Bioluminescence and Chemiluminescence

Bioluminescence and Chemiluminescence

Current Status

Proceedings of the VIth International Symposium on Bioluminescence and Chemiluminescence, Cambridge, September 1990

Edited by
Philip E Stanley and Larry J Kricka

JOHN WILEY & SONS
Chichester · New York · Brisbane · Toronto · Singapore

Other Wiley Editorial Offices

John Wiley & Sons, Inc., 605 Third Avenue,
New York, NY 10158-0012, USA

Jacaranda Wiley Ltd, G.P.O. Box 859, Brisbane,
Queensland 4001, Australia

John Wiley & Sons (Canada) Ltd, 22 Worcester Road,
Rexdale, Ontario M9W 1L1, Canada

John Wiley & Sons (SEA) Pte Ltd, 37 Jalan Pemimpin   05-04,
Block B, Union Industrial Building, Singapore 2057

Library of Congress Cataloging-in-Publication Data available.

British Library Cataloguing in Publication Data:

International Symposium on Bioluminescence and
Chemiluminescence (6th 1990 Cambridge)
  Bioluminescence and chemiluminescence: current
status.
  I. Title   II. Stanley, Philip   III. Kricka, Larry
574.19

  ISBN 0 471 92993 X

Printed in Great Britain by
Biddles Ltd., Guildford, Surrey

Preface

This volume contains the proceedings of the VIth International Symposium on Bioluminescence and Chemiluminescence held at Cambridge, England between 10–13 September 1990.

Previous and successful Symposia in this series have been held in Brussels (1978), San Diego (1980), Birmingham (1984), Freiburg (1986) and Florence (1988). Each meeting has had its own flavour but we have seen a transition in their contents as time has passed. Thus in the early meetings there were only very minor components concerning immunoassays and DNA probes whereas today they represent major ones. Likewise we have seen a spectacular increase in cellular luminescence contributions. At this meeting we have for the first time a session on imaging, a technique which allows workers to measure and locate light sources in two dimensions such as microtitre plates, blots and cell cultures.

The 1990 Marlene DeLuca Memorial Prize, generously sponsored by Laboratorium Prof. Dr. Berthold, was awarded to Mr Michael White for his paper on the "Applications of direct imaging of firefly luciferase expression in single intact mammalian cells using highly sensitive charge-coupled device cameras".

We wish to express our sincere thanks to the Scientific Organizing Committee: T.P. Whitehead (London) Chairman, R.E. Ansorge (Cambridge), A.K. Campbell (Cardiff), I.A. Cree (Dundee), J.G.M. Hastings (Birmingham), C.E. Hooper (Surrey), L.J. Kricka (Philadelphia, USA), F. McCapra (Sussex), B.J. McCarthy (Leeds), G.S.A.B. Stewart (Nottingham), G.H.G. Thorpe (Birmingham) and A. Townshend (Hull)

and to the International Advisory Committee:

W.R.G. Baeyens (Belgium), T.O. Baldwin (USA), F. Berthold (W. Germany), E. Cadenas (USA), M. Ernst (W. Germany), J.W. Hastings (USA), L.J. Kricka (USA), A. Lundin (Sweden), W.D. McElroy (USA), J.C. Nicolas (France), M. Pazzagli (Italy), A. Roda (Italy), E. Schram (Belgium), H. Schroeder (USA), P.E. Stanley (UK), A. Tsuji (Japan), N.N. Ugarova (USSR), W.G. Wood (Germany) and J. Woodhead (UK).

Editors Note: These Proceedings were compiled from camera ready manuscripts which were not subjected to peer review so as to speed publication. The Editors do not take responsibility for the scientific content of the papers.

Acknowledgements

We gratefully acknowledge the sponsorship of the following companies who generously supported the Symposium by their involvement in various ways:

Amersham International PLC (Little Chalfont, UK)
Amgen, Inc (Thousand Oaks, CA, USA)
Analytical Luminescence Laboratory (ALL) (San Diego, CA, USA)
Astromed Limited (Cambridge, England)
Berthold Instruments (UK) Limited (St Albans, UK)
BioOrbit UK Limited (Hungerford, UK)
Biotrace Limited (Bridgend, Wales, UK)
Byk-Sangtec Diagnostica GmbH & Co. KG (Dietzenbach, West Germany)
Ciba-Corning Diagnostics Limited (Halstead, UK)
Dynatech Laboratories (Billingshurst, UK)
ELA Technologies (Athens, GA, USA)
Enzymatix Limited (Cambridge, UK)

Fujirebio Inc (Tokyo, Japan)
ICN-Flow (High Wycombe, UK)
Hamamatsu Limited (Enfield, UK)
Hi-Tech Scientific Limited (Salisbury, UK)
Image Research Limited (Cambridge, UK)
John Wiley & Sons, Publishers (Chichester, UK)
Labsystems (UK) Limited (Basingstoke, UK)
Polaroid Corporation (Cambridge, MA, USA)
Sankyo Co. Limited (Tokyo, Japan)
Steptech Instrument Services Limited (Stevenage, UK)
Strategic Technologies International (Mundelein, IL, USA)
Tropix, Inc (Bedford, MA, USA)
Turner Designs (Sunnyvale, CA, USA)
Wright Instruments Limited (Enfield, UK)

Laboratorium Prof. Dr. Berthold (Wildbad, W. Germany) for donation of the Marlene DeLuca Memorial Prize and for Marlene DeLuca Bursaries to permit certain delegates to attend.

Cambridge Research & Technology Transfer Ltd for organizing the Symposium.

The cover was designed by Paula Davis of Ellis Design, Cambridge, England.

Table of Contents

Session II: IMMUNOASSAY AND DNA PROBES

Table of Contents

Session III: CELLULAR LUMINESCENCE

Table of Contents

Table of Contents   xiii

Industrial Seminar: ELA TECHNOLOGIES, INC

Table of Contents

Industrial Seminar: BERTHOLD INSTRUMENTS

Session I

GENETIC ENGINEERING

MOLECULAR BIOLOGY OF BIOLUMINESCENCE

E.A. Meighen

*Department of Biochemistry, McGill University,
3655 Drummond St., Montreal, Quebec,
Canada, H3G 1Y6*

INTRODUCTION

Luciferases that catalyze the light emitting reaction have been isolated from a wide variety of organisms. Although O_2 is a common component of all bioluminescence reactions, the substrates or luciferins can have quite diverse structures for the different luminescent systems. DNA coding for luciferases have been isolated and sequenced from luminescent systems with luciferins whose structures can be classified into three basic types (Fig.1)

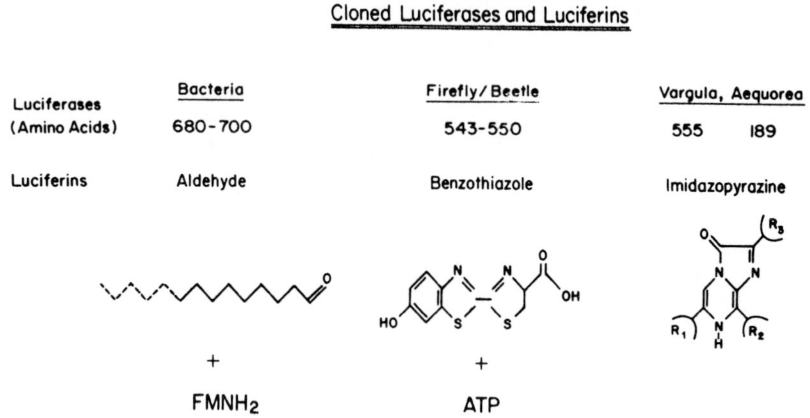

Cloned Luciferases and Luciferins

	Bacteria	Firefly/Beetle	Vargula, Aequorea	
Luciferases (Amino Acids)	680-700	543-550	555	189
Luciferins	Aldehyde	Benzothiazole	Imidazopyrazine	

+

FMNH₂

+

ATP

Fig. 1 Structures of luciferins from cloned luciferases.

The substrates for luciferases from marine and terrestrial bacteria are $FMNH_2$ and a long chain aldehyde generally of eight carbons or longer. The enzyme has a molecular mass of 77 kDa and is composed of two nonidentical subunits, α and β (40 and 37 kDa, respectively), coded by the *lux*A and *lux*B genes (1).

Firefly or click beetle luciferases are composed of polypeptides of 62 kDa (2). These luciferases catalyze the ATP dependent decarboxylation of luciferin, which is a benzothiazole derivative. Photoproteins or luciferases isolated from the jellyfish, *Aequorea*, or the ostracod, *Vargula* (formerly *Cypridina*) both have luciferins with an imidazopyrazine nucleus with quite different substituents (R1,R2,R3) (3). The mass of the polypeptide coded by the cDNA for the *Vargula* luciferase is about three times that of the Aequorea photoprotein. All three luminescent systems catalyze emission of light between 480 and 600 nm with the firefly/beetle system emitting light at the higher wavelengths (green to orange).

The DNA coding for the luciferases has been transferred into a wide range of different organisms and cells as reporters or sensors of gene expression and/or cellular function (4,5). The number of different applications has been increasing at a remarkable rate as researchers in other fields recognized the high sensitivity of the assay, the absence of endogenous activities, the speed and simplicity of the measurement of light, and the linear relationship between light and amount of luciferase over a wide range in the luminescence assays making quantitative determination of the amount of enzyme much simpler than in most if not all radiometric and spectrophotometric assays. The expression of the luciferase genes in cells and organisms is dependent on many factors including efficiency of gene transfer, copy number of the luciferase genes, promoter proficiency, processing, stability and translation of the luciferase mRNA, and stability and folding of the protein product to form the functional enzyme. Moreover availability of the substrates is just as important as the level of functional luciferase in generating the luminescence signal.

At present, the DNA coding for the firefly or bacterial luciferases has been used in most applications, with the cDNA for firefly luciferase being primarily used in eukaryotic systems and the lux genes for bacterial luciferase in prokaryotic systems. However, both luciferases can be expressed in prokaryotic and eukaryotic systems and bacterial luciferase has been expressed at very high levels in eukaryotic systems particularly in systems (plants or insects) that are expressed at temperatures of 30°C or lower.

RESULTS AND DISCUSSION
Bacterial luciferases and organization of the bacterial lux genes
A number of different bacterial luciferases as well as firefly and click beetle luciferases have been cloned. Table 1 lists strains from three different genera, *Vibrio*, *Photobacterium* and *Xenorhabdus*, including terrestrial and marine bacteria for which both the *lux*A and *lux*B genes for luciferase have been cloned and the nucleotide sequence determined (6). Differences in amino acid sequences of up to 46% exist between the *lux*A gene products (α subunits) and up to 56% between the *lux*B gene products (β subunits). A similar degree of variation in amino acid sequence has been observed between firefly and click beetle luciferases (2). For the bacterial luminescent system, the genes (*lux*CDE) responsible for synthesis of the aldehyde substrate have been cloned for most if not all of the strains listed in Table 1 and have been used to supply the aldehyde directly inside prokaryotic cells(4). As there is clearly a wide choice of luciferases genes

Table 1

CLONED AND SEQUENCED LUCIFERASE GENES[a]

Species	Strain	luxCDE[b]
X. luminescens	ATCC 29999	+
V. harveyi	B392	+
V. fischeri	MJ-1	−
	ATCC 7744	+
P. leiognathi	54	−
	721	−
	ATCC 25521[c]	+
P. phosphoreum	NCMB 844[c]	+

[a] Only includes strains for which both luxA and luxB have been sequenced. Significant differences exist in amino acid sequence (15-56%) between all luciferase genes except the two V. fischeri strains and two of the P. leiognathi strains (54 and 721). Specific references given in (6).
[b] Strains for which the luxCDE genes have been sequenced.
[c] unpublished data.

producing luciferases with different properties, the opportunity exists to select the most suitable system for a particular application.

The bacterial bioluminescent reaction and the pathways for supplying the substrates (FMNH$_2$ and aldehyde) are outlined in Fig. 2. Long chain aldehyde can be supplied by generating fatty acids from activated acyl groups in the cell (e.g. acyl-ACP) which are then reduced and activated by the enzymes (transferase, synthetase and reductase) coded by the luxCDE genes of the bacterial bioluminescent system. FMNH$_2$ can be regenerated by the NAD(P)H dependent reduction of FMN by oxidoreductases present in luminescent and other bacteria.

In all bacterial lux systems so far analyzed (Table 1), the lux genes for luciferase (AB) and for aldehyde synthesis (CDE) are present and organized as depicted in Fig. 3 with the direction of transcription from left to right. The luciferase genes are adjacent (luxAB) flanked by the genes for aldehyde synthesis with the luxCD genes immediately upstream and the luxE gene immediately downstream or separated by another lux gene. Only the five genes (luxCDABE) are required for light emission in bacteria as there is an adequate supply of FMNH$_2$ and the activated acyl precursors for aldehyde synthesis.

There are however significant differences in these systems in the organization of other lux genes involved in regulatory and/or unknown functions. Between luxB and luxE, a gene designated as luxF and homologous to luxB (30% identity) is present in P. phosphoreum and two of the P. leiognathi strains (54 and 721) but not in the other strains (1).

Downstream and closely linked to luxE in all marine bacteria so far investigated but not in the terrestrial bacteria,

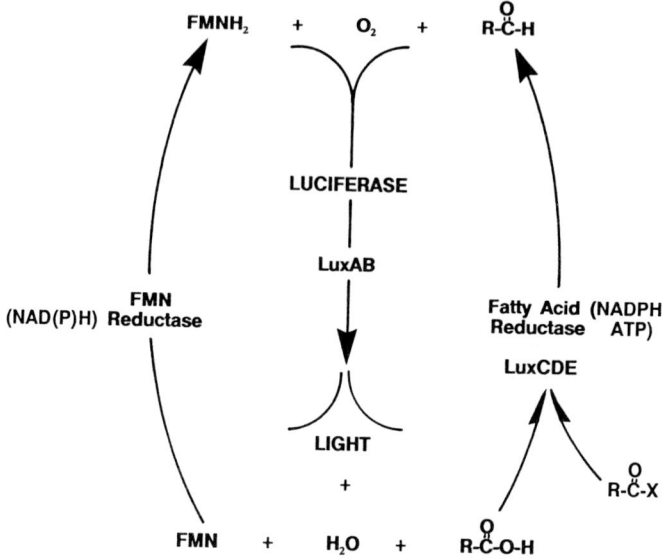

Fig.2. Bioluminescence reaction and regeneration of substrates.

Xenorhabdus, is a new gene designated as *lux*G(7). An additional
gene further downstream, *lux*H, is also present in *V. harveyi* but
not in *V. fischeri* or *P. leiognathi* (ATCC 25521). The functions
of these downstream genes are unknown but they appear not to be
directly required for luminescence but may be related to the
physiology or environmental niche of the marine bacteria.

Upstream of *lux*C, regulatory genes (*lux*IR) are present only
in the *V. fischeri* lux systems whereas the other strains that
have been investigated contain extensive AT rich regions with
only short open reading frames (40 codons). Regulatory lux genes
have however been detected at other loci in the *V. harveyi*
genome. In spite of the differences in lux gene organization and

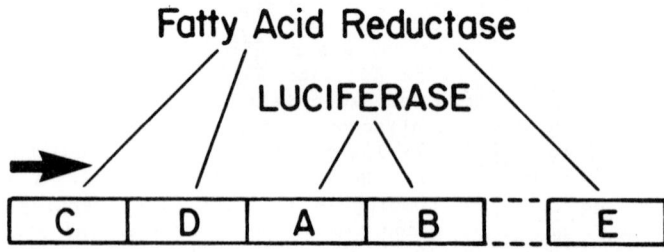

Fig. 3. Organization of the *lux*CDABE genes in luminescent
bacteria. Arrow indicates direction of transcription.

location of the regulatory genes, expression of most but not all bacterial luminescent systems appear to be induced with cellular growth.

Application of the bacterial *lux* genes. The transfer of lux genes into bacteria has been used to monitor gene expression as well as a sensor of the metabolic functions of the bacteria and the effect of different reagents (e.g. antibiotics) on cellular function. The bacterial lux genes have been transferred to over 24 different species of bacteria including both gram negative and gram positive bacteria (4,5) (Table 2). At present, primarily the lux genes from *V. harveyi* and *V. fischeri* have been used. As sufficient $FMNH_2$ is present in the bacteria, aldehyde must be added. Decanal has generally been added to cells containing the *V. harveyi* luxAB genes and dodecanal to cells with the *V. fischeri* luxAB based presumably on the *in vitro* specificities of the respective luciferases. However, the luminescent response of *E. coli* containing the recombinant *lux*AB genes from either *V.* fischeri or *V. harveyi* clearly shows a preferred response for the shorter chain aldehydes, in particular nonanal and to a lesser extent decanal (Fig. 4). As the *V. harveyi lux* system appears to be more stable particularly at higher temperatures, application of the *V. harveyi lux*AB genes may be preferred.

In some cases, the lux genes for aldehyde synthesis (*lux*CDE) have also been transferred so that it is not necessary to add exogenous aldehydes (Table 2). Most studies in this regard have used the *V. fischeri lux*CDABE genes (4) as they can be readily transferred as a single fragment whereas suitable restriction sites are not available for the *V. harveyi lux*CDABE system to transfer as a single fragment. However, the X. luminescens lux

Table 2

BACTERIA CONTAINING RECOMBINANT LUX GENES

Lux Genes	Genera	Aldehydes Added[a]
*Vf lux*CDABE	*Agrobacterium, Bacillus, Erwinia, Escherichia, Pseudomonas, Rhizobium, Xanothomas, Xenorhabdus*	None
*Vf lux*AB	*Anabaena, Citrobacter, Klebsiella, Lactobacillus, Lactococcus, Listeria, Salmonella, Shigella, Staphylococcus, Streptococcus*	Dodecanal
*Vh lux*CDABE	*Escherichia, Photobacterium, Vibrio Xenorhabdus*	None
*Vh lux*AB	*Agrobacterium, Anabaena, Bacillus, Bradhyrhizobium, Pseudomonas, Rhizobium, Streptomyces*	Decanal

[a] Aldehydes added to give *in vivo* luminescence.

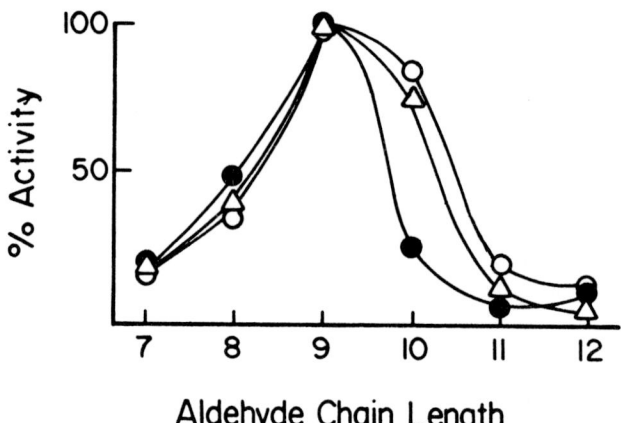

Fig.4. Luminescent response of *E. coli* containing the recombinant *lux*AB genes from *V. fischeri* (●), *V. harveyi* (○) and *X. luminescens* (△).

system may be the most useful as all five genes are enclosed within a 7 kbp EcoRI fragment and the luciferase is even more stable than *V. harveyi* luciferase with a half time for inactivation of over 3 hrs. at 45°C (6).

<u>Expression in eukaryotic systems</u>. As expression in eukaryotic systems requires in most instances one promoter per gene, the application of the bacterial lux system was limited until the luciferase genes were fused to form a monocistronic gene. A variety of fused *V. harveyi luxA-B* genes have been created in which coded for the α and β subunits linked by a polypeptide. Fig. 5 shows the structure of one such fused polypeptide. As a consequence, the fused bacterial luciferase has been expressed in a number of different eukaryotic systems including plant and insect cells as well as mammalian cells (Table 3). The level of expression is quite high in these systems and reaches levels at least as high if not higher than that obtained using the cDNA for firefly luciferase particularly in cells grown at temperatures of 30°C or lower.

Fused lux A-B

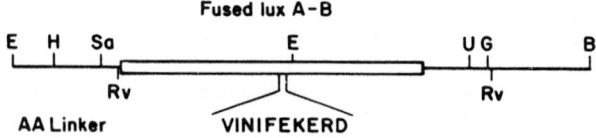

Fig. 5. Structure of a fused <u>lux</u>A-B gene. Restriction sites are; BamHI(B), BglII (G), EcoRI (E), EcoRV (Rv), HindIII (H), PvUII (U), SalI (Sa).

TABLE 3

**EXPRESSION OF BACTERIAL AND FIREFLY
LUCIFERASES IN DIFFERENT EUKARYOTIC SYSTEMS[a]**

	Luciferase[b]	
	Bacterial	Firefly
Mammalian Cells		
Mouse fibroblasts	0.06[c]	–
Monkey kidney cells	–	0.04
Tobacco plants	0.4	0.06
Virus-infected cells		
Insect (baculovirus)	60[d]	–
Monkey (vaccinia virus)	–	0.8
Yeast	16	–
In vitro	$\mu g/h/mg$ RNA	
Reticulocyte lysate	0.3	2
Mouse ascites extract	–	120

[a]The levels of expression are highly dependent on vector, host and experimental conditions and can only be compared qualitatively.
[b]Luciferase (in μg) with the same specific activity as the native enzyme necessary to account for the highest level of activity per mg of extracted protein (8-13).
[c]Unpublished data, J. Bell. Maxiumum levels detected at temperatures below 30°.
[d]Unpublished data, C. Richardson.

The highest level of expression of the fused luciferase is found in insect cells infected with baculovirus containing the luxAB gene inserted in the viral vector in place of the gene coding for the polyhedron capsid. The level of expression of the proteins is as high as that of another reporter gene, placed under a different promoter in the same vector and can easily be detected by protein staining (C. Richardson, unpublished data). As little as a single cell needs to be extracted to be detected using a continuous assay that provides stable light emission (<0.5% change/min) for three hours or longer. Consequently luminescence can readily be measured in a scintillation counter or detected on film.

ACKNOWLEDGEMENTS
Supported by grants from the Medical Research Council of Canada. I would like to thank Joyce Herron for her patience and skill in typing this manuscript.

REFERENCES

1. E.A. Meighen, Enzymes and genes from the *lux* operons of bioluminescent bacteria. Annual Review of Microbiology 42, 151-176(1988).

2. K.V. Wood, Y.A. Lam, H.H. Seliger, and W.D. McElroy, Complementary DNA coding click beetle luciferases can elicit bioluminescence of different colors. Sci. 244, 700-702(1989).

3. E.M. Thompson, S. Nagata, and F.I. Tsuji, Cloning and expression of cDNA for the luciferase from the marine ostracod Vargula hilgendorfii. Proc. Natl. Acad. Sci. 86, 6567-6571(1989).

4. J.J. Shaw, and C.I. Kado, Development of a *Vibrio* bioluminescence gene-set to monitor phytopathogenic bacteria during the ongoing disease process in a non-disruptive manner. Biotechnology 4, 60-64(1986).

5. S.A.A. Jasmin, A. Ellison, S.P. Denyer and G.S.A.B. Stewart, *In vivo* bioluminescence: A cellular reporter for research and industry. J. Biolumin. and Chemilumin. 5, 115-122(1990).

6. R. Szittner, and E. Meighen, Nucleotide sequence, expression and properties of luciferase coded by *lux* genes from a terrestrial bacterium. J. Biol. Chem. In Press (1990).

7. E. Swartzman, C. Miyamoto, A. Graham, and E. Meighen, Delineation of the transcriptional boundaries of the *lux* operon of *Vibrio harveyi* demonstrates the pPresence of two new *lux* genes. J. Biol. Chem. 265, 3513-3517(1990).

8. M. Boylan, J. Pelletier, and E. Meighen. Fused bacterial luciferase subunits catalyze light emission in eukaryotes and prokaryotes. J. Biol. Chem. 264, 1915-1918(1989).

9. J.R. De Wet, K.V. Wood, M. DeLuca, D.R. Helinski, and S. Subramani. Firefly luciferase gene: Structure and expression in mammalian cells. Mol.and Cell. Biol. 7, 725-737(1987).

10. C. Koncz, N. Martini, R. Mayerhofer, R.Z. Koncz-Kalman, H. Körber, G.P. Redei, and J. Schell. High-frequency T-DNA-mediated gene tagging in plants. Proc. Natl. Acad. Sci. U.S.A.86, 8467-8471(1989).

11. D.W. Ow, K.V. Wood, M. Deluca, J.R. DeWet, D.R. Helinski, and S. H. Howell. Transient and stable expression of the firefly luciferase gene in plant cells and transgenic plants. Science. 234, 856-59(1986).

12. J.F. Rodriguez, D. Rodriguez, J.R. Rodriguez, E.B. McGowan, and M. Esteban. Expression of the firefly luciferase gene in vaccinia virus: A highly sensitive gene marker to follow virus dissemination in tissues of infected animals. Proc. Natl. Acad. Sci. USA 85, 1667-1671(1988).

13. G.D. Kutuzova, E.A. Skripkin, N.I. Tarasova, N.N. Ugarova, and A.A. Bogdanov. Synthesis and pathway of luciola mingrelica firefly luciferase in Xenopus laevis frog oocytes and in cell-free systems. Biochimie. 71, 579-583

THE ORIGIN OF BEETLE LUCIFERASES

Keith V. Wood

Promega Corporation, 2800 Woods Hollow Rd., Madison WI 53711

INTRODUCTION

It is said that to know truly of the world, one must know also of its past. This maxim apparently applies also to our understanding of beetle luciferases. The chemistry of beetle luminescence, studied mostly of the North American firefly, *Photinus pyralis*, is believed to be well established. The components of the reaction had been identified by the 1960's, and by 1978 the dioxetane mechanism was firmly established (1). It is generally accepted that beetle luciferases work by combining ATP and luciferin to produce an activated intermediate, which is then oxidized to produce a photon. Little new information about the enzymatic mechanism has been obtained in the past decade.

I propose that coenzyme A is also a substrate of the beetle luciferases that has been largely neglected in their study. The importance of coenzyme A was realized through investigating the evolutionary ancestry of these enzymes. Comparisons of protein primary structures have revealed that the luciferases are related to coenzyme A ligases. Experimental evidence shows that coenzyme A participates in the luminescent reaction, and that this participation is mediated by the reactive thiol group of coenzyme A. It is likely that coenzyme A forms a thiol ester with the carboxylate of luciferin during catalysis. The mechanistic consequences of this interaction are not yet clear.

MATERIALS AND METHODS

Sequence alignments and hydropathy plots: Sequence alignments were done manually aided by GenePro software, Riverside Scientific (Seattle). Hydropathy plots were done using the method of Kyte and Doolittle (2); 15 residue values were averaged for each point.

Enzyme assays: Luciferin and firefly luciferase were from JBL Scientific (San Luis Obispo). Coenzyme A (lithium salt) was from Pharmacia; dethioCoA (lithium salt) was from Sigma. Measurements of luminescence were made on a Turner Model 20 luminometer. Buffer for luminescence assays was 30 mM Tricine, 8 mM Mg^{2+} (magnesium carbonate), 10 mM DTT, 0.2 mM EDTA, pH 7.8. Luciferase at 10 mg/ml was diluted 10,000-fold into buffer containing 10% glycerol and 1 mg/ml BSA. For assay, 10 μl of diluted enzyme were mixed with 200 μl buffer; luminescence was initiated by injection of 100 μl buffer containing 3 mM luciferin, 1.5 mM ATP, and various concentrations of coenzyme A and/or dethioCoA.

RESULTS AND DISCUSSION

Beetle luciferases are likely to be young as a class of enzymes, arising near or after terrestrial invasion by animals (about 400 million years ago). This seems true since the luminescence chemistry of beetles is not found in any of the luminous marine arthropods. Since most other luciferase systems studied are of marine organisms, the beetle luciferases could be the most recently evolved of the studied systems. Given the recency of these enzymes, the question of their enzymatic origins is particularly interesting.

The amino acid sequences of several beetle luciferases have become available through cloning and sequencing of their respective cDNA sequences (3, 4, 5). Judged by the amino acid replacement rate in these sequences, the beetle luciferases are evolving at an unusually high rate. The rate per 100 amino acids has been estimated to be 23 to 31 replacements per 100 million years (6) The comparible rates for hemoglobins and cytochrome c are 12 and 4 replacements per 100 million years, respectively. At this rate, if applied uniformly to the amino acid sequences, little similarity could be expected between the primary structures of luciferases and other enzymes with common ancestry of more than 1 billion years ago.

Common ancestry has been found, however, between beetle luciferases and two other enzymes unrelated to luminescence. One of these enzymes, 4-coumarate:CoA ligase, is central to the phenylpropanoid metabolism of plants (7). Phenylpropanoids yield many specialized compounds such as lignin, chemicals for defense against microbes and animals, colors and fragrances to attract animals, and possibly chemicals for protection from ultraviolet radiation. The structural relationship between luciferases and 4-coumarate:CoA ligase is especially surprizing since plants and animals have diverged from common ancestry more than 1 billion years ago.

The other enzyme found to have common ancestry is long-chain acyl-CoA synthetase (8). This enzyme, isolated from rat liver, is involved in the metabolism of fatty acids through β-oxidation and also in synthesis of other cellular lipids. Though this enzyme is also derived from an animal, it is structurally much less similar to luciferase than the 4-coumarate:CoA ligase of plant origin (Fig. 1). This suggests that

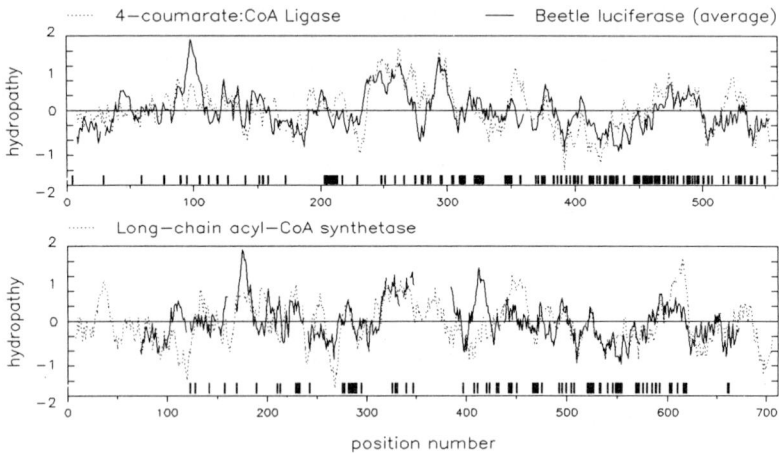

Fig. 1 Comparison of hydropathy plots for beetle luciferases (solid line) and 4-coumarate:CoA ligase (dotted line, top) and long-chain acyl-CoA synthetase (dotted line, bottom). Vertical bars along the abscissa indicate the positions of amino acids that are identical in all sequences of each plot.

long-chain acyl-CoA synthetase is decedent from this lineage of enzymes before the divergence of plants and animals.

Relative similarities of these enzymes to luciferases were determined by alignment with sequences derived from beetles of three genera, *Photinus pyralis* (firefly), *Luciola cruciata* (firefly), and *Pyrophorus plagiophthalamus* (click beetle). Alignment of the beetle luciferase sequences shows identity at 40% of their amino acid positions. Of these identities, 56% and 35% are identical with the aligned sequences of 4-coumarate:CoA ligase and long-chain acyl-CoA synthetase, respectively (for positions not conserved between the luciferases, the average identities are 16% and 11%, respectively). Thus, much of the sequence that is identical between luciferases was maintained from before their ability to produce light.

The relative similarity of luciferases to the CoA-utilizing enzymes is also revealed in their hydropathy plots (Fig. 1). Since hydropathy is a major factor in the stability of protein tertiary structure, these plots are expected to reflect that structure. Hydropathy plots of the enzymes were superimposed based on the alignment of their amino acid sequences. Gaps were placed in the plots where the sequence alignments dictated. For the luciferases, the hydropathy values at each amino acid position of their aligned sequences were combined to give an averaged hydropathy plot.

Similarity between the plots for the luciferases and 4-coumarate:CoA ligase is immediately apparent. Over the full length of the sequences, the plots show common general trends. This is true even in regions where the sequence homology is not especially high, for example between amino acid positions 1 and 200. The comparison between plots for luciferases and long-chain acyl-CoA synthetase shows less similarity. As expected, similarity is greatest in regions with the most sequence identity. For both CoA-utilizing enzymes, similarities with the luciferases are greatest near the carboxy termini.

These similarities strongly suggest that coenzyme A was important also to the ancestry of luciferase. Similarity is also evident in the chemistries of these enzymes. Both 4-coumarate:CoA ligase and long-chain acyl-CoA synthetase catalyze formation of a thiol ester with the carboxyl group of their respective substrates. This is done by first using ATP to form an enzyme-bound acyl-AMP intermediate. With the luciferases, such an acyl-AMP intermediate of luciferin is believed to be the chemical species that is oxidized for photon production. By analogy, the ancestor of luciferases

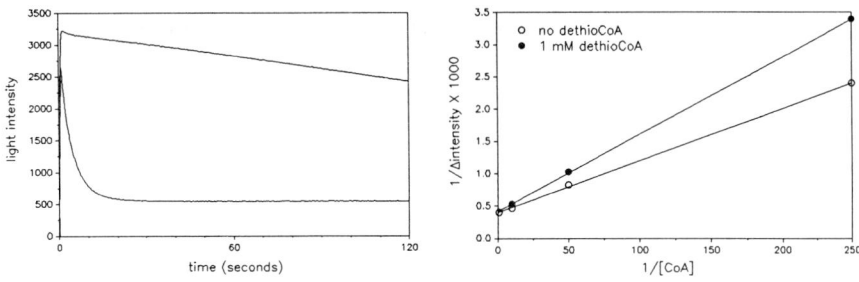

Fig. 2 Enhancement of enzyme turnover by coenzyme A is shown on the left. The lower plot shows the kinetics of light emission in absence of coenzyme A; the upper plot is an identical reaction with 1 mM CoA added. Inhibition by 1 mM dethioCoA is shown in a reciprocal plot on the right. "Δintensity" is the increase of light intensity at steady-state (25 s) caused by the addition of coenzyme A.

may similarly have catalyzed formation of a thiol ester of luciferin (or rather the ancestor of luciferin).

What of the possibility that coenzyme A is still involved with the chemistry of beetle luminescence? Experiments in the late 1950's showed that addition of coenzyme A to a reaction with purified luciferase reduced the characteristic self inhibition of activity (Fig. 2)(9). With use of a non-oxidizable substrate analog, dehydroluciferin, it was shown that coenzyme A could indeed form a thiol ester bond with the carboxyl group. It was also known that coenzyme A is abundant in the light organs of fireflies. However, the biological relevance of these observations was not clear.

In light of the evolutionary evidence, it is now apparent that coenzyme A plays a specific role in the production of light. It can be shown that the activity of coenzyme A is mediated specifically through its thiol group by using the analog, dethioCoA. DethioCoA is identical to coenzyme A without the sulfur atom. When added to the luminescence reaction in absence of coenzyme A, the dethioCoA has no effect. However, dethioCoA specifically inhibits the increased luminescence caused by coenzyme A. On a reciprocal plot, the inhibition by dethioCoA displays competitive kinetics with respect to the activity of coenzyme A. Thus, dethioCoA binds to the same site on the enzyme as coenzyme A, but without the sulfur atom it lacks activity. By the demonstrated activity of coenzyme A with dehydroluciferin, and by analogy to the coenzyme A chemistries of the related enzymes, it is very likely that activity mediated by the sulfur atom is through formation of a thiol ester with luciferin.

Thus, beetle luciferases are apparently coenzyme A ligases. The difference between these and other coenzyme A ligases is that luciferases are also oxygenases. Because the tertiary structures of the luciferases are probably quite similar to 4-coumarate:CoA ligase, the oxygenase probably is a new activity that has evolved into an existing enzymatic structure. The role of coenzyme A in the chemistry of luciferase is not obvious since light production *in vitro* can proceed in its absence. However, with further investigation of its action, many of our concepts about the mechanism of luciferase may change. For example, it is now clear that the "flash" of light traditionally witnessed of the enzyme in test tubes does not relate to the flash of fireflies in the wild.

REFERENCES

1. J. Wannlund, M. DeLuca, K. Stempel, and P.D. Boyer, Use of ^{14}C-Carboxyl-Luciferin in Determining the Mechanism of the Firefly Luciferase Catalyzed Reactions. *Bioch. Biophys. Res. Commun.*, **81**, 987-992 (1978).
2. J. Kyte and R.F. Doolittle, A Simple Method for Displaying the Hydropathic Character of a Protein. *J. Mol. Biol.*, **157**, 105-132 (1982)
3. J.R. de Wet, K.V. Wood, M. DeLuca, D.R. Helinski, and S. Subramani, Firefly Luciferase Gene: Structure and Expression in Mammalian Cells. *Molec. Cell. Biol.*, **7**, 725-737 (1987).
4. T. Masuda, H. Tatsumi, and E. Nakano, Cloning and Sequence Analysis of cDNA for Luciferase of a Japanese Firefly, *Luciola cruciata*. *Gene*, **77**, 265-270 (1989).
5. K.V. Wood, Y.A. Lam, H.H. Seliger, and W.D. McElroy, Complementary DNAs Encoding Click Beetle Luciferases Can Elicit Bioluminescence of Different Colors. *Science*, **244**, 700-702 (1989).
6. K.V. Wood, Luciferases of Luminous Beetles: Evolution, Color Variation, and Applications, Ph.D. Dissertation (Department of Chemistry, University of California, San Diego).
7. J. Schroder, Protein Sequence Homology Between Plant 4-Coumarate:CoA Ligase and Firefly Luciferase. *Nucl. Acids Res.*, **17**, 460 (1989).
8. H. Suzuki, Y. Kawarabayasi, J. Kondo, T. Abe, K. Nishikawa, S. Kimura, T. Hashimoto, and T. Yamamoto, Structure and Regulation of Rat Long-chain Acyl-CoA Synthetase, *J. Biol. Chem.*, **265**, 8681-8685 (1990)
9. R.L. Airth, W.C. Rhodes, and W.D. McElroy, The Function of Coenzyme A in Luminescence, *Biochim. Biophys. Acta*, **27**, 519-532 (1958).

ENGINEERING OF A BACTERIAL LUCIFERASE ALPHA–BETA FUSION PROTEIN WITH ENHANCED ACTIVITY AT 37°C IN *ESCHERICHIA COLI*

A. Escher, D.J. O'Kane[1], and A.A. Szalay

University of Alberta, Plant Molecular Genetics, 6-30 Medical Science Building, Edmonton, AB T6G 2H7, Canada and [1]Department of Biochemistry, University of Georgia, Athens, GA 30602, USA

INTRODUCTION

We have previously reported the construction of a gene encoding an $\alpha\beta$ fusion of the subunit polypeptides of the luciferase from *Vibrio harveyi* strain MAV (1). This fusion luciferase (Fab9) is active as a monomer with light emitting properties similar to wild type enzyme when synthesized at low temperatures (15°C–23°C). In contrast, synthesis at 37°C results in a marked decrease in activity ($\approx 2\%$ of wild type) due to temperature sensitive folding of the fusion polypeptide. The *fab9* gene has been used as a marker of gene expression in a variety of organisms (2,3). However the temperature sensitivity of Fab9 protein has excluded its effective use in mammalian systems. In order to obtain a luciferase fusion with increased activity at 37°C, we have cloned and sequenced the *luxA* and *luxB* genes encoding a thermostable luciferase from thermotolerant *Vibrio harveyi* strain CTP5 (unpublished data).

Here we describe the construction of a *luxAB* open reading frame encoding a full–length $\alpha\beta$ fusion luciferase with enhanced activity at 37°C, and we compare the activity of MAV wild type luciferase with CTP5 wild type luciferase as well as MAV fusion luciferase with CTP5 fusion luciferase in *Escherichia coli* cells grown at different temperatures.

MATERIAL AND METHODS

Strains and Plasmids. *E. coli* strain HB101 was used for cloning and strain DLT101 for expression of *lux* constructs (1). All constructs were cloned in vector pT_7T_319 under transcriptional control of the T_7 promoter Ø 10. Plasmids pLX203ab and pLX703 fab9 have been previously described (4,1). pLX 203ab carries a 2.2 kb DNA fragment containing the *luxA* and *luxB* genes from *V. harveyi* strain MAV; pLX703fab9 carries a 2.2 kb DNA fragment containing the MAV *luxAB* gene fusion *fab9*; $pLX208a_cb_c$ carries a 2.3 kb DNA fragment containing the *luxA* and *luxB* genes from *V. harveyi* strain CTP5 (thereafter named $luxA_c$ and $luxB_c$ respectively); $pLX708fa_cb_c9$ carries a 2.3 kb DNA fragment containing the CTP5 $luxA_cB_c$ gene fusion fa_cb_c9. Gene fa_cb_c9 was constructed by ligating a 1.2 kb SalI–EcoRI DNA fragment, containing the $luxA_c$ gene fused in frame with a DNA sequence encoding a peptide linker identical to the one linking the α and β polypeptide moities of Fab9, with a 1.1 kb EcoRI–BamHI DNA fragment encoding $luxB_c$. Recombinant DNA techniques and analysis of gene products were performed as previously described (1).

<u>Bioluminescence Assay</u>. Light emission was measured *in vivo* using a
video image analyzer (Hamamatsu Argus–100). *E. coli* DLT101 cells
transformed with different *lux* constructs were grown at 37°C to mid–
log phase in Luria–Bertani (LB) medium (5) in presence of 100 µg/ml
ampicillin. Optical densities were measured and adjusted to 7Klett
by dilution in LB broth. Twenty microliter of each culture was
applied on four nitrocellulose papers using a blotting apparatus. The
papers were then put onto LB plates (100 µg/ml ampicillin) and
incubated at 23, 30, 37, and 42°C for 3 hours. Decanal was then
applied uniformly on the lid of each plate and photon emission was
measured immediately for 15 s at 100% sensitivity of the photon
counting tube.

RESULTS AND DISCUSSION
Expression of *lux* Constructs in *E. coli*.
Both the *luxAB* and *luxA$_c$B$_c$* genes code for α and β subunit polypeptides
of molecular weight 40 kDa and 36 kDa respectively. However the CTP5
luciferase has 5 amino acid substitutions in the α subunit polypeptide
and 7 amino acid substitutions in the · β subunit polypeptide
(unpublished data). This may explain the difference in mobility
observed in a 10% SDS polyacrylamide gel between the MAV and CTP5
luciferase subunit polypeptides even after boiling samples for 4 min
in presence of 2% SDS (Fig. 1A). The substitutions may cause
strengthening of protein secondary structures, especially in the α
subunit polypeptide of CTP5 luciferase. Fig. 1B shows that the
difference in migration rate is also visible between the luciferase
fusions Fab9 and Fa$_c$b$_c$9, which have identical amino acid sequence
except for the 12 substitutions stated above.

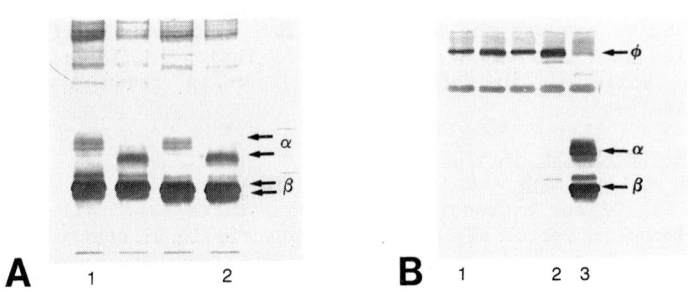

Fig. 1. Immunoblot analysis of protein crude extracts from *E.
coli* DLT101 cells transformed with different *lux* constructs.
Antibodies raised against denatured α and β luciferase subunit
polypeptides were used to detect the presence of α, β, and
fusion proteins.
Fig. 1A.   Cells transformed with pLX203ab (lane 1), or
pLX208a$_c$b$_c$ (lane 2) were grown at 42°C.   Fig. 1B. Cells
transformed with pLX703fab9 (lane 1), pLX708fa$_c$b$_c$9 (lane 2), and
pLX203ab (lane 3) were grown at 37°C.

The increased amounts of CTP5 wild type and fusion luciferase synthesized relative to MAV wild type and fusion luciferase is in part due to the presence of an additional 100 nucleotides in the 5' untranslated region of CTP5 luciferase transcripts from constructs pLX208a$_c$b$_c$ and pLX708fa$_c$b$_c$9 and which results in doubling the amount of luciferase synthesized. This was confirmed by comparing at different temperatures the luciferase amounts and activities of MAV wild type luciferase gene constructs with or without the additional 100 nucleotides (not shown). Subsequently all values of *in vivo* bioluminescence obtained from cells transformed with pLX208a$_c$b$_c$ or pLX708fa$_c$b$_c$9 were divided by two.

In vivo Activities of Luciferase Enzymes.

There is no significant difference in bioluminescence between cells transformed with *luxAB* or *luxA$_c$B$_c$* and grown at 23, 30, and 37°C (Table 1). In contrast there is a more than 10 fold increase in light emission from cells synthesizing the thermostable CTP5 luciferase at 42°C. Comparison of *in vitro* activities of the same enzymes purified at 23°C shows a 2–3 fold activity increase of CTP5 luciferase at 42°C (6). The difference observed between *in vivo* and *in vitro* measurements implies a temperature sensitive folding of MAV wild type luciferase at 42°C. Whether the improved folding of CTP5 wild type luciferase at 42°C is solely due to its primary structure or is a result of interactions with heat shock proteins remains to be determined.

Table 1. Ratios of *in vivo* bioluminescence measurements done on *E. coli* cells expressing different *lux* constructs and grown at increasing temperatures.

	23°C	30°C	37°C	42°C
LuxA$_c$B$_c$.[a] LuxAB	0.9 ± 0.2	0.8 ± 0.2	0.8 ± 0.1	12 ± 2
Fa$_c$b$_c$9[b] Fab9	1.2 ± 0.2	2.8 ± 1.6	27 ± 5	c

[a]*In vivo* activity of wild type CTP5 luciferase relative to wild type MAV luciferase.
[b]*In vivo* activity of CTP5 fusion luciferase relative to MAV fusion luciferase.
[c]No light was detected from *E. coli* cells expressing *fab9* and grown at 42°C by using the protocol described in material and methods.

In contrast cells transformed with CTP5 gene fusion *fa$_c$b$_c$9* display a marked increase in photon emission relative to cells transformed with MAV luciferase gene fusion *fab9* when grown at 30°C and above (Table 1). This may be explained in view of the extreme temperature sensitivity of fusion luciferase Fab9. Any amino acid substitution causing improved polypeptide folding at elevated temperature will show its effect at a lower temperature in the fusion luciferase than in the wild type heterodimeric enzyme.

At 37°C fusion luciferase fa$_c$b$_c$9 has roughly 30% of wild type activity (Table 2). This makes gene *fa$_c$b$_c$9* better suited for use as a marker of gene expression in mammalian systems and provides a basis for the further engineering of a luciferase fusion protein with improved activity at elevated temperature.

Table 2. *In vivo* activities of luciferase fusion protein Fab9 and Fa_cb_c9 synthesized at different temperatures and expressed as a percentage of the activity of their respective wild type enzyme synthesized at the same temperatures.

	23°C	30°C	37°C	42°C
Fab9[a] LuxAB	55% ± 10	20% ± 10	2% ± 1	c
Fa_cb_c9[b] LuxA$_c$B$_c$	70% ± 12	60% ± 13	30% ± 10	2% ± 1

[a]*In vivo* activity of MAV fusion luciferase relative to wild type MAV luciferase.

[b]*In vivo* activity of CTP5 fusion luciferase relative to wild type CTP5 luciferase.

[c]No light was detected from *E. coli* cells expressing *fab9* and grown at 42°C by using the protocol described in material and methods.

REFERENCES

1. A. Escher, D.J. O'Kane, J. Lee A.A. Szalay, Bacterial luciferase $\alpha\beta$ fusion protein is fully active as a monomer and highly sensitive *in vivo* to elevated temperature. Proc. Natl. Acad. Sci. USA, 86, 6528–6532 (1989).

2. W.H.R. Langridge, F.J. Fitzgerald, C. Koncz, J. Schell and A.A. Szalay, The dual promoter of *A. tumefacians* mannopine synthease genes is regulated by plant growth hormones. Proc. Natl. Acad. Sci. USA, 86, 3219–3223 (1989).

3. A. Escher, D.J. O'Kane, J. Lee, W.H.R. Langridge and A.A. Szalay, Construction of a novel functional bactaerial luciferase by gene fusion and its use as a gene marker in low light video image analysis. Proceedings of the SPIE's 33rd Annual International Symposium on Optical and Optoelectronic Applied Science and Engineering, 1161, 230–235 (1989).

4. O. Olsson, C. Koncz and A.A. Szalay, The use of the *luxA* gene of the bacterial luciferase operon as a reporter gene. Mol. Gen. Genet. 215, 1–9 (1988).

5. T. Maniatis, JE.F. Fritsch and J. Sambrook, In Molecular cloning: a laboratory mannual, Cold Spring Harbor Lab., Cold Spring Harbor, p. 440 (1982).

6. D.J. O'Kane, B.G. Gibson, J. Lee, A.A. Szalay and J.E. Wampler, Isolation of a thermostable bacterial luciferase from thermotolerant *Vibrio harveyi* isolates. In Lux gene symposium, p. 17 (1989).

CLONING THE INDIVIDUAL COMPONENTS OF THE FATTY ACID REDUCTASE GENES *LUXC, D* AND *E* IN *E. COLI*: BIOLUMINESCENCE WITHOUT A COMPLETE REDUCTASE COMPLEX

P.J. Hill and G.S.A.B Stewart

Department of Applied Biochemistry and Food Science
Nottingham University, Sutton Bonington,
Leicestershire, LE12 5RD, UK

INTRODUCTION

Bacterial luciferase catalyses the flavin-mediated oxidation of a long chain fatty aldehyde, tetradecanal, with concomitant emission of blue green light. This aldehyde is produced by a fatty acid reductase complex, which is composed of three different polypeptides; acyl transferase (*luxD* product) which releases tetradecanoic acid from acyl-ACP into the luminescence system; acyl protein synthase (*luxE* product) which in the presence of ATP activates the fatty acid; acyl protein reductase (*luxC* product) which is responsible for the transfer of the acyl group from the synthase and its subsequent NADPH dependent reduction in order to release free tetradecanal (1). An outline of the steps involved in synthesis of aldehyde for the bioluminescent reaction is shown in Figure 1.

The object of this study was to clone the *luxC*, *luxD* and *luxE* genes from *Vibrio fischeri* ATCC7744, independently of the regulatory region *luxI* and *luxR* which lie upstream of *luxC*, and of the luciferase genes *luxA* and *luxB* which lie between *luxD* and *luxE* (1).

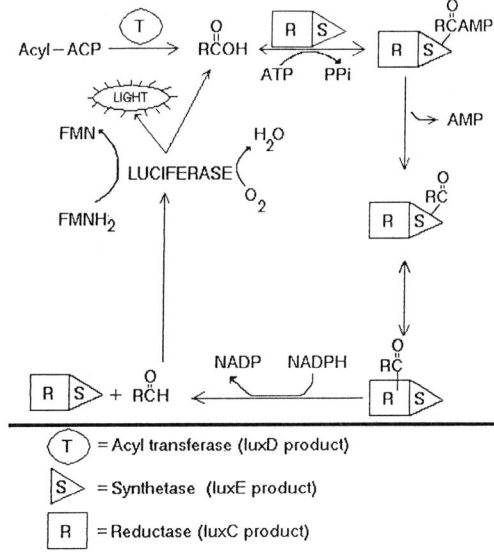

Figure 1. The reactions of the fatty acid reductase complex.

MATERIALS AND METHODS

Bacterial strains and growth media. *Vibrio fischeri* ATCC7744 was purchased from the National Collections of Industrial and Marine Bacteria (Aberdeen AB9 8DG). *E. coli* strains were grown in Lennox medium (LB broth & AL plates)(2). *V. fischeri* was grown on marine agar (Difco). To select for pBR328 and its derivatives, ampicillin was added at 50mg/l. Chloramphenicol at 30mg/l was used for selection of pSB247.

Cloning experiments. Restriction endonuclease reactions and ligations were performed as described by Maniatis *et al.* (2). Ligation products were used to transform *E. coli* as detailed by Park *et al.* (3).

Polymerase chain reaction. PCR was performed as described by Saiki *et al.* (4) using Taq DNA polymerase purchased from Amersham International. The amplification reactions were carried out in a Hybaid intelligent heating block; the DNA was denatured at $90^{\cdot}C$ for 2 minutes, annealed at $50^{\cdot}C$ for 2 minutes and polymerase extended at $70^{\cdot}C$ for 10 minutes, for 25 cycles, with a final extension at $70^{\cdot}C$ for 15 minutes to ensure completion of all strands.

Primers for PCR. Primers were synthesised using a Milligen Cyclone DNA synthesiser. Primers were designed with the aid of previously published sequences (5), with a minimum of 20 base homology to ensure accurate hybridisation. Unique restriction sites were incorporated into the 5' end of each primer to facilitate cloning of the PCR product. In an effort to circumvent problems of expression of *luxE*, primers were designed in order that the *luxE* start codon could be placed after *luxD* in the precise position that the *luxA* start codon occupies in the wild type *lux* operon. Coding strand sequences (CSS) non-coding strand sequences (NCSS) and primer sequences (Primer) are defined in Table 1.

In vivo luciferase assays. Bioluminescent colonies on solid media were visualised using an Hamamatsu Argus 100 VIM 3 photon imaging camera. Bioluminescence from broth cultures was measured in a Turner Designs 20e luminometer.

TABLE 1. PCR Primers used.

Primer #1. *luxC* coding strand, *EcoRI* and *NruI* sites underlined, *luxC* start codon in italics
CSS   ATT TTA AAT ACT AAG TAT ATT ATA GGG GAA ATA *ATG* AAT
Primer ATT TT<u>G AAT</u> <u>TCT</u> <u>CGC</u> GATATT ATA GGG G

Primer #2. *luxD* non-coding strand, *EcoRV* site underlined, *luxA* start codon in italics
CSS   AGG AAT AGA GT*A TG*A AGT TTG GAA ATT
NCSS TCC TTA TCT CAT TCA AAC CTT TAA
Primer TCC TTA T<u>CT</u> <u>ATA</u> GCT TCA AAC CTT TAA

Primer #3. *luxE* coding strand, *NdeI* site underlined, *luxE* start codon in italics
CSS   GAG GGG ATG GT*A TG*A CTG TTC ATA
Primer GAG GGG AT<u>C</u> <u>ATA</u> <u>TGA</u> CTG TTC

Primer #4. *luxE* non-coding strand, *SmaI* and *NcoI* sites underlined, *luxE* stop codon in italics
CSS   GAT *TAA* GTT ATG ATT GTT GAT GGC AGA GTT TCA
NCSS CTA ATT CAA TAC TAA CAA CTA CCG TCT CAA AGT
Primer   AA TA<u>G</u> <u>GGC</u> <u>CCG</u> <u>GTA</u> <u>CC</u>G TCT CAA AG

RESULTS AND DISCUSSION

The *luxE* fragment obtained from the chromosome of *V. fischeri* ATCC7744 via PCR using primers #3 and #4, was digested with the restriction enzymes *NdeI* and *NcoI* and ligated to an equivalent digest of pSB219 (6), thus providing *luxE* expression under the control of *trp*poL. The ligation was used to transform *E. coli* JM107 containing a compatible plasmid, pSB247, constructed by cloning the thermostable *Vibrio harveyi luxA/B* from pSB226 (6) into pSU8 (7) as an *EcoRI/SmaI* fragment. Transformants which were both ampicillin and chloramphenicol resistant were assessed for their ability to elicit a bioluminescent phenotype without the addition of exogenous aldehyde, using the photon video camera. Bright colonies were confirmed in their construction by miniprep analysis and were found to produce approximately 10^4-fold less light without aldehyde than when decanal was added exogenously. These clones were unable to produce any light in the absence of exogenous aldehyde at temperatures above 30°C, indicating that as with other gene products of the *V. fischeri lux* operon such as *luxR* (Hill and Stewart, unpublished data) and *luxA* and *luxB* (9), the polypeptide coded for by *luxE* is heat labile.

The *luxCD* fragment obtained via PCR with primers #1 and #2 was digested with the restriction enzymes *EcoRI* and *EcoRV* and ligated to an equivalent digest of pBR328. The ligation was used to transform *E. coli* JM107[pSB247] with transformants exhibiting both ampicillin and chloramphenicol resistance being screened for light production. Very few transformants were obtained, with only three showing a bioluminescent phenotype without the addition of exogenous aldehyde. Bioluminescence from these clones had an intensity similar in magnitude to the *luxE* clone discussed above. On restriction enzyme analysis of the bioluminescent clones, it was found that not all of the *luxD* gene was present, at least 300bp being absent from the 3' end. Subsequent attempts at cloning a complete *luxCD* fragment have failed. However, the digestion of the above PCR product with *BclI*, which provides two fragments (*luxCD'*, containing *luxC* and the first 30 bases of *LuxD*, and *luxD'* which contains the remaining *luxD* sequence), has allowed the successful cloning into pBR328 of *EcoRI/BclI* and *BclI/EcoRV* fragments respectively, with neither being able to elicit a bioluminescent response in *E. coli* JM107[pSB247].

In parallel with the above, the equivalent *EcoRI/BclI* fragment was excised from one of the above bioluminescent clones having *luxC* and a truncated *luxD*. Such a fragment, recloned into pBR328, again failed to provide any bioluminescent phenotype in *E. coli* JM107[pSB247].

In conclusion, *luxE* alone can provide a weak bioluminescent phenotype when expressed in *E. coli* harbouring an independent clone of *luxA/B*. *LuxC* requires at least a portion of *luxD* beyond the *BclI* site, to provide a similar weak bioluminescent phenotype. The inability, to date, to clone an intact *luxCD* or *luxCDE* in a multicopy plasmid in *E. coli*, suggests that such constructs may be lethal; a factor supported by observing that the only previously reported cloning of *luxCD* in *E. coli*, included the highly efficient transcriptional regulatory elements *luxR* and *luxI* (8). The weak bioluminescence obtained from the above *lux* constructs may help to elucidate the *in vivo* as opposed to *in vitro* mechanism of aldehyde synthesis.

ACKNOWLEDGEMENTS
We wish to thank the Agricultural and Food Research Council for funding.

REFERENCES
1. E.A. Meighen, Enzymes and genes from the *lux* operons of bioluminescent bacteria. Ann. Rev. Microbiol. 42,151-176, (1988).
2. T. Maniatis, E.F. Frisch and J. Sambrook, Molecular cloning: A laboratory manual, Cold Spring Harbor Laboratory, Cold Spring Harbor, N.Y. (1982).
3. S.F. Park, D.A. Stirling, C.S.J. Hulton, I.R. Booth, C.F. Higgins and G.S.A.B. Stewart, A novel, non-invasive promoter probe vector: Cloning of the osmoregulated *proU* promoter of *Escherichia coli* K12. Mol. Microbiol. 3, 1011-1023 (1989).
4. R.K. Saiki, D.H. Gelfand, S. Stoffel, S.J. Scharf, R. Higuchi, G.T. Horn, K.B. Mullis and H.A. Erlich, Primer-directed enzymatic amplification of DNA with a thermostable DNA polymerase. Science 239,487-491 (1988).
5. T.O. Baldwin, J.H. Devine, R.C. Heckel, J.-W. Lin and G.S. Shadel, The complete nucleotide sequence of the *lux* regulon of *Vibrio fischeri* and the *luxABN* region of *Photobacterium leiognathi* and the mechanism of control of bacterial bioluminescence. J. Biolumin. Chemilumin. 4, 326-341 (1989).
6. P.J. Hill, S. Swift and G.S.A.B. Stewart, PCR based gene engineering of the *Vibrio harveyi lux* operon and the *Escherichia coli trp* operon provides for biochemically functional native and fused gene products. Mol. Gen. Genet. In press (1990).
7. E. Martinez, B. Bartolome and F. De La Cruz, pACYC184-derived cloning vectors containing the multiple cloning site and *lacZ alpha* reporter gene of pUC8/9 and pUC18/19 plasmids. Gene 68, 159-163 (1988).
8. C. Miyamoto, M. Boylan, L. Cragg and E. Meighen, Comparison of the *lux* systems in *Vibrio harveyi* and *Vibrio fischeri*. J. Biolumin. Chemilumin. 3, 193-199 (1989).
9. G.N. Sakharov, A.D. Ismailov and V.S. Danilov, Temperature dependences of the reaction of bacterial luciferase. Biokhimiya 53, 891-898 (1988). (Original article).

ENGINEERING PROTEIN KINASE RECOGNITION SITES INTO FIREFLY
LUCIFERASE AND AEQUORIN

G. Sala-Newby and A.K. Campbell

Department of Medical Biochemistry
University of Wales College of Medicine,
Heath Park, Cardiff, CF4 4XN, UK

INTRODUCTION

A universal feature of eukaryotic cells is the ability of
physiological agonists, drugs and pathogens, acting at the plasma
membrane, to activate or injure chemical reactions and structures
within the cell. Binding or insertion of the initiating agent to
the plasma membrane results in the generation of signals such as
$Ca2+$, IP_3 or cyclic nucleotides which then induce covalent
modifications to intracellular proteins, which are responsible for
provoking the end response (1,2). The end response include cell
movement, secretion, transformation, division, cell defence and
death, either via apoptosis or lysis. Their are two major
problems one conceptual, the other technical, which have, so far,
prevented the elucidation of the complete sequence from plasma
membrane to end response. When measurements are carried out on
cell populations or on tissue homegenates it appears that the
passage from signal to protein to response is smooth and
continuous. However, when examined at the level of individual
cells, it can be seen that each cell crosses a series of
thresholds or "rubicons, which must be crossed in the correct
sequence and at the right place if the cell is to produce the
appropriate end response (3). The time and magnitude of each
"rubicon" vary from cell to cell. Thus in order to elucidate the
pathway for cell activation or injury it is essential that methods
are available to measure, locate and manipulate the signals,
covalent modification of proteins and the end response in
individual, live cells.

A new strategy has been developed using the polymerase
chain reaction (PCR), to engineer bioluminescent proteins to
measure signals i.e. cyclic nucleotides, IP_3, GTP, ATP and
covalent modifications i.e. phosphorylation, such that a change in
light intensity or colour enables a particular parameter to be
quantified (4). Benzothiazole luciferases occur in luminous
beetles and are 60KDa, euglobular proteins composed of
approximately 550 amino acids, and catalyse a reaction which can
emit green, green-yellow, yellow, orange or red light (5). In
contrast imidazolopyrazine luciferases and photoproteins emit blue
light with maximum at about 440 to 490 nm (2).

The aim of the work reported here was to engineer the protein kinase A recognition heptapeptide, kemptide LRRASLG (6), on to the N and C terminus of firefly luciferase and aequorin DNA. A valine at position 217 in firefly luciferase was converted to an arginine by a two base change, resulting in a putative protein kinase A site, RRFS. A T7 promoter was incorporated at the 5', enabling the activity of new protein to be assessed following formation of mRNA and translation *in vitro* (7).

MATERIALS AND METHODS
Firefly luciferase and aequorin DNA: Firefly luciferase (Photinus pyralis) cDNA was isolated by cloning from a cDNA library (8). Aequorin cDNA was isolated by reverse transcriptase PCR (9) using poly A$^+$ mRNA (Aequorea kind gift from Dr. C.C. Ashley, Oxford) and 5' sense primer + T7 promoter TAATACGACTCACTATAGGGAGAATGGTCAAGCTTACATCAGACTTCGAC and the 3' antisense primer GAATTCTTAGGGGACAGCTCCACCGTA. Genomic DNA was also isolated from firefly tails and Aequorea by standard methods, and amplified using the same primers for aequorin cDNA, and primers CACCTAATACGACTCACTATAGGGAGAATGGAAGACGCCAAAAAC 5' sense, and TCATCGCTGAATACAGTTAC 3' antisense (non-coding region) for firefly c or g DNA (Fig. 1.)

DNA	exp	gel	DNA	exp	gel
1.Plasmid-A10.6.1		5370	8.FFT7,Nterm mutant	689	692
2.salI fragment	2400	2485	9.FF mutant +T7	1703	1698
3.FF,full length	1682	1641	10.aequorin	573	589
4.FF+T7 promoter	1703	1718	11.aequorin +T7	603	631
5.FF-36bp C term.	1646	1622	12.aequorin rePCR	573	596
6.FF+T7-36bp	1667	1660	13.primers alone	50	<200
7.FF,Cterm mutant	1040	1059	14.no PCR		

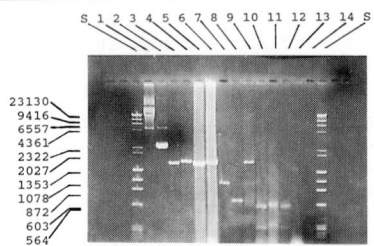

Fig.1

Incorporation of protein kinase sites: Normal reactions consisted of 25 cycles of 1 min at 94°C, 1 min at 55°C, 2 min at 72°C + 5 sec extension on each cycle in 10mM TRIS pH8.3, 50mM KCl, 2mM

$MgCl_2$ 0.01% gelatin (w/v), 0.2mM of each of the four
deoxyribonucleoside triphosphates, 5 x 10^{-4}M each primer and 8 U/ml
Ampli Taq DNA polymerase, followed by 30 min at 37°C with 400 U/ml
Klenow fragment of DNA polymerase. Incorporation of kemptide at
the N terminus and mutation of residue 217 to R in firefly
luciferase involved two step PCR procedures. In the latter the
cDNA was prepared in two pieces in PCR Stage 1 using
AGAACTGCCTGCCGCAGATTCT CGCA and TGCGAGAATCTGCGGCAGGCAGTTCT and
then linked together in PCR Stage 2 which required only 7-10
cycles (Fig. 1).

Assays In vitro transcription using T7 RNA polymerase to form
capped mRNA, and then translation in rabbit reticulocyte lysate
to form firefly luciferase or apo aequorin was as previously
described (7,8), as were the assays for luciferase and
reactivation of Ca^{2+}-activated photoprotein.

RESULTS AND DISCUSSION
Transcription-translation of cDNA versus g DNA: Firefly luciferase
g DNA has six introns (10) and the PCR product was approx. 296 bp
longer for the g than cDNA. Both formed mRNA of the correct
length, but only mRNA from the cDNA translated to form active
luciferase, as expected. However aequorin c and g DNA were the
same length on agarose gels (ca 600 bp) and both formed active
aequorin from mRNA, suggesting that aequorin g DNA has no introns.
Characterisation of PCR products: Successful PCR reactions were
confirmed by three criteria (a) correct length of ethidium
bromide stained bands from agarose gel electrophoresis (Fig. 1)
(b) formation of ^{32}P mRNA of correct length on glyoxal
polyacrylamide electrophoresis (8) (c) translation of mRNA to ^{35}S
protein and measurement of luminescent activity (Fig. 2).

The activity of bioluminescent protein per g RNA varied
considerably. Mutation of V217 to R resulted in some 80% loss in
activity, but addition of kemptide to the C terminus caused little
loss. Similarly addition of kemptide to the N terminus of
aequorin resulted in full retention of Ca^{2+}-activated
photoprotein.

Attempts to demostrate a change in pI between the various
recombinant luciferase, using isoelectric focussing, were
unsuccessful, because of artifactual bands generated from the
focussing procedure. However the major band for recombinant and
extracted luciferase had the same pI (6.6). Measurement of the
ratio of chemiluminescence at 603/545nm of extracted, recombinant
full length, and recombinant containing RRXS or kemptide at the C
terminus, luciferase showed no apparent difference at pH 7.8.
Furthermore, under conditions where kemptide was phosphorylated
30% no effect on the ratio of light emission at 603/545nm between
the various luciferases was detected, in spite of differences in
specific activity (Fig. 2).

If a change in intensity or colour of these mutated
bioluminescent proteins, or "rainbow" proteins" (4) can be
detected when phosporylation occurs, as has been demonstrated for

chemically formed kemptide-luciferase (9), then it should be possible, for the first time, to monitor protein phosphorylation in live cells.

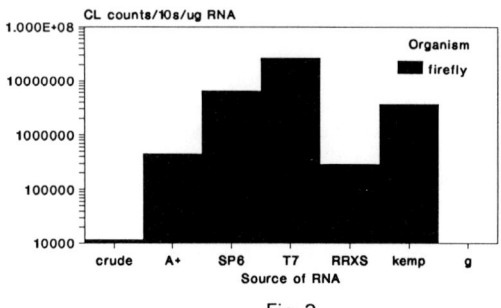

Fig. 2

ACKNOWLEDGEMENTS
We thank the MRC and AFRC for financial support.

REFERENCES
1.   A.K. Campbell. Intracellular calcium: its universal role as regulator. Wiley, Chichester (1983).
2   A.K. Campbell. Chemiluminescence: principles and applications in biology and medicine. Horwood/VCH, Chichester and Weinheim (1988).
3.   A.K. Campbell. The Rubicon Hypothesis: a quantal framework for understanding the molecular pathway of cell activation and injury in calcium and oxygen radicals in cell injury in press. C.J. Duncan (Ed) SEB Monograph, Cambridge University Press, Cambridge (1991).
4.   A.K. Campbell. Rainbow protein. British Patent Application 8916806.6 (1989).
5.   K.V. Wood, Y.A. Larn and W.D. McElroy. Complementary DNA coding click beetle luciferases can elicit bioluminescence of difference colours. Science 244, 700-702 (1989).
6.   J.L. Maller, B.E. Kemp and E.G. Krebs. In vitro phosporylation of a synthetic peptide substrate of cyclic AMP-dependent protein kinase. Proc. Natl. Acad. Sci. USA 75 248-251 (1978).
7.   A.K. Campbell, G. Sala-Newby, P.J. Aston, N. Kalsheker, Y. Kishi and O. Shimomura. From Luc and Phot genes to the hospital bed. J. Biolum. Chemilum. 5, 131-139 (1990).
8.   G.Sala-Newby, N. Kalshekar and A.K. Campbell. Removal of twelve C-terminal amino acids from firefly luciferase abolishes activity. Biochem. Biophys. Res. Commun. in press (1990).
9.   T.M. Jenkins, G. Sala-Newby and A.K. Campbell. Measurement of protein phosphorylation by covalent modification of firefly luciferase. Biochem. Soc. Trans. 18, 463-464 (1990).
10.   J.R. DeWet, K.V. Wood, M. DeLuca, D.R. Helinski and S. Subramani. Firefly luciferase gene = structure and expression in mammalian cells. Mol. Cell. Biol. 1, 725-735 (1987).

DIFFERENCES IN THE REGULATORY SIGNALS
CONTROLLING EXPRESSION OF THE *VIBRIO LUX* SYSTEMS

E. Swartzman, J.G. Cao, C. Miyamoto,
A. Graham, and E. Meighen

*Department of Biochemistry, McGill University,
Montreal, Quebec, Canada H3G 1Y6*

INTRODUCTION

Our understanding of the mechanisms involved in the regulation of bacterial luminescence has increased greatly over the last number of years. The genes responsible for light production in *Vibrio harveyi* and *Vibrio fischeri* have been identified and expressed in *Escherichia coli*(1,2). A comparison of the *lux* operons (Fig. 1) shows that while the two marine bacteria have common structural genes, the mode of *lux* gene regulation must differ. *V. harveyi* and *V. fischeri* share five structural *lux* genes, *luxA-E*, that are located in the same order within the two lux operons: the *luxA* and *B* genes code for the α and ß subunits of luciferase, and the *luxCD* and *E* genes encode the polypeptides of the fatty acid reductase complex that is required for aldehyde synthesis. Recently, two new *V. harveyi lux* genes, *luxG* and *H*, have been identified and sequenced(3). Although *luxG* and *luxH* are located immediately after *luxE* and are within the *lux* transcriptional unit, it is not known what function they have in bioluminescence. Analysis of the *V. fischeri* DNA downstream of *luxE* has shown that this system also contains the *luxG* gene, but lacks *luxH*(unpublished results). In contrast to *V. harveyi*, the *V. fischeri* regulatory *lux* genes are found linked to the structural genes. LuxI, believed to be responsible for autoinducer production, is cotranscribed with and located immediately upstream of *luxCDABEG*. LuxR is found adjacent to *luxI*, but is transcribed in the opposite direction and encodes a regulator protein that is thought to bind autoinducer and stimulate transcription of the right *lux* operon (*luxICDABEG*). Although no analogous *luxI* or *luxR* genes have been found linked to the *V. harveyi lux* structural genes, a gene having *lux* regulatory properties has been located at an unlinked locus(4). It is apparent then that the gene organization of *lux* regulation differs in these two *Vibrio* species. In this study, the molecular mechanisms controlling bioluminescence in *V. harveyi* and *V. fischeri* are compared and contrasted.

MATERIALS AND METHODS

Lysates of *Photobacterium phosphoreum* NCMB844, *V. fischeri* ATCC 7744, *V. harveyi* B392 and *E. coli* RR1 were prepared according to a procedure provided by Lee Wall(personal communication). Mobility shift assays were performed according to procedures previously described(5), using 0.5xTBE as the electrophoresis buffer. The DNA probe used in the assay was obtained from a region upstream of the *V. harveyi lux* promoter(3) and end-labelled with α-[^{32}P]dATP and the large fragment of DNA polymerase I.

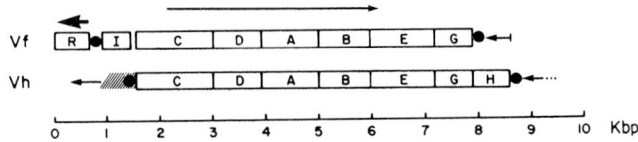

Fig. 1 Gene organization of the *lux* operons of *V. fischeri* and *V. harveyi*. The *V. fischeri lux* system contains two operons(top). The right operon contains *luxICDABE* and *G*, while the left operon contains *luxR*. The *V. harveyi lux* operon contains *luxCDABEG* and *H*(bottom). The arrows indicate direction of transcription.

RESULTS AND DISCUSSION

Several factors involving the regulation of the *Vibrio lux* systems have already been studied, including transcription initiation and termination. The promoter for the *V. harveyi lux* operon has been located in front of *luxC*, while the promoter for the *V. fischeri* right *lux* operon is found just upstream of *luxI*. The *V. fischeri* inducer gene is therefore part of the right *lux* operon and, together with the *luxR* gene product, is responsible for the positive feedback loop controlling autoinduction of luminescence. The genetics of autoinduction in *V. harveyi* has yet to be defined. Transcription terminators have been found for the *lux* operon of *V. harveyi* and the right *lux* operon of *V. fischeri*(Fig. 2). In *V. harveyi*, the terminator is located just after *luxH* and precedes a terminator for

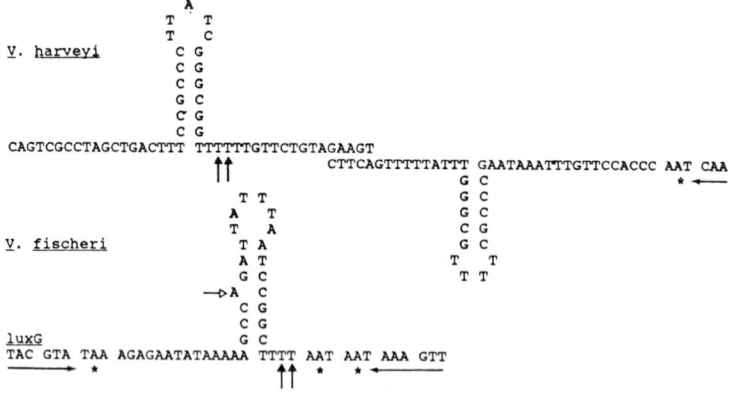

Fig. 2 Transcription terminators of the *V. harveyi* and *V. fischeri lux* systems. Filled arrowheads indicate transcription termination sites of *lux* DNA. An empty arrowhead points to the terminator site of gene transcribed in opposite direction.

a gene transcribed in the opposite direction. In contrast, the terminator for the *V. fischeri* right *lux* operon is located after *luxG* as well as a gene transcribed in the opposite direction, and has been shown to function bidirectionally (unpublished results). Although quite rare, bidirectional terminators are found in some prokaryotic and eukaryotic systems, but it is not known what effect they may have on gene expression.

Light induction in *V. harveyi* and *V. fischeri* is cell-density dependent and is responsive to a small molecule, termed autoinducer, that is produced by the cells and diffuses freely across the cell wall. The autoinducers of both *V. harveyi* and *V. fischeri* have been purified and shown to have very similar chemical structures(6,7). The autoinducer of *V. fischeri* has been identified as N-(ß-ketocaproyl) homoserine lactone, while that of *V. harveyi* is N-(ß-hydroxybutyryl) homoserine lactone(Fig. 3). Although both are composed of an amino acid linked by amide bond to a fatty acid metabolite, the autoinducers are species-specific and cannot be interchanged. Consequently, the molecular mechanisms controlling light emission in the two bacteria must differ.

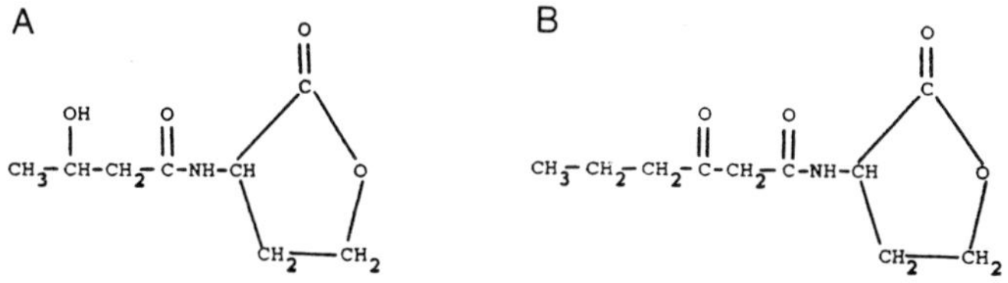

Fig. 3 Autoinducers of the *V. harveyi* and *V. fischeri lux* systems.   (A)   *V.   harveyi*   autoinducer   N-(ß-hydroxybutyryl) homoserine   lactone.   (B)   *V.   fischeri*   autoinducer   N-(ß-ketocaproyl) homoserine lactone.

As regulatory proteins directly affecting *V. harveyi* light induction have yet to be identified, we attempted to search for proteins that would bind near the *V. harveyi lux* promoter. This was accomplished with mobility shift assays using an insert of DNA from just upstream of the promoter and cell extracts. As can be seen in Fig. 4, only protein extracted from *V. harveyi* caused the DNA fragment to migrate slower in the native polyacrylamide gel, indicating specific DNA-protein interactions. No mobility shift could be seen when protein from *V. fischeri, E. coli* or *P. phosphoreum*, a different genera of luminescent bacteria, was used in the assay. The protein that is binding near the *V. harveyi lux* promoter is species-specific and hence may be involved in *lux* gene regulation.

Examination of all aspects of light production in *V. harveyi* and *V. fischeri* reveals some interesting contrasts. Although the five known structural genes (*luxABCD* and *E*) are conserved, the genes controlling *lux* regulation have diverged. The 5' and 3' ends of the *lux* operons differ, including transcription initiation and termination sites. The species specific autoinducers are very similar in structure and stimulate light production in the same

manner, but cannot be interchanged.   Finally, protein in extracts
of *V. harveyi*, but not *V. fischeri*, bind to DNA near the *V. harveyi*
lux promoter.   Further work comparing and contrasting the *lux*
regulatory mechanisms in *V. harveyi* and *V. fischeri* will lead to a
better understanding of gene regulation in luminescent bacteria.

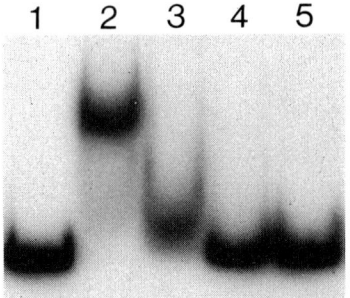

Fig. 4   *V. harveyi* DNA-Protein Binding. A ^{32}P-labelled DNA
fragment located just upstream of the *V. harveyi lux* promoter was
used in the mobility shift assay with protein extracted from *V.
harveyi*(lane 2), *E. coli*(lane 3), *V. fischeri*(lane 4) and *P.
phosphoreum*(lane 5).   Lane 1 contains the probe without any added
protein.

ACKNOWLEDGMENTS
This work was supported by Grant MA-7672 from the Medical Research
Council of Canada.

REFERENCES
1. C.M. Miyamoto, D. Byers, A.F. Graham, E.A. Meighen, Expression
   of Bioluminescence in *Escherichia coli* by recombinant *Vibrio
   harveyi* DNA J. Bacteriol. 169, 247-53 (1987).
2. J. Engebrecht, K.H. Nealson, M. Silverman, Bacterial
   Bioluminescence: Isolation and Genetic Analysis of Functions
   from *Vibrio fischeri*. Cell 32, 773-81 (1983).
3. E. Swartzman, C. Miyamoto, A. Graham, E. Meighen, Delineation
   of the Transcriptional Boundaries of the *lux* Operon of *Vibrio
   harveyi* Demonstrates the Presence of two New *lux* Genes. J.
   Biol. Chem. 265, 3513-517 (1990).
4. M. Martin, R. Showalter, M. Silverman, Identification of a
   Locus Controlling Expression of Luminescence Genes in *Vibrio
   harveyi*. J. Bacteriol. 171, 2406-2414 (1989).
5. R. J. Rolfes, H. Zalkin, *Escherichia coli* Gene *purR* Encoding a
   Repressor Protein for Purine Nucleotide Synthesis. J. Biol.
   Chem. 263, 19653-19661 (1988).
6. J.-G. Cao, E.A. Meighen, Purification and Structural
   Identification of an Autoinducer for the Luminescence System of
   Vibrio harveyi. J. Biol Chem. 264, 21670-21676 (1989).
7. A. Eberhard, A.L. Burlingame, C. Eberhard, G.L. Kenyon, K.H.
   Nealson, N.J. Oppenheimer, Structural Identification of
   Autoinducer of *Photobacterium fischeri* luciferase. Biochemistry
   20, 2444-2449 (1981).

EXPRESSION OF FUSED BACTERIAL LUCIFERASE IN MAMMALIAN CELLS

Stacy Costa, Elizabeth Douville, John Bell, and *Edward Meighen

*Departments of Biochemistry and Medicine, University of Ottawa,
451 Smyth Road, Ottawa, (Ontario, Canada K1H 8M5),
and *Department of Biochemistry, McGill University,
3655 Drummond Street, Montreal, (Quebec, Canada H3G 1Y6)*

INTRODUCTION

The expression of firefly luciferase in eukaryotic cells has stimulated much interest in the use of light emission as a highly sensitive marker of gene expression (1). Bacterial luciferase, which catalyzes the oxidation of $FMMH_2$ and a long chain aldehyde resulting in the emission of light, is a heterodimeric enzyme (α,β) composed of two nonidentical subunits (α and β) coded by luxA and luxB (2). Generation of a fused monocistronic bacterial luciferase gene (3) has now made it possible to express the bacterial luciferase genes under a single promoter in eukaryotic cells. The fused gene(luxA-B) contains the coding regions(luxA and luxB) for both subunits(α,β) of luciferase linked together by a short peptide connecting the carboxyl terminal of the α subunit to the amino terminal of the β subunit. Luciferase activity comparable to the wild type heterodimeric enzyme ($\alpha\beta$) has been observed for the fused luciferase. Fused luciferases have been expressed not only in bacterial cells but also in eukaryotic systems including yeast (3) and plant (4) cells as well as *in vitro* in a reticulocyte translation system (3).

The present experiments show that the fused bacterial luciferase can be expressed in mouse fibroblast cells. Moreover, by complementation of an altered cell line containing a defective retroviral genome, it was possible to general lux+ virions that can be used for highly efficient infection of mammalian cells.

MATERIALS AND METHODS

Cells ands Plasmids. Psi-2 NIH 3T3 cells are a helper cell line identical to normal mouse fibroblasts(NIH 3T3) except they possess a defective retroviral genome(psi-2) such that the viral RNA cannot be packaged into virions. However, all the proteins needed for the virions can be synthesized. The plasmid (pWE β-actin) containing the psi-2 packaging locus was a gift from R. Mulligan and P. Gros.

Electroporation. Plasmid constructs(60 μg) after ethanol precipitation were mixed with cells released by trypsinization from 175 cm^2 plates into 1.0 ml of α-MEM (serum free) for 10 min. The mixture was then transferred to a cuvette and electroporated at 500 microfarads and 300 volts. The cells were then plated in

MEM with 10% serum (50:50 newborn:fetal calf serum) and screened
48 hours later for G418 resistance (1 mg/mL).
Luciferase Assays. Extracts from mammalian cells were obtained
by freeze thawing and mild sonication followed by centrifugation
of the cellular debris. Extract (<400 μl) was added to 1.0 ml
of 0.05 M Na/K phosphate buffer, pH 7.0, containing 0.2% bovine
serum albumin and 0.002% decanal followed by injection of 1.0 ml
of 50 μM $FMNH_2$. Activity is given by the maximum light intensity
in light units (LU) where one LU corresponds to 1 x 10^{10}
quanta/sec. Protein concentrations were measured by the BioRad
assay.

RESULTS and DISCUSSION

Construction of a plasmid that could express luciferase in
mammalian cells was accomplished by insertion of an EcoRV fragment
coding for the fused bacterial luciferase gene(luxA-B) after the
β-actin promoter in the pWE plasmid (Fig. 1).

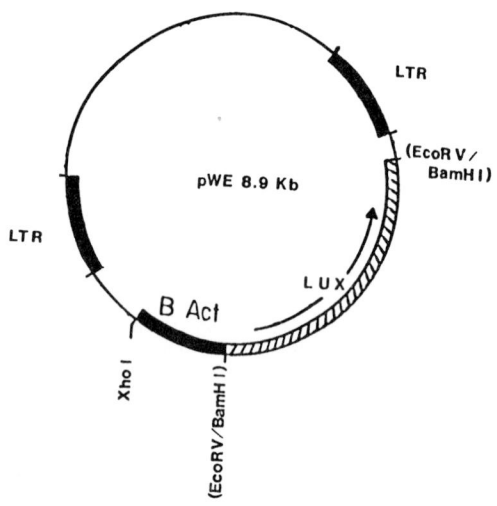

Figure 1. Construction of a luxA-B[+] plasmid for expression in
mammalian cells. The pWE-luxA-B plasmid was constructed by
ligating an EcoRV DNA fragment (2.3 kbp) starting 3 nucleotides
before and ending 309 nucleotides after the initiation and
termination codons of a fused Vibrio harveyi luxA-B gene (3),
respectively, with T4-DNA ligase into the pWE plasmid between the
long terminal repeat sequences (LTR) after restriction of the
vector with BamHI and blunt ending with the Klenow fragment of
DNA polymerase.

This plasmid also contains the gene conferring resistance to neomycin. Transfected cells containing the fused luciferase gene can be readily selected by their resistance to specific aminoglycosides conferred by the cointegration of the luxA-B gene and the neomycin gene into the host genome.

After electroporation, cells were grown for 48 hours and screened for the expression of the neomycin gene by their resistance to the aminoglycoside, G418. Resistant colonies were passaged for several generations at 37°C before growth at various temperatures (Table 1). Luciferase activity in lysates from the 3T3 cells was highly dependent on the temperature of growth with highest activity in extracts of cells grown at the lowest temperature.

Table 1

EFFECT OF GROWTH TEMPERATURE ON LUCIFERASE
ACTIVITY IN FIBROBLASTS[a]

Activity (LU/mg x 10^3)	T
0.3	37 C
2.5	30 C
30	24 C

Virions containing the fused luciferase gene were produced by the lux+ NIH 3T3 psi-2 cells. These virions composed of the retroviral proteins but containing RNA transcribed from the pWE-luxA-B DNA could be used for infection of normal 3T3 fibroblasts. Analysis of G418 resistant colonies demonstrated that luciferase activity could be detected in lysates from most clones although the level of activity varied over a wide range (Table 2).

The present results have demonstrated that the fused luciferase can be expressed in mammalian cells and in particular in mouse fibroblast cell lines. We also produced virions containing the fused luxA-B viral RNA and used these to infect normal fibroblast cells. Analysis of individual G418 resistant colonies demonstrated that most of the infected fibroblasts emitted light although there was considerable variation in level due presumably to integration of the luxA-B gene at different sites in the mouse genome. This approach appears to have considerable potential as an efficient method for introduction of the luxA-B genome into mammalian cells.

Although relatively high levels of luciferase activity could be extracted from cells grown at low temperatures (24° or 30°C) the level of activity in cells grown at 37°C was somewhat disappointing. Recent experiments have indicated that the folding and/or stability of the fused luciferase in E.coli cells grown at higher temperatures (i.e. 37°C) is extremely poor compared to the native enzyme formed via interaction of the unlinked α and β polypeptides. In contrast, the relative activities of the fused and native bacterial luciferase are much more comparable in cells

Table 2

LUCIFERASE LEVELS IN FIBROBLAST CELLS
INFECTED WITH LUX+-VIRIONS

Activity (LU/mg x 10^3)	No. of Colonies*
< 1	2
1- 4	7
5-30	4
106	1
1560	1

---

*Normal 3T3 fibroblasts were infected with lux+-virions giving approximately 1000 colonies with an average luciferase activity of 0.025 LU/mg. Analysis of 15 individual colonies demonstrated a wide range of luciferase levels.

grown at lower temperatures (5).   Construction of a fused luciferase with improved folding and/or stability may be necessary for the use of bacterial luciferase as a reporter of gene expression in mammalian cells at 37°C.

ACKNOWLEDGEMENTS
Supported by grants MA-7672 and MA-9852 from the Medical Research Council of Canada.   We wish to thank Rosza Szittner for her invaluable assistance during the course of these experiments.

REFERENCES
1. D.W. Ow, K.V. Wood, M. Deluca, J.R. DeWet, D.R. Helinski and S.H. Howell, Transient and stable expression of the firefly luciferase gene in plant cells and transgenic plants. Science. 234, 856-859(1986).
2. E.A. Meighen, Enzymes and Genes from the Lux Operons of Bioluminescent Bacteria. Annual Review of Microbiology 42, 151-176(1988).
3. M. Boylan, J. Pelletier and E. Meighen, Fused Bacterial Luciferase Subunits Catalyze Light Emission in Eukaryotes and Prokaryotes. J. Biol. Chem. 264, 1915-1918(1989).
4. O. Olsson, A. Escher, G. Sandberg, J. Schell, C. Koncz and A. Szalay, Engineering of monomeric bacterial luciferases by fusion of luxA and luxB in Vibrio harveyi. Gene 81, 335-347 (1989).
5. A. Escher, D.J. O'Kane, J. Lee and A. Szalay, Bacterial luciferase αβ fusion protein is fully active as a monomer and highly sensitive in vivo to elevated temperature. Proc. Natl. Acad. Sci. USA 86, 6528-6532(1989).

THE CLONING AND EXPRESSION OF *LUXAB* IN *LISTERIA MONOCYTOGENES*

S.F. PARK, U. NISSEN AND G.S.A.B. STEWART

Department of Applied Biochemistry and Food Science,
Nottingham University, Sutton Bonington,
Leicestershire, LE12 5RD, UK

INTRODUCTION

In vivo bioluminescence requires a functional intracellular biochemistry and, consequently, can be used as a rapid monitor of microbial viability and stress operating in complex environments and in real-time (1). Such studies need not be confined to the limited number of inherently bioluminescent bacterial species since normally "dark" terrestrial microorganisms which have been engineered to both contain and express the (*luxAB*) genes from marine bacteria also emit light. Although a number of Gram positive organisms, including *Bacillus subtilis* and Lactic acid bacteria (2,3) have been engineered to a bioluminescent phenotype, there are as yet no reports of bioluminescent derivatives of Gram positive pathogens.

The Gram-positive bacterium *Listeria monocytogenes* is a facultative intracellular pathogen responsible for a variety of diseases of medical and vetinarian importance. In recent years this organism has been increasingly associated with food contamination, causing illness and death (4). The object of this study, therefore, was to construct a bioluminescent derivative of *L. monocytogenes* to enable developments such as described above to be initiated for this organism. The utility of bioluminescent *L. monocytogenes* for monitoring biocide action and sub-lethal cell injury is discussed.

MATERIALS AND METHODS

Bacterial strains and growth media *L. monocytogenes* 23074 (serotype 4b) was purchased from the American Type Culture Collection. Strains were grown in Brain Heart Infusion broth (Oxoid) at $30^{\circ}C$ and 150 rpm or on BHI agar plates. To select for pSB292 and its derivatives, chloramphenicol was included at 5 mg/l.

Cloning experiments Chromosomal DNA from *L. monocytogenes* was isolated as described previously (5). Restriction reactions and ligations were performed as described by Maniatis *et al* (6). The products of ligation were used to transform *L. monocytogenes* after dialysis against TE (10 mmol/l Tris-HCl, pH 7.9; 1mmol/l EDTA) on VS membranes (Millipore) using the method of Park and Stewart (7).

Luciferase assays When cells were grown on solid media, bioluminescent colonies were visualised using a Hamamatsu Argus 100 VIM 3 photon imaging camera following the addition of dodecanal to the petridish lid. *In vivo* bioluminescence from broth cultures was assessed, following the addition of 0.001 vol. 1% dodecanal in ethanol, using a luminometer (Turner Designs).

Biocide assay Cells of bioluminescent *L. monocytogenes* were grown up to mid-exponential phase in BHI containing chloramphenicol. Aliquots of 0.1ml (10^7 cells) were then added to wells in a microtitre plate containing equal volumes of BHI with appropriate biocide concentrations. Cells were incubated at $30^{\circ}C$ for 15 min and bioluminescence assessed using the photon counting camera after the addition of 0.001 vol. 1% dodecanal solution to the individual wells.

RESULTS AND DISCUSSION

To use bioluminescence in studies of *L. monocytogenes* a *lux* expression vector was needed to transform the organism. To this end the broad host range promoter probe pSB292 was constructed (unpublished data). This vector is based on the plasmid pCK1 (8) and contains a promoterless copy of the *luxAB* genes from *Vibrio fischeri* as marker genes (Fig.1). Expression of *luxAB* and the acquisition a bioluminescent phenotype in *L. monocytogenes* was facilitated by the introduction of Listerial promoters into the upstream multicloning site of pSB292. Chromosomal DNA fragments, generated by restriction with the enzymes *Alu*I, *Hae*III and *Rsa*I were cloned into the compatible *Sma*I site in pSB292. The products of the ligation were used to transform *L. monocytogenes* and of the resulting 3×10^3 transformants, 43 were found to be bioluminescent when visualized by the VIM 3 camera. These constructs were shown subsequently to contain pSB292 derivatives in which expression of the *luxAB* genes was under the control of putative Listerial promoters. Fig. 2 shows a strain of *L. monocytogenes*, containing one such plasmid, imaged by its own bioluminescence.

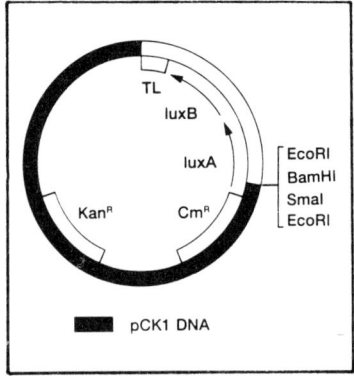
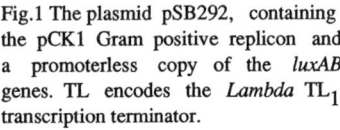

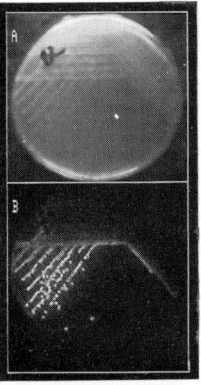

Fig.1 The plasmid pSB292, containing the pCK1 Gram positive replicon and a promoterless copy of the *luxAB* genes. TL encodes the *Lambda* TL$_1$ transcription terminator.

Fig.2 Light emitting Listeria: *Listeria monocytogenes* 23074 containing an expression competent derivative of pSB292, viewed under external illumination (A) and imaged by its own bioluminescence (B).

The utility of a bioluminescent strain of *L. monocytogenes* can be demonstrated readily. Fig. 3 shows a simple growth curve in which bacterial growth was monitored either by following absorbance at 600nm or by measuring *in vivo* bioluminescence. It demonstrates clearly that bioluminescence correlates to bacterial growth until the approach of stationary phase when it begins to decay, presumably as cellular biochemistry is perturbed. The ability of cellular bioluminescence to reflect the biochemical status of the cell can be utilized to provide a rapid assay for the activity of biocides against *L. monocytogenes*. Fig. 4 shows the prototype of such a test where bioluminescent cells were exposed to various concentrations of different biocides. A reduction in cellular bioluminescence is dependent on both the concentration and the activity of particular antimicrobial agents. These results are consistent with those previously obtained with bioluminescent *Escherichia coli* (9).

The potential of these new constructs for exploring growth *in situ* in real food systems and for monitoring the direct effect of food preservation conditions on this important pathogen is considerable.

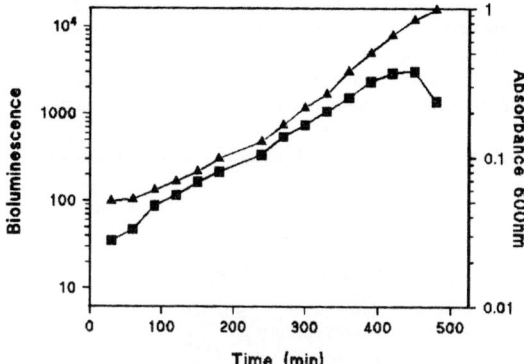

Fig.3 Growth of *L. monocytogenes* followed by absorbance at 600nm (▲) and bioluminescence (■).

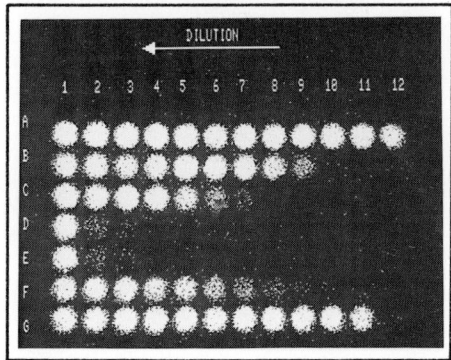

Fig.4 Testing biocide activity against *L. monocytogenes* by measuring cellular bioluminescence. Each well of the microtitre plate contained 10^7 bioluminescent cells. The initial biocide concentration in the wells of column 12 was 1% (v/v) and decreased in twofold dilutions from right to left. The wells in column 1 and row A were controls which contained no biocide. Biocides used were: (B), Chloroxylenol; (C), Chlorhexidine gluconate; (D), 4° Ammonium Formulation 1; (E), 4° Ammonium Formulation 2; (F), Hypochlorite formulation; (G), Hypochlorite.

ACKNOWLEDGEMENTS
We wish to thank the Agricultural and Food Research Council for funding SFP and providing the Hamamatsu Argus 100 VIM 3 photon counting camera.

REFERENCES
1. G.S.A.B. Stewart, *In vivo* bioluminescence: new potentials for microbiology. Lett. Appl. Microbiol. 10, 1-8 (1990).
2. M. Karp, Expression of bacterial luciferase from *Vibrio harveyi* in *Bacillus subtilis* and in *Escherichia coli*. Biochim. et Biophys. Acta 1007, 84-90 (1989).
3. K.A. Ahmad and G.S.A.B. Stewart, The production of bioluminescent Lactic acid bacteria suitable for the rapid assessment of starter culture activity in milk J. Appl. Bacteriol. In press (1990).
4. J. Lovett, *Listeria monocytogenes*, In Foodborne Bacterial Pathogens, P. Doyle (Ed), Marcel Dekker Inc. New York, p. 284-311 (1989).
5. D.G. Pitcher, N.J. Saunders and R.J. Owen, Rapid extraction of bacterial genomic DNA with guanidium thiocyanate. Lett. Appl. Microbiol. 9, 151-156 (1989).
6. T. Maniatis, E.F. Fritsch and J. Sambrook, Molecular cloning: a laboratory manual, Cold Spring Harbor Laboratory, Cold Spring Harbor, N.Y. (1982).
7. S.F. Park and G.S.A.B. Stewart, High efficiency transformation of *Listeria monocytogenes* by electroporation of penicillin treated cells. Gene In press (1990).
8. M.J. Gasson and P.H. Anderson, High copy number plasmid vectors for use in lactic streptococci. FEMS Microbiol. Lett. 30, 193-196 (1985).
9. S.A.A. Jassim, A. Ellison, S.P. Denyer and G.S.A.B. Stewart, *In vivo* bioluminescence: a cellular reporter for research and industry. J. Biolumin. Chemilumin. 5, 115-122 (1990).

COMPLEMENTATION *IN VIVO* OF THE *LUX*-SPECIFIC
FATTY ACID REDUCTASE SUBUNITS
FROM DIFFERENT LUMINESCENT BACTERIA

S. Ferri, R. Soly, C. Miyamoto and E. Meighen

*Department of Biochemistry, McGill University,
Montreal, Quebec, H3G 1Y6, Canada*

INTRODUCTION (1)

Light emission in luminescent bacteria is catalyzed by luciferase which oxidizes long chain aldehydes to the corresponding fatty acids. The aldehyde substrate for the luciferase reaction is supplied by the fatty acid reductase complex which is made up of acyl-CoA reductase, acyltransferase and acyl-protein synthetase subunits coded by the *lux*C, D and E genes, respectively. The transferase (*lux*D) can hydrolyze acyl-ACP and acyl-CoA and is believed to be responsible for diverting myristic acid from the fatty acid biosynthetic pathway to the luminescent pathway. The synthetase subunit (*lux*E) activates fatty acids to produce an acylated enzyme intermediate. In the presence of NADPH, the reductase subunit (*lux*C) catalyzes the reduction of the acylated synthetase or acyl-CoA to produce the aldehyde substrate for the luciferase reaction.

Although attempts have been made to purify the fatty acid reductase complex from various luminescent bacteria only that of *Photobacterium phosphoreum* has been purified to homogeneity. In the present study, *lux*C, D and E genes from *P. phosphoreum* and *Vibrio harveyi* were respectively transferred into M42 (2), M17 and A16 (3), which are reductase, transferase and synthetase mutants of *V. harveyi*.

MATERIALS AND METHODS

Genes were mobilized into *V. harveyi* mutants (ampicillin resistant) as previously described (4) using *Escherichia coli* MM294 with recombinant pMGM110 (streptomycin resistant) and *E. coli* MM294 containing the conjugative plasmid pRK2013 (kanamycin resistant). Selection was performed on LB agar containing ampicillin (100 μg/ml), kanamycin (30 μg/ml) and streptomycin (25 μg/ml). The selected colonies were grown in LB medium with the same antibiotic content at 27°C in a rotating incubator. *In vivo* luminescence was measured using a photomultiplier calibrated with the light standard of Hastings and Weber ; 1 LU = 7 x 10^8 quanta per second.

All expressed genes were inserted downstream of an inducible *V. harveyi* promoter located in the cloning vector pMGM110, derived from pKT230 (4). DNA fragments containing the *V. harveyi* (B392) and *P. phosphoreum* (NCMB844) *lux*C, D and E genes as shown in Fig. 1 were mobilized into the respective *V. harveyi* mutant hosts to produce the

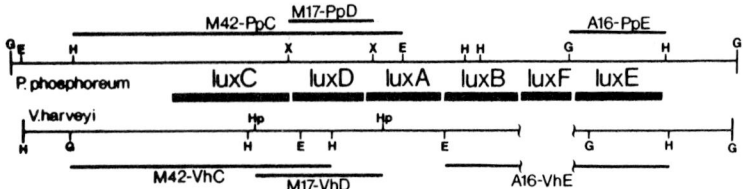

Fig. 1: Restriction map of *P. phosphoreum* and *V. harveyi*
lux DNA with fragments inserted into pMGM110 to produce the
indicated mutants. E, *Eco*RI; G, *Bgl*II; H, *Hin*dIII; Hp, *Hpa*II
(only two are shown); and X, *Xba*I.

following cell lines: M42 containing V. harveyi (M42-VhC) and *P.*
phosphoreum (M42-PpC) *lux*C; M17 containing the *V. harveyi* (M17-VhD)
and *P. phosphoreum* (M17-PpD) *lux*D; and A16 containing *V. harveyi*
(A16-VhE) A and *P. phosphoreum* (A16-PpE) *lux*E.   *V. harveyi* wild-
type and mutant cells containing the pMGM110 vector with no insert
were also used as control.

RESULTS AND DISCUSSION
Complementation of M42 with the *lux*C gene is shown in Fig. 2.   The
*V. harveyi lux*C gene restored light to slightly higher than wild-
type levels while the *P. phosphoreum* gene allowed luminescence to
reach 5% wild-type levels.   Nevertheless, the *P. phosphoreum* gene

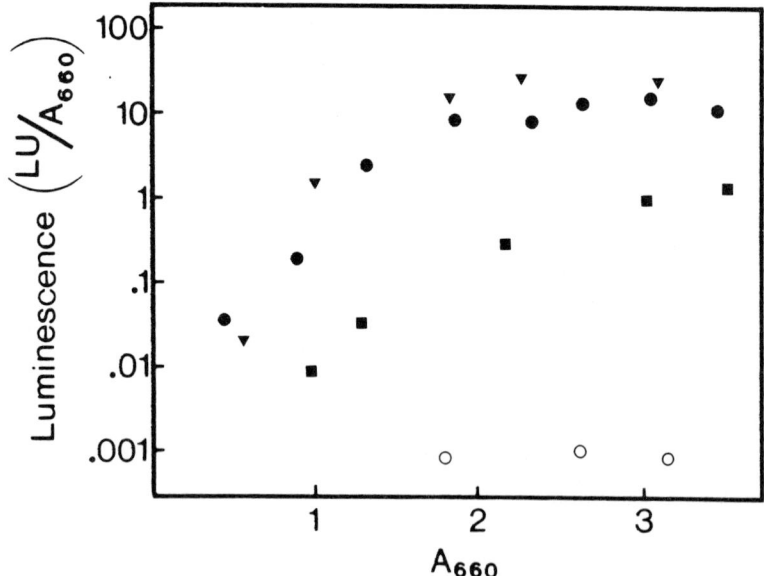

Fig. 2: *In vivo* luminescence (LU/A_{660}) as a logarithmic
function of cell growth (A_{660}) for M42 cell lines: M42 (○),
M42-PpC (■), M42-VhC (▼) and the *V. harveyi* control (•).

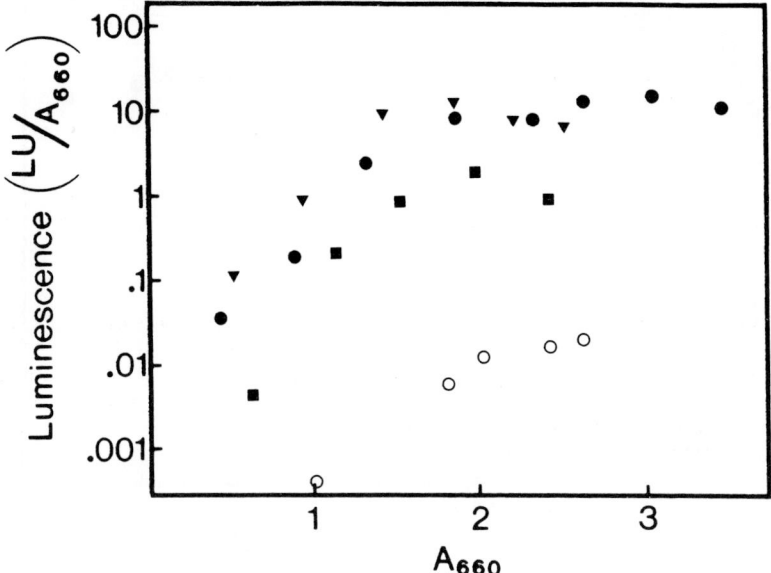

Fig. 3: In vivo luminescence (LU/A_{660}) as a logarithmic function of cell growth (A_{660}) for M17 cell lines: M17 (∘), M17-PpD (∎), M17-VhD (▼) and the V. harveyi control (•).

caused a 10^3-fold stimulation of light emission over the mutant. Addition of *lux*D to this mutant results in no increase in luminescence (not shown).

Complementation of M17 with *lux*D is shown in Fig. 3. The *V. harveyi lux*D restored wild-type levels of luminescence while the *P. phosphoreum* gene caused a 200-fold stimulation of light relative to the mutant to reach 10% of wild-type levels. Furthermore, addition of exogenous myristic acid to these cells restored wild-type levels of light. Addition of *lux*C to these mutants results in no complementation (not shown).

Complementation of A16 with *lux*E is shown in Fig. 4. In this case, the *P. phosphoreum* gene could restore wild-type luminescence levels while with the V. harveyi gene, the cells could attain 20% of wild-type light emission.

The three *V. harveyi* mutants M42, M17, and A16 can be specifically complemented by the *V. harveyi* and *P. phosphoreum lux*C, D and E genes, respectively. These results confirm that M42, M17 and A16 are mutations of the reductase, transferase and synthetase subunits of fatty acid reductase (2,3). The synthetase (*lux*E) seems to complement the most successfully in this system. It is also interesting that proteins from *P. phosphoreum* can replace *V. harveyi* subunits in the fatty acid reductase complex to restore activity even though these proteins have only approximately 60% amino acid sequence identity (unpublished data).

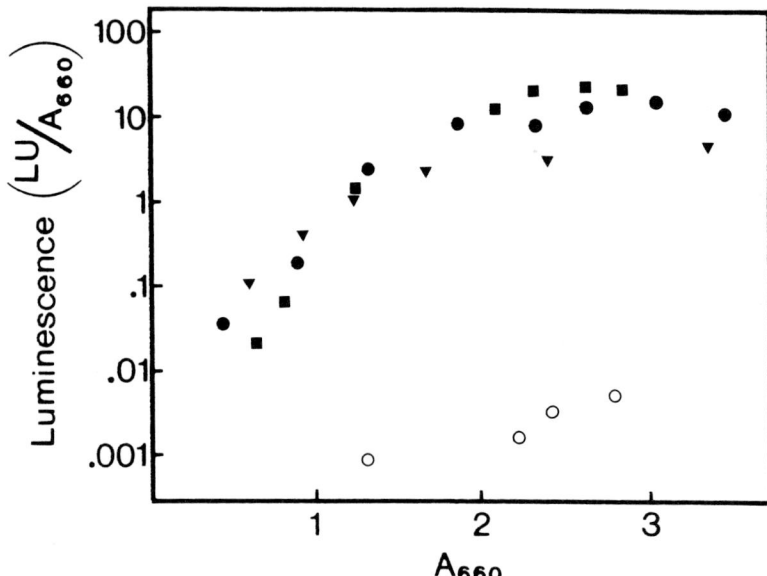

Fig. 4: In vivo luminescence (LU/A$_{660}$) as a logarithmic
function of cell growth (A$_{660}$) for A16 cell lines: A16 (○),
A16-PpE (■), A16-VhE (▼) and the V. harveyi control (•).

Gene transfer is a powerful tool that can be extended to other
luminescent bacteria to characterize dark mutants by identifying
genes which can complement the mutations. This technique can also
be used to study protein subunit interactions by examining the
properties of hybrid complexes without the necessity of purifying
the proteins.

ACKNOWLEDGMENTS
This work was supported by a grant (MT-4314) from the Medical
Research Council of Canada.

REFERENCES
1. E. Meighen, Enzymes and genes from the *lux* operons of
   bioluminescent bacteria, <u>Ann. Rev. Microbiol.</u> <u>42</u>,151-176(1988).
2. D. Byers and E. Meighen, Inhibition of *Vibrio harveyi*
   bioluminescence by cerulenin: *in vivo* evidence for covalent
   modification of the reductase enzyme involved in aldehyde
   synthesis, <u>J. Bacteriol.</u> <u>171</u>, 3866-3871 (1989).
3. L. Wall, D. Byers and E. Meighen, *In vivo* acylation of proteins
   in *Vibrio harveyi*: identification of proteins involved aldehyde
   production for bioluminescence, <u>J. Bacteriol.</u> <u>159</u>, 720-724
   (1984).
4. C. Miyamoto, E. Meighen and A. Graham, Transcriptional
   regulation of lux genes transferred into Vibrio harveyi, <u>J.
   Bacteriol.</u> <u>172</u>, 2046-2054 (1990).

OSMOREGULATION OF BIOLUMINESCENCE EXPRESSION IN P. PHOSPHOREUM IS RELATED TO GYRASE ACTIVITY

Haruo Watanabe[1], Humio Inaba[1,2], and
J. Woodland Hastings[3]

[1]
Biophoton Project,
Research Development Corporation of Japan,
2-1-1, Yagiyama-minami, Taihaku-ku, Sendai 982 Japan
[2]
Research Institute of Electrical Communication,
Tohoku University,
2-1-1, Katahira, Aoba-ku, Sendai 980 Japan
[3]
Department of Cellular and Developmental Biology,
Harvard University,
16 Divinity Avenue, Cambridge, MA 02138 U.S.A.

INTRODUCTION

In the marine luminous bacterium Photobacterium phosphoreum (strain 496), cell growth is normal in media with either 1% NaCl or 3% NaCl, but no bioluminescence occurs in the 1% NaCl medium. However, its development can be triggered in a log phase culture by increasing the NaCl to 3%. As previously shown (8), this development of luminescence is blocked by inhibitors of protein synthesis and RNA polymerase, and also by inhibitors of DNA gyrase(9). In mutants resistant to nalidixic acid, growth was quite normal, while bioluminescence was equal or even greater than in the wild type. In the presence of added nalidixic acid, there was a pronounced and specific further increase in the magnitude of luminescence development. However, this increase was not attributable to increases in luciferase. It has been found that the addition of tetradecanal induces bioluminescence in cells growing in 1% NaCl medium. Also, the shift of NaCl concentration in the medium to 3% results in an increase in intracellular potassium ions. These results suggest that aldehyde synthesis genes may be controlled by the status of DNA negative supercoiling (1,2,5).

MATERIALS AND METHODS

Wild type Photobacterium phosphoreum (strain 496) cells were cultured in liquid medium, as described previously (8). Shifting 1% to 3% NaCl in media was carried out by the addition of solid NaCl (8).

Nalidixic acid (nal) resistant strains (nal-r) were isolated by a method similar to that of Puga and Tessman (6), as described elsewhere (8).

Intracellular ion concentrations were determined with dried cells by atomic absorption analysis, and expressed as concentration (mmol/l) in cells, based on the method of Shultz and Solomon (7). The measurements of cell density, in vivo luminescence, and luciferase activities were described elsewhere (3,8). Tetradecanal was dissolved to 0.1% in 50 mmol/l phosphate buffer, pH 7.0 containing 0.5% BSA. An appropriate amount of aldehyde solutions was added for measurements of the luminescence intensity in vivo.

RESULTS AND DISCUSSION

The effect of nal on growth and the development of luminescence in the wild type strain of P. phosphoreum was examined. At a nal concentration where growth is only partially inhibited (1 μg/ml), luminescence is 2000 times less; at 5 μg/ml, growth is strongly inhibited, while luminescence is only barely and transiently detectable (data not shown). The changes in luminescence are not paralleled by the activities of extractable luciferase (Table 1). The large increase in the in vivo luminescence per cell (lum/OD) during growth (over 1000 fold from hr 6 to hr 10) is similar to the autoinduction described in other species of bacteria, where a large concomitant increase in luciferase activity occurs. In this strain, however, the specific luciferase activity only increases by less than a factor of three over the same time. Inhibition of the development of the in vivo bioluminescence by nal is also not reflected in the luciferase activity. In vitro specific activities at hr 10, after 4 hrs exposure to nal, were decreased by less than 50% despite a more than 1000 fold inhibition of the development of luminescence (Table 1).

Mutants capable of growth on nal were also characterized. At low concentrations (10 μg/ml or less) where the growth of the wild type is still strongly inhibited, the growth of nal-r is essentially unaffected, and it is only partially inhibited at higher concentrations, up to 100 ug/ml (data not shown). At the same time, the development of luminescence is

	6 h		10 h			
	wt	nal'-52	wt		nal'-52	
			nal (μg/ml)			
			0	25	0	25
O.D.	0.40	0.20	1.16	0.60	0.68	0.62
In vivo						
Luminescence	0.61	0.13	1980.	0.60	632	6090
Lum/OD	1.5	0.65	1710.	1.0	929	9820
In vitro						
Luciferase	1.22	1.05	9.4	3.3	2.5	2.6
Luciferase/OD	3.1	5.2	8.1	5.5	3.7	4.2

Table 1. Effects of nal added at hr 6 on growth(OD),luminescence, and luciferase in the wild type (wt) and mutant nal-r(52).

actually stimulated by nal, maximally at concentrations of 25 µg/ml; lower levels of luminescence, but still higher than the control, are expressed at higher nal. These effects of nal on bioluminescence in the nal-r mutants are also not paralleled by changes in cellular luciferase, as shown in Table 1, where there is a ten fold stimulation in the development of the _in vivo_ luminescence by added nal, but essentially no difference in the extractable luciferase activity.

(A)	OD	lum.	Na	K	Mg	Ca
1% NaCl (5hr) (170mM)	0.215	0.00	51.8	126.5	19.6	<0.51
3% NaCl (6hr) (490mM)	0.331	2.65	221.4	321.6	14.4	<0.50

(B)	OD	lum.	Na	K	Mg	Ca
a) 3%→ 3%	2.175	24100	75.5	601.7	17.4	<0.73
b) 3%→ 1%	1.065	0.98	63.2	189.1	19.2	<0.53
c) 1%→ 3%	1.643	1725	211.6	377.0	16.5	<0.71
d) 1%→ 1%	1.160	0.11	57.4	169.9	19.6	<0.53

Table 2. Cell density (OD), luminescence and intracellular ions (mmol/l) in the wild type cultured in media with different NaCl concentrations. (A) Culture in 1% NaCl at hr 5 was shifted to 3% NaCl, and measured again at hr 6. (B) Precultures in 1% or 3% NaCl for 24 hrs were transfered to fresh media and cultured for 12 hrs, when measurements were made.

When the NaCl concentration in the medium was shifted from 1% (170 mmol/l) to 3% (490 mmol/l) at hr 5, the intracellular Na and K ions in the wild type strain increased (measured at hr 6), concomitant with the development of luminescence (Table 2(A)); but Mg and Ca ions did not change so much. With cells precultured for 24 hrs in either 1% or 3% NaCl medium, cell density, luminescence and intracellular ion concentrations were compared 12 hrs later, after shifts up and down in NaCl concentration (Table 2(B)). In cultures where luminescence developed, intracellular K ion concentrations were higher than 300 mmol/l.

It was found that tetradecanal added at the time of measurement of luminescence evoked the luminescence of cells growing in a 1% NaCl medium (data not shown). In the presence of nal, luminescence failed to develop following shift to 3% NaCl in the wild type strain. However, the addition of tetradecanal can evoke luminescence. In 1% NaCl, luminescence is still suppressed in nal-r, but the addition of tetradecanal also induced luminescence.

The supercoiling and superhelical tension status of the bacterial chromosome may have profound effects on transcription (1,2,5). In P. phosphoreum, inhibitors of gyrase known to decrease supercoiling decrease luminescence, while nal treatment of nal-r, which is known to increase supercoiling, causes an increase in light emission. The salt induction of luminescence by NaCl might be similarly mediated: an increase in salt is known to increase supercoiling, at least in plasmids (4). In the wild type strain, the shift of the NaCl concentration in the medium to 3% caused an increase of the intracellular K ions to 300 mmol/l or more, concomitant with the development of luminescence. It is possible that an increase in the concentration of intracellular K ions may activate DNA gyrase, and/or change DNA conformation to be suitable for transcription of aldehyde synthesis genes.

REFERENCES

1. K. Drlica, Biology of bacterial deoxyribonucleic acid topoisomerase. Microbiol. Rev., 48, 273-289(1984).
2. M. Gellert, DNA topoisomerases. Annu. Rev. Biochem., 50, 879-910(1981).
3. J.W. Hastings, T.O. Baldwin, and M.Z. Nicoli, Bacterial luciferase: assay, purification and properties. Methods in Enzymology, 57, 136-152(1978).
4. C.F. Higgins, C.J. Dorman, D.A. Stirling, L. Waddel, I.R. Booth, G. May, and E. Bremer, A physiological role for DNA supercoiling in the osmotic regulation of gene expression in S. typhimurium and E. coli., Cell, 52, 569-584(1988).
5. S. Javanovich and J. Lebowitz, Estimation of the effect of coumermycin A on Salmonella typhimurium promoters by using random operon fusions. J. Bact., 169, 4431-4435(1988).
6. A. Puga and I. Tessman, Mechanism of transcription of bacteriophage S13. II. Inhibition of phase-specific transcription by nalidixic acid. J. Mol. Biol., 75, 99-108(1973).
7. S.G. Schultz and A.K. Solomon, Cation transport in Escherichia coli: I. Intracellular Na and K concentrations and net cation movement. J. Gen. Physiol., 45, 355-369(1961).
8. H. Watanabe and J.W. Hastings, Expression of luminescence in Photobacterium phosphoreum: Na regulation of in vivo luminescence appearance. Arch. Microbiol., 145, 342-346 (1986).
9. H. Watanabe and J.W. Hastings, Expression of bacterial luminescence is stimulated by nalidixic acid resistant mutant. Arch. Microbiol., 154, 239-243, (1990).

IMMUNOLOCALIZATION OF LUCIFERASE IN LUMINOUS BACTERIA PREPARED BY FAST-FREEZE FIXATION

M.T. Nicolas, P. Colepicolo*, G. Nicolas,
J. W. Hastings** and J.M. Bassot

CNRS, Laboratoire de Bioluminescence et Technologie
Appliquée, 105 boulevard Raspail, 75006 Paris, France.
*Instituto de Quimica, Cidade Universitaria, CP. 20.780,
01498, Sao Paulo, Brasil.
**Cellular and Developmental Biology, Harvard University,
16 Divinity Avenue, Cambridge, MA 02138, USA.

INTRODUCTION
Among the different groups of bioluminescent organisms luminous bacteria are one of the best understood with regard to their biochemistry, ecology, genetics and molecular biology (1) (2). However, little is known about the ultrastructure and the localization of the components of the luminescent reaction. In order to study this aspect, we have used antibodies directed against bacterial luciferases to localize the enzymes by immunogold labeling (IGL) (3).

The major technical problem is to preserve both ultrastructure and antigenicity. Liquid chemical fixatives can produce precipitation, masking, and displacement of the antigens. An alternative is to immobilize physically the specimen by fast-freeze fixation (FFF). Followed by freeze substitution (FS), this technique allows IGL on sections and gives remarkably sharp images. This article will review the results obtained with several species of luminescent bacteria and show unexpected localizations.

MATERIALS AND METHODS
The marine luminescent bacteria Vibrio harveyi, V. albensis, V. cholerae, V. molerae and the pathogenic luminous human bacterium Xenorhabdus luminescens were grown on solid medium. A small gelose sample bearing colonies was slammed onto a polished copper block cooled by liquid helium at -260°C (4). FS was done at -90°C, to replace water in its solid state by a solvent (acetone) with or without fixative agents added. After warming to room temperature, the samples were embedded in epoxy (Epon) or acrylic (LR White) resins. The IGL was done on sections, (3). The grids were first incubated with a polyclonal antibody against luciferase of V. harveyi or

against luciferase of X. luminescens, raised in a rabbit.
Then a goat antirabbit antibody with 10 nm gold
particles attached was applied (5)(6)(7).

RESULTS AND DISCUSSION

The Luminescent Vibrio: As already described in E. coli
prepared by FFF-FS (8), the ultrastructure was excellent
and revealed features completely new compared with images
obtained after liquid chemical fixation. The nucleoid was
not precipitated in the center of the cell but extended
up to the plasma membrane in an intricate meshwork.
Mesosomes were never observed. The cell wall was sharply
defined with the successive layers characteristic of gram
negative bacteria. In V. harveyi, dense patches 50-100 nm
in diameter, often aggregated in polyglobular clusters,
were present at the periphery of the cells.
In all the Vibrio studied the IGL was intracellular, in
the ribosome rich areas. In V. harveyi it was
particularly concentrated on the patches (5) (Fig. 1).
Moreover, induction of bioluminescence during growth was
nicely paralleled by increase in the IGL (6).
The labeling over the patches could be indicative of a
functional stage and could correspond to the "electron
opaque formations" previously observed (9). Special
functions in the bacteria might be confined to specific
areas, even though there is no limiting membrane. This is
reminiscent of the differentiation of the scintillons in
Gonyaulax; they occur first free in the cytoplasm and
become secondarily associated with a membrane (10).

The pathogenic Xenorhabdus luminescens: From the
morphological point of view, the appearances of both the
nucleoid and the cell wall were very much the same as
described for Vibrio. However, numerous flagella and
several types of cytoplasmic inclusions (bundles of
filaments, clusters of dense material) were observed.
The most striking feature was the IGL of luciferase
(7). Gold particles occurred first within the cells,
especially in cortical areas rich in ribosomes, but not
significantly over any kind of the cellular inclusions.
Labeling occurred also outside the cells, on the
irregular cell coat or its detached fragments (Fig.2).
Alignments of gold particles crossing the cell capsule
were often observed. This surprising localization,
extending beyond the limits of the cells, strongly
suggests that a translocation of luciferase occurs at
precise channels located along the capsule. However, the
biological significance of the export of the luciferase
visualized here by IGL is enigmatic and requires further
studies.
In summary, FFF-FS provides an excellent preservation
of the ultrastructure and of the antigenicity. The two

examples reported here open up new insights in bacterial
bioluminescence: luciferase localizes at precise
intracellular sites in V. harveyi; it is excreted
extracellularly in X. luminescens.

ACKNOWLEDGMENTS
    Thanks to D. Touret and D. Gache for her technical
assistance. Our work was supported by CNRS/NSF (160059),
NSF DMB 8616522 and NIH GM19536 grants.

REFERENCES
1.   K. Nealson, and J.W. Hastings, The luminous bacteria.
In: The Prokaryotes, 2nd Ed., A. Balows, H.G. Trüper, M.
Dworkin, W. Harder and K.H. Schleifer (Eds), Springer-
Verlag (New York) (in press).
2.   T. Meighen, Enzymes and genes from the lux operons of
bioluminescent bacteria. Ann. Rev. Microbiol. 42, 151-175
(1988).
3.   J. de Mey, Colloidal gold probes in immunocyto-
chemistry. In:Immunocytochemistry, J.M. Polak and S. Van
Noorden (Eds), Wright-PSG (Bristol, UK), p.82-112 (1983).
4.   J. Escaig, New instruments which facilitate rapid
freezing at 83K and 6K. J. Microscopy 125, 221-229 (1982)
5.   M.T. Nicolas, J.M. Bassot and G. Nicolas, Immunogold
labeling of luciferase in the luminous bacterium Vibrio
harveyi after fast freeze fixation and different freeze-
substitution and embedding procedure. J. Histochem.
Cytochem. 37, 663-674 (1989).
6.   P. Colepicolo, M.T. Nicolas, J.M. Bassot and J.W.
Hastings, Expression and localization of bacterial
luciferase determined by immunogold labeling. Arch.
Microbiol. 152, 72-76 (1989).
7.   M.T. Nicolas, P. Colepicolo, J.W. Hastings and J.M.
Bassot, Translocation of luciferase by the bioluminescent
bacterium Xenorhabdus luminescens revealed by immunogold
labeling after fast-freeze fixation. J. Cell. Sci.
(submitted)
8.   J.A. Hobot, W. Villiger, J. Escaig, M. Maeder, A.
Ryter and E. Kellenberger, Shape and fine structure of
nucleoids observed on sections of ultra-rapidly frozen
and cryosubstituted bacteria. J. Bacteriol. 162, 960-971
(1985).
9.   S.E. Medvedera and M.V. Salnicov, Morphology and
ultrastructure of marine luminous bacteria. In:
Bioluminescence in the Pacific, J.Gitelson and J.W.
Hastings (Eds), USSR Pacific Sci.Congr.,p.213-218 (1982).
10.   M.T. Nicolas, G. Nicolas, C.H. Johnson, J.M. Bassot
and J.W. Hastings, Characterization of the bioluminescent
organelles in Gonyaulax polyedra (dinoflagellates) after
fast-freeze fixation and antiluciferase immunogold
staining. J. Cell Biol. 105, 723-735 (1987).

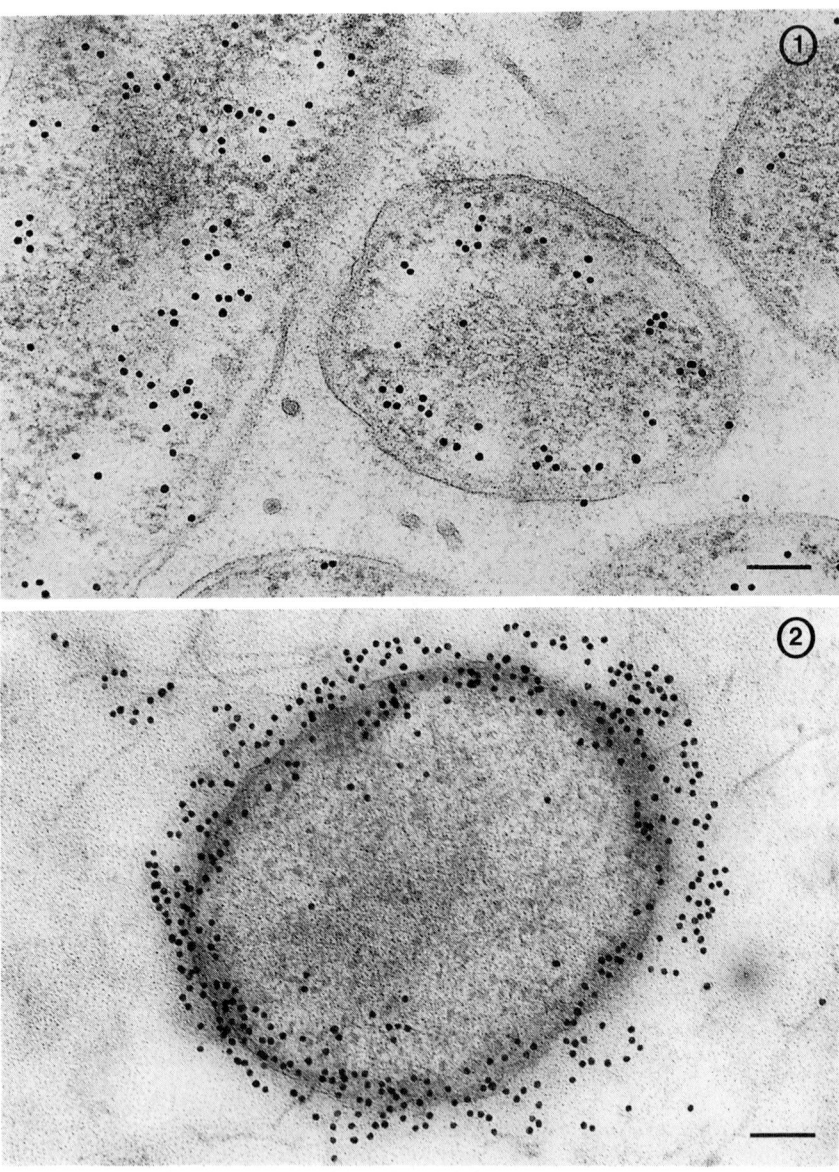

Figure 1. <u>Vibrio harveyi</u>. FFF, FS in acetone-osmium,
Epon, antiluciferase IGL. Bar: 0.1 μm. Gold particles
are strictly intracellular, mostly over the patches.
Figure 2. <u>Xenorhabdus luminescens</u>. FFF, FS in acetone-
osmium, LR White, antiluciferase IGL. Bar: 0.1 μm.
Gold particles are both intra and extracellular.

CHARACTERIZATION OF A NEW
BIOLUMINESCENT CELLULAR MODEL
OF RESPONSE TO ESTROGENS.

M. Pons, D. Gagne, J. C. Nicolas and M. Mehtali [1].

INSERM Unité 58, 60, Rue de Navacelles, 34090 Montpellier France and
[1] Transgene S.A., 11 rue de Molsheim, 67000 Strasbourg-Cédex, France.

INTRODUCTION

The role of estrogens in the growth of hormone-dependent tumors is well established. Estrogens regulate the expression of specific genes through the interaction of the estrogen-receptor complex with specific nucleotide sequences termed estrogen-responsive elements (EREs) (1). As a consequence of this interaction, there are changes in the synthesis of specific RNAs and proteins involved in the regulation of cell proliferation, differentiation and physiological function. The search for potent and specific new antiestrogens able to inhibit or reverse tumor growth is still very urgent, but this research is impeded by the complex technology that their screening requires. Therefore, we investigated whether a new model based upon recent genetic engineering progress would be more efficient. Using an estrogen receptor-positive breast cancer cell line, MCF-7, we established a stable estrogen-dependent expression system by transfecting these cells with the pVit-tk-Luc plasmid. In this plasmid, the 5' flanking region of the Xenopus vitellogenin A2 gene (Vit) which contains an ERE, controls the firefly luciferase structural gene (Luc) (2). In such transfected cells, estrogenic activity is easily and quickly followed by measuring the luciferase activity ; compared to conventional tests, such a model offers many advantages such as simplicity, sensitivity, and rapidity of measurement as well as low cost.

MATERIALS AND METHODS
Chemicals, materials, cell culture

Origins of steroids and of products used in cell culture have been described elsewhere (3). Luciferin was purchased from Sigma. The single photon counting camera "Argus 100" was obtained from Hamamatsu. Construction of the pVit-TK-Luc plasmid and the use of pAG60 plasmid (conferring resistance to the agent G418) have been previously described (3). MCF-7 cells were grown in monolayers using minimal essential medium without phenol red (MEM Gibco) and supplemented with fetal calf serum free of endogeneous steroid as described elsewhere (3).
Cloning of the responsive stable transfectants : establishment of a permanent cell line

The stable cell line, MVLN-15-C7 was established by cotransfection of MCF-7 cells with the plasmids pVit–tk-Luc and pAG-60, then by selection of the resistant clones in the presence of G 418, and finally selected by a sub-cloning step performed on positive cultures using a single photon detecting camera (3).
Luciferase assay in cell-free conditions

At the end of the cell incubation with the various compound(s) (24, 48 h), the culture media was removed and cells extensively washed with cold PBS. Cells in luminescence buffer were then disrupted, and the corresponding luciferase activity was determined by the integration value (15 s integration time, arbitrary units) of the luminescence peak following the injection of luciferin (3).
Luciferase assay in whole cells : use of a single photon detecting camera

At the end of incubation of the cells with estradiol (or other compounds), the culture medium was renewed with a fresh medium containing 3×10^{-4} mol/l luciferin. Cell flasks can be immediately analyzed with a single photon counting camera to measure the luminescence. Images can be printed (on a Video Graphic Printer UP-811 from Sony) and/or stored on disk.

RESULTS AND DISCUSSION

Until now the screening tests to characterize the estrogenicity or antiestrogenicity of new molecules generally required several days whether the tests were carried out in established cell lines or in animals. In order to break free from this cumbersome technology we have investigated, with the help of genetic engineering techniques, a new model able to be used as an early screening test. Our results suggested that stable transfection of the pVit-tk-Luc plasmid in MCF-7 cells could be a convenient model to screen new estrogen or antiestrogen molecules.

In agreement with other results obtained with an analogous plasmid (pVit-tk-CAT) (4), pVit-tk-Luc was able to respond to estradiol by expressing, in stably transfected MCF-7 cells, the firefly luciferase in a dose-dependent manner with a EC50 value lower than 10^{-10} mol/l. This value corresponds to that of a physiological effect mediated by the estrogen receptor. The maximum response was reached at 10^{-10}-10^{-9} mol/l estradiol (corresponding to an average of 22,500 molecules of luciferase induced per cell) and was approximately 4-5 times higher than the basal level (vehicle alone) (Fig. 1). Moreover hydroxytamoxifen (OH-Tam), a typical antiestrogen, when present in 1000-fold excess, was able to inhibit the estradiol-induced response. This inhibition was concentration-dependent (3) and could be reversed by 1 μmol/l estradiol. Such results were obtained whether the luminescence was measured in cell-free experiments (luminometer) (3), or in intact cells as shown in Fig 2.

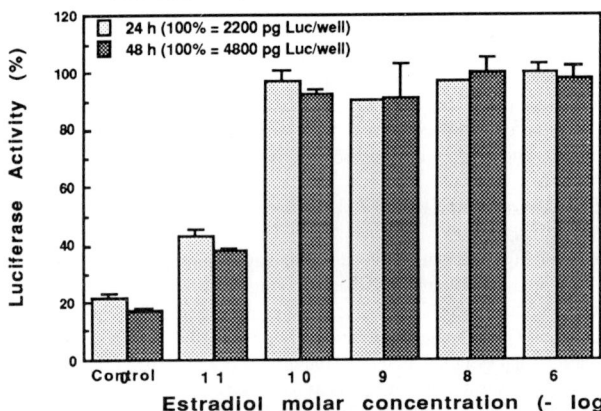

STIMULATORY EFFECT OF ESTRADIOL

Fig. 1 : Dose response curve of the MVLN-15-C7 cells incubated for 24 h or 48 h with estradiol. Results are expressed as a percentage of luciferase activity measured per Petri dish (mean ± SEM of triplicates).The 0 % value represents the luminometer background value obtained in the absence of luciferase (8 to 9 arbitrary units) whereas the 100 % values represent the maximum intensity obtained in the presence of estradiol (2260 and 4630 arbitrary units for 24 and 48 h incubation time, respectively).

Estrogen induction of firefly luciferase in our model allows a quick and easy bioluminescent measurement : after the cells are harvested and sonicated, the measurement of the luciferase activity lasts 15 s per test-tube. This rapid and simple measurement (compared to hours or days for the CAT assay) is also highly sensitive since some 10^5-10^6 stably transfected cells were enough to obtain an easily measurable inducible activity (compared to 20-40x10^6 cells needed for the measurement of estradiol promoted induction of progesterone receptors in MCF-7 cells. This result

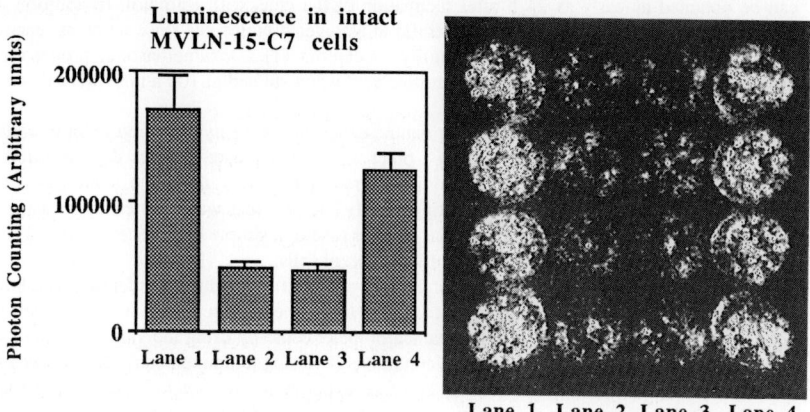

Fig. 2 : Estrogen dependent luminescence in intact MVLN-15-C7 cells. The cells cultured in a 24-well multidish (four wells per condition) were treated for 24 h with 1 nmol/l estradiol (lane 1), with 1 µmol/l OH-Tam alone (lane 2) or in presence of 1 nmol/l estradiol (lane 3), or of 1µmol/l estradiol (lane 4). After an exposure time of 10 min, images of the wells obtained by the single photon detecting camera were shown on the right side and the corresponding photon countings (in arbitrary units) were from lane 1 to lane 4 : 171239 ± 26145 ; 48737 ± 5800 ; 47462 ± 5000 ; 124959 ± 12563 respectively.

Kinetics of the appearance or disappearance
of induced luciferase

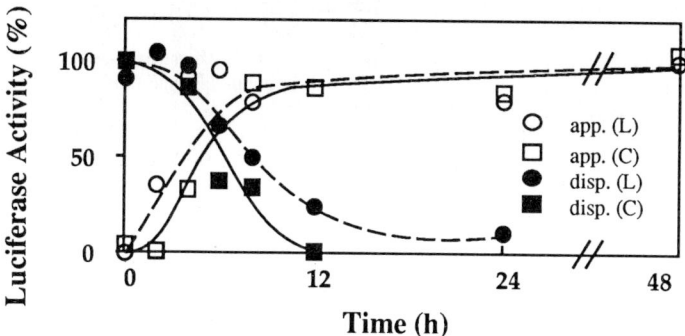

Fig 3 : Kinetics of luciferase appearance (induction by estradiol) and of disappearance (in the presence of OH-Tam). MVLN-15-C7 cells were incubated at various times with different concentrations of estradiol (appearance (app)) or incubated for 24 h with 1 nmol/l estradiol and then incubated for various times in the presence of 1 µmol/l OH-Tam (disappearance (disp)). Luciferase was determined in intact cell (video camera (C)) and in cell-free (luminometer (L)) experiments. Results are expressed in %, with 100% being the maximal level of luminescence due to estradiol.

can be obtained as early as 24 h after incubation of the cells with estradiol. In addition, this hormone dependent luminescence was specific to estrogenic molecules since no cross-reactivity appeared with ligands of the "superfamilly" receptors (1) (Dexamethasone, progesterone, testosterone, calcitriol, aldosterone, and retinoic acid when studied at 100 nmol/l were devoid of any agonist property (3)).

As illustrated in Fig 2, estrogen dependent luminescence can be easily measured in intact cells by use of a single photon detecting camera ; other results not presented here demonstrated the reversibility of the phenomenon. By adding estradiol or OH-Tam to the same cells, they successively become luminescent or extinguished. The fact that we could easily measure the luminescence in intact cells without cellular damage (use of a simple luminescence buffer) was a particulary helpful tool in the subcloning of luminescent cells.

Kinetics of appearance and disappearance of the estrogen dependent induced luciferase was followed in MVLN-15-C7 cells. This experiment was performed either in intact cell or in cell-free conditions. As shown in Fig 3, results obtained in intact cells (by using the video camera) are in agreement with those obtained in cell-free conditions (by using a luminometer). This showed that the experimental conditions we used in cell-free experiments do not lead to any significant degradation of luciferase activity. The response of the cells to estradiol is quick, since they need to be incubated with the hormone for only about 8-10 h before observing the full level of induction ; this level then persists for at least 24-48 h. Disappearance of the luminescence, due to the antiestrogenic effect of OH-Tam on cells previously incubated with estradiol then with OH Tam, suggested a relatively high stability of the luciferase enzyme in these cells since the half-life time of the activity was about 8 to 10 hours. These last observations together with the high sensitivity of the method confirm the advantages of this new cellular model to screen estrogen or antiestrogen activities.

REFERENCES.

1. M. Beato, Gene regulation by steroid hormones. Cell 56, 335-344 (1989).

2. J.R. De Wet, K.V. Wood, M. De Luca, D.R. Helinski and S. Subramani, Firefly luciferase gene : structure and expression in mammalian cells. Mol. Cell. Biol. 7, 725-737 (1987).

3. M. Pons, D. Gagne, J.C. Nicolas and M. Mehtali, A new cellular model of response to estrogens : a bioluminescent test to characterize (anti)estrogen molecules. BioTechniques (in press) vol 9, n° 4 (1990).

4. L. Klein-Hitpass, M. Schorpp, U. Wagner and G.U. Ryffel, An estrogen-responsive element derived from the 5' flanking region of the Xenopus vitellogenin A2 gene functions in transfected human cells. Cell 46, 1053-1061 (1986).

IRON-REGULATION IN PHOTOBACTERIUM PHOSPHOREUM BY THE FUR PROTEIN OF ESCHERICHIA COLI

K. Knöchelmann, P. Lümmen[1] and U.K. Winkler

Lehrstuhl für Biologie der Mikroorganismen, Ruhr-Universität,
D-4630 Bochum 1, FRG, and
[1]Hoechst AG, Pflanzenschutz-Forschung Biochemie,
Postfach 800320, D-6230 Frankfurt 80, FRG

INTRODUCTION

Bioluminescence of P. phosphoreum has been shown to be controlled by the iron concentration of the growth medium: luciferase synthesis was induced at significantly lower cell densities under iron-restricted conditions as compared to iron-rich cultures. Furthermore, three iron-regulated outer membrane proteins (IROMP) appeared under iron limitation only (8). Presumably these proteins are part of a high-affinity iron uptake system (2).
In E. coli iron-regulated genes are controlled by the Fur protein acting as a repressor in the presence of Fe^{2+}, Mn^{2+} or Co^{2+}. A consensus sequence for Fur binding at promoter/operator regions of iron-regulated genes has been published (3).
The objective of our research project was to study the genetic regulation of iron-responsive genes in P. phosphoreum NCMB7 by the fur gene product of E. coli.

MATERIALS AND METHODS

Culture conditions for P. phosphoreum, measurement of in vivo bioluminescence, and chemical mutagenesis has been described previously (7). Mutants with deregulated IROMP gene expression were selected according to the $MnCl_2$ method of Hantke (6). Standard protocols were applied for the isolation of chromosomal and plasmid DNA, cleavage with restriction enzymes, ligation of DNA fragments, and transformation of competent E. coli cells (9). Labeling of DNA with the digoxigenin kit, blotting of electrophoretically separated DNA fragments, and hybridization of labeled probes essentially followed the manufacturer's recommendations (Boehringer manual). Recombinant plasmids derived from broad host-range vector pKT230 (1) and plasmid pMH15 (5) were conjugationally transferred to P. phosphoreum using the helper plasmid pRK2013 (4).

RESULTS AND DISCUSSION

By application of a selection method published for E. coli (6) fifteen genetically independent Mn^{2+} resistant mutants were isolated from P. phosphoreum NCMB7. SDS-electrophoretic analysis of outer membranes prepared from cells grown under iron-rich conditions proved three iron-regulated outer membrane proteins (IROMP) to be de-repressed (30, 70 and 77 kDa). So a common transcriptional regulation of the IROMP genes could be postulated.
Synthesis of iron-chelating compounds (siderophores) was also uncoupled from the iron concentration in the medium at least in some mutants (data not shown) indicating an iron-regulon in Photobacterium. The P. phosphoreum mutants were designated fur (ferric uptake regulation) to stress the phenotypical similarity with corresponding E. coli mutants (6). In some fur mutants of P. phosphoreum luciferase synthesis was induced at lower cell densities compared to wild-type cells even in the presence of excess iron (8).

A recombinant plasmid pKK9 (13.0 kb) was constructed by ligating the large BamHI/HindIII fragment of pKT230 with a 2.2 kb fragment of pMH15 (5) containing the E. coli fur^+ gene (Fig. 1). After conjugative transfer into fur mutants of P. phosphoreum transconjugants were obtained and grown in the presence of 250 µg/ml streptomycin for several generations. Plasmid DNA was isolated and hybridized with the digoxigenin-labeled 2.2 kb E. coli fur^+ fragment after Southern transfer (Fig. 2). The results clearly showed that pKK9 was replicated in the photobacterial host.

Mutant L1010 (fur) was chosen to investigate the effect of the cloned fur^+ gene on IROMP expression. L1010 and L1010 (pKK9) were grown in complex medium LM, in LM supplemented with ferric ammonium citrate, and in the presence of the strong synthetic iron chelator ethylendiamine-di-(o-hydroxyphenyl)acetic acid (EDDA). Expression of all three IROMP in L1010 was significant under iron-rich conditions with a slight increase when EDDA was present in the culture broth as revealed by SDS gel electrophoresis (Fig. 3B) and Western blotting (Fig. 3A). Transconjugant L1010 (pKK9) showed the wild-phenotype with IROMP synthesized only

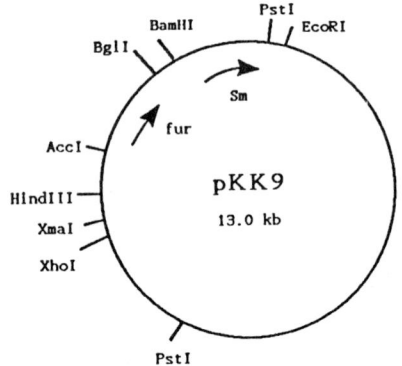

Fig. 1: Restriction map of plasmid pKK9

under iron limitation (Fig. 3, d-f). From these complementation experiments two conclusions were drawn: (i) the E. coli fur^+ gene was expressed in P. phosphoreum, and (ii) the fur^+ gene product was functionally active as a repressor of iron-regulated genes in P. phosphoreum. As a consequence structural homology between the Fur repressors and between the recognition sequences (Fur consensus) of E. coli and Vibrionaceae should be expected. Furthermore, Fur has been shown to control the iron-regulated hemolysin expression in Vibrio cholerae (10).

Since no Fur-specific consensus sequence located in the regulatory regions of bacterial lux operons has been reported so far, there is no evidence for direct transcriptional control of bioluminescence by iron. An alternative explanation of the iron effect on luciferase synthesis may be the facilitated uptake of autoinducer via iron transport systems. Investigations regarding the regulation of bioluminescence in the L1010 (pKK9) strain and the cloning of the photobacterial fur+ gene are under way.

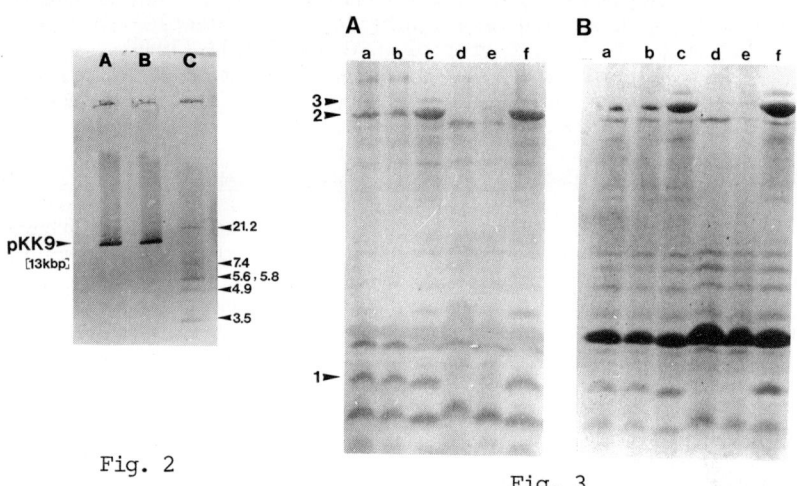

Fig. 2

Fig. 3

Fig. 2: Hybridization of plasmid DNA isolated from P. phosphoreum transconjugants L1010 (pKK9) [lane A] and L1055 (pKK9) [lane B] with the 2.2 kb E. coli fur+ fragment. The plasmid DNA was linearized with BamHI before gel electrophoresis, blotting and hybridization. Lane C: λ/EcoRI marker (kb) also labeled with digoxigenin.

Fig. 3: Western blot (A) and SDS-gel (B) of outer membrane proteins from P. phosphoreum fur mutant L1010 [lane a-c] and transconjugant L1010 (pKK9) [lane d-f]. Growth of bacteria in complete medium LM [lane a, d], LM + 50 μg/ml ferric ammonium citrate [lane b, e], and LM + 20 μM EDDA [lane c, f].

REFERENCES

1.   M. Bagdasarian, R. Lurz, B. Rückert, F.C.H. Franklin, M.M. Bagdasarian, J. Frey and K.N. Timmis, Specific-purpose plasmid cloning vectors II. Broad host range, high copy number, RSF1010-derived vectors, and a host-vector system for gene cloning in Pseudomonas. Gene 16, 237-247 (1981).
2.   A. Bagg and J.B. Neilands, Ferric uptake regulation protein acts as a repressor, employing iron(II) as a cofactor to bind the operator of an iron transport operon in Escherichia coli. Biochemistry 26, 5471-5477 (1987).
3.   S.B. Calderwood and J.J. Mekalanos, Confirmation of the fur operator site by insertion of a synthetic oligonucleotide into an operon fusion plasmid. J. Bacteriol. 170, 1015-1017 (1988).
4.   D.H. Figurski and D.R. Helinski, Replication of an origin-containing derivative of plasmid RK2 dependent on a plasmid function provided in trans. Proc. Natl. Acad. Sci. USA 76, 1648-1652 (1979).
5.   K. Hantke, Cloning of the repressor protein gene of iron-regulated systems in Escherichia coli K12. Mol. Gen. Genet. 197, 337-341 (1984).
6.   K. Hantke, Selection procedure for deregulated iron transport mutants (fur) in Escherichia coli K12: Fur not only affects iron metabolism. Mol. Gen. Genet. 210, 135-139 (1987).
7.   P. Lümmen and U.K. Winkler, Bioluminescence of outer membrane defective mutants of Photobacterium phosphoreum. FEMS Microbiol. Lett. 37, 293-298 (1986).
8.   P. Lümmen, U. Winkler and M.R.W. Brown, Iron affects bioluminescence and outer membrane composition of Photobacterium phosphoreum (Abstract). J. Biolum. Chemilumin. 2, 230 (1988).
9.   T. Maniatis, E.F. Fritsch, J. Sambrook, Molecular Cloning. A Laboratory Manual. Cold Spring Harbor Lab., New York (1982).
10.   J.A. Stoebner, S.M. Payne, Iron-regulated hemolysin production and utilization of heme and hemoglobin by Vibrio cholerae. Infect. Immun. 56, 2891-2895 (1988).

Acknowledgements. This investigation was supported by a grant (Lu 412/1-1) of the Deutsche Forschungsgemeinschaft. We thank Dr. K. Hantke, Tübingen, for supplying plasmid pMH15.

Escherichia coli strains MM294, MM294 (pKT230), and MM294 (pRK 2013) were kindly provided by Dr. E.A. Meighen, Montreal.

THE DETERMINATION OF THE Ca^{2+}-FLUX AND A CORRELATED CHEMILUMINESCENCE IN ELECTROPORATED HUMAN LEUCOCYTES

H. Schmidt, V. S. Malinin[1], U. Graba and
A. B. Putvinski[1]

Research Centre Biotechnology, Berlin, 1017,
Alt-Stralau 62
[1]Institute of Radioengineering & Electronics,
Academy of Sciences of the USSR
K. Marx av. 18 Moscow, GSP-3, 103907, USSR

INTRODUCTION

V. S. Malinin et al. (1) found, that high voltage pulses induce a chemiluminescent response of human blood leucocytes in the presence of luminol, which depends both on the amplitude of the pulse and on the calcium concentration of the incubation medium. Several authors demonstrated a dependence of the cell activation from the Ca^{2+}-concentration (2, 3). This paper concerns the quantitative measurement of the Ca-influx into leucocytes during electroporation by short electric pulses (T= 50 us) in the amplitude range between
1 und 6 kV/cm using the radioactive Isotope ^{45}Ca.

MATERIALS AND METHODS

The isolation of leucocytes from human blood was followed by sedimentation of erythrocytes in Dextran T500, centrifugation of the leucocyte-rich supernatant in a Ficoll-Verografin gradient (1,078 g/cm^3) and hypo-osmotic lysis of residual erythrocytes. The cells were resuspended in HBSS and stored at 5°C. The number of cells was counted under a microscope for each solution (1E+06-1E+07 cells/vial). For evaluation of the cell volume we measured the K-concentration at a flame-spectrometer. The 1 ml samples contained additional 50 umol/l luminol and various amounts of labelled Ca-solution. They were stirred and thermostated at 37°C in a specially constructed chemiluminometer (1).
After electroporation and luminescence measuring the samples were centrifugated and 2 times washed with a EDTA containing puffer. The cells were lysed with 0,5 ml Triton X100 and the Ca-activity of this suspension was measured in 5 ml liquid dioxane scintillator. The desintegration rates were determined in a Rack-Beta-

Spectral-Spectrometer applying a stored ^{45}Ca-quench
curve. The Henks balance salt solution (HBSS) contained
136 mmol/l NaCl, 5 mmol/l KCl, 0,4 mmol/l Na$_2$HPO$_4$,
0,4 mmol/l KH$_2$PO$_4$, 5 mmol/l D-glucose, 10 mmol/l
HEPES at pH 7,4. The liquid scintillator contained 7
g PPO, 0,3 g POPOP and 100 g naphthalene for 1 l
dioxane. The activity concentration of the applied Ca-
solution was nearly 1E+05 dpm/100 ul (aqueous ^{45}CaCl$_2$-
solution from Amersham, 74 MBq/ml, β-energy (max.) =
257 keV).

RESULTS AND DISCUSSIONS
The Ca-flux into leucocytes during electroporation was
investigated as a function
- of the HV-amplitude of a single electric field pulse
- of the number of single pulses with constant
  amplitude
- and of the Ca-concentration in the incubation
  solution.
For interpreting the Ca-flux during electroporation we
determined also the half life of generated membrane
pores by addition of the labelled Ca-solution at
different times after pulsing and measuring the
radioactivity contents of the cells. The CL-dates are
concerned to the recorded maximum in relative units,
which we got 2 - 2,5 minutes after pulsing (1). As
demonstrated in Fig. 1 the Ca-content of the cells and
the CL-signal grows with HV-amplitudes greater then 1 -
1,5 kV/cm. While the CL-signal reaches at 5 kV/cm
respectively 7,3 umol/l Ca-uptake a maximum, the Ca-
content grows nearly linearly till a concentration
of 10 umol/l.

FIG.1	FIG.2
Dependence of Ca-Influx (a) and CL (b)	**Dependence of Ca-Influx (a) and CL (b)**
from the pulse amplitude	**from the number of HV-pulses**

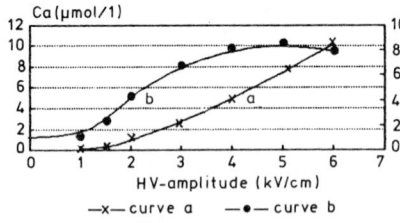

—x—curve a   —•— curve b

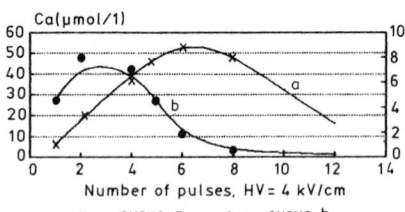

—x— curve a   —•— curve b

In Fig. 2 we used a constant HV-pulse amplitude of
4 kV/cm and realized the Ca-uptake by applying a
different number of pulses in a distance of 1
minute. Under these conditions the Ca-content grows

linearly till 56 umol/l. The decreasing Ca-content   of
the cells after 6 times pulsing originates in an
increasing number of destroyed cells as we could see
under a microscope. In this experiment we received the
maximum of CL at a Ca-concentration of 17,5 umol/l.
Because the relation between the number of cells
in Fig. 2 and Fig. 1 is 1,3E+07 : 5E+06 = 2,6, we can
conclude, that for a maximum in CL a Ca-concentration
of nearly 1,4E-06 umol/l* cell is necessary. The
maximum of the measured Ca-uptake in this experiment
corresponds to 4,3E-06 umol/l* cell.
We demonstrate in Fig. 3 for a physiologicaly relevant
Ca-concentration range of the cell solution, that the
Ca-content of the cells after electroporation grows
linearly till 30 umol/l in 3E+06 cells (1E-05 umol/l*
cell) at a pulse amplitude of 6 kV/cm. Comparing this
single pulse experiment with the multiple pulse
electroporation at 4 kV/cm-amplitudes (Conc.$_{max.}$=4.3*
E-06 umol/l* cell),we measure a maximal Ca-content in a
4 mmol/l-Ca-incubation solution of 17,5 umol/l (5,8E-06
umol/l* cell).   This result means, that leucocyte
electroporation is not reversible after multiple
pulsing for 6 times. The change of the CL in this
experiment confirms a CL-maximum at 1,4 E-06 umol/l*
cell.

FIG.3
Dependence of the Ca-content in cells from the Ca-concentration of the solut.

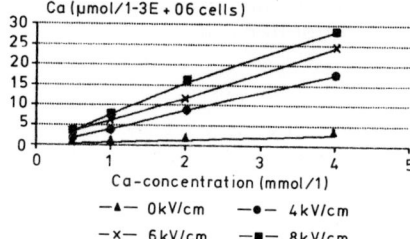

—▲— 0kV/cm   —●— 4kV/cm
—x— 6kV/cm   —■— 8kV/cm

FIG4
Determination of the mean lifetime of membrane pores

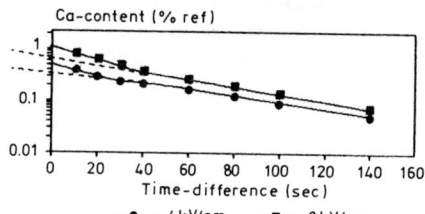

—●— 4kV/cm   —■— 8kV/cm

Malinin et al. (1) interpreted the Ca-ion influx into
the cells by reversible electroporations with a
trigger-like reaction, which finally at higher field
strengths leads to an increase in the diameter of the
evolving pores. They determined by a CL-measurement
method, that
the mean lifetime of the membrane pores was affected by
the amplitude of electric pulses.We also investigated
the mean lifetime of leucocyte-membrane pores by
additing the labelled Ca-solution at different time

intervals after pulsing. We analyzed 9 decay curves   by
exponential   regression   method.   The   results,
demonstrated   in Fig. 4, lead to the   conclusion,   that
pores with mean lifetimes of T_1= 25 sec and T_2= 50   sec
exist   in   leucocyte membranes. But we did not   find   a
significant   dependence from the pulse   amplitude.   The
great   variance   of   mean   lifetimes   for   one   pulse
amplitude is caused primarily by the different membrane
properties   of different leucocytic samples. While   the
mean   correlation coefficient of all regressions   r   =
0,988   had a standard deviation of$\pm$ 0,014, we   got   for
instance for the 4 kV-amplitude an average T_1-value   of
23,2 $\pm$ 4,6   and T_2 of 46,2 $\pm$ 6,5 and for   the   6   kV-
amplitude T_1 = 26,6 $\pm$ 3,5 and T_2= 57 $\pm$ 19,1.   These
deviations   correspond   to the   deviation   of   K-
concentrations   in the applied cell cultures ([K]   =
0,35   mmol/l $\pm$ 0,13) and the mean   quotient   of   Ca/K-
concentration   ([Ca/K] = 0,799$\pm$0.22). The   extrapolated
cutting   points   for   t = 0   should be related   to   the
number of pores, which belong to the mean lifetimes T_1
and T_2 . For HV-pulses of 4 kV/cm we evaluated n(T_1)
/n(T_2) = 1,2 and for HV = 6 kV/cm pulses n(T_1)/n(T_2) =
1,66   respectively. We   concluded,   that   with   the
augmentation   of the electric field the number of T_1 -
pores   increase. This correlation   would   explain   the
increasing   Ca-uptake   with growing HV-pulses,   but   it
should   be   repeatedly examined with   single   identical
cell cultures.

ACKNOWLEDGEMENTS
For helpful supporting and discussions we are obliged
 to Mrs. Dr.Schultze, Prof. Dr. Liebscher, and Prof.
Yu.   A. Vladimirov.

REFERENCES
1.   V.S. Malinin, V.S. Sharov, A.V. Putvinsky   et   al.,
Chemiluminescent   reactions   of phagocytes   induced   by
electroporation, Bioelectrochemistry and Bioenergetics,
22, 37 - 44 (1989)
2. K T. Hartiala, J.G. Scott, M.K. Villjanen and K.   E.
O.Akerman,   Biochem. Biophys. Res. Commun. 144, p.   794
(1987)
3.   T Hamachi, M. Hirata and T. Koga,Biochim.   Biophys.
Acta, 889, p. 136 (1986)
4.A. Boyum, Scand. J. Immunol., 15, p. 9 (1976)

THE EFFECT OF EXTERNAL INDUCER ON <u>E.COLI</u>
CONTAINING THE LUX OPERON OF <u>V.FISCHERI</u> WITH A <u>LUXI</u>
DELETION DURING DIFFERENT STAGES OF GROWTH.

Y.Y.Adar,*, J. Kuhn** and S. Ulitzur*

Dept. of Food Engineering and Biotechnology* and Faculty
of Biology**, Technion-Institute of Technology,
Haifa 32000, Israel

INTRODUCTION
The transcription of the cloned <u>lux</u> system of <u>Vibrio
fischeri</u> in <u>E.coli</u> is regulated by both a specific
autoinducer and the <u>luxR</u> protein (1-2). Recently we have
shown that a part of these components, the transcription
of the <u>lux</u> system, is positively controlled by HtpR
protein (σ^{32}) and negatively controlled by LexA protein
(3-4). According to the models suggested by Nealson <u>et
al</u>. (1) and Engebrecht <u>et al</u> (2), the primary regulation
of bioluminescence in <u>V.fischeri</u> involves the
interaction of the LuxR protein and the autoinducer
produced by the enzyme encoded by <u>luxI</u>. The LuxR-
autoinducer complex stimulates transcription of the
promoter of the right operon. Such stimulation increases
the level of the autoinducer which, in turn, results in
a positive feed-back loop and an exponential increase in
luminescence. According to this model the autoinduction
of the luminescence system is solely dependent on the
accumulation of the autoinducer in the cells. It was
thus expected that the level of luminescence of the <u>luxI</u>
deleted mutant would be directly dependent on the
concentration of externally added inducer and that the
response of the cells to the inducer would not be
dependent on the growth phase of the culture. In the
present communication we show that the responsiveness of
<u>luxI</u> deleted mutant cells to externally added inducer is
largely dependent on the growth phase. Cells from the
late exponential phase of growth showed over 100-fold
more light with a given concentration of inducer than
that of cells from early logarithmic stage of growth.
HtpR-mutant of the <u>luxI</u> deleted cells showed 1000 times
less light with a given concentration of the inducer
than the HtpR+ isogenic strain, and their luminescence
was not dependent on the growth phase.

MATERIALS AND METHODS
<u>Bacterial strains</u>: <u>E.coli</u> W3110 was used as the wild
type strain. <u>E.coli</u> MC4100 and KY1603 R.40 are isogenic

htpR+ and htpR- strains (5).

Plasmids: The luxI deleted mutant pW21A was obtained
from Biolume Ltd., Haifa, Israel. This plasmid is a
derivative of the plasmid pChv1 (3) that was deleted in
the TaqI region. In the absence of externally added
inducer, cells harbouring this plasmid are very dim. The
inducer of V.fischeri was synthesized according to the
procedure described by Eberhard (6). The cells were
grown in Luria Broth (LB) consisting of 10g tryptone, 5g
yeast extract and 5g NaCl in 1 l water. Luminescence was
determined as previously described (3).

RESULTS AND DISCUSSION
Fig. 1 shows the effect of 100 pg/ml of the synthetic
inducer of V.fischeri on the kinetics of light
development in W3110/pW21A cells that were taken from
different stages of growth. It can be seen that the rate
of specific luminescence development increases along
with the increase in cell concentration in the growth
medium. Cultures showing optical densities (O.D.-600)
above 0.5 gave over 100-fold more light per cell than
culture coming from earlier stages of growth. Specific
luminescence enhancement by externally added inducer on
cells taken from different stages of growth is
summarized in Figure 1B. A similar phenomenon has
recently been shown by us with E.coli cells harbouring
the whole intact lux system of V.fischeri (3). We
attribute this phenomenon to the increase in HtpR
activity in the starving cells that are accumulated in
the culture towards the end of the logarithmic phase of
growth. This assumption was further strengthened in this
study. HtpR- cells harbouring the plasmid pW21A almost
did not respond to the addition of 100 pg/ml of the
inducer and their luminescence was less than 0.1% of the
light developed by HtpR+ cells under the same
conditions. Moreover, the activity of the inducer was
not dependent on the growth phase from which the cells
were taken for the assay (data not shown). These results
clearly indicate that the dramatic increase in specific
luminescence during late stages of exponential phase of
growth results from the combined effect of the increase
in both the HtpR and the inducer activities. It appears
that the HtpR activity by itself is responsible for at
least a 100-fold increase in specific luminescence
during induction. The mode of action of the HtpR protein
has not yet been elucidated. However, it is clear that
the HtpR protein is active only in the presence of the
inducer, probably as a transcriptional factor.

Fig.1. The effect of <u>V.fischeri</u> inducer on the
kinetics of light development at different stages of
growth cycle of <u>E.coli</u> W3110/pW21A cells.
A. <u>E.coli</u> W3110/pW21A cells were grown overnight with
shaking in LB medium at 30° C. The culture was diluted
1000-fold in fresh LB and the cells were incubated with
shaking at 30° C. At different times aliquots were
removed for optical density determination. Samples
showing an OD600 of more than 0.1 were diluted properly
in LB to obtain this density. One ml samples were placed
in vials containing the <u>V.fischeri</u> synthetic inducer
(100 pg/ml). The vials were incubated at room
temperature (24° C) and their luminescence was determined
every 30 minutes. The figure shows the kinetics of
specific luminescence (LU/O.D.) at different optical
densities.
B. The luminescence enhancement factor: the ratio
between the specific light intensity (LU/O.D.) after 60
minutes of incubation with the inducer to the light
intensity at time zero is given for each optical density
of W3110/pW21A shown in Figure 1A.

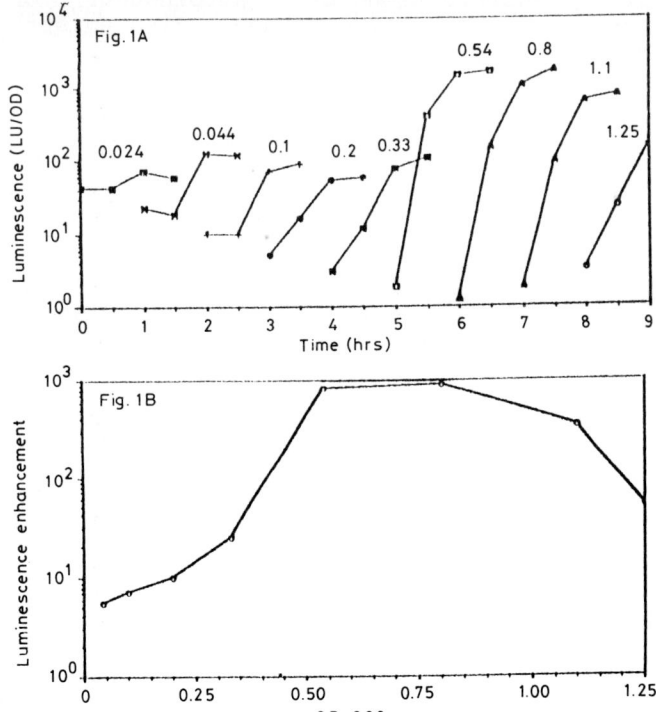

REFERENCES

1. K.H.Nealson,   T.   Platt   and   J.W.Hastings.
    Cellular control of the synthesis and activity of
    the bacterial luminescent system. J. Bacteriol.104,
    313-322 (1970).
2. J.   Engebrecht,   K.   Nealson and M. Silverman.
    Bacterial luminescence: isolation and genetic
    analysis of functions from Vibrio fischeri. Cell
    32, 773-781 (1983).
3. S. Ulitzur and J. Kuhn. The transcription of
    bacterial luminescence is regulated by Sigma 32.
    J.Biolum. and Chemilum. 2, 81-93 (1988).
4. S.Ulitzur. The regulatory control of the bacterial
    luminescence system - a new view. J.Biolum. and
    Chemilum. 4, 317-325 (1989).
5. N.Kusukawa and T.Yura. Heat shock protein GroE of
    E.coli: key   protective roles against thermal
    stress. Genes and Development 2, 874-882 (1988).
6. A.Eberhard,   A.L.   Burlingame,   C.Eberhard, G.L.
    Kenyon,   K.H. Nealson and N.J. Oppenheimer.
    Structural identification of autoinducer of
    Photobacterium fischeri luciferase. Biochemistry
    20, 2444-2449 (1981).

AMPLIFIED BIOLUMINESCENCE ASSAYS

G.K.Athwal, E.A.El-Kholy and A.E.G.Cass

Centre for Biotechnology,Imperial College of Science,Technology & Medicine, London SW7 2AZ, UK

INTRODUCTION

Luminescent immunoassays have developed in response to the need for sensitive alternatives to radioimmunoassays and are available in a variety of configurations (1). Although there have been several reports of immunoassays based upon bacterial luciferase this particular enzyme suffers from the disadvantage of being very susceptible to inactivation by typical protein cross-linking reagents thus making conjugation difficult (2). On the positive side bacterial luciferase offers a high quantum yield, long lived emission and ready availability. An alternative approach to improving the sensitivity of (non-luminescent) enzyme immunoassays has been via enzyme amplification (3). In its original configuration the assay involves a two step process; in the first step the enzyme label is used to generate a product during a fixed time incubation. Under these circumstances the amount of product generated is proportional to the amount of enzyme present. The product of this step is then amplified through a second enzymic recycling reaction where each molecule of product is recycled many times to yield the final signal. In its most developed form this scheme works with alkaline phosphatase as the first enzyme acting on $NADP^+$ to generate NAD^+ as the product. The NAD^+ then recycles between oxidised and reduced forms using alcohol dehydrogenase/ethanol and diaphorase/a tetrazolium salt. The final product is a formazan dye.

Ultimately the sensitivity of this type of assay is determined by the specificity of the amplifying reagents for the product rather than the substrate of the first enzyme, that is the background signal in the absence of the first enzyme. In an attempt to develop a high sensitivity immunoassay based around bacterial luciferase we have explored the use of nucleotide pyrophosphatase (NPP) as the first enzyme and luciferase as the amplifying enzyme. The scheme envisages the conversion of FAD to FMN by the NPP followed by the FMN dependent turnover of bacterial luciferase in the presence of oxidoreductase, NADH and decanal.

MATERIALS AND METHODS

<u>Enzymes and Antibodies</u>. Luciferase and oxidoreductase from *Vibrio harveyii* were isolated using published procedures (4). Nucleotide pyrophosphatase from *Crotalus atrox* was purified from either a crude venom or from a partially purified enzyme both from Sigma. The purification in both cases was by FPLC using a MonoQ column. Bovine IgG and sheep anti(bovine IgG) were from Sigma.

<u>Other Reagents:</u> All other reagents were from Sigma and the FAD was analysed by thin layer chromatography and HPLC.

<u>Enzyme Assay Procedures:</u> Luminescence assays were performed in an LKB 1250 luminometer with a 50mmol/l phosphate buffer pH 7 containing 0.2% bovine serum albumin (BSA) and 1µg/ml decanal. Other conditions are given in the text. The output from the luminometer was displayed on a y-t chart recorder and the plateau intensity was used as a measure of the signal. NPP was assayed spectrophotometrically at 405nm with 5 mmol/l thymidine-5'-monophosphate-p-nitrophenyl ester at pH 7.4 in 0.1mol/l Tris buffer. Incubations of NPP with FAD were carried out in the same buffer and 10µl of this solution added to 1ml of the luciferase solution.

<u>Conjugation Reactions:</u> NPP was conjugated to bovine IgG using p-benzoquinone or glutaraldehyde by standard methods (5).

<u>Immunoassays:</u> Competitive immunoassays were carried out in coated tubes following established protocols and employing BSA as a blocking agent.

RESULTS AND DISCUSSION

NPP from *Crotolus atrox* was purified to homogeneity by ion exchange chromatography and the purified enzyme had a molecular weight of 25kD as judged from SDS gels and a specific activity of 5U/mg in the colorimetric assay. This purified material was used in all subsequent experiments.

In order to determine the working conditions for the NPP assay we explored the effects of varying concentrations on the overall light output in the amplified assay. The optimisation of the assay is complicated by the observations that excess FMN inhibits bacterial luciferase and that FAD can also act as both a poor substrate and an inhibitor of the same enzyme. We observed that the light output from a coupled luciferase assay is proportional to the FMN concentration upto 5μmol/l and above that concentration is decreased, consistent with excess FMN acting as an inhibitor. At this concentration of FMN, FAD is inhibitory in concentrations greater than 250μmol/l. Interestingly the inhibition by FAD at high concentration of FMN is additive suggesting that the FAD competes at the catalytic FMN site but not the inhibitory one. When non-inhibiting concentrations of FAD are used in the NPP-luciferase coupled assay the light intensity increases linearly with incubation time.

Similarly with a fixed incubation time the light intensity increases linearly with NPP concentration, Fig 1.

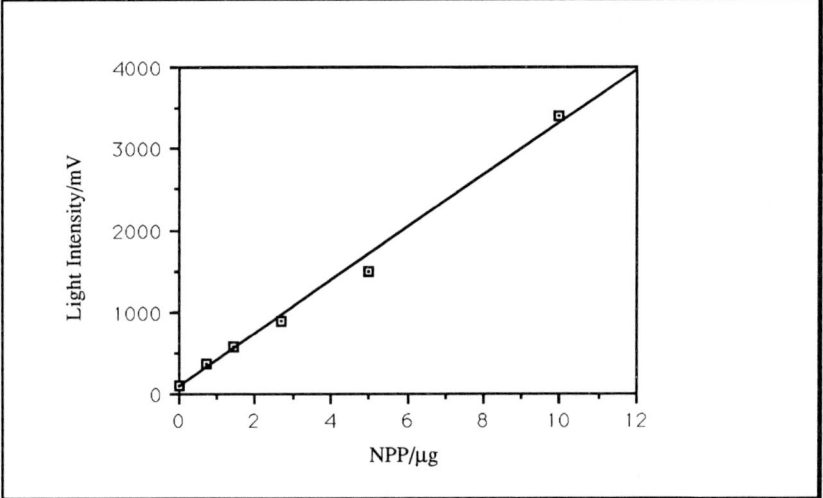

Fig 1. Dependence of light intensity on NPP in the NPP-luciferase coupled assay. 250μmol/l FAD was incubated with variable amounts of NPP for 10 minutes, then 0.1ml was added to the luciferase-oxidoreductase assay mixture and the steady state light intensity recorded

The effect of FAD concentration is more complex because of the inhibitory action of this nucleotide on the luciferase reaction discussed above. It is not therefore surprising that the FAD dependence shows a peak response, Fig 2:

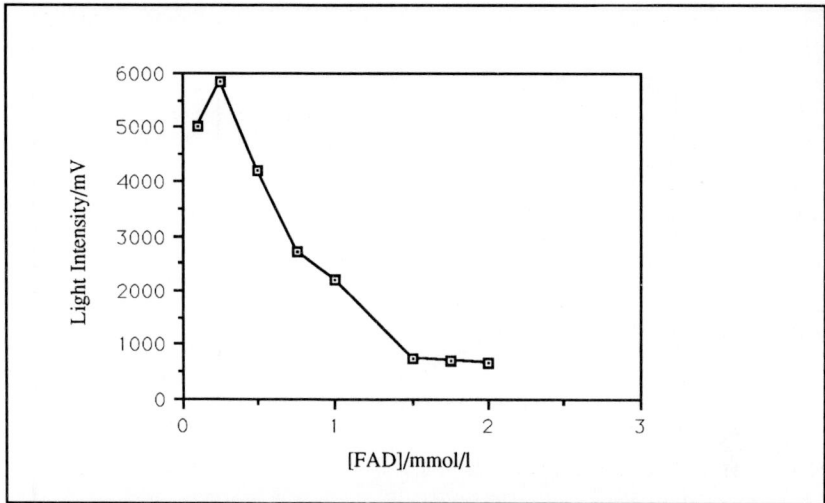

Fig 2. Dependence of light intensity on FAD in the NPP-luciferase coupled assay. Variable concentrations of FAD were incubated with NPP for 10 minutes, then 0.1ml was added to the luciferase-oxidoreductase assay mixture and the steady state light intensity recorded

In Figs 2and 3 above it can be seen that that the intercept on the ordinate is non zero implying some luciferase turnover even in the absense of FMN. We thought at first that this might be due to contaminating FMN in the FAD samples, thin layer chromatography on silica plates showed considerable contamination of the FAD by riboflavin but no evidence of FMN. Similarly reverse phase HPLC on a C_{18} column again revealed considerable riboflavin content but no FMN. Finally FAD purified by HPLC showed exactly the same behviour as unpurified material. FAD has been reported to be a substrate for bacterial luciferase (6) and we have shown that the oxidoreductase can reduce FAD with NADH. Despite these observations we have found that the background signal varied with different preparations of luciferase suggesting that it can be supressed by appropriate manipulation of the luciferase to oxidoreductase ratios.

At the lowest background we have so far achieved the detection limit for NPP is 0.6pmol for a signal to background ratio of 2:1 and an icubation time of 10 minutes. This is approximately three orders of magnitude better than the colorimetric assay. We have not yet systematically explored the effect of the composition of the luciferase system on this value although we have observed that increasinsing the NADH concentration from 0.1 mmol/l to 1mmol/l improves the overall dose- response behaviour but doesn't affect the background signal.

NPP is also active against NADH, hydrolysing it to NMN and adenine, however control experiments have shown that this activity does not interfere with the luciferase assay and we attribute this to the short measurement time (1-2 min) involved. In addition because the optimum pH's for both the NPP and the luciferase reactions are comparable there is no need to adjust the pH between the two parts of the experiment. Both of these properties are in contrast to the alkaline phosphatase assay where the first reaction has to be stopped and a pH change made before the amplification step.

NPP was conjugated to bovine IgG using standard glutaraldehyde or p-benzoquinone chemistries, in the former case the reaction resulted in almost total loss of NPP activity, whereas the latter gave an 80% retention of activity. The conjugates were separated from unreacted NPP and IgG by FPLC gel filtration on Superose-6 and used in a competitive immunoassay. The dose-response curve for bovine IgG is shown in Fig 3 and the detection limit is 1.5 µg. We are currently exploring the use of this system in other configurations.

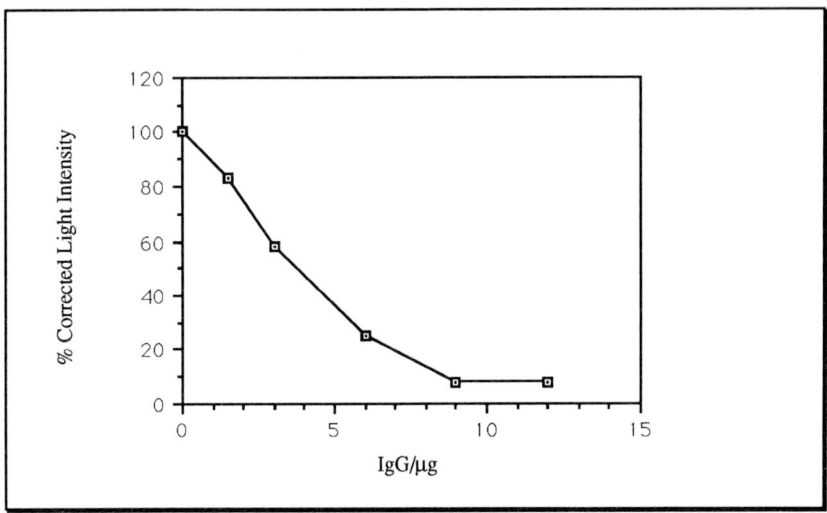

Fig 3: Competitive Immunoassay employing solid phase sheep anti(bovine IgG) and NPP
labelled bovine IgG

REFERENCES
1.   F.Kohen,M.Pazzagli,M.Serio,J.De Boevers and D.Vandekerckhove,Chemiluminescence
and Bioluminescence Immunoassay in Alternative Immunoassays W.P.Collins (Ed) Wiley,
Chichester, p.103-121 (1985)
2.   J.Wannlund and M.Deluca Bioluminescent Immunoassays:Use of Luciferase-Antigen
Conjugates for Determination of Methotrexate and DNP in Bioluminescence and
Chemiluminescence M.Deluca and W.D.McElroy (Eds), Academic Press, New York (1981)
3.   D.L. Bates, Enzyme Amplification in Diagnostics. Trends Biotech. 5, 204-209 (1987)
4.   J.W.Hastings, T.O.Baldwin and M.Z.Nicoli, Bacterial Luciferase:Assay, Purification
and Properties Methods in Enzymology LVII 135-152 (1978)
5.   P.Tijssen Practice and Theory of Enzyme Immunoassays, Elsevier, Amsterdam (1985)
6.   G.Mitchell and J.W.Hastings The effect of Flavin Isomers and Analogues upon the
Color of Bacterial Bioluminescence. J.Biol.Chem. 244,2572-2579 (1969)

Session II
IMMUNOASSAY AND DNA PROBES

NOVEL CHEMILUMINESCENT ADAMANTYL 1,2-DIOXETANE ENZYME SUBSTRATES

I. Bronstein, R. R. Juo, J. C. Voyta and B. Edwards

TROPIX, Inc., 47 Wiggins Avenue, Bedford, MA 01730

INTRODUCTION

1989 marked the twentieth anniversary of Kopecky and Mumford's synthesis of the first 1,2-dioxetane (1). Since this seminal event, a vast amount of research and development has been dedicated to understanding the basic chemistry of dioxetanes and exploiting their unique properties. In 1972, Wynberg's group reported the synthesis of adamantylideneadamantyl 1,2-dioxetane, notable for its high thermal stability (decomp. above 165°C) (2). Wynberg's elegantly simple synthesis, via photooxygenation of adamantylideneadamantane, capitalized on the lack of allylic protons which precluded competing ene reactions. The resulting 1,2-dioxetane product also demonstrated that two adamantyl groups did not sterically inhibit peroxide formation. In 1977, McCapra and coworkers described the synthesis of an unsymmetrically substituted adamantyl 1,2-dioxetane, (9-(2-adamant-ylidene)-N-methylacridan 1,2-dioxetane), which, when heated, generated chemiluminescence attributed exclusively to the excited singlet state of N-methylacridone (3). In the same year, McCapra proposed that dioxetanes which contain electron donating substitutents decompose via charged intermediates in an electron transfer process (4). Later work, originating from Adam's laboratories, detailed the syntheses and stabilities of several unsymmetrically substituted adamantyl 1,2-dioxetanes. They concluded that the stabilization mechanism for these compounds is quite complex, depending on conformational isomerism as well as other factors (5). Subsequently, Adam suggested the application of chemically functionalized 1,2-dioxetanes as potentially useful chemotherapeutic agents (6).

The first important application of 1,2-dioxetanes to clinical analysis was described by Hummelen *et al.* in 1987 (7). Suitably derivatized adamantylideneadamantyl 1,2-dioxetanes were utilized as labels in thermochemiluminescent immunoassays. Further molecular modification, in our laboratory and elsewhere, produced enzyme-activated adamantyl substituted 1,2-dioxetane substrates which, when coupled with the existing ELISA test formats, led to the development of highly sensitive assays for a variety of clinically important analytes (8). Currently, a new generation of dioxetane substrates and enhancers has

been developed which exhibit improved properties in a number of bioassay formats.   In this paper we describe recent progress in adamantyl
1,2-dioxetane enzyme substrate technology and its applications.

DISCUSSION

AMPPD (disodium 3-(4-methoxyspiro[1,2-dioxetane-3,2'-tricyclo-
[3.3.1.1.^{3,7}]decan]-4-yl)phenyl phosphate, CAS#124951-96-8), was
developed as a substrate for alkaline phosphatase with detection
capabilities superior to currently available colorimetric, bioluminescent or
fluorimetric substrates. The stable 1,2-dioxetane can be conveniently
stored and used as an aqueous solution with a half-life of one year at
room temperature.

Figure 1. AMPPD (R = H)

Dephosphorylation of AMPPD generates the metastable phenolate anion
(AMP$^-$D), which decomposes while emitting light. This enzyme-mediated
chemiluminescence has been successfully incorporated in sensitive,
alkaline phosphatase-based protein and nucleic acid probe assays (9).

Design of 1,2-Dioxetane Enzyme Substrates.   Close examination of a
series of 1,2-dioxetane properties has permitted the intelligent design of
substrates with specific properties ,e.g., thermal stability, ease-of-use, and
intense chemiluminescent signal. The basic design of 1,2-dioxetane
substrates includes several components as shown in Fig. 2.   The
presence of a single adamantyl group is sufficient to provide the
necessary thermal stability in our asymmetric 1,2-dioxetanes. Scission of
the peroxide bond releases approximately 100 kcal/mol to populate
excited states of an aryl carbonyl fragment, from which emission occurs.
This weak oxygen-oxygen bond (23 kcal/mol) thus acts as an "energy
storehouse." Dioxetane decomposition is initiated by enzymatic cleavage
of a protecting group on the phenoxy oxygen. For example, alkaline
phosphatase dephosphorylates AMPPD to initiate chemiluminescence.
The orientation of the aryl alkoxy substituent to the dioxetane attachment
point on the phenyl ring is referred to as "disjoint" and is key to the
performance of the compound as an enzyme substrate. Two other
important features include 1) the alkoxy substituent on the dioxetane
which enables efficient photooxygenation of the enol ether precursor and

2) a disaggregation moiety on the adamantyl ring which will be discussed later in this paper.

Figure 2. Schematic Design of 1,2-Dioxetane Substrates

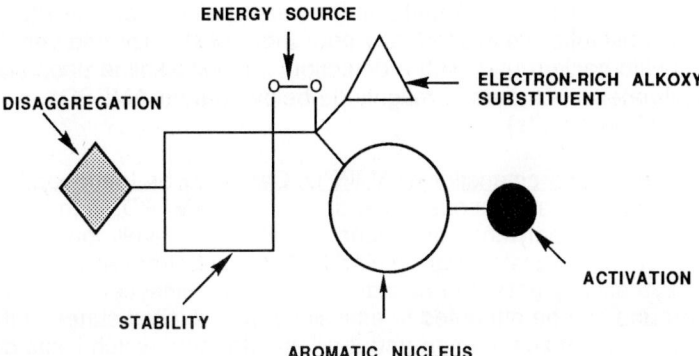

Disjoint (odd) versus Conjugated (even) Substitution Patterns. It was reported in 1982 that a *meta*, or disjoint, isomer pattern in *m*-amino-diphenyl dioxene-1,2-dioxetane exhibited a higher total chemilumi-nescence lifetime, greater stability, and a reasonable efficiency for the population of singlet states in a polar environment (10). We found that in naphthalene-substituted dioxetane systems, the position of the donor (O⁻) relative to the point of the attachment to the dioxetane ring, affects the characteristics of the excited ester product. *(An odd pattern is one in which the donor group's point of attachment to the aryl ring in relation to the dioxetane ring's point of attachment to the aryl ring is such that the total number of aryl ring carbon atoms separating these points, including the atoms at the point of attachment, is an odd whole number).* The odd substituted naphthalene dioxetane phosphates (2,7- 1,3- and 1,6- isomers) exhibited longer half-lives for their respective oxyanions (11). In addition, the odd-patterned dioxetanes produced greater levels of chemiluminescence, lower backgrounds, and bathochromically shifted emissions compared to the even-substituted 2,6-naphthalene compound.

Theoretical AM1 molecular orbital calculations revealed that the greatest amount of charge transfer from donor to acceptor occurred in the odd substituted oxynaphthoic acid model systems. The phenomenon of disjoint substitution suggests that dioxetane decomposition occurs along several reaction coordinates which strongly depend on the aromatic substitution pattern.

Our investigations have shown that an odd substitution pattern also leads to the preferred dioxetane properties in phenyl-substituted dioxetanes. Again, AM1 molecular orbital calculations for the 1,3-carboxyphenolate, and 1,4-carboxyphenolate anions reveal that the HOMO-LUMO gaps are 7.14 eV and 7.97 eV for the disjoint and conjugated systems respectively.

The corresponding net charge transfers are 0.77 eV (disjoint) and 0.63 eV (conjugated). On the basis of these results, excited states with greater charge transfer character are expected to lead to chemiluminescent systems with greater excitation yields, lower nonspecific backgrounds, and more stable intermediates preceding the formation of the excited state ester products. We found that the *para* analogue of AMPPD (i.e., 1,4 phenyl substitution) exhibited very poor thermal stability and very high nonspecific background, with a detection limit for alkaline phosphatase approximately six orders of magnitude below that for AMPPD (unpublished results).

Performance Characteristics of AMPPD: Can They be Improved? We have observed that aqueous buffer solutions of AMPPD, in the presence and absence of polymeric enhancers, exhibit a relatively long delay before reaching constant light emission. This substrate also gives some nonenzymatically activated background chemiluminescence. This background can be attributed to emissions from excited states of the methyl *m*-oxybenzoate anion and 2-adamantanone, which limits detection sensitivity. The long delay leading to steady state light emission, and the accompanying "enhanced" backgrounds, are due to the amphiphilic nature of the molecule. This inherent property leads to the aggregation of AMPPD and its dephosphorylated anion. Background signal produced from any nonenzymatically generated excited state emitter is amplified, since the hydrophobic environment of the aggregate stabilizes the emitter. In order to minimize these aggregation effects, we designed a new group of derivatized adamantyl 1,2-dioxetane phosphates with improved analytical performance.

Substituted 5-Adamant-2'-ylidene-1,2-Dioxetane Enzyme Substrates. The guiding premise in the design of this new class of dioxetane substrates was a modification of the hydrophobic adamantyl group in order to prevent molecular self-assembly. This "new generation" of derivatized adamantyl-substituted 1,2-dioxetane substrates (R = OH, OMe, Cl, Br, I) is synthesized by coupling 5-substituted adamantan-2-ones with α-methoxy-α-phenyl methane phosphonate esters, using a Wittig-Horner reaction. The resulting phenolate enol ethers are subsequently phosphorylated and photooxygenated to yield the desired modified dioxetanes.

5-Substituted-adamantyl 1,2-Dioxetanes: Anion Half-lives and Thermal Backgrounds. Table 1 shows that derivatization of the adamantyl portion of the dioxetane results in faster decomposition of the intermediate dioxetane anion produced by alkaline phosphatase catalyzed dephosphorylation; thus, steady state emission is attained more rapidly. Furthermore, this molecular modification also has a significant impact on the thermal or "nonspecific" background of the dioxetane.

Table 1. Half-lives and Thermal Backgrounds
for Various Dioxetane Phosphates

DIOXETANE	BACKGROUND AT 0.4 mM (TLU)	HALF-LIFE OF ANION * (min)
AMPPD	1.83	2.10
MeO-AMPPD	0.71	1.52
HO-AMPPD	0.83	1.19
Cl-AMPPD	0.83	0.96
Br-AMPPD	0.78	1.01
I-AMPPD	0.46	0.93

TLU - Turner Light Units
* Anion produced by the addition of 0.764 picomoles of alkaline phosphatase to 40 nanomoles of AMPPD and R-AMPPD in 0.1M DEA, 1mM $MgCl_2$, pH 10.0.

Chemiluminescence Kinetics on Nylon Membranes. We compared the kinetics of light emission from AMPPD and Cl-AMPPD on a neutral nylon membrane. Figure 3 shows the time course of chemiluminescence emission from Cl-AMPPD, which is dramatically faster compared to AMPPD. This feature in turn, permits detection of lower DNA concentrations in shorter times.

Figure 3. Chemiluminescence Kinetics on Nylon Membranes
AMPPD vs Cl-AMPPD

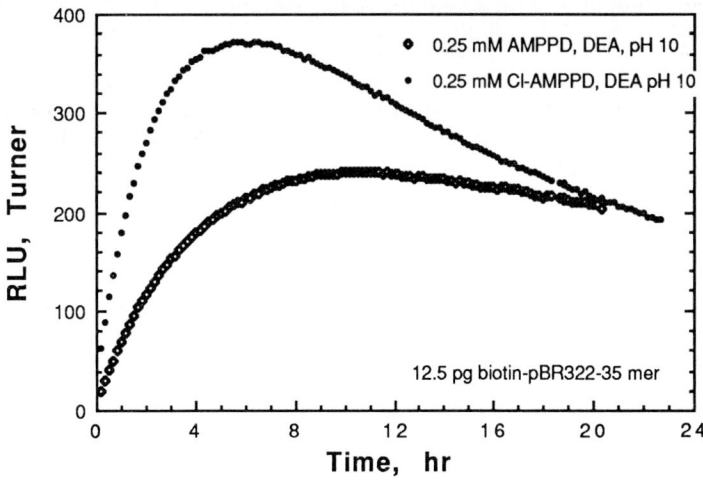

Alkaline Phosphatase Detection Limits. The concentrations of alkaline phosphatase which can be detected in solution with various dioxetane phosphate substrates (0.1 M diethanolamine containing 1 mM $MgCl_2$ at pH 10), using a Labsystems Luminoscan microtiter plate luminometer, are summarized in Table 2. The data is expressed as the enzyme concentration levels determined at twice the background, after 5 and 20 minute incubations. Again, as our results indicate, the derivatized dioxetane substrates provide increased sensitivities of alkaline phosphatase detection in shorter time periods.

Table 2. Detection Limits for Alkaline Phosphatase with Different Chemiluminescent 1,2-Dioxetane Substrate/Enhancer Systems

DIOXETANE	ENHANCER 1		ENHANCER 2	
	5 MIN	20 MIN	5 MIN	20 MIN
AMPPD	4.0	1.4	3.5	1.1
HO-AMPPD	3.0	0.8	1.0	0.3
Cl-AMPPD	1.0	0.6	0.8	0.4
Br-AMPPD	2.6	0.5	0.4	0.3

*Detection limits expressed in femtomoles /liter

TSH Immunoassay. We also compared the performance of AMPPD and HO-AMPPD in the detection of thyroid stimulating hormone (TSH). Figures 4A and 4B show plots of Relative Light Units (RLU) versus TSH concentration, determined in the Berthold LB952 luminometer. The 5 s integral RLU measurements were performed at 6, 12, 18, 24, 30, 40, 50, and 60 minutes.   Figures 4A and 4B indicate that 5-hydroxyadamantyl 1,2-dioxetane is superior in its performance due to the rapid establishment of a reliable dose-response curve. This feature is the result of faster kinetics leading to steady state emission which, as shown in Fig. 5, enables faster detection in assays using HO-AMPPD. It is important to note that while the derivatized dioxetanes exhibit absolute chemiluminescent signal levels that are lower or equal to AMPPD, their nonspecific background levels are also lower, resulting in greater signal-to-noise ratios.

Figures 4A and 4B. Chemiluminescent Signal vs. TSH Concentration
A Comparison of AMPPD and HO-AMPPD in Emerald

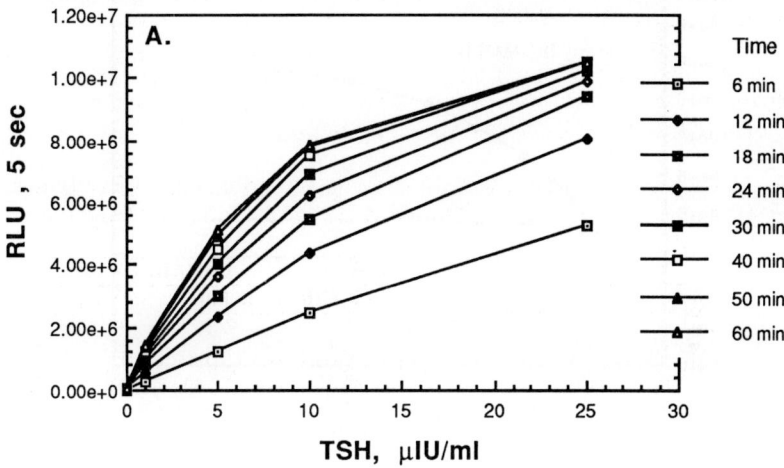

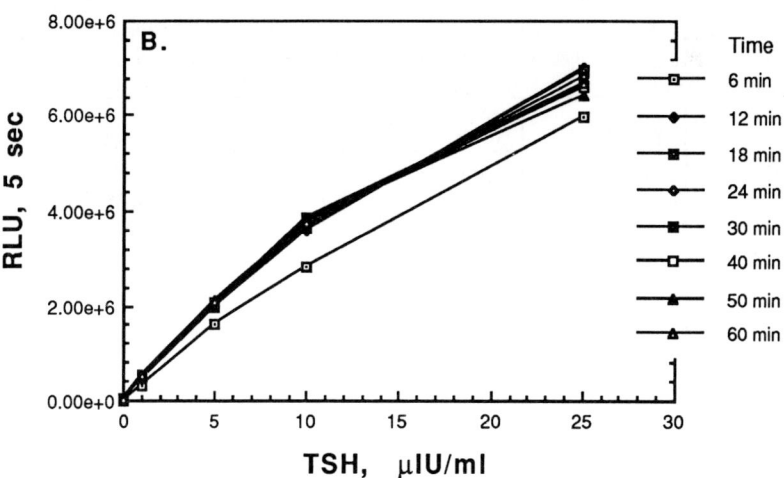

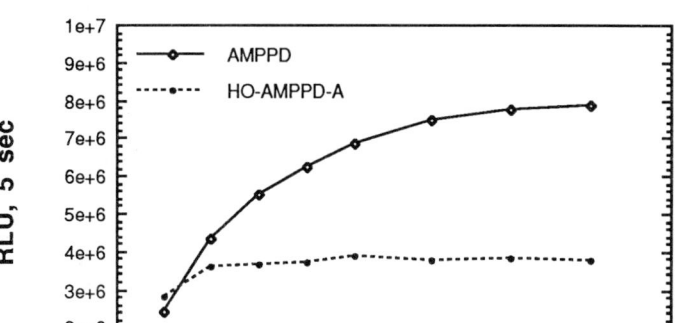

Figure 5. Chemiluminescent Kinetics
A Comparison of AMPPD and HO-AMPPD

<u>Signal/Noise for Various Dioxetane Substrates.</u> Table 3 compares
signal-to-noise levels for the various dioxetane phosphates determined
at two incubation times, in different enhancing solutions. Nearly all
derivatized dioxetanes exhibit considerably higher signal-to-noise levels

Table 3. Chemiluminescent Signal/Noise Levels
for Substituted AMPPD Derivatives

ENHANCER	TIME (min)	R = H	R = OH	R = Cl	R = Br
NONE	5	21	55	82	77
	20	34	73	98	92
SAPPHIRE 1	5	159	229	577	383
	20	364	305	827	566
EMERALD 1	5	109	307	583	149
	20	282	386	964	915
SAPPHIRE 2	5	192	466	1319	688
	20	434	643	2221	1019
EMERALD 2	5	113	247	839	315
	20	291	299	1280	503

All enhancers used at 1mg/ml in 0.1M DEA, pH 10.
Alkaline phosphatase concentration = 2.83×10^{-13} M
Measurements performed in a Berthold LB915T Luminometer.

when compared to the parent compound. Specifically, it is remarkable that in the presence of the more efficient polymeric enhancer (Sapphire 2), the signal-to-noise ratio measured for Cl-AMPPD is nearly an order of magnitude greater than that observed for AMPPD under identical conditions. For the 5-substituted adamantyl 1,2-dioxetanes, two isomeric forms exist in which the substituent is either *syn* or *anti* to the dioxetane ring. In the case where R = OH, these isomers can be separated *via* reverse-phase HPLC. The data listed in Table 3 corresponds to the faster emitting HO-AMPPD isomer.

Conclusions. We have developed a new generation of 5-substituted adamantyl 1,2-dioxetane substrates which exhibit improved performance features in a range of applications. Steady state emission is reached more rapidly and nonspecific backgrounds are lower than that achieved with the parent dioxetane, AMPPD. The mechanisms underlying these phenomena have not been fully explored. In general,a lower degree of aggregation in these compounds is certainly one of the key factors. However, dipole effects and electronic perturbation induced by some of the adamantyl substituents cannot be ruled out at this time. With the advent of this new class of substrates, faster and more sensitive protein and DNA detections are now possible.

ACKNOWLEDGEMENTS

The authors greatly appreciate the support and contributions of Tropix scientists in assembling this manuscript. In particular, the help from Alison Sparks was invaluable during critical technical discussions. We also thank Gary Thorpe for his contributions in the TSH experiments.

REFERENCES

1. K. R. Kopecky and C. Mumford, Can. J. Chem. 47, 709 (1969).
2. J. H. Wierynga, J. Strating and H. Wynberg, Adamantylidene-adamantane Peroxide: A Stable 1,2-Dioxetane. Tetrahedron Lett. 2, 169-172 (1972).
3. F. McCapra, I. Beheshti, A. Burford, R. A. Hann and K. A. Zaklika, Singlet Excited States from Dioxetan Decomposition. J. C. S. Chem. Commun. 944-946 (1977).
4. F. McCapra, Alternative Mechanism for Dioxetan Decomposition. J. C. S. Chem. Commun. 946-948 (1977).
5. W. Adam, L. Encarnacion and K. Zinner, Thermal Stability of Spiro[adamantane-[1,2]dioxetanes]. Chem. Ber. 116, 839-846 (1983).
6. W. Adam, V. Bhushan, T. Dirnberger and R. Fuchs, Functionalized 1,2-Dioxetanes as Potential Chemotherapeutic Agents: The Synthesis of Dioxetane-Substituted Carbamates. Synthesis. 330-332 (1986).

7.   J. C. Hummelen, T. M. Luider and H. Wynberg, Functionalized
          Adamantylideneadamantane 1,2-Dioxetanes: Investigations
          of Stable and Inherently Chemiluminescent Compounds as a
          Tool for Clinical Analysis. Pure Appl. Chem. 59, 639-650
          (1987); J. C. Hummelen, A. C. Brouwer, T. M. Luider, F. van
          Bolhuis and H. Wynberg, Xanthenylideneadamantane 1,2-
          Dioxetane. Preparation, Properties and its Potential Use as an
          Inherently Thermochemiluminescent Label. Tetrahedron Lett.
          29, 3137-3140 (1988).
8.   A. P. Schaap, R. S. Handley and B. P. Giri, Chemical and
          Enzymatic Triggering of 1,2-Dioxetanes. 1: Aryl Esterase-
          catalyzed Chemiluminescence from a Naphthyl Acetate-
          Substituted Dioxetane. Tetrahedron Lett. 28, 935-938 (1987);
          A. P. Schaap, M. D. Sandison and R. S. Handley, Chemical
          and Enzymatic Triggering of 1,2-Dioxetanes. 3: Alkaline
          Phosphatase-catalyzed Chemiluminescence from an Aryl
          Phosphate-substituted Dioxetane. Tetrahedron Lett. 28, 1159-
          1162 (1987); I. Bronstein, B. Edwards and J. C. Voyta, 1,2-
          Dioxetanes: Novel Chemiluminescent Enzyme Substrates.
          Applications to Immunoassays. J. Biolumin. Chemilumin. 4,
          99-111 (1989); G. Thorpe, I. Bronstein, L. J. Kricka, B. Edwards
          and J. C. Voyta, Chemiluminescent Enzyme Immunoassay of
          α-Fetoprotein Based on an Adamantyl Dioxetane Phenyl
          Phosphate Substrate. Clin. Chem. 35, 2319-2321 (1989); I.
          Bronstein and P. McGrath, Chemiluminescence Lights Up.
          Nature. 338, 599-600 (1989).
9.   I. Bronstein and J. C.Voyta, Chemiluminescent Detection of
          Herpes Simplex Virus I DNA in Blot and In-Situ Hybridization
          Assays. Clin. Chem. 35, 1856-1857 (1989); I. Bronstein, J. C.
          Voyta, K. G. Lazzari, O. J. Murphy, B. Edwards and L. J. Kricka,
          Rapid and Sensitive Detection of DNA in Southern Blots with
          Chemiluminescence. BioTechniques. 8, 310-314 (1990);
          R. Tizard, R. L. Cate, K. L. Ramachandran, M. Wysk, J. C. Voyta,
          O. J. Murphy and I. Bronstein, Imaging of DNA Sequences
          with Chemiluminescence. Proc. Natl. Acad. Sci. USA. 87,
          4514-4518 (1990); I. Bronstein, J. C. Voyta, K. G. Lazzari,
          O. J. Murphy, B. Edwards and L. J. Kricka, Improved
          Chemiluminescent Detection of Alkaline Phosphatase.
          BioTechniques. 9, 160-161 (1990).
10.   S. Gagnon, Dissertation, Wayne State University, 1982.
11.   B. Edwards, A. Sparks, J. C. Voyta and I. Bronstein, Unusual
          Luminescent Properties of Odd- and Even-substituted
          Naphthyl-derivatized Dioxetanes. J. Biolumin. Chemilumin. 5,
          1-4 (1990); B. Edwards, A. Sparks, J. C. Voyta, R. Strong, O. J.
          Murphy and I. Bronstein, Naphthyl Dioxetane Phosphates:
          Synthesis of Novel Substrates for Enzymatic Chemilumines-
          cent Assays. J. Org. Chem. 55, in press.

RAPID AND SENSITIVE LUMINESCENT METHODS
FOR NUCLEIC ACID DETECTION

Denise V Pollard-Knight

Pollards Wood Laboratories, Amersham International plc, Nightingales Lane, Chalfont St Giles, Bucks HP8 4SP, UK

Nucleic acid hybridisation to detect specific sequences is one of the most widely used molecular biology techniques. The nucleic acid probe must be detected following hybridisation. Labelling is most commonly performed with radioactivity, for example ^{32}P or ^{3}H. Significant advances have been made in the development of simple assay formats that can be used for hybridisation analyses. The convenience of these techniques, and the sensitivity ($<10^{-18}$ moles) implicit in using radioactive labels, has resulted in their application to basic research problems. For example, the organisation of the human genome, the detection of genetic diseases, and the detection of microbial, plant and viral pathogens. Although radioactive labels continue to be used in research laboratories, the short half-life, disposal and safety considerations, and the necessity to perform long autoradiographic exposures, which can vary from a few hours for phosphorus-32 to up to a month for tritium, has resulted in the search for suitable non-radioactive labels and detection techniques. Initially, alternatives were based on enzymes with a colorimetric or fluorescent end-point. These methods, however, do not have the sensitivity required, for example, for analysis of single-copy genes in human genomic DNA. More recently, novel chemiluminescent detection methods, which are beginning to approach the sensitivity of radioactive labels, have been devised. These methods use both enzyme and chemical labels.

Applications of the most widely used chemiluminescent detection systems will be reviewed. Factors that influence the use of a particular luminescent detection method are discussed, and the impact of probe and target amplification techniques on the requirement for high sensitivity non-radioactive signal-generation system is briefly considered.

FORMAT FOR ASSAY

Both the assay methodology and the detection technology are closely related. Assay formats therefore need to be considered. The main types of hybridisation analysis are described below.

1. <u>Sequence Detection</u>

    This is the simple analysis of the presence or absence of a sequence in a sample. In a spot or dot blot, the extracted nucleic acid sample is fixed to a nitrocellulose or nylon membrane by baking, ultraviolet light (uv), crosslinking or alkali treatment. The presence of a specific nucleic acid sequence on the membrane is determined by hybridisation with a labelled probe. Alternatively, a sandwich hybridisation format may be used with microtitre plates or plastic beads as the solid support.

Another development is the 'reverse' dot blot. In this
approach, DNA oligonucleotide probes are immobilised onto a
nylon support via a poly-dT tail by ultraviolet irradiation.
The immobilised probes are hybridised to a nucleic acid target
from solution.

The presence of a recombinant plasmid or lambda bacteriophage
carrying a DNA sequence of interest may be identified by two
other techniques known as colony and plaque hybridisations.
Colony hybridisations require the transfer of bacteria from a
plate onto a membrane and then hybridisation with a labelled
probe. In plaque hybridisation, DNA present in bacteriophage
plaques is bound to a membrane by applying the membrane to the
surface of a plate followed by baking or uv crosslinking as
described above. Hybridisation with a labelled probe is then
used to identify positive plaques.

2.   Sequence Identification with Molecular Weight Estimation

The presence of a particular sequence and its' size in
relation to other sequences or standards may be detected using
Southern or Northern blots. In the case of DNA, this is known
as the Southern blot(1), and for RNA it is termed a Northern
blot. The extracted DNA is first digested with restriction
enzymes, the fragments are electrophoretically separated and
then transferred onto a nitrocellulose or nylon filter:
digestion is not required for a Northern blot. A labelled
probe is then hybridised to the target nucleic acid on the
membrane.

3.   Sequence Identification with Tissue or Cellular Localisation

This technique involves annealing of a labelled probe to
target sequences immobilised in cytological and histological
preparation(2). In situ hybridisation therefore detects the
presence of a specific sequence and where it is localised in
tissues or cells.

LUMINESCENT DETECTION METHODS

For the many applications of hybridisation analyses, a range of
detection limits or 'sensitivities' are required of the detection
method. The standard for Southern and dot blots has been the
detection of a single copy gene in a 10 μg sample of human genomic
DNA. For a DNA probe of 1 kilobase (kb) this is 3.3 pg, 1×10^{-18}
moles or 3×10^6 molecules. This figure does not take into account
any target losses that may occur, particularly on nitrocellulose.
Although these losses are less on nylon membranes, the detection
limit is often limited by high backgrounds.

For in situ hybridisation, the aim is detection of a unique gene in
a single cell with a resolution of less than 1 μm, which, on
average, is 1000 nucleotides from a total human genome of 3×10^9
nucleotides. Resolution, therefore, also determines the choice of
label.

The detection limits required for the different applications of
hybridisation analysis are as follows:

Application	Range of Detection Limit

Southern or DNA dot blots

Single copy gene detection in 10 μg of:

1. human genomic DNA — 5×10^{-18} moles
2. *Drosophila* genomic DNA — 1 to 5×10^{-17} moles
3. *E.coli* genomic DNA — 5 to 50×10^{-17} moles

Northern or RNA dot blots

1. Detection of low abundance mRNA — 1 to 50×10^{-18} moles
2. Plaque and colony hybridisation — 5 to 50×10^{-17} moles

***In situ* hybridisation**

Detection of:

1. Single copy gene in 1 human cell — approx. 10×10^{-24} moles
2. Amplified sequences or transcripts — approx. $2\text{-}10 \times 10^{-21}$ moles

For comparison, using ^{32}P-labelled nucleotides and a 1 kb probe labelled to a specific activity of 10^9 counts/min/μg, an experienced user will achieve a detection limit of 30 fg or 2×10^4 molecules when probing a Southern blot. This normally requires an overnight autoradiographic step: the resulting autoradiogram of a nylon membrane will have a strong signal and low background.

Several major advances in the development and application of luminescent detection methods have occurred in the past five years. Some of these are based upon alkaline phosphatase and horseradish peroxidase catalysed reactions, but chemically triggered chemiluminescent compounds have also been prepared. The major advantages of a luminescent end-point include a permanent record, eg film, and the possibility for repeated exposures to achieve the optimum signal-to-noise ratio if the light signal is of sufficient duration. The signal can be quantified by appropriate instrumentation.

Horseradish Peroxidase-catalysed Enhanced Chemiluminescence

The thousand-fold enhancement, by substituted phenols, of the chemiluminescent signal from the oxidation of luminol catalysed by horseradish peroxidase was first described by Whitehead et al(3). The mechanism of enhancement is not fully understood, but this has not prevented its widespread application. The degree of enhancement and duration of the light output depends upon the type of peroxidase, the enhancer, and the chemiluminescent substrate.

In nucleic acid detection, we and others have used enhanced chemiluminescence to detect biotin-labelled(4) and horseradish peroxidase-labelled(5,6) DNA and RNA probes. The light is emitted with a quantum yield of 0.01 at 428 nm, it decays over a period of hours and may be detected on blue-sensitive X-ray or Polaroid film. Less than sixty minute exposures of blots are typically required(4,7). A cooled charge-collection device (CCD) camera or luminometer can also be used for quantification. A detection limit of 0.1×10^{-18} moles can be achieved; however, a more realistic figure for an inexperienced user is 1 to 5×10^{-18} moles. A detailed description of the application of this technique can be found elsewhere(8). We have also recently used enhanced chemiluminescence and horseradish peroxidase oligonucleotide probes to detect single base mis-matches and for DNA fingerprinting.

Alkaline Phosphatase-catalysed Chemiluminescence

Luciferin-O-phosphate   A new substrate for alkaline phosphatase, luciferin-O-phosphate, has now been synthesised(9). The enzyme cleaves the phosphate group to release luciferin, which is oxidised by firefly luciferase to produce light at 525 nm. This is potentially a very sensitive assay system, since it combines the high catalytic rate of alkaline phosphatase for this substrate ($k_{cat} = 1000$ sec^{-1}) and the high quantum yield of the luciferase reaction (Q = 0.9). A family of luciferin substrates for other enzymes, including ß-galactosidase and an esterase, have also been prepared, but the alkaline phosphatase substrate appears to be the most sensitive. About 3×10^{-20} moles of IgG can be detected in an immunoassay(9). Using streptavidin-alkaline phosphatase to detect a biotinylated DNA probe, 7×10^{-15} moles of a 4.3 kb sequence have been detected on nitrocellulose dot blots(10). We have detected 50×10^{-18} moles of a plasmid DNA in a Southern blot using luciferin-O-phosphate, a biotinylated probe and a streptavidin-alkaline phosphatase conjugate. Non-enzymatic hydrolysis of the luciferin-O-phosphate in solution, however, has been observed to result in high backgrounds.

1,2-Dioxetanes   One of the most innovative chemiluminescent systems to be developed is the enzymatic triggering of 1,2-dioxetanes. Dioxetanes, which decompose to generate chemiluminescence, were first prepared 20 years ago. The synthesis of 'triggerable' derivatives, however, greatly facilitated their application to detection systems. The most useful dioxetane for nucleic acid detection is currently a phenylphosphate derivative, which is a substrate for alkaline phosphatase ($k_{cat} = 1000$ sec^{-1})(11). On removal of the phosphate group, the unstable dioxetane intermediate decomposes with emission of light at 495 nm and a quantum yield of 1.3×10^{-5} in aqueous medium. This quantum yield can be increased four hundred-fold to 4.8×10^{-3} by addition of a detergent-solubilised derivative of fluorescein, which participates in inter-molecular energy transfer. The light is now emitted at fluorescin wavelength of 528 nm and may be detected on green-sensitive X-ray film, Polaroid film, a CCD camera or a luminometer. Less than 1000 molecules of alkaline phosphatase can be detected in a photon-counting luminometer with a signal-to-background ratio of six(11).

A pH-dependent delay in attaining the steady state light-output is observed with the dioxetane assay. This results from the long lifetime ($t_{1/2}$ = 37 minutes at pH 9.1) of the dephosphorylated intermediate in the presence of detergent micelles; this necessitates longer exposures of nitrocellulose blots to film than for other chemiluminescent systems. It is not yet clear how diffusion of this long-lived micelle-solubilised intermediate on the blots is prevented.

Clyne et al(12) have applied the dioxetane assay to a DNA-probe sandwich assay for Chlamydia on a microtitre plate. This substrate has also been used with a biotinylated oligonucleotide probe and streptavidin-alkaline phosphatase conjugated bound to nylon. In a one minute exposure, 8×10^{-17} moles of target were detected. Similarly, Bronstein et al(13) have detected 4.39×10^4 molecules of hepatitis B virus 'core antigen' DNA and 1×10^{-19} moles of plasmid DNA in a dot blot, using an oligonucleotide probe directly labelled with alkaline phosphatase and dioxetane-phosphate. We have applied it to Southern blots and have demonstrated a detection limit of 0.2×10^{-18} moles with a biotinylated 0.4 kb DNA probe and a streptavidin-alkaline phosphatase complex, followed by an overnight exposure of a nitrocellulose blot to X-ray film(14). The light can be detected on blots over at least seven days without addition of fresh substrate. This allows variation of the exposure-time of the film to achieve the optimum signal-to-background ratio. Shorter exposures of less than two hours are sufficient to achieve equivalent detection limits on nylon membranes. We have also used oligonucleotide probes directly labelled with alkaline phosphatase to detect plasmid DNA on nylon dot blots and for fingerprinting.

Other Luminescent Systems

A bioluminescent assay coupling glucose-6-phosphate dehydrogenase to bacterial luciferase and an oxidoreductase system has been applied to DNA detection. The light emitted at 495 nm can be detected as described for enhanced chemiluminescence. When coated beads, rather than a membrane, were used 10×10^{-18} moles of target DNA were detected on a luminometer.

The potential of acridinium esters as a chemiluminescent detection system has been known for some time. Light is emitted by treatment of the ester with an alkaline solution of hydrogen peroxide, which cleaves the molecule to form an unstable intermediate; this decays to produce light with a high quantum efficiency. Acridinium esters have been attached to DNA oligonucleotide probes. In particular, a

methyl acridinium ester has been used.   In the presence of base and
hydrogen peroxide, a cyclo-oxetane ring intermediate is formed.
This rapidly converts to an excited N-methylacridine that emits
light on relaxing tot he ground state.   Arnold et al(15) showed that
5×10^{-19} moles of the purified chemiluminescent probes can be
detected. In a homogeneous assay procedure, 10^{-16} to 10^{-17} moles of
target sequences were detectable.   By introducing a solid-phase
separation, the detection limit was increased to 10^{-17} to 10^{-18}
moles of target sequence.   Although it is simpler than other
systems, the light is produced over a few seconds, and sensitivity
is limited by the geometry of detection and the concentration of
ester available for reaction.

The acridinium ester method involves no amplification.   A chemical
system has been developed by McCapra(16) that results in
amplification.   It is based upon labelling nucleic acid probes with
a photochemical sensitiser. This can be activated by light of the
appropriate wavelength and, if photobleaching does not occur, a
steady-state concentration of the excited form of the dye may occur.
Singlet oxygen is generated from molecular oxygen and, in the
presence of a suitable olefin, a dioxetane results.   On gentle
heating the dioxetane emits light.   This technique is potentially
very sensitive since many dioxetane molecules can be accumulated per
initial photochemical sensitiser present in the hybrid.   Alternative
methods of triggering the light emission will undoubtedly be
devised.

Amplification Strategies

The use of enzymes for signal generation can be considered an
amplification method, since many product molecules are produced per
enzyme molecule.   Alternative approaches to nucleic acid detection
that remove some of the requirement for highly sensitive signal-
generation systems involve amplification of the target or probe
molecules.   Some of these have been combined with a luminescent end-
point.

In an elegant technique a series of oligonucleotide probes
complementary to different parts of the target sequence and with
single-stranded overhangs are used(5).   The single-stranded
overhangs are used to capture the probe/target hybrid onto a bead
and to label the hybrid by further hybridisation with another
oligonucleotide, which is crosslinked to horseradish peroxidase or
to alkaline phosphatase.   This format results in many enzyme-
labelled oligonucleotides bound per initial probe molecule
hybridised, thereby increasing the signal density. Detection of 6 x
10^4 molecules of Hepatitis B virus(5) and fifteen different types of
Chlamydia serovars(12) has been demonstrated with horseradish
peroxidase and alkaline phosphatase labels, respectively, using
chemiluminescent end-points.   This method can be applied to
detection of any target sequence by simply changing the two initial
oligonucleotide probes, and it therefore represents a generic
approach.

Target Amplification

The polymerase chain reaction (PCR) is a relatively rapid method for
generating greater than a million-fold increase in the number of an
RNA or DNA sequence(17).   The PCR product can be analysed directly
with an ethidium bromide stain after electrophoresis in agarose gel,

but hybridisation of probe sequences to dot blots of the product is used to improve the specificity of analysis.

One of the major disadvantages of the PCR technique is the difficulty of quantification. This may be necessary, for example, for the detection of Hepatitis B virus where the number of virus particles in a blood sample is used to monitor the state of an infection. This problem has not been fully solved. Approaches being taken include amplification of standards at the same time as the sample, and estimation of the amplification efficiency by comparing the signal from the amplified product on slot or dot blots with signal from dilutions of an unamplified standard.

At Amersham we have combined PCR with detection by enhanced chemiluminescence. This allows a reduced number of PCR cycles and a rapid detection step: a sensitivity of less than 10^{-20} moles of target nucleic acid has been achieved. PCR will continue to have a major impact on the use of non-radioactive methods for many applications of nucleic acid hybridisation. In some instances, it has removed the necessity for a hybridisation step, in others, a non-radioactive detection system of equivalent sensitivity to radioactivity is no longer required.

Conclusions and Future Perspectives

Non-radioactive nucleic acid detection was first described over six years ago. Since then, many new labelling systems and novel detection methods have emerged.

Development of some of these methods into commercial kits has made them accessible to more laboratories. The chemiluminescent detection systems have improved performance compared with the colorimetric techniques, and they represent the first non-radioactive methods to approach the sensitivity of radioactive labels. The use of a particular luminescent system depends on the assay requirements. Where a rapid result is required, but extremely high sensitivity is not necessary, enhanced chemiluminescence a horseradish peroxidase label or the acridinium esters can be used. If very high sensitivity is essential, but the assay time is not critical, the alkaline phosphatase-dioxetane system may be used.

Future advances in chemiluminescent assays may develop from chemical, rather than enzymatic amplification techniques, that can be used in rapid, simple and sensitive nucleic acid probes.

REFERENCES

1. E M Southern, Detection of specific sequences among DNA fragments separated by gel electrophoresis J. Mol. Biol **98**, 503-517 (1975)
2. A T Haase, M Brakic, L Stowring and H Blum, Detection of viral nucleic acids by in situ hybridisation Meth. Virol.7, 189-226 (1984)
3. T P Whitehead, G H G Thorpe, T J N Carter, C Groucutt and L J Kricka, Enhanced luminescence procedure for sensitive determination of peroxidase-labelled conjugates in immunoassay Nature **305**, 158-159 (1983)
4. J A Matthews, A Batki, C Hynds, L J Kricka, Enhanced chemiluminescent method for the detection of DNA-dot hybridisation assays Anal. Biochem. **151** 205-209 (1985)

5.   M S Urdea, J A Running, T Horn, J Clyne, L Ku and B D Warner, A
     novel method for the rapid detection of specific nucleotide
     sequences in crude biological samples without blotting or
     radioactivity: application to the analysis of hepatitis B virus
     in human serum *Gene* **61**, 253-264 (1987)
6.   M S Urdea, B D Warner, J A Running, M Stempien, J Clyne and
     T Horn, A comparison of non-radioisotopic hybridisation assay
     methods using fluorescent, chemiluminescent and enzyme-labelled
     synthetic oligo-deoxyribonucleotide probes *Nucl. Acid Res.* **16**,
     4937-4956 (1988)
7.   D Pollard-Knight, C A Read, M J Downes, L A Howard,
     M R Ledbetter, S A Pheby, E McNaughton, A Syms and M A W Brady,
     Non-radioactive nucleic acid detection by enhanced
     chemiluminescence using probes directly labelled with
     horseradish peroxidase, *Anal. Biochem.* **185**, 84-89 (1990)
8.   I Durrant, L C A Benge, C Sturrock, A T Devenish, R Howe, S
     Roe, M Moore, G Scozzafava, L M F Proudfoot, T C Richardson and
     K G McFarthing, The application of enhanced chemiluminescence
     to membrane-based nucleic acid detection *Biotechniques* **8**, 564-
     568 (1990)
9.   W Miska and R Geiger, Synthesis and characterisation of
     luciferin derivatives for use in bioluminescence-enhanced
     enzyme immunoassays *J. Clin. Chem. Clin. Biochem.* **25**, 23-30
     (1987)
10. R Hauber and R Geiger, A sensitive bioluminescent-enhanced
     detection method for DNA dot-hybridisation *Nucl. Acid Res.* **16**,
     1213 (1988)
11. A P Schaap, H Akhavan and L J Romano, Chemiluminescent
     substrates for alkaline phosphatase: application to
     ultrasensitive enzyme-linked immunoassays and DNA probes, *Clin.
     Chem.* **35**, 1863-1864 (1989)
12. J M Clyne, J A Running, M Stempien, R S Stephens, H Arhavan-
     Tafti, A P Scaap and M S Urdea, A rapid chemiluminescent DNA
     hybridisation assay for the detection of *Chlamydia trachomatis*,
     J. Biolumin. Chemilumin. **4**, 357-366 (1989)
13. I Bronstein, J C Voyta and B Edwards, A comparison of
     chemiluminescent and colorimetric substrates in a hepatitis B
     virus DNA hybridisation assay *Anal. Biochem.* **180**, 95-98 (1989)
14. D Pollard-Knight, A C Simmonds, A P Schaap, H Akhavan and
     M A W Brady, Non-radioactive DNA detection of Southern blots by
     enzymatically triggered chemiluminescence *Anal. Biochem.* **185**,
     353-358 (1990)
15. L J Arnold Jr, P W Hammond, W A Wiese and N C Nelson, Assay
     formats involving acridinium-ester labelled DNA probes
     Clin.Chem. **35**, 1588-1594 (1989)
16. F McCapra, D Watmore, F Sumun, A Patel, I Beheshti, K
     Ramakrishnan and J Branson, Luminescent labels for immunoassay
     - from concept to practice *J. Biolumin. Chemilumin.* **4**, 51-58
     (1989)
17. R K Saiki, S Scharf, F Faloona, K B Mullis, G T Horn,
     H A Erlich and N Arnheim, Enzymatic amplification of ß-globin
     genomic sequences and restriction analysis for the diagnosis of
     sickle-cell anaemia *Science* **230**, 1350-1354 (1985)

ENHANCED CHEMILUMINSCENT IMMUNOASSAYS FOR ENVIRONMENTAL MONITORING

G.W. Aherne, A. Hardcastle, P. Saleem and N. England

Department of Biochemistry, University of Surrey, Guildford, Surrey, GU2 5XH, UK

INTRODUCTION

Many chemicals are used in our everyday lives and recent concern, expressed in both the scientific and lay press on the fate and persistence of these chemicals, has highlighted the need for improved methods for monitoring the environment. Environmental analysis is required to assess the prevalance of chemical micropollution, to monitor the extent of pollution over a time period and as an aid to prevent further contamination. Increased monitoring of the environment is also required to comply with various national and international regulations.

Because of the large number of chemicals in daily use in eg agriculture, horticulture, manufacturing, the analytical challenge is immense. Readily available, sensitive and specific assays which are also cost-effective are required, at least for the most widely used and/or most toxic compounds. Because of the different sample matrices involved in environmental monitoring eg soil, air filters, water, food, the assays must also be robust and adaptable. In addition, occupational exposure studies may require the analysis of biological samples such as plasma and urine. It is only relatively recently that the potential for immunoassay in environmental monitoring has been considered (1,2).

The micropollution of water supplies has been of concern for some time especially in areas where a high proportion of drinking water is obtained from river water. Monitoring of water supplies is normally carried out using conventional analysis such as HPLC, GC, and GC-MS. Although these assays are sensitive and specific they require extensive sample extraction techniques and are often time consuming. As an alternative to chromatographic methods immunoassays have many attributes to offer especially in monitoring programmes. They combine specificity and sensitivity with ease of use, large sample throughputs, minimal sample cleanup and also offer the possibility of developing assays which can be used away from traditional analytical laboratories. This paper describes the use of enhanced chemiluminescent endpoints for the detection of pesticides in water using immunoassays and adaptations of these assays for occupational and air monitoring.

In 1985 the legal limits for pesticides in water were set by the Council of the European Communities in the Drinking Water Directive (3). The Maximum Admissable Concentration (MAC) for any single pesticide is 100ng/l and for total pesticides is 500ng/l. In a survey carried out between July 1985 and June 1987 (4) 369 breaches of the MAC were recorded. More than 70% of these breaches were due to the presence of triazine herbicides (atrazine, simazine and propazine). The triazine herbicides are widely used to control weed growth on railway lines, roadways, carparks etc and have been shown to be particularly persistant in soil and ground water. It was decided therefore to use the triazine herbicides as a model for the development of immunoassays suitable for monitoring water samples. Enhanced chemiluminescence (5) was chosen as the end-point because of the potential for high sensitivity and the robust nature of the prolonged light signal produced.

MATERIALS AND METHODS

A triazine antiserum was raised in a sheep to an immunogen prepared by conjugating 2-chloro-4-isopropyl amino-6-carboxy pentylamino s-triazine to porcine thyroglobulin using an N-hydoxysuccinamide active ester technique(6). Suitable antisera were identified using C^{14}-atrazine in a radioimmunoassay and were purified using DEAE chromatography. A horse radish peroxidase (HRPO)-hapten conjugate (molar ratio 1:20) was prepared using N-hydroxysuccinamide and purified on a Sephadex G100 column (25x1.5cm).

The competitive enhanced chemiluminesent immunoassay (ECLIA) was set up in polystyrene tubes (Clinicon) coated overnight with 300ul antiserum diluted 1:2000 in 0.07M sodium barbitone pH 9.6. Following washing in 1% Tween 20/distilled water, atrazine standards (prepared in non-immunoreactive tap water) or water samples (100ul) and HRPO conjugate (50ul) diluted 1:5000 in phosphate buffered saline with gelatin and Tween 20 (PBSGT) were added to the tubes. The tubes were incubated in an iced water bath for 2h and washed again. 300ul enhancement reagent (Amerlite, Amersham International plc) was automatically added to 24 tubes in one batch using an LKB Wallac 1251 luminometer (time taken 2 min) and then the light intensity produced in each tube was measured (time taken 2 min).

The feasibility of adapting this assay format for biological samples and air monitoring and of using a portable camera luminometer to produce results away from the laboratory were investigated.

RESULTS AND DISCUSSION

<u>Atrazine ECLIA</u> Cross-reactivity studies showed that the antiserum produced in response to the immunogen was specific to the triazine class of herbicides. However, results were expressed as "immunoreactive triazine" because of the spectrum of cross-reactions observed with different triazines. Atrazine crossreacted by 100%, propazine by >100%, simazine by 15%, prometryne by 37.5% and terbutryn by 1%.

The sensitivity of the atrazine standard curve depended on the concentration of both the coating antiserum and the HRPO conjugate as well as the incubation conditions. An incubation time of 2h in an iced water bath was found to give reproducible and sensitive curves. A loss of assay sensitivity was observed if the incubation period was prolonged. The standard curve ranged from 25-500 ng/l with a mean CV across the standards of 9.6% (n=5). The theoretical sensitivity of the curve (2SD inhibition of the binding at zero analyte concentration) was less than 50ng/l. Atrazine could be quantitatively recovered (83-112%) from tap water over a range of concentrations (60-240ng/l) with a mean CV of 20.2%. The CV of sample measurement was less than 25%. Immunoreactive triazine concentrations measured in authentic water samples correlated with those obtained by HPLC (r=0.915; y=45.6+0.512x, n=27) although the HPLC results were approximately double the immunoassay results.

A survey of local water sources (R. Wey, lakes, ponds) and tap water showed that immunoreactive triazines were frequently present. Out of 126 water samples, 63 had triazine levels greater than the MAC, 14 gave values within 10% of the MAC and 59 were below the MAC. In the samples obtained from the lakes on the University campus (Fig.1) levels as high as 400ng/l were found during a

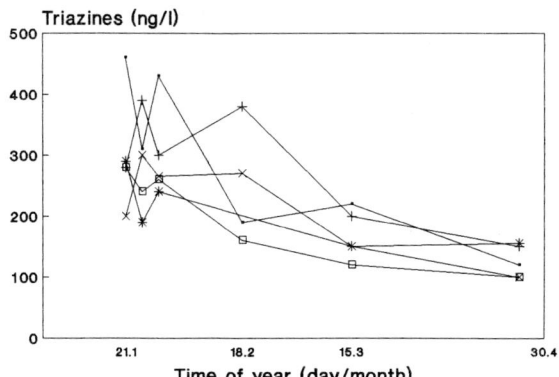

Fig.1. Immunoreactive triazine in water from 5 sampling points in the University lakes.

period of heavy winter rainfall. These levels declined to
approximately 100ng/l later in the year. The presence of
triazines in the lake was attributed to the past use of
herbicides in nearby ornamental areas and on flat roofs
and to the proximity of the lake to railway lines etc. A
similar pattern of triazine levels were found in the
River Wey which runs through the centre of Guildford
where there is municipal use of triazines. Only 1 of 20
tap water samples had triazine immunoreactivity which
exceeded the MAC (114ng/l).

Adaptation of the ECLIA to biological samples
Occupational exposure to triazine herbicides has been
linked to an increased risk of developing cancer (7). It
may therefore be appropriate to measure the concentration
of triazines in the plasma or urine of exposed workers.
There was a marked increase in the antibody binding (Bo)
when plasma was added to the triazine ECLIA. This was
found to be due to a matrix effect on the assay rather
than an effect on the end-point measurement. Reproducible
standard curves for the measurement of triazines in
plasma were produced by preparing the standards in normal
plasma and re-optimising the concentrations of reagents.
At an antibody coating dilution of 1:8000 and a HRPO
conjugate dilution of 1:15000 a sensitivity of <0.1ng/ml
was obtained. Recovery of atrazine "spiked" into normal
plasma was 110% at 0.4 and 0.6ng/ml (CV=15% and 14%
respectively). If standards were prepared in urine the
assay could also be used to detect and measure atrazine
in urine at concentrations of <0.5ng/ml. This illustrates
how readily the ECLIA can be adapted to different types
of sample whilst maintaining sensitivity.

ECLIA for air monitoring One of the main problems
encountered in environmental analysis is the variety of
sample types it may be necessary to analyse. Sample
extraction time eg for extraction of analytes from soil
or food would negate some of the attractions of
immunoassay even if simple extraction procedures were
carried out. The feasibility of using immunoassay as an
alternative to HPLC for air monitoring was investigated
using the atrazine ECLIA.

The occupational exposure level for atrazine is
10mg/m^{3}time weighted average for an 8h working day. Known
amounts of atrazine (which reflected this limit and the
solubility of atrazine) deposited onto air filters were
extracted using methanol. The methanol extracts were
subjected to analysis by HPLC or ECLIA. The HPLC standard
curve ranged from 0-20ug/ml and the extracts were
analysed without dilution. On the other hand because of
the high sensitivity of the immunoassay it was necessary
to dilute the extracts at least 10000 times before assay.
The recovery of atrazine was quantitative for both HPLC
and ECLIA over a range of 10-180 ug spiked onto the
filter. However, the precision of measurement was only
3-5% for HPLC compared to 6-17% for the immunoassay. The

poor precision associated with the immunoassay was
probably mainly due to the large sample dilutions used.
Thus, sensitive immunoassays may not be suitable for all
types of environmental analyses. For air monitoring
purposes it may be more appropriate to capitalise on the
sensitivity of immunoassay by reducing the air sampling
times from the standard 2h period.
ECLIA for paraquat Provided suitable antisera are
available immunoassays should be applicable to a range of
different groups of pesticides. For example, an ECLIA for
paraquat has also been established using a sheep antibody
and a similar assay format to that described for the
triazines. For water analysis a limit of detection of
10ng/l was obtained from the standard curve. Although
paraquat is not a problem as far as water micropollution
is concerned, the availability of a rapid sensitive assay
such as this may be valuable to monitor the disappearance
of paraquat following accidental spillage.

   The injestion of paraquat either deliberately or
accidentally is often fatal. The clinical management of
paraquat poisoned patients is facilitated by a knowledge
of the plasma concentrations. The ECLIA for paraquat was
adapted for plasma measurements. At the high dilutions of
plasma used (>1/100) no matrix effect was evident. The
sensitivity of the standard curve was 0.2ng/ml which is
equivalent to a concentration of 20ng/ml in undiluted
plasma.

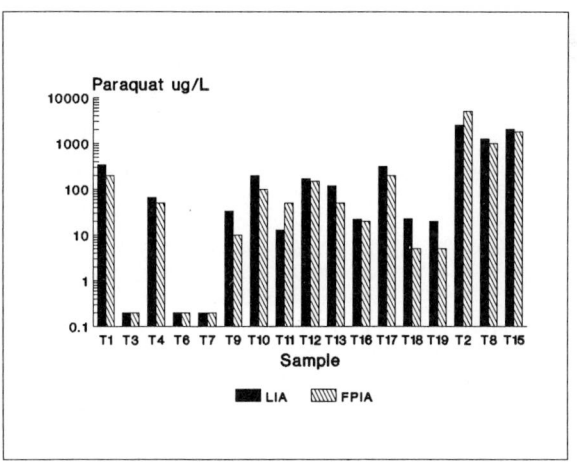

Fig.2. Comparison of plasma paraquat
results.

Results obtained with the assay correlated well (Fig.2)
with those obtained with the fluorescence polarisation
immunoassay (FPIA) (8) normally used in clinical

situations.   In its present form the ECLIA took 2-3 hours
to complete but it would be possible to achieve a   more
rapid   assay using different   reagent   concentrations
without significant loss of sensitivity.
Photographic endpoints for environmental immunoassays.One
of the advantages of using enhanced luminescent endpoints
in   immunoassays   is   the   possibility   of   obtaining
photographic records of results where semi-quantitative
and   qualitative   results are appropriate   (9).   The
feasibility of using a camera luminometer and simple
procedures   applicable   to   "field"   conditions   was
investigated.   Soft   microtitre plates were   coated with
diluted   antibody   solutions   and washed with a water/
Tween solution using a wash bottle.   The   assays   were
carried   out   as   before but reduced dilutions of   HRPO
conjugates and   an incubation period of only 30 min   at
room   temperature   were used.   At the end of   the assay
enhancement   reagent was added and the plate   exposed   to
Polaroid high speed film (612) in the Microlite camera
luminometer (Dynatech). Exposure of the film could   take
place at any time from 2-30 min following addition of
enhancement solution. It was found that the concentration
of   reagents and photographic exposure time could be
optimised to produce an endpoint, ie either light or dark
exposure,   at a predetermined concentration. The atrazine
screening assay has been described previously (10)   and
results obtained with the paraquat assay are shown in
Fig.3.

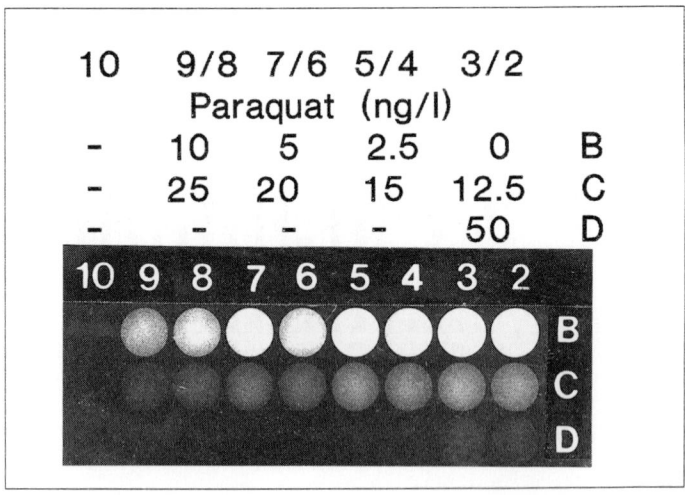

Fig.3. Camera luminometer assay of
paraquat in water.

    A   selection of water samples and   atrazine standards
were assayed under "field" conditions using   the   camera

luminometer and individuals were asked to score the
results as above, equal to or below the MAC by comparing
the brightness of the sample well on the photograph with
the 100ng/l standard wells. An atrazine solution of
50ng/l was clearly observed to be below the MAC but
observers had more difficulty in distinguishing the
150ng/l standard from the MAC. Authentic water samples
containing immunoreactive triazine levels below the MAC
(as determined previously in the ECLIA) were mostly
considered to be negative with respect to the 100ng/l
standard. The mean percentage correct score was 83%
(n=12, ie 6 scorers, each sample in duplicate). Samples
containing 200ng/l triazine or greater were 98% correctly
scored as being darker than the 100ng/l standard ie above
the MAC.

Summary ECLIA techniques have been developed for the
triazine herbicides and paraquat in water. The
sensitivity of each assay is <100ng/l – the legal limit
for pesticides in drinking water - and they represent a
simple and rapid means of detecting samples which are in
breach of the MAC. The assays described should not be
viewed as alternatives to more conventional
chromatographic assays but as complementary to them. In
our laboratory the assays have been carried out in tubes
but they can be easily transferred to the more amenable
microtitre plate format when the appropriate equipment is
available.

The immunoassays have been adapted for use with other
types of sample matrices ie. plasma, urine and air
filters. At the moment one of the factors preventing more
widespread use of immunoassays in environmental analysis
is that a wide range of antibodies to analytes of
interest is not available. Also the various Registration
and Regulatory Authorities are only just beginning to
recognise the validity of results obtained by
immunoassay.

One exciting application of ECLIA, however, is the
possibility of producing a permanent record of assay
results using a camera luminometer. The precise
analytical situations in which rapid, semiquantitative or
qualitative results are required must be defined and the
requirement for sample extraction considered. The ability
to distinguish samples which are positive for an analyte
or group of analytes at or above a threshold
concentration may greatly facilitate the increased
monitoring of water supplies required by EC regulations
and may also have applications in other areas of
environmental analysis.

ACKNOWLEDGEMENTS
The authors wish to thank the Water Research Centre and
the Health and Safety Executive for financial support and
Dr S. Dawling for supplying the plasma samples for
paraquat analysis.

REFERENCES

1. G.W. Aherne, Immunoassays in environmental analysis. Anal. Proc. 24, 140-141 (1987).
2. J.M. Van Emon, J.N. Seiber and B.D. Hammock, Immunoassay techniques for pesticide analysis. Analytical Methods for Pesticides and Plant Growth Regulators Vol.XVII. p.217-261 (1989).
3. The Drinking Water Directive. Official Journal, N229/1, Directive 80/778/EEC, (1980).
4. A. Lees and K. McVeigh, An investigation of pesticide pollution in drinking water in England and Wales. Friends of the Earth, London, (1988).
5. G.H.G. Thorpe, L.J. Kricka, S.B. Morley and T.P. Whitehead, Phenolic enhancers of the horseradish peroxidase-luminol-hydrogen peroxide reaction: Application in luminescence monitored enzyme immunoassays. Clin.Chem. 31, 1335-1341 (1985).
6. A. Gadow, H. Fricke, C.J. Strasburger and W.G. Wood, Synthesis and evaluation of luminescent tracers and hapten-protein conjugates for use in luminescence immunoassays with immobilised antibodies and antigens. J.Clin.Chem.Clin.Biochem. 22, 337-347 (1984).
7. A. Donna, P. Crosignani, F. Robutti et al., Triazine herbicides and ovarian epithelial neoplasms. Scand. J.Work Environ.Health 15, 47-53 (1989).
8. D.L. Colbert and R.E. Coxon, Paraquat measured in serum with the Abbott TDx. Clin.Chem. 34, 1948 (1988).
9. L.J. Kricka and G.H.G. Thorpe, Photographic detection of chemiluminescent and bioluminescent reactions. Methods in Enzymology 133, 404-420 (1986).
10. A. Hardcastle, G.W. Aherne and G.H.G. Thorpe, Photographic screening procedure for triazine herbicides in water using an enhanced luminescent immunoassay. In Watershed '89. The Future for Water Quality in Europe Vol.II, D. Wheeler, M.L. Richardson and J.W. Bridges (Eds), Pergamon Press, Oxford, p.293-296 (1989).

HYDROPHILIC ACRIDINIUM-9-CARBOXYLIC ACID DERIVATIVES USED AS LABELS IN LUMINESCENCE IMMUNOASSAYS

A. Mayer, E. Schmidt, T. Kinkel, P. Molz,
S. Neuenhofer and H.J. Skrzipczyk

Hoechst AG, D-6230 Frankfurt 80, FRG

INTRODUCTION

During the last years interest in chemiluminescent probes for molecular recognition has rapidly increased (1,2). In luminescence immunoassays a major progress was initiated by the introduction of acridinium-9-carboxylic acid derivatives as labels (3). While the stability of the first acridinium esters was not suffucent for commercial use (4), efforts in several laboratories yielded optimised acridinium labels with significantly enhanced stability (4-7). Meanwhile chemiluminescent immunoassays being at least as sensitive as radioactive ones are available (1,2).

Of the compounds synthesized in our laboratories we chose a representative acridinium-9-(N-sulphonyl)carboxamid label with good overall properties for the development of routine immunoassays. Such a system is now commercially available. An example for an immunoassay is presented below.

A label should not significantly alter the physico-chemical properties of the molecule to which it is attached (2). In fact the inherent hydrophobicity of the mentioned acridinium-9-(N-sulphonyl)carboxamides may decrease the solubility of labelled compounds in aqueous buffers. The same is true for most acridinium esters (4). This effect may arise especially in labelling hydrophobic haptens, e.g. thyroxine. Proteins are less susceptible to solubility changes by labelling. Actually the coupling of acridinium esters to haptens was not successful in most cases (2). Hydrophilic acridinium ester derivatives for liposome encapsulation have been reported (4). However, these materials do not have a reactive group for coupling to antibodies.

In order to have tailor-made labels for every purpose, we decided to synthesize hydrophilic compounds. A major requirement was to retain the good characteristics of our routine label such as high stability and fast kinetics of light emission. We expected advantages especially in the field of hapten assays.

MATERIALS AND METHODS

Preparation and purification of labels: All reagents are available commercially or can be prepared according to literature procedures. The experimental details of synthesis are published in the patent literature (6,8). Purification was achieved by medium or high pressure reversed phase chromatography.

<u>Determination of emission kinetics:</u>   The measurements using anti-
bodies labelled with the different acridinium derivatives were performed on a Berthold LB 952 luminometer using 50 µl sample volume.

RESULTS AND DISCUSSION
Features of acridinium-9-(N-sulphonylcarboxamide labels 1 and the
resulting effects in luminescence immunoassay are listed below.
Features: fast light emission; high light yield; high stability.
Effects: short measuring time; high sensitivity; long shelf life.

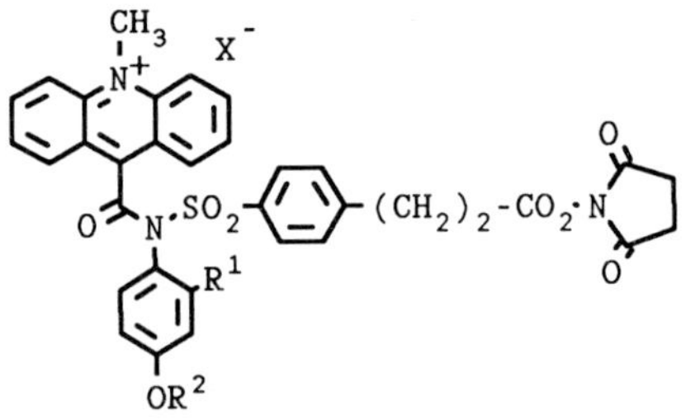

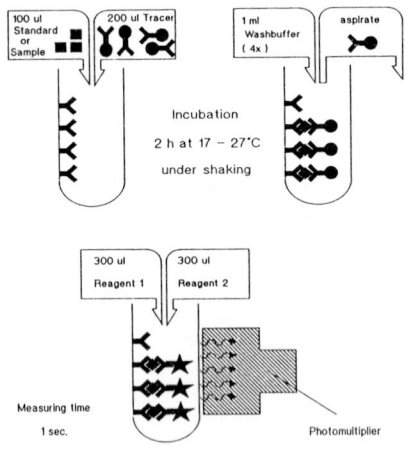

Comp.	R^1	R^2
1a	H	CH$_3$
1b	OCH$_3$	CH$_3$
1c	H	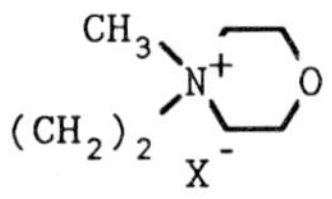 (CH$_2$)$_2$ N$^+$(CH$_3$) morpholine X$^-$

Fig. 1. Structure of some acridinium-9-(N-sulphonyl)-
                carboxamide labels

Six assays for thyroid and tumor diagnostics on this basis are now
commercially available. Others will follow soon. A schematic assay
protocol for TSH is shown in Fig. 2.

Fig.2. BeriLux hTSH
Assay principle

Reagent 1: 0.5% H$_2$O$_2$
in HNO$_3$ (0.1 mol/l)

Reagent 2: NaOH
(0.25 mol/l)

As outlined before, hydrophobic labels may affect the tracer solubility in aqueous buffers. Indeed thyroxine conjugates with compound 1a showed adsorptive interactions with hydrophobic surfaces.
The development of hapten assays with acceptable properties was not
possible.
        The complex structure of acridinium-9-(N-sulphonyl)carbox-

amid labels allows consideration of numerous possibilities to increase hydrophilicity by introduction of polar or ionic substituents. These possibilities, however, are reduced by the condition
to retain all good characteristics of label **1a**.
According to our studies, substituent changes in the phenylring attached to the amide nitrogen has clear effects on stability and
emission kinetics. Thus the donor group in 4-position is necessary.
A second methoxy group in 2-position (Fig. 1, compound **1b**) decreases the emission kinetics (Fig. 3). A small increase in stability is observed. For labels of type **1b** a reading time of only one
second, as used with compound **1a**, would be too short for high sensitivity and precision.

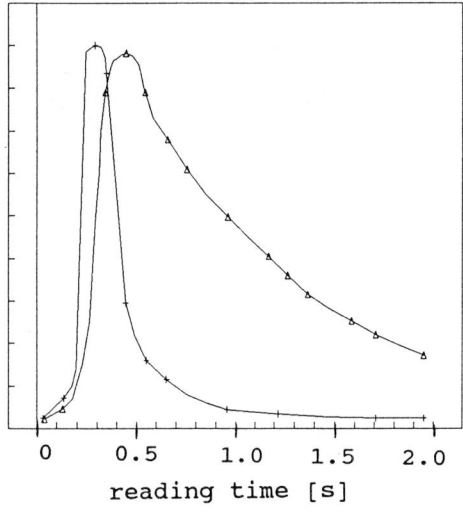

Fig.3. Emission kinetics

of label **1a** (+) and **1b** (Δ)

reading time [s]

Hence we decided to maintain the basic structure of compound **1a** and
to exchange the methyl group in 4-position for an appropriate substituent. We chose the morpholinoethyl group because in the final
methylation step of the acridine nitrogen the more nucleophilic
nitrogen atom of the morpholino substituent should be also methy-
lated, thus generating a second hydrophilic ammonium-centre. The
synthesis of label **1c** was performed according to Fig. 4.
    Compound **1c** is much more soluble in aqueous buffers than
label **1a**. While **1a** can be extracted into dichloromethane easily,
extraction of label **1c** from water is only possible if a lipophilic
counterion is present. First results using hapten conjugates
labelled with compounds **1a** or **1c** show higher solubility and immune
reactivity for the latter.
    These results suggest, that the potential of acridinium-
9-(N-sulphonyl)carboxamides has not yet entirely been exploited. It
seems possible to prepare tailor-made labels for special purposes
by changing substitution pattern.

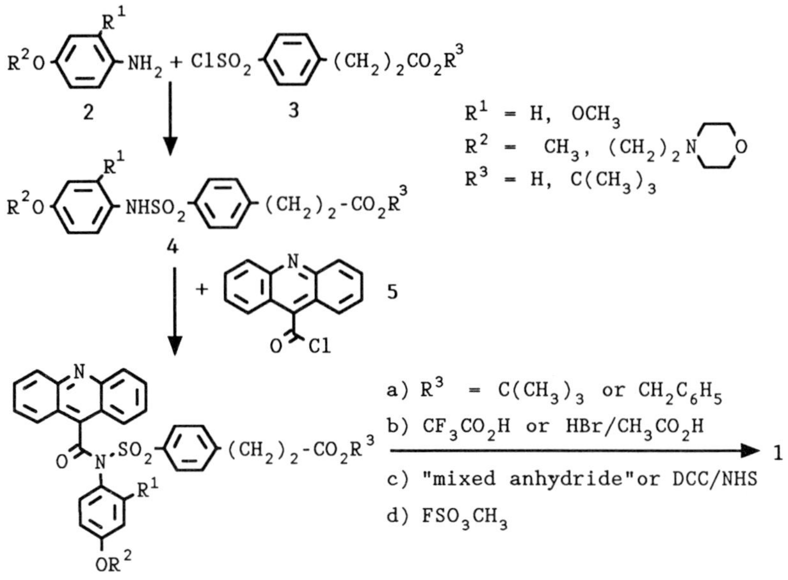

Fig.4. Synthesis of acridinium-9-(N-sulphonyl)carboxamide
labels

REFERENCES

1.   W.G. Wood, Routine Luminescence Immunoassays-Dream or Re-
        ality. J. Biolumin. Chemilumin. 4, 79-87 (1989).
2.   A. K. Campbell, An alternative to radioactive labels. In
        Chemiluminescence, A.K. Campbell (Ed.), VCH, Weinheim;
        Horwood, Chicester, Chap. 8 (1988).
3.   I. Weeks, I. Behesti, F. McCapra, A.K. Campbell and J.S.
        Woodhead, Acridinium esters as high-specific activity la-
        bels in immunossay. Clin. Chem. 29, 1974-1979 (1983).
4.   S.-J. Law, T. Miller, U. Piran, C. Klukas, S. Chang and J.
        Unger, Novel Poly-substituted Aryl Acridinium Esters and
        their Use in Immunoassay. J. Biolumin. Chemilumin. 4,
        88-98 (1989).
5.   T. Kinkel, H. Lübbers, E. Schmidt, P. Molz and H.J.
        Skrzipczyk, Synthesis and Properties of New Luminescent
        Acridinium-9-carboxylic Acid Derivatives and their Appli-
        cation in Luminescence Immunoassays (LIA). J. Biolumin.
        Chemilumin. 4, 136-139 (1989).
6.   P. Molz, H.J. Skrzipczyk, H. Lübbers, H. Strecker, G.
        Schnorr and T. Kinkel, German Patent Application DE-A-36
        28 573 (1986); European Patent Application EP-A- 257 541
        (1987).
7.   P.G. Mattingly, L.G. Bennett, US Patent Application US-A-
        921979 (1986); European Patent Application EP-A-273 115.
8.   T. Kinkel, P. Molz, E. Schmidt, G. Schnorr and H.J.
        Skrzipczyk, German Patent Application DE-A-38 05 318
        (1988); European Patent Application EP-A-330 050 (1989).

CHEMICAL AND ENZYMATIC TRIGGERING OF 1,2-DIOXETANES: STRUCTURAL EFFECTS ON CHEMILUMINESCENCE EFFICIENCY

A. P. Schaap, R. DeSilva, H. Akhavan-Tafti and R. S. Handley

Department of Chemistry, Wayne State University, Detroit, Michigan 48202 USA

INTRODUCTION

A series of stable 1,2-dioxetanes has been prepared which can be chemically or enzymatically triggered to produce chemiluminescence. The chemiluminescence is generated by removal of a protecting group X from an aryl-OX group to form an unstable aryloxide intermediate which spontaneously decomposes by an intramolecular electron-transfer process to produce light. The efficiency of the chemiluminescence is strongly dependent on structural effects. Treatment of dioxetane **1d** (X = SiMe$_2$t-Bu) with fluoride ion in DMSO generates intense blue chemiluminescence with an efficiency of 25%. The chemiluminescence efficiencies for the reaction of **3d** and **4d** with fluoride, in contrast, are only 0.002 and 0.004% respectively in DMSO. These results led us to predict that other "meta-type" dioxetanes might also exhibit chemiluminescence with high efficiencies. We report herein a study of the chemiluminescent decomposition of dioxetanes **1a-c** and **2a-c** triggered by base or the enzymes alkaline phosphatase and ß-galactosidase.

MATERIALS AND METHODS

All dioxetanes used in this study were synthesized in our laboratory or were supplied by Lumigen, Inc., Detroit, MI 48202. Dioxetane **1b** is available from Lumigen under the tradename Lumigen™PPD in a ready-to-use formulation with an enhancer as Lumi-Phos™530. Solutions of enzymes were prepared fresh daily from stock solutions stored at 4°C. Chemiluminescence efficiencies were determined using luminometers built in this laboratory and are relative to the efficiency of 25% for the reaction of dioxetane **1d** with fluoride in DMSO. The latter was measured relative to the luminol standard. Chemiluminescence decays were measured on either the above luminometer or a Turner TD-20e luminometer. Chemiluminescence spectra were recorded using a Spex Fluorolog spectrofluorimeter run in a ratio mode to compensate for the decrease in light intensity during the scan. Fluorescence quantum yields are expressed relative to quinine sulfate.

RESULTS AND DISCUSSION

In 1987 we reported the discovery of highly stable adamantyl-substituted dioxetanes which could be triggered by the action of an appropriate chemical agent or enzyme to give chemiluminescence (1-3). In the first reported example, treatment of dioxetane **4a** with base converted a dioxetane which is stable for several years at room temperature into the unstable aryloxide form which decomposes within minutes with emission of bluish light from the anion of methyl 6-hydroxynaphthoate.

a X = H
b X = PO$_3$Na$_2$
c X = ß-D-galactoside
d X = Si(CH$_3$)$_2$t-Bu

Chemiluminescence from the action of aryl esterase on the acetate-protected form of **4a** marked the first example of the enzyme-catalyzed chemiluminescent decomposition of a dioxetane. Chemiluminescence efficiencies for these reactions were < 0.01%. The corresponding meta- and para-substituted phenyl dioxetanes **1a,d** and **3a,d**, while equally stable, show dramatically different chemiluminescence efficiencies. Comparison of the fluoride-triggered decomposition of **1d** and **3d** (X = Si(t-Bu) Me$_2$) in DMSO revealed the para-substituted compound to be weakly luminescent (CL efficiency = 0.002%) while the meta- isomer generates brilliant blue luminescence (CL efficiency = 25%) (2).   The prediction that similarly high chemiluminescence efficiencies might be achieved by an analogous "meta-type" naphthyl-substituted dioxetane has proved to be correct. Treatment of dioxetane **2a** with potassium t-butoxide in DMSO produces bright blue chemiluminescence with 8% efficiency (Table 1) which exactly matches the fluorescence of the anion of methyl 3-hydroxy-1-naphthoate.

Table 1. Chemiluminescence efficiencies for fluoride-triggered decomposition of dioxetanes **1d-4d** in DMSO.

	1	2	3	4
CL (%)	25	8	0.002	0.004
Φ_F	0.44	1.0	9.0×10^{-4}	0.55
CE (%)	57	8	0.2	0.007

A direct comparison of the base-triggered decomposition of the two dioxetanes with a 1,3 substitution pattern **1a** and **2a** in aqueous solution at pH 12 and in 1 mmol/l CTAB, pH 12 shows that the naphthyl-dioxetane decomposes four times more slowly. Given the high fluorescence quantum yield of the anion of methyl 3-hydroxy-1-naphthoate in aqueous solution (0.04 in 2-amino-2-methanol-1-propanol ("221") buffer, pH 9.6 or aq. NaOH, pH 12 and 0.63 in 1 mmol/l CTAB, pH 12) we were led to prepare the phosphate and ß-galactoside derivatives **2b-c** in order to evaluate their use as chemiluminescent enzyme substrates for alkaline phosphatase and ß-galactosidase, respectively. With roughly comparable rates of decomposition of the aryloxide form of dioxetanes **1a** and **2a** and a significantly more fluorescent cleavage product it was thought that dioxetanes **2b-c** might prove to be more sensitive reagents for the detection of these two enzymes. Decomposition of dioxetane **2b** in 0.75 mol/l 221 buffer with alkaline phosphatase, in fact, shows a 5.8-fold greater chemiluminescence efficiency than **1b** (Table 2).   Inclusion of 1 mmol/l CTAB causes a further increase in chemiluminescence efficiency of 7 times for **2b** but addition of the same fluorescent enhancer (4) as is employed with **1b** in Lumi-Phos™ 530 at the same concentration leads to no further increase in chemiluminescence efficiency.   The highest chemiluminescence efficiency found for the naphthyl dioxetane is, therefore, only one-ninth the value for **1b** in the optimized Lumi-Phos™ 530 formulation.

Table 2. Comparison of the properties of phenyl and naphthyl "meta-type" dioxetanes.

	Half-lives of hydroxydioxetanes* at 37°C (min)		Chemiluminescence efficiencies (%) of phosphatedioxetanes	
	1a	**2a**	**1b**	**2b**
0.75 mol/l 221 buffer, pH 9.6	3.0	8.7	1.3×10^{-3}	7.5×10^{-3}
+ 1 mmol/l CTAB	30	99	1.3×10^{-2}	5.5×10^{-2}
+ enhancer			0.48	5.5×10^{-2}

* Generated via rapid dephosphorylation of **1b** and **2b** with alkaline phosphatase.

The combined effect of the lower chemiluminescence efficiency and slower rate of decomposition of the hydroxy dioxetane makes **2b** substantially less sensitive for the detection of alkaline phosphatase than **1b** (Lumigen™ PPD). Figure 1 shows a plot of the rate of grow in and decay of the chemiluminescent signal from Lumi-Phos™ 530 and from dioxetane **2b** (3.3×10^{-4} mol/l) in the same buffer solution with and without 1 mmol/l CTAB generated by the addition of 5×10^{-11} mol/l of alkaline phosphatase. The curve for Lumi-Phos™ 530 is shown at 10% of the actual intensity values for clarity. The measured detection limits of alkaline phosphatase with **2b** and Lumi-Phos™ 530 are 5.5×10^{-18} mol and 1.6×10^{-21} mol, respectively. The decay of the curves results from substrate depletion at this relatively high enzyme concentration.

Fig. 1 Time profile of the chemiluminescence from alkaline phosphatase-initiated decomposition of **1b** and **2b**.

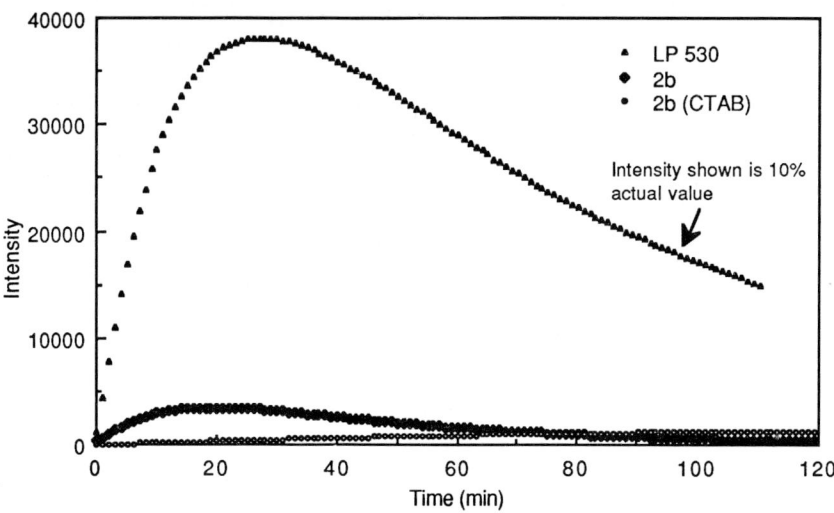

Chemiluminescent detection of ß-galactosidase with the ß-D-galactoside-protected dioxetane **1c** (Lumigen™GPD, commercially available as Lumi-Gal™530) compared to **2c** again demonstrates that the phenyl dioxetane gives a significantly lower level of detection . In a preliminary evaluation, the naphthyl dioxetane is turned over by the enzyme more slowly and the hydroxy form decomposes at a slower rate resulting in a long ramp-up time to plateau intensity. The slow kinetics combined with the lower chemiluminescence efficiency make **2c** a much less sensitive reagent than **1c** for detecting ß-galactosidase. Dioxetane **1c** in Lumi-Gal™ 530 provides a highly sensitive system for the one-step detection of ß-galactosidase. Enzyme levels as low as 5 x 10^{-19} mol of ß-galactosidase can be readily detected. The light intensity at the plateau is proportional to enzyme concentration over a range of at least 4 orders of magnitude.

Fig. 2   Time profile of the chemiluminescence from detection of 5 x 10^{-11} mol/l of ß-galactosidase at 37°C using Lumi-Gal™530.

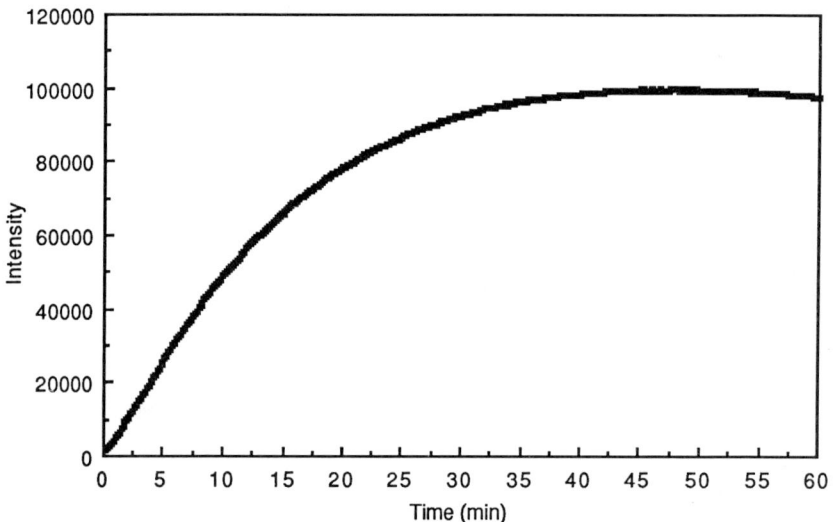

<u>REFERENCES</u>
1. A.P. Schaap, R.S. Handley, and B.P.Giri, Chemical and Enzymatic Triggering of 1,2-Dioxetanes.
    1: Aryl Esterase-catalyzed Chemiluminescence from a Naphthyl Acetate-substituted Dioxetane.
    <u>Tetrahedron Lett.</u> <u>28</u>, 935 (1987).
2. A.P. Schaap, T.-S. Chen, R.S. Handley, R. DeSilva and B.P. Giri, Chemical and Enzymatic
    Triggering of 1,2-Dioxetanes. 1: Fluoride-induced Chemiluminescence from a Naphthyl Acetate-
    substituted Dioxetane. <u>Tetrahedron Lett.</u> <u>28</u>, 1155 (1987).
3. A.P. Schaap, M.D. Sandison and R.S. Handley, Chemical and Enzymatic Triggering of 1,2-
    Dioxetanes 3: Alkaline Phosphatase-catalyzed Chemiluminescence from an Aryl Phosphate-
    ubstituted Dioxetane. <u>Tetrahedron Lett.</u> <u>28</u>, 1159 (1987).
4. A.P. Schaap, H. Akhavan, and L.J. Romano, Chemiluminescent Substrates for Alkaline
    Phosphatase: Application to Ultrasensitive Enzyme-Linked Immunoassays and DNA Probes. <u>Clin.</u>
    <u>Chem.</u> <u>35</u>(9), 1863-1864 (1989).

TECHNICAL CHALLENGES IN THE DEVELOPMENT OF THE CIBA CORNING ACS:180 BENCHTOP IMMUNOASSAY ANALYZER

E. Krodel, J. Boland, G. Carey and M. Kwiatkowski

Ciba Corning Diagnostics Corp., 333 Coney St., E. Walpole, MA 02032 USA

INTRODUCTION

Automation of immunoassays should provide rapid and reliable test results to allow timely clinical intervention as well as savings in manual labor. The ACS:180™ analyzer automates immunoassays employing paramagnetic particles as the solid phase and chemiluminescence as the measurement technology. The time from application of the sample to the first result is 15 minutes, allowing throughput of as much as 180 tests per hour. Developing a robust system for rapid assays required parallel design efforts in assay and instrument development. The instrument required assays with common incubation intervals. The assays required an instrument with a variety of flexible protocols using a single process track. In this paper, we describe the instrument system and discuss the strategies we used to produce a system with a broad test menu, high throughput, and reliable operation.

MATERIALS AND METHODS

Instrument description A simplified diagram of the instrument is shown in Fig. 1. Each test is performed in a discrete reaction vessel that also serves as the cuvette for the chemiluminescence measurement. Cuvettes are loaded by the operator into a bin on the front left of the instrument. The instrument orients the cuvettes and places them in the linear process track. The temperature of the process track is controlled by thermal electric devices which serve to preheat the cuvettes before sample addition as well as to maintain a constant reaction temperature for precision.

The sample and reagent carousels are located on the front of the instrument, opposite the fluid delivery system. The carousels rotate to place sample or reagents under the probes for aspiration. Since the solid phase reagent consists of particles in suspension, it is necessary to prevent the particles from settling. This is performed by the reagent carousel which agitates the solid phase bottles and the solid phase containers which have internal baffles that aid in the agitation of the solid phase reagent.

The fluid delivery system contains four probes mounted on robotic arms with three degrees of freedom. One probe is dedicated to aspirating the sample from a primary tube or sample cup and delivering the sample to the cuvette in the track. After addition of sample (and any pretreatment reagent, if the test requires it), the cuvettes progress through several positions before the first reagent is added. During this time period, any pretreatment reagent can act on the sample. Reagent additions can occur at three time points which are immediately following pretreatment, at 2.5 minutes after pretreatment, or 5 minutes after pretreatment. The total incubation time with reagents as the cuvettes advance along the process track is 7.5 minutes. Because the incubation interval is only 7.5 minutes, the reagents must reach the proper temperature before being dispensed into the cuvettes. This is accomplished by aspirating the reagents by reagent probes and drawing the reagents back into a heat exchanger utilizing stainless steel coils surrounded by a constant temperature fluid.

After incubation is complete, the cuvette moves into the wash station. Stationary magnets (composed of neodinium and boron) are mounted on the side of the process track so that particles are pulled to the cuvette wall. Assay supernate is aspirated by a probe and wash liquid is delivered from a fixed port mounted above the process track. Depending upon the protocol selected, the wash liquid may be aspirated and a second wash dispensed. Final aspiration occurs by a second probe after an additional 20 seconds of residence time of the wash liquid in the cuvette.

Particles are resuspended in 300 µl of a solution of 0.5% H_2O_2 in 0.1 N nitric acid. The cuvette is then transferred from the process track to the luminometer via an elevator mechanism.

The luminometer chamber is a rotary housing with six wells. The photomultiplier tube, which operates in the photon counting mode, is located in front of the third well. A fixed injection port is mounted above this well and delivers 300 µl of 0.25 N NaOH, which initiates the chemiluminescence reaction. The light emitted is measured for 5 seconds after base injection. In the next well, the cuvette contents are aspirated and transferred to the waste container. The cuvette advances to the fifth well where it is ejected from the luminometer into a bag for later disposal by the operator.

The light emitted from the cuvette is acquired in 10 millisecond intervals. The relative light units from each interval are summed over the 5 second collection period. Dark counts are measured before the start of the chemiluminescent reaction and are subtracted to correct for thermal drift of the photomultiplier tube. The analyte concentration in the sample is then calculated from a standard curve. A master curve for each lot reagents is generated at the time of manufacture and entered into the computer contained in the instrument. Calibration is performed by measuring two standards whose values are used to adjust the stored master curve.

<u>Reagents</u>   Solid phase for the hCG assay is prepared by covalently coupling a purified sheep polyclonal anti-hCG antibody to paramagnetic particles. Tracer antibody is a dimethylacridinium ester labeled monoclonal anti-hCG antibody. Fifty µl of sample and 100 µl of tracer (20 ng antibody) are incubated in the instrument for 5 minutes. No sample pretreatment is required. Solid phase reagent (42 µg antibody immobilized on 200 µg paramagnetic particles in 450 µl) is then dispensed and incubated for an additional 2.5 minutes. Particles are captured and washed twice prior to generation of the chemiluminescence signal.

RESULTS AND DISCUSSION
Key features of both assays and instrument subsystems were exploited to deliver rapid tests. The use of a particulate solid phase offers enhanced assay kinetics compared to coated tubes or microtiter plates. The high surface area of paramagnetic particles selected allows immobilization of large amounts of antibody,

thereby driving the kinetics of analyte capture. Treatment of the solid phase after coupling of the antibody by incubation in a buffer with a high concentration of non-immune IgG reduces non-specific binding of tracer antibody to 0.05%.

Instrument subsystems were also developed to allow rapid assays. Since increased temperature drives reaction kinetics, a thermally controlled environment was provided for incubation. Magnets with high field strength and high field gradient were chosen. Challenges in luminometer design that led to the implementation of the rotary housing were the need to keep stray light to a minimum, while still allowing entry and exit of the reaction vessels at a high enough rate for the desired throughput.

hCG assay results The accuracy of the test was demonstrated by correlation with Magic Lite hCG (Ciba Corning Diagnostics, Medfield, MA). For 172 patient samples, the correlation coefficient was 0.98. Slope and intercept of the regression line were 1.08 and 1.03 mIU/mL, respectively. Parallelism was evaluated by linear dilution of patient serum samples of known concentrations. Dilution recoveries for dilutions from 1:2 to 1:6 averaged 101% for 3 patient samples. Similarly, recoveries of hCG from 3 human patient samples supplemented with a known amount of hCG averaged 105%. Sensitivity was assessed by dilution of serum samples and comparison of the observed dose to zero. Doses of 0.4 to 0.6 mIU/mL were statistically different from zero. Within run and total precision of dose was assessed over a 5 week period. Five determinations of dose (in replicates of 3) were made on control materials, giving the results shown below:

| | %CV of dose | |
Dose, mIU/mL	Within Run	Total
13.9	3.74	4.77
124.8	3.42	4.7
329.1	2.69	7.4

The range of hCG levels determined on 53 non-pregnant females was 0 - 7.2 mIU/mL, with a mean of 0.9 mIU/mL.

In summary, design strategies for the ACS:180 analyzer were employed in both chemistry and instrumentation to allow development of a simple, yet flexible, system that delivers rapid results. The performance of hCG assay reported here demonstrates that when these strategies are employed, the result is precise and accurate patient values.

ACKNOWLEDGEMENT
We thank the entire Oberlin and Walpole ACS:180 development team for their dedication in developing the system. We thank Eleanor Ling, John Kilroy and Kim Beals for the hCG data and Robert Hemmond for the illustration.

Figure 1. ACS:180™ System Layout

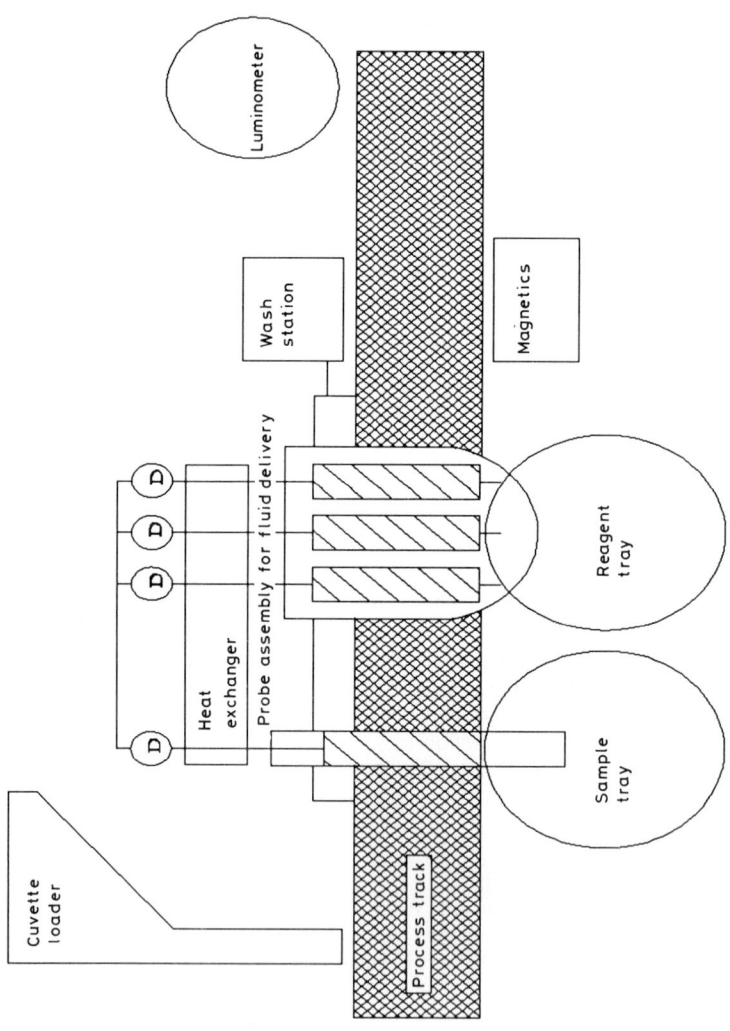

A NOVEL HOMOGENEOUS IMMUNOASSAY FOR ESTRONE-3-GLUCURONIDE BASED ON ANTIBODY PROTECTED-EFFECTS ON THE ENHANCEMENT OF CHEMILUMINESCENCE BY NaOH

J.B. Kim, O.J. Kwon and G. Barnard[1]

Department of Animal Products Science, Kon-Kuk University, Seoul, Korea,
[1]Department of Chemistry, City University, London, UK

INTRODUCTION

One of the great advantages of non-isotopic immunoassay systems over conventional radioimmunoassay (RIA) is that it is possible to develop homogeneous systems which do not require a physical separation step. Potentially, homogeneous assays are simpler, quicker, more precise and easily automated. The basis of homogenous assays is the finding that the properties of the label are altered upon binding to the antibody and that the signals generated by the antibody-bound and free moieties can be discriminated (1). Based upon the antibody-enhanced chemiluminescence, the homogeneous chemiluminescence immunoassays (CIA) for progesterone (2), cortisol (3) and estradiol-16α-glucuronide (4), were developed. However, we report here a new homogeneous system based on a different principle from antibody-enhanced chemiluminescence. We have exploited the finding that on addition of sodium hydroxide, the antibody bound chemiluminescence label is protected from the effect of sodium hydroxide on the enhancement of chemiluminescence for a short period prior to dissociation. This has enabled the development of a novel homogeneous CIA for estrone-3-glucuronide (E_1-3-Glu).

MATERIALS AND METHODS

Chemiluminescence marker conjugate and antibodies to E_1-3-Glu

E_1-3-Glu was coupled to aminoethylethyl-isoluminol (AEEI) as described (5). Both polyclonal and monoclonal antibodies to E_1-3-Glu-BSA were kindly donated from Dr. F.Kohen, in the Weizmann Institute of Science, Rehovot, Israel.

Assay protocol

One hundred µl of diluted urine or 100µl of standard (range 1.56 to 400pg/100µl of immunoassay buffer) were added in duplicate to flat bottomed polystyrene tubes (44.2 x 9.3mm). Subsequently, 100µl of E_1-3-G-6-AEEI (500pg/100µl assay buffer) and 100µl of suitably diluted antiserum (1:3000, v/v) were added and the mixture was incubated at room temperature for 1 hr. Two hundred µl of 2N sodium hydroxide and 100µl of microperoxidase solution (100µl/ml) were added to the assay tube, which was placed in the luminometer. The chemiluminescence reaction was initiated and the light emitted was measured with a LKB Luminometer Model 1250 in association with an LKB 1223 Databox.   An automatic dispenser was used to inject the solution of hydrogen peroxide (Hook and Tucker Instruments, Ltd., Surrey, UK). The signal was integrated over 10s.

RESULTS AND DISCUSSION
The effects of anti-E_1-3-Glu antiserum (polyclonal and monoclonal) on the light emission of E_1-3-Glu-AEEI in the presence of NaOH (2N) was investigated by incubating serial dilutions of the antiserum (range 1:200 to 1:100,000, v/v) with 100µl of E_1-3-Glu-AEEI (50µg) in the presence and absence of 100pg of authentic E_1-3-Glu. The results are shown in Fig. 1. As the concentrations of antiserum increased, the light yields decreased, which indicates that the chemiluminescence of antibody bound E_1-3-Glu-AEEI conjugate is protected from the effect of NaOH on the enhancement of chemiluminescence. But this effect was reduced by unlabeled E_1-3-Glu that competed with labeled conjugate to bind to the combining site of antibody. One interesting thing is that the mode of polyclonal antibody-protected effect is different from that of monoclonal antibody.   This difference may be due to the differences of antibody combining sites in size, structure and conformation. In this case, combining site of monoclonal antibody recognize both E_1-3-Glu and AEEI while that of polyclonal antibody recognize probably only part of E_1-3-Glu-AEEI conjugate as represented in Fig. 2.

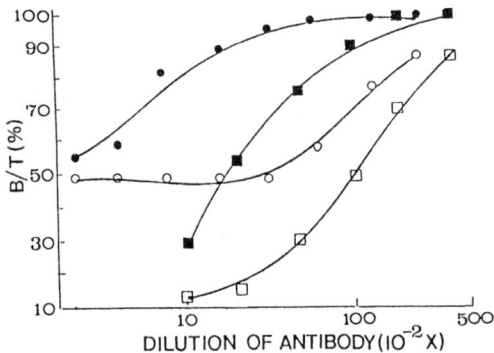

Fig. 1. Dilution curves of polyclonal (O) and monoclonal (□) antibodies with 100pg of authentic E_1-3-Glu (●,■)

Based on the finding mentioned above, calibration curves were obtained at different dilutions of the monoclonal antibody (range : 1:10,000 to 1:400,000, v/v). The results are shown in Fig. 3.   Fig. 4. shows a typical calibration curve (mean ± SD ; 5 replicates) for E_1-3-Glu as determined with appropriate dilution (1:50,000) of monoclonal antibody.   The minimum concentration of E_1-3-Glu that could be significantly distinguished from zero (mean - 2SD) was 8pg/tube, which is as sensitive as those obtained by other heterogeneous methods such as RIA and ELISA.

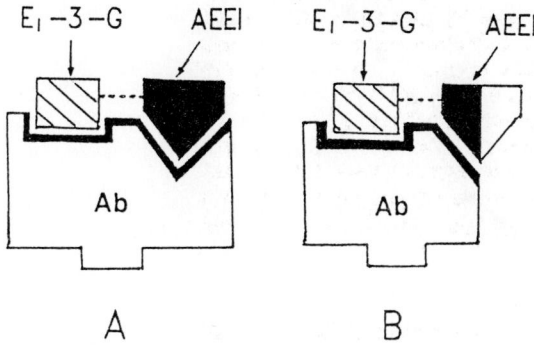

Fig. 2. Schematic representation which shows antibody combining sites in recognizing ligand(E_1-3-Glu), bridge and chemiluminescence marker (AEEI):
A recognition of both E_1-3-Glu and AEEI;
B recognition of E_1-3-Glu and partial AEEI.

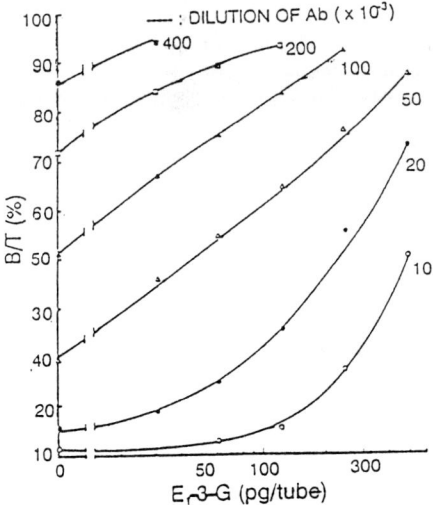

Fig. 3. Calibration curves at different antibody dilutions

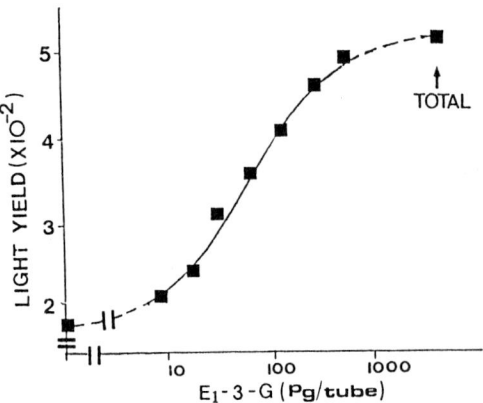

Fig. 4. Calibration curves of E_1-3-Glu in homogeneous CIA.

It will be very interesting to see if this novel homogeneous CIA can be applied to all kinds of hormones to be assayed. There appear to be, however, no general rules with respect to the type of quenching of chemiluminescence obtained when the label is bound to the antibody.

REFERENCES
1.   A.K. Campbell, R.A. Roberts and A. Patel, Chemiluminescence energy transfer: a technique for homogenous immunoassay. in: Alternative Immunoassays. W.P. Collins (Ed), John Wiley and Sons, Chichester, p. 1-9 (1987).

2.   F. Kohen, M. Pazzagli, J.B. Kim and H.R. Lindner, An assay procedure for plasma progesterone based on antibody enhanced chemiluminescence. Febs. Letts. 104, 201-205 (1979).

3.   F. Kohen, M.Pazzagli, J.B. Kim and H.R. Lindner, An immunoassay for plasma cortisol based on chemiluminescence. Steroids. 36, 421-437 (1980).

4.   F. Kohen, J.B. Kim, G. Barnard, H.R. Lindner, An assay for urinary estriol-16 α-glucuronide based on antibody-enhanced chemiluminescence. Steroids. 36, 405-419 (1980) .

5.   D.A. Weerasekera, J.B. Kim, G.J. Barnard, W.P. Collins, F. Kohen and H.R. Lindner. Monitoring ovarian function by a solid-phase chemiluminescence immunoassay. Acta Endocr. Copenhagen, 101, 254-263 (1982).

CHEMILUMINESCENT ENZYME IMMUNOASSAY OF HUMAN-GROWTH-HORMONE BASED ON ADAMANTYL DIOXETANE PHENYL PHOSPHATE SUBSTRATE

S. Albrecht, H.Ehle[1], K. Schollberg, R.Bublitz[1] & A. Horn[1]

Institute of Clinical Chemistry of the Medical Academy Dresden and [1]Institute of Biochemistry of the Friedrich-Schiller-University Jena, Germany

INTRODUCTION

The growth hormone (hGH, somatotropin STH) the most abundant hormone in the anterior pituitary, is a polypeptide of M.W. 21500 containing 191 aminoacids and two intramolecular disulfide bridges. It bears a marked structural resemblance to prolactin and to placental lactogen. During most of the day, the plasma concentration of hGH in normal adults remains stable and relative low, often below the level of detectability of most assays.

Adults and children show a marked rise in hGH-secretion 60-90 min after the onset of sleep.

The plasma levels are primarity a function of pituitary secretion. The secretion is believed to be under the control by hGHRH and somatostatin, as well as modulated by other factors, such as somatomedins. The release of the two hypothalamic factors is controlled by the higher centres of the brain.

The determination of hGH (base-line and following stimulation) is indispensable for the differential diagnosis of disturbed growth. We evaluated the substrate adamantyl methoxy 1,2-dioxetane phenyl phosphate (AMPPD, Fa. Tropix)[1] for the alkaline phosphatase label used in a sandwich-ELISA of hGH. Detection limit for hGH in the presence of an amplifier was 5 pmol/L. Intra- and interassay precision in cases of normal and pathological serum-samples in comparison to a hGH-RIA will be discussed.

MATERIAL AND METHODS

Luminometry: Chemiluminescence (6s-integral) was measured with a Berthold LB 9502 Luminometer.

Coating procedure: Luminometer-tubes (75x12 mm, Fa. Sarstedt, Germany) were incubated 4 hours with 100 μL hGH-antibody-solution (8 μg/mL in 50 mmol/L carbonate buffer, pH 9,6). The polyclonal antibodies (rabbit) were isolated by affinity chromatography with hGH-Sepharose 4B. Two wash steps were then performed using PBS-Tween (0,137 mol/L NaCl; 1,47 mmol/L KH_2PO_4; 20,43 mmol/l Na_2HPO_4; 2,68 mmol/L KCl; 0,02% NaN_3; 0,05% (v/v) Tween 20).

Chemiluminescent substrate: AMPPD and the amplifier "Sapphire" (λ_{max} (Fluor.)= 466 nm) are available from Tropix Inc. (Bedford).

Methods comparison: The competitive radioimmunoassay hGH-RIA (Isocommerz GmbH, Leipzig, Germany) was used for the methods comparison.

Assay procedure: The anti-hGH-coated tubes were incubated 4 hours with 100 μL sample or standard (1:3 diluted with PBS-Tween) and washed two times with PBS-Tween, following by the next incubation step with 100 μL enzyme-antibody-conjugate (AP-anti-hGH-conjugate, prepared by using the glutardialdehyde-method, 1-3 IU/mL).After 4 hours the   tubes were washed two times with PBS-Tween and Aqua bidest.

The tubes were filled with 200 μL of substrate buffer (0,05 mol/L Na_2CO_3/NaHCO$_3$; 0,001 mol/L $MgCl_2$) containing the "Sapphire" amplifier and loaded into the Berthold Clinilumat LB 9502. We initiated the chemiluminescent reaction by injecting 300 μL of AMPPD (0,4 mmol/L in substrate buffer). The signal was recorded as 6s-integral every 10 min during the first hour. A constant light emission was found between 30 and 60 min. We used the 30 min value for the calculation of the assay.

RLU [counts per 6 s] x 10^{-3}

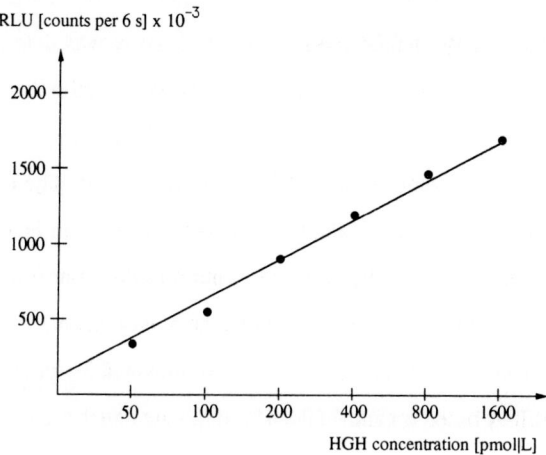

Fig.1 HGH-Assay - standard curve

HGH - RIA [pmol/L]

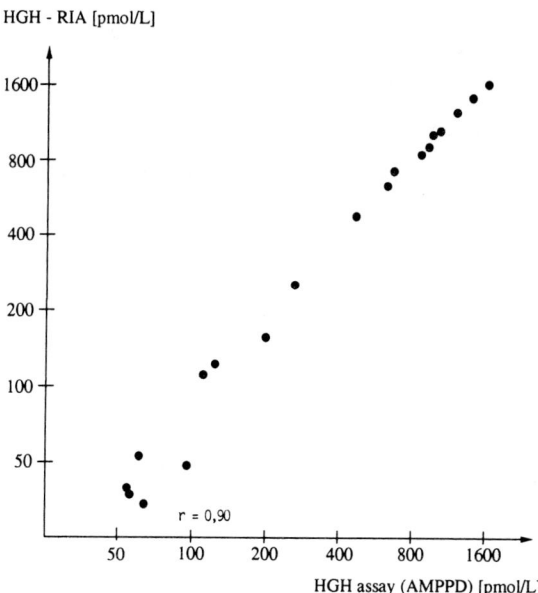

Fig.2 Correlation between HGH-RIA and chemiluminescent assay results of 20 serum samples

RESULTS AND DISCUSSION

The relative linear standard curve in the range of 25 to 1600 pmol/L hGH is shown in Fig.1. The sensitivity of this luminescent hGH-assay was determined to be 5 pmol/L when defined as the mean of the zero standard + 2SD (n=10). The intra- and interassay precision was found to be 10-9% (50 pmol/L, n=15; 800 pmol/L, n= 15) and 14-10% (50 pmol/L, n=15; 800 pmol/L, n=15). The correlation bet- ween the chemiluminescent assay and the hGH-RIA is shown in Fig.2.

The correlation between the radiometric and chemiluminescent results for the ran- ge of patient hGH values measured was satisfactory (r=0,9; n=20). The sensitivity was improved 10-fold in comparison to the RIA-assay and 3-fold in comparison to a ultramicro-ELISA based on anti-hGH-AP-conjugate and 4-methylumbelliferylp- hosphate as substrate .[2]

Further improvements of the chemiluminescent version of the assay will require optimization of substrate concentration, incubation time, coating procedure and amplifier concentration. Then it will be possible to determine hGH in human urine, where the hGH-concentration is only one p.c. of serum samples.

REFERENCES

[1] I. Bronstein & L.J. Kricka: "Clinical Application of Luminescent Assay for En- zymes and Enzyme Labels"
J.Clin.Lab.Anal. 3 (1989) 316-322
[2] H. Ehle, K. Hagner, S. Richter, R. Bublitz, A. Horn & G. Jahreis: "Ultramikro- ELISA zur hGH-Bestimmung im Kapillarblut"
Proceedings of the 6. Symposium für klinische Endokrinologie, Friedrich Schiller Universität Jena 1989

CHEMILUMINESCENT ASSAY OF ALKALINE PHOSPHATASE USING ASCORBIC ACID 2-PHOSPHATE AS SUBSTRATE AND ITS APPLICATION TO CHEMILUMINESCENT ENZYME IMMUNOASSAYS

M. Maeda, A. Tsuji, K.H. Yang[1] and S. Kamada[2]

School of Pharmaceutical Sciences, Showa University, Tokyo, Japan,
[1]Department of Animal Reproduction, Kon-Kuk University, Soul, Korea
and [2]Tosoh Co., Tokyo, Japan

INTRODUCTION

Recent, highly sensitive bioluminescent and chemiluminescent methods for alkaline phosphatase (ALP) have been reported(1 - 3). We have also developed a sensitive chemiluminescent method for ALP using NADP as substrate and NADH chemiluminescent reaction (4). Although the sensitivity of the method was high, the procedure was slightly complicate as compared with other methods. In this study we have developed a new chemiluminescent method for ALP using ascorbic acid 2-O-phosphate as substrate. Enzymatic dephosphorylation produces ascorbic acid which reacts with lucigenin to emit light. The sensitivity of this method was examined and compared with other methods using adamantyl 1,2-dioexatane phosphate (1,2), luciferin-O-phosphate (3) or NADP (4) as substrate. We have also applied this method to the enzyme immunoassay using ALP as label.

MATERIALS AND METHODS

INSTRUMENTS A Luminometer UPD-8000 (Meidenshya Co., Tokyo) and a luminescence Reader (Aloka Co., Tokyo) were used for measuring chemiluminescence.

REAGENTS Alkaline phosphatase (ALP, EC 3.1.3.1), $NADP^+$ and malic enzyme were purchased from Boeringer Mannheim _ Yamanouchi Co. (Tokyo). 3-(2'-spiroadamantane)-4-methoxy-4-(3"-phosphoryloxyphenyl)-1,2-dioxetan (AMPPD) was gifted by Fuji Revio Co., (Tokyo). Ascorbic acid 2-O-phosphate was obtained from Wako Chemical Co., (Osaka). Lucigenin (N,n'-dimethyl-9,9'-bisacridinium dinitrate) and other chemicals were obtained from Tokyo Chemical Industry Co. (Tokyo). 17α-Hydroxyprogesterone, anti-17α-hydroxyprogesterone antiserum and double antibody coated beads were the same ones as reported in the previous paper (4).

ASSAY PROTOCOL

(1) CL assay of ALP using ascorbic acid 2-O-phosphate as substrate : To the assay tube, each $50\mu l$ if ALP solution and 1×10^{-2} M ascorbic acid 2-O-phosphate solution in 0.1 M Tris buffer (pH 9.8) were added. After incubation for 60 min at room temperature, $500\mu l$ of 6×10^{-4} % lucigenin solution in 0.1 M KOH containing 1×10^{-3} M Triton X-100 was added and the chemiluminescence was measured for 10 sec after 15 sec waiting time by a luminometer.

(2) CL assays of ALP using $NADP^+$ and AMPPD : Both methods were performed according to the reports.

(3) CL EIA of 17α-hydroxyprogesterone (17-OHP) using ALP as label : 17-OHP-ALP conjugate was prepared by using 4-(2-carboxymethylthio)-17-OHP and ALP by the activated ester method. To each assay tube add serially $100\mu l$ each of the 17-OHP standard solution or sample solution, diluted anti-17-OHP antiserum $(1 : 1.6 \times 10^5)$ and diluted 17-OHP-ALP solution $(1 : 4 \times 10^4)$, mix well, then add one bead coated with second antibody (affinity purified goat anti-rabbit IgG) and incubate at 4^oC overnight. After washing the bead three times with isotonic saline (1 ml each), assay 17-OHP-ALP conjugate bound to the bead by the above chemiluminescent method.

RESULTS AND DISCUSSION

CL assay of ALP

Lucigenin is one of the classical organic CL reagents, its CL being produced by addition of either H_2O_2 or organic reducing compounds to an alkaline solution (5). Various reducing sugars react with lucigenin to emit light. We developed a CL assay of invertase using sucrose as substrate, in which enzymatic hydrolysis of non-reducing sucrose produces reducing glucose and fructose and emits light by reaction of lucigenin (6). However, the sensitivity of the method was lower than those of other CL methods. Ascorbic acid gives the most intensive CL with lucigenin among reducing compounds. Recently, ascorbic acid 2-O-phosphate was synthesized and used as additives of cosmetics and foods. We have examined utility of ascorbic acid-2-O-phosphate as the substrate for ALP. The reaction mechanism is shown in Fig.1. Ascorbic acid, produced upon dephosphorylation reacts with lucigenin in alkaline condition to emit light.

Various factors affecting on CL intensity, such as reagent concentration, reaction temperature, reaction time, were examined. Finally, the optimal assay conditions are decided as described in the Assay protocol (1). As shown in Fig. 2, the linearity of working curve ranged from 10^{-14} M to 10^{-11} M ; CV% less than 6.4%.

Recently, Bronstein (1) and Schaap (2) developed a new CL substrate for ALP, AMPPD, and Miska (3) reported D-luciferin-O-phosphate as new BL substrate for ALP, respectively. We have also developed CL assay of ALP using $NADP^+$/ethanol/alcohol dehydrogenases and NADH-CL reaction (4)/ or enhanced CL reaction (7). Detection limits of ALP obtained by these methods are summarized in Table 1. The detection limit was 1×10^{-19} mol/assay by 60 min assay. The sensitivity of the present method is slightly lower than those of the AMPPD method and the $NADP^+$ - enhanced enzyme cycling method. But the cost of reagents is lower than the AMPPD method and the assay procedure is more simple than the $NADP^+$- enhanced enzyme cycling method.

CL EIA for 17-OHP

The sensitivity of EIAs for steroids is markedly influenced by the combination of antibody and enzyme-labeled antigen : that is, "homologous" and "heterologous" combinations. According to the previous results of various EIAs developed in our laboratory, the bridge heterologous combinations was used in this study. The monoclonal antibody prepared by using 4-(2-carboxyetylthio)-17-OHP-BSA conjugate and the 17-OHP-ALP conjugate prepared by using 4-(2-carboxymethylthio)-17-OHP and ALP were used. After examination of various conditions, the dilutions of anti-17-OHP antibody and 17-OHP-ALP conjugate were 1 : 1.6 x 10^5 and 1 : 4 x 10^4, respectively. The typical working curve CL EIA for 17-OHP using ALP as the label enzyme is shown in Fig.3. The measurable range was from 0.5 to 128 pg/assay and the detection limit was 0.5 pg/assay, corresponding to 1.0 femtomol. The coefficients of variation at each 17-OHP level were in the range from 0.4 to 9.3%.

Figure 1. Principle of CL assay of alkaline phosphatase

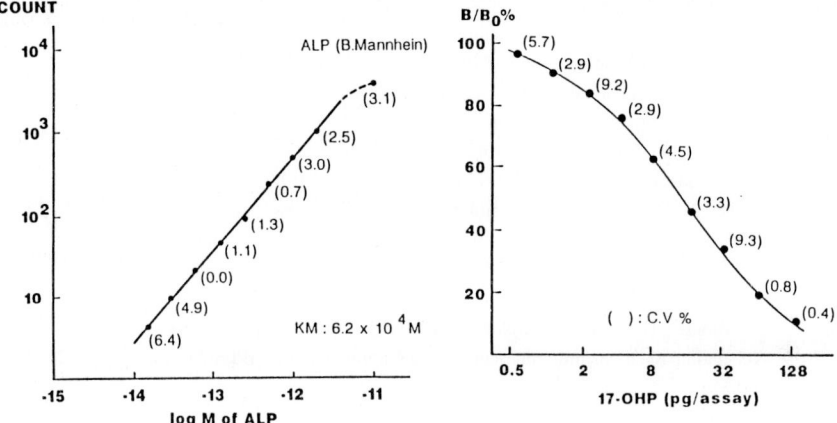

Figure 2. Working curve of ALP   Figure 3. Working Curve of 17-OHP

Table 1. Comparison of alkaline phosphatase detection limits
obtained by various methods

Method	Substrate	Detection limit (mol/assay)
NADP[+]	NADP[+]/EtOH/ADH/1-MPMS/IL/m-POD	2×10^{-18}
Enhanced NADP[+]	NADP[+]/Enzyme cycling	3.6×10^{-20}
AMPPD	AMPPD	1×10^{-20}
Bioluminescence	D-Luciferin-O-phosphate	1×10^{-19}
Present method	Ascorbic acid 2-O-phosphate	1×10^{-19}

ADH : Alcohol dehydrogenase, 1-MPMS : 1-Methoxy-5-methylphenathinium methylsulfate,
IL : Isoluminol, m-POD : Microperoxidase, AMPPD : 3-(2'-spiroadamantane)-4-methoxy-4-
(3"-phosphoryloxy)phenyl-1,2-dioxetane disodium salt

REFERENCES
1.   I.Bronstein, B.Edwards, and J.C.Voyta, 1,2-Dioxetanes: novel chemiluminescent enzyme
        substrates. Applications to immunoassays. J. Biolumi. Chemilumi., 4, 99 - 111 (1989).
2.   A.P.Schaap, M.Sandison, and R.S.Handley, Chemical and enzymatic triggering of 1,2-
        dioxetanes. 3: alkaline phosphatase-catalyzed chemiluminescence from an aryl phosphate -
        substituted dioxetane. Tetrahedron Lett., 28, 1159 - 1162 (1987).
3.   W.Miska, and R.Geiger, Luciferin derivatives in bioluminescence - enhanced enzyme
        immunoassays. J. Biolumi. Chemilumi., 4, 119 - 128 (1989).
4.   M.Maeda, H.Arakawa, and A.Tsuji, Chemiluminescent assay of various enzyme activities
        and its application to enzyme immunoassays. J. Biolumi. Chemilumi., 4, 140 -148 (1989).
5.   M.Maeda, and A.Tsuji, Chemiluminescent flow injection analysis of biological compounds
        based on the reaction of lucigenin. Anal. Sci., 2, 183 - 189 (1989)
6.   S.Shimazu, H.Arakawa, M.Maeda, and A.Tsuji, Chemiluminescent assay of invertase and
        its application to chemiluminescent enzyme immunoassays. Bunseki Kagaku, 37, 123 -128
        (1988)
7.   M.Kawamoto, H.Arakawa, M.Maeda, and A. Tsuji, Chemiluminescent assay of alkaline
        phosphatase using enzyme cycling enhanced method. Meeting abstracts of 110th Annual
        meeting of the Pharmaceutical Society of Japan, Sapporo, August 21 -23.3, p.256 (1990)

FULLY-AUTOMATED ANALYZER FOR CHEMILUMINESCENT ENZYME IMMUNOASSAY

K.SEKIYA, Y.SAITO, T.IKEGAMI, M.YAMAMOTO, Y.SATO, M.MAEDA[1] and A.TSUJI[1]

Sankyo, Co., Ltd., Tokyo Japan.
[1]School of Pharmaceutical Sciences
Showa Univrersity, Tokyo, Japan

INTRODUCTION

Recent progress of nonisotopic immunoassay in the field of clinical chemistry is remarkable. Various fully-automated analyzers have also been developed for turbidimetric, latex and enzyme immunoassays, and become to be used in the clinical laboratories. We have developed chemiluminescent enzyme immunoassay (CLEIA) based on the CL assay of glucose oxidase (GOD) (1, 2). In this study, we have developed a fully-automated analyzer for CLEIA using GOD as the label enzyme, and evaluated the usefulness of the instrument for clinical application.

MATERIALS AND METHODS

REAGENTS Luminol was purchased from Tokyo Chemical Industry Co., Ltd., Glucose Oxidase(GOD) from Toyobo Co., Ltd., microperoxidase (m-POD) from Sigma Chemical Co., Ltd., hydrogen peroxide from Wako Pure Chemical Co., Ltd, N-succinimidyl-6-maleimidohexanoate from Dojindo Laboratories, and Ultragel from Pharmacia-Japan Co., Ltd.

APPARATUS As showing Fig.1, the fully-automated CLEIA analyzer is composed of the main part (W=1170, D=770 and H=1250mm) and the operation part (W=430, D=770 and H=1250mm). The main part is constituted from three compartments. Reagents, buffers, effluents, power source, compressor and motor-driver are accommodated in the lower compartment. In the middle, total 500 tubes coated with antibody (25 tubes per one tray) are housed and additional 168 tubes for dilution can be installed. In the upper compartment, the reaction part and detection part are housed. The part for accommodation of chips and installation of samples are set in this part. The system uses disposable chips for the collection of samples in order to avoid the carry-over of samples. Chips and samples are accommodated in the upper compartment.

The CL reaction is carried out in a cell of an optical integrating sphere. The CL intensity is counted by time integration for 10 sec from the injection of the CL reagent into the sample by two photo multipliers (PM). Two types of PMs (Hamamatsu Photonics R1104, R1546) are used in the system; one is for high sensitivity (detection of low CL intensity) and the other for low sensitivity (detection of high CL intensity). Blockdiagram for detection system as shown in Fig.2. This system makes it possible to simultaneously measure low and high CL, and provides very wide dynamic determination range. Every reagent supplied for this system is bar-coded, which enables the allocation of reagents corresponding to respective measuring automatically.

Main specifications are shown as below :

Principle of measurement: single-line discrete system

Simultaneous measuring channels : Max 20 channels of randomaccess per one sample

Installed samples : 100 samples (additional) and 10 samples in emergency

Continuously measuring samples : max 2000 tests

Through put: 120 tests/h

Measuring mode : 45 minute assay, 30 minute assay and 180 minute assay
Time for testing : about 45 min till the first test result
Sample volume : 30 - 100μl (by each channel), 10 - 100μl (when diluted)
Temperature control system : plate-heater method, at 37° C
Detection system: Two PMs, high and low sensitivity
Other functions : automated dilution and re-testing system, cooling system of reagents, preventing
system from evaporation of sample

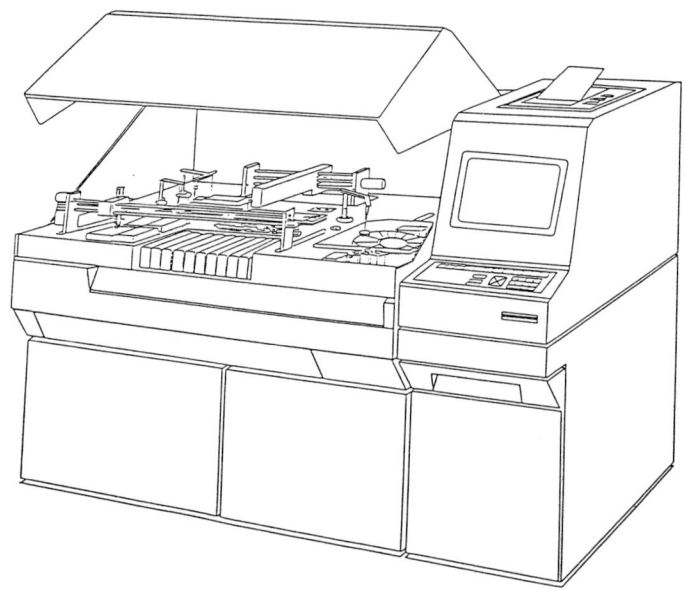

Fig.1 Fully-automated CLEIA analyzer

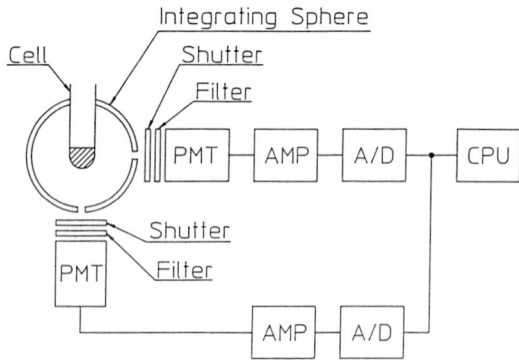

Fig.2 Blockdiagram for Detection System

RESULTS AND DISCUSSION

GOD is an enzyme which produces H_2O_2 during a preincubation period that is appropriately selected in order to increase the sensitivity of the system. GOD oxidizes glucose to yield H_2O_2 which can be measured by CL reaction of luminol/m-POD system. The working curve of GOD by the present analyzer ranges from 6.2×10^{-18} to 1.0×10^{-12} mol/assay.

From the result of examination on various factors, the assay protocol was decided as mentioned above. A typical standard curve of AFP ranges from 0.5 to 1,000 ng/ml, as shown in Fig.3. The detection limit was 0.048 ng/ml. The assay ranges, detection limits and reproducibility of analysis are listed in Tab.1.

All assays were not interfered by fat, hemoglobin, and free and conjugated bilirubin at their normal concentrations. The correlation between the results obtained by this system and conventional assay kits are as follows; AFP: Y=1.006X + 6.101 (r=0.995, n=75), CEA: Y=0.988X-1.146 (r=0.979, n=150). The reproducibility was tested with use of control serum. The results of intra assay and inter assay are shown in Tab.2. Other measuring items are FER, PAP, HCG, TSH, T_4, T_3, TBG, FT_4, LH, FSH, PRL, IgE, Insulin, Miosin light chain.

The CL method was generally considered to have high sensitivity but poor reproducibility. The present system enables to obtain the reproducibility of more or less 1% of CV by the combination of the integrating sphere and the time integration on detection. The characteristic of the system compared with conventional detectors is in the wide dynamic range, which can be obtained by two photo multiplier system. The system can pick-up the samples of over-ranged, automatically dilute the samples and carry out re-measurement. These features of the system are considered to contribute greatly to the cost-reduction and rationalization of clinical laboratory tests.

ACKNOWLEDGEMENT

We express our hearty thanks to Professor Gomi of Showa University Hospital who helped us tabulate the data.

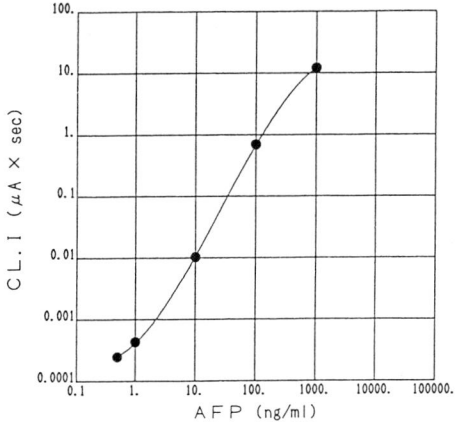

Fig.3 Standard Curve of AFP

Table 1 Assay range and detection limit

Analyte	Assay range	Detection limit*
AFP	0.5 - 1000 ng/ml	0.048 ng/ml
CEA	0.5 - 1000 ng/ml	0.086 ng/ml

* 0 + 2SD

Table 2 Intra assay and inter assay

	Intra assay (n=30)			Inter assay (n= 30)		
	X (ng/ml)	S.D.	C.V.%	X (ng/ml)	S.D.	C.V.%
AFP	30.3	0.6	2.0	31.2	1.3	3.8
	153.6	6.5	4.2	145.3	7.1	4.9
CEA	4.0	0.2	5.1	4.0	0.3	7.5
	23.3	1.0	4.3	24.4	1.4	5.8

REFERENCES

1.   H.Arakawa, M.Maeda and A.Tsuji, Chemiluminescence enzyme immnoassay of 17α-hydroxyprogesterone using glucose oxidase and bis (2,4,6-trichlorophenyl) oxalate - fluorescent dye system. Chem. Pharm. Bull., 39, 3036-3039 (1982)

2.   M.Maeda and A.Tsuji, Enzymatic immunoassay of α-fetoportein, insulin and 17-α-hydroxy-progesteron based on chemiluminescence in a flow-injection system. Anal. Chim. Acta, 167, 241-248 (1985)

3.   A.Tsuji, M.Maeda and H.Arakawa, Chemiluminescent enzyme immunoassay. Anal. Sci., 5, 497-506 (1989)

DEVELOPMENT OF ENZYME - MEDIATED CHEMILUMINESCENCE
IMMUNOASSAYS FOR HUMAN LUTEINIZING HORMONE (LH)
AND HUMAN FOLLICLE STIMULATING HORMONE (FSH)

G. Odstrchel, E. Guthrie, J. Mauricio, D. Garcia
and V. Mahant

Nichols Institute Diagnostics,
San Juan Capistrano, CA 92675, USA

INTRODUCTION

The hormones LH and FSH are glycoproteins of approximately 30,000
MW that are secreted by the pituitary gland. They are comprised
of two protein chains. The alpha subunits are identical and the
beta subunits are responsible for the biological and immunological
specificity of these protein hormones. The assessment of these
hormone levels is important in men in that in concert with testosterone they maintain spermatogenesis and in women they are
critical in the proper assessment of the menstrual cycle. Proper
assessment of these levels permit diagnosis and treatment of
several irregularities in the reproductive cycle and in recent
years have become very important tests in the area of in-vitro
fertilization.

MATERIALS AND METHODS

This manuscript describes the development of LH and FSH assays in
a non-isotopic luminescent format. The assays are based on the
use of a biotinylated specific monoclonal capture antibody and
another monoclonal antibody that is conjugated with the enzyme
calf intestine alkaline phosphatase. A streptavidin coated ¼"
polystyrene bead is added to complete the assay. After incubation
(1.0 hour) the bead is washed and the substrate, phenylphosphate-
substituted dioxetane is added to the tube containing the bead.
Upon dephosphorylation by alkaline phosphatase the substrate
becomes destabilized and produces light. The bead is incubated for
an additional thirty (30) minutes and at precise time intervals the
relative light units (RLU) are measured. A standard curve of RLU's
versus concentration of standards is constructed and the LH and FSH
values obtained from this curve.

RESULTS AND DISCUSSION

Use of the enzyme mediated chemiluminescent technology yields
assays for LH and FSH that are very sensitive (0.1 mIU/mL). These
assays are unaffected by high dose hook effects and can be done
with a 1.0 hour incubation time. Correlation of these assays with
reference methods was excellent (r= .95-998, slope .92-.98).
Chemiluminescent technology offers an alternative to isotopic IRMA
tests without sacrificing the quality of the results.

LH/FSH ASSAY PROCEDURE

a. Pipet 50 µL of standard or patient sample

b. Pipet 200 µL of reagent B (conjugate and
    biotinylated antibody solution)

c. Vortex

d. Add one avidin-coated bead

e. Incubate at room temperature for 1 hour on
    rotator (170 rpm)

f. Wash the bead
    Aspirate reaction mixture
    Add 2 x 2 mL saline/azide wash solution

g. Add 200 µL luminescent substrate

h. Read after 30 minutes in luminometer

TYPICAL LUMINESCENT LH STANDARD CURVE

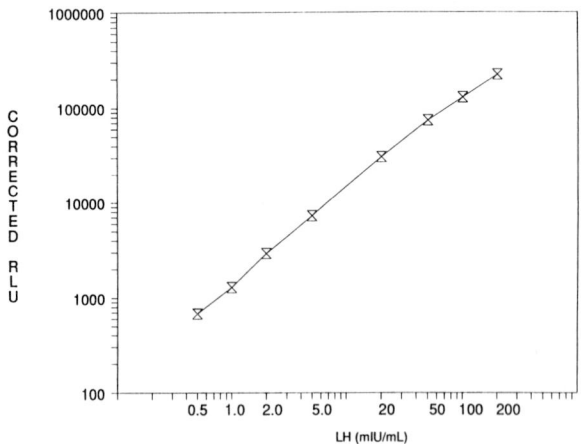

TYPICAL LUMINESCENT FSH STANDARD CURVE

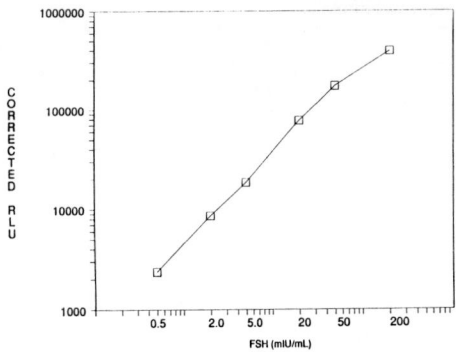

LH PATIENT CORRELATION

LUMINESCENT ASSAY vs AMERLITE-LH

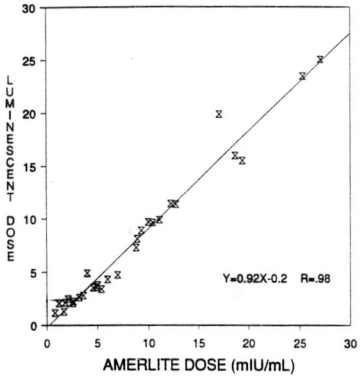

FSH PATIENT CORRELATION

LUMINESCENT ASSAY vs ALLEGRO-IRMA

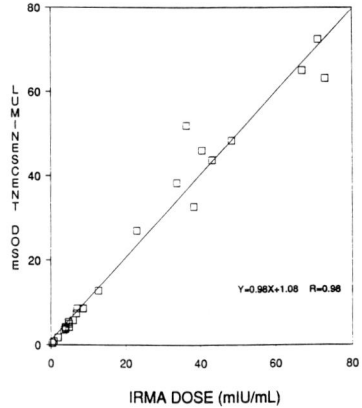

LH ASSAY PERFORMANCE

INTRA-ASSAY VARIATION

Mean Value (mIU/ml)	N	%CV
3.1	10	4.8
54.7	10	4.5

SENSITIVITY

Standard	N	Detectable Dose
0	10	.1 mIU/mL

FSH ASSAY PERFORMANCE

INTRA-ASSAY VARIATION

Mean Value (mIU/mL)	N	%CV
4.3	20	4.3
41.1	20	5.0

SENSITIVITY

Standard	N	Detectable Dose
0	24	.1 mIU/mL

SUMMARY

SPECIFICATION	LH	FSH
Sensitivity	0.1 mIU/mL	0.1 mIU/mL
Intra-assay %CV	≤ 4.8	< 5.0
Sample Size	50 µL	50 µL
High Dose Hook	>20,000 mIU	>20,000 mIU
Incubation Time	1.0 hr.	1.0 hr.
Substrate Incubation	0.5 hr.	0.5 hr.
Correlation	Amerlite LH (n = 31)	Allegro IRMA (n = 25)
- Coefficient	.98	.95
- Slope	.92	.98

SYNTHESIS AND COMPARISON OF
SUBSTITUTED CHEMILUMINESCENT ACRIDINIUM DERIVATIVES
INCLUDING A NOVEL ACRIDINIUM AMIDE

M.L. Grayeski, T.J. Novak and A.G. Mohan[1]

*Department of Chemistry, Seton Hall University, South Orange, New Jersey
07079 and [1]State of New Jersey, Department of Health, Public Health and
Environmental Labs., P.O. Box 361, Trenton, New Jersey 08625-0361*

INTRODUCTION

The use of chemiluminescent acridinium haptens as
labels in immunoassays is attracting increased
interest because of the possibility of high
incorporation ratios of efficient acridinium salts with
high retention of immunological activity (1). Our
research interests involve using acridiniums for a
variety of applications including DNA hybridization
probes and HPLC detection as well as immunoassays.
Since these applications have different requirements
for the measurement process in terms of
chemiluminescence efficiency and kinetic behavior, we
began investigating the influence of substitution at
the 9-position of the acridinium nucleus on the
chemiluminescence intensity under a variety of aqueous
reaction conditions. A series of esters of 10-
methylacridinium-9-carboxylic acid has been prepared
with aryl substitution including p-t-butylphenyl, 2-,
3-monochlorophenyl, 2,4-dichlorophenyl and 2,4,6-
trichlorophenyl. A new chemiluminescent amide,
10-methyl-9-carboxyl (N-trifluoromethylsulfonyl-n-
butyl) acridinium amide trifluoromethane sulfonate
(BATQ), was also synthesized. These are compared to
aryl esters which have been examined previously: 4-
chlorophenyl, p-acetylphenyl, 2-napthylphenyl and
phenyl (2).

MATERIALS AND METHODS

<u>Synthesis.</u> Acridine-9-carboxylic acid was prepared by
the procedure of Newman and Powell using chlorobenzene
as the reaction solvent (3). Esterification and
quaternization with methyltriflouromethanesulfonate
were performed as previously described (1). N-
butyltrifluoromethane
sulfonamide was prepared according to Tseng (4).
Amidation and quaternization were done using the
procedure of Molz and Skrzipczyk (5).

Solution preparation. Working aqueous solutions of the
acridinium derivatives were prepared at a concentraton
of approximately 8x10(-9) mol/l by dilution of a stock
solution in acetonitrile with dilute pH 3.0 nitric acid
solution. Hydrogen peroxide solutions were prepared by
dilution of a 30% solution with 0.05 mol/l phosphate
buffered at various pH levels. Hydrogen peroxide
concentrations ranged from 20umol/l to 1 mol/l with
lower concentrations used at higher pH levels. For each
pH value investigated, 3 levels of hydrogen peroxide,
each differing by an order of magnitude, were used.
Chemiluminescence measurements. Chemiluminescence was
initiated by injection of 50ul of the aqueous solution
of the acridinium derivative into 200ul of the buffered
peroxide solution. The intensity was monitored using a
Turner Designs model TD 20e luminometer interfaced to
an AT&T PC6300 computer via a DT 2808 A/D converter
(Data Translation, Marlboro, Mass.). Data reduction was
performed using ASYSTANT+ (Asyst Software Technologies,
Rochester, NY).

RESULTS AND DISCUSSION

Chemiluminescence efficiency was compared at various
reaction pH levels by measuring total emission
normalized for acridinium concentration (fig. 1). Under
the conditions chosen, the reaction of the BATQ
derivative resulted in an intensity greater than all of
the aryl esters investigated. BATQ exhibited an
intensity which was twice that of the phenyl ester at
pH 11. The phenyl ester is presented graphically
because it exhibited the highest efficiency of the
array of esters previously examined (2). It is
interesting to note that aryl esters having lower pKa
values displayed the lowest intensity at pH levels
greater than 10 (fig. 2) which would appear to
contradict expectations based upon accepted mechanistic
considerations (2). However further experimentation
revealed that the reaction conditions under which
luminescence occurs have important ramifications in
terms of efficiency comparisons. The conditions
employed here use a relatively low concentration of
peroxide which results in a chemiluminescence rate
slower than competing hydrolysis. Inductive effects of
the 9-position substituent which enhance light
producing pathways may also enhance base mediated dark
pathways as well. Hydrolysis, a dark reaction, competes
with chemiluminescence pathways. This was verified by
comparing relative intensities using various peroxide
concentrations at constant reaction pH. At pH 9, the
intensity of the phenyl ester was 2.5 fold greater than
that of the p-acetylphenylester using 1 mmol/l
peroxide as oxidant. However by increasing the peroxide
concentration to 70 mmol/l,   thus enhancing the

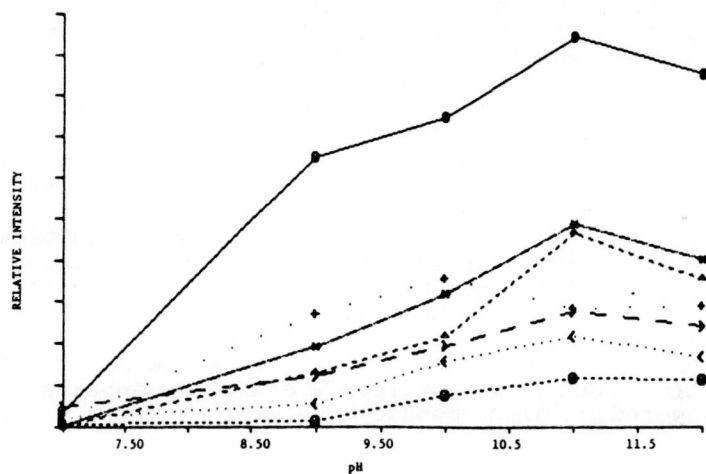

Figure 1. Relative chemiluminescence intensity expressed as total integrated area normalized for acridinium concentration of various 9-substituted acridinium derivatives as a function of reaction pH. The hydrogen peroxide concentration used was 1 mmol/l. 8 BATQ, * Phenyl Ester, ^ p-Tertbutyl phenyl Ester, + 3-Chlorophenyl Ester, > 2-Chloro phenyl Ester, < 2,4-Dichlorophenyl Ester, 0 2,4,6-Trichlorophenyl Ester.

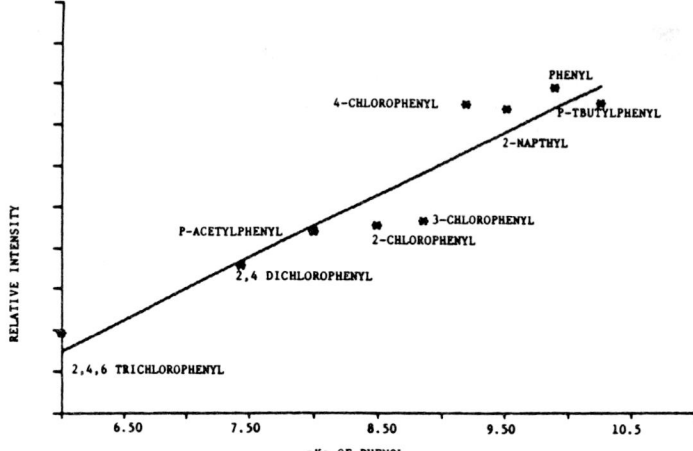

Figure 2. Relative chemiluminescence intensity expressed as total integrated area normalized for acridinium concentration of 9-substituted aryl acridinium esters as a function of leaving group pKa using 1 mmol/l hydrogen peroxide as oxidant at pH 11.

emission rate, the intensity of the p-acetylphenylester was twice that observed for the phenylester as would be expected based upon consideration of pKb values of the conjugate leaving group.. This behavior is also confirmed by noting the reversal of relative intensities of the phenyl and 3-chloro derivatives as reaction pH was lowered (fig.1). At lower pH ranges, emission competes more favorably with hydrolysis for the 3-chlorophenylester. It should be noted that the BATQ derivative exhibited higher intensities at all peroxide concentrations and pH values, albeit in many cases slower emission kinetics. These observations may be attributed to minimization of known dark reaction pathways (1,2) because of steric hindrance afforded by the trifluoromethylsulfonyl group to attack by nucleophiles at the 9-position of the acridinium ring. However, reactive intermediate formation leading to excited states appears to be favored over pseudo-base production.

REFERENCES
1.   I. Weeks, I. Beheshti, F. McCapra, A.K. Campbell, and J.S. Woodhead, Acridinium Esters as High Specific-Activity Labels in Immunoassay. Clin. Chem. 29, 1474-1479 (1983).
2.   F. McCapra, Chemical Mechanisms in Bioluminescence. Acc. Chem. Res. 9, 201-208 (1976).
3.   M.S. Newman, and W.H. Powell, Synthesis and Properties of 4,5-Dimethyl acridine and 1,4,5,8- (1961).
4.   S.S. Tseng, A.G. Mohan, L.G. Haines, L.S. Vizcarra, and M.M. Rauhut, Efficient Chemiluminescence from Reactions of N-(Trifluoromethyl sulfonyl)oxamides with Hydrogen Peroxide and Fluorescers. J.Org.Chem. 44, 4113-4116 (1979).
5.   P. Molz, and H.J. Skrzipczyk, Chemiluminescent Acridine Derivatives, Their Preparation and Use. Ger.Pat. App. DE 362 8573 (1988).

NOVEL ENHANCED LUMINESCENCE IMMUNOASSAYS FOR
HAPTENS AND MACROMOLECULES

M.J.O'Sullivan, J.K.Horton, J.E.James, O.M.Evans and S.Swinburne

*Amersham International plc, Cardiff Laboratories, Whitchurch,
Cardiff, CF4 7YT, Wales, UK*

INTRODUCTION

Many different non-isotopic tracers have been evaluated over the
last two decades(1). One of the more popular approaches has been
the application of enzymes(2). Some of these tracers, especially
those based on horseradish peroxidase (HRP), have provided enzyme
immunoassays (EIA) with good sensitivity. Unfortunately
conventional EIA still has a significant number of disadvantages,
mainly associated with the steps required to quantitate the
enzyme activity. A second popular approach has been the use of
luminescent detection systems(3). HRP can catalyse luminol
oxidation with light emission, but the intensity is low and its
duration is short. A dramatic advance was the discovery that the
light output in the HRP catalysed reaction could be increased
more than 1000 fold by the presence of enhancers such as firefly
luciferin and p-iodophenol. The reaction produces very intense
light which is emitted as a continuous glow(4). Enhanced
luminescence immunoassays have been developed for adenosine
3',5'-cyclic phosphate (cAMP), guanosine 3',5-cyclic phosphate
(cGMP), interleukin-1α (IL-1α) and human chorionic gonadotrophin
(HCG). The performance of these assays has been compared with
both conventional EIA's and RIA's. The cAMP and IL-1α assays
will be described in detail in this paper.

MATERIALS AND METHODS

Preparation of the cAMP-HRP conjugate. A cAMP-HRP conjugate was
prepared via a stable N-hydroxy-succinimide ester intermediate
(5). The ester was prepared by reaction of the carboxylic acid
group of 2'-O-monosuccinyladenosine-3':5'-cyclic monophosphate
(Sigma) in the presence of 1-ethyl-3(3-dimethylaminopropyl)
-carbodiimide hydrochloride (Sigma) and was stable at -20°C for
at least one year. The activated ester was dissolved in 1ml of
dry DMF and 25μl containing 0.25μmol was added to 0.5ml of
0.1mol/1 phosphate buffer pH8.0 containing HRP (Sigma type XII
affinity purified, 1mg, 0.025μmol). The reaction mixture was
vortex mixed and allowed to stand at room temperature for 4
hours. Low molecular weight components of the reaction mixture
were removed by gel filtration on a short Sephadex G-25 column
(Pharmacia) and the conjugate stored at 4°C in the presence of 1%

bovine serum albumin.

Protocol for the cAMP Enhanced Luminescent Assay. cAMP antisera
(rabbit) was diluted 1:10,000 in 0.1mol/l phosphate buffer pH7.4
containing 1% BSA. This antiserum was from the same pool as
supplied in Amersham's cyclic AMP assay kit (RPA 509). The cAMP
tracer was diluted 1:2000 in phosphate buffer containing 1% BSA.
The standard (64pmol) supplied in Amersham's cAMP RIA kit was
reconstituted with 2ml of distilled water and serially diluted to
produce an appropriate set of diluted standards. The standards
were then acetylated using the protocol described in the RIA pack
leaflet except that 25μl rather than 50μl of acetylation reagent
was employed. The assays were performed in opaque microtitre
wells precoated with anti-rabbit Ig. These were standard
production plates which are routinely used in Amersham's
progesterone ELIA kit. To perform the assay, acetylated
standards (50μl) and diluted antiserum (100μl) were added to the
opaque microtitre plates, mixed and left for 2 hours at 4°C.
Diluted tracer (100μl) was then added, mixed and the plates held
at 4°C for an additional hour. The plates were then washed four
times with phosphate buffer containing 0.1% Tween 20. The
enhanced luminescence was measured using Amersham's standard
signal reagents and buffer in an Amersham Research Luminometer.
The procedure for the conventional EIA was as described above
except that HRP activity was measured using o-phenylenediamine.
The reaction was terminated after 30 min., the solutions
transferred to a clear plate and the optimal density at 492 read
using a Titertek Multiscan MCC/340. The cAMP ^{125}I based RIA was
performed using the Amersham kit.

Protocol for the Interleukin-1α Assay. Opaque microtitre plates
were coated with 200μl per well of mouse monoclonal anti-IL-1α
(DiaNippon) at a 1:2000 dilution in 0.1mol/l carbonate buffer
pH9.5. The plates were left overnight at 4°C, washed four times
with 0.05mol/l phosphate buffer pH7.4 containing 0.1% Tween 20.
The plates were then blocked with 1% w/v skimmed milk powder in
phosphate buffer (PB) and washed as above.IL-1α standard (32 to
0.062fmol per well, 100μl) in PB was added to the coated wells.
The plates were held for 1 hour at room temperature and then
washed x4 with PB containing 0.1% Tween 20. Rabbit anti-IL-1α
(100μl in PB, diluted 1:4000) was added to the wells. After 1
hour at room temperature the plates were washed as described
above. Donkey anti-rabbit HRP conjugate (100μl, diluted 1:8000
in PB) was added to each well. After two hours the plates were
washed as described above. HRP activity was measured either by
enhanced luminescence as described previously or by using OPD.
The OPD reaction was terminated after 5 minutes, and the coloured
solutions transferred to clear plates for reading. The IL-1α
standard and antiserum were from Amersham Interleukin-1α assay
system (RPA.528). The anti-rabbit HRP conjugate was a standard
Amersham catalogue item.

RESULTS AND DISCUSSION
Standard curves for the cAMP assay, ELIA and RIA are virtually

identical as shown in Fig.1. For clarity the assay replicates
are not shown but the replicate precision of the two techniques
is comparable. The RIA incorporates a straightforward magnetic
solid phase (Amerlex-M) separation system. However this system
is arguably less convenient than the microtitre plate based ELIA.
When plotted in the same fashion, the conventional HRP EIA is
again virtually superimposable over the ELIA. However, to obtain
satisfactory optical density readings (~0.5 optical density units
for the zero standards) a minimum of 30 min. is required when
using OPD as substrate, even though OPD is one of the better HRP
substrates from the point of view of colour development. In our
hands colour development using OPD is 3x and 20x faster than when
using TMB and ABTS. Maximum signal with the enhanced
luminescence reagents is reached within 2 minutes and remains
relatively constant for the next 20 minutes. In addition there
is no requirement to terminate the reaction with sulphuric acid.

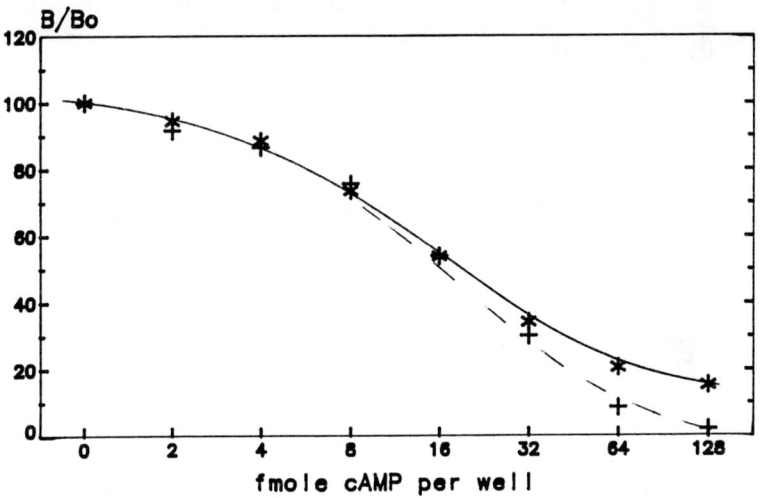

Fig. 1 Comparison of cAMP standard curves: *RIA   +EIA

IL-1α assay. The standard curve for the enhanced luminescence
assay is shown in Fig. 2. The assay has a working range from
0.125 to at least 32fmol/well. The conventional EIA has a range
from 0.25 to 16fmol/well. The sensitivities of the ELIA and EIA
are approximately 0.05 and 0.1fmol/well respectively. Amersham's
IL-1α [125I] assay system covers the range 0.5 to 16fmol/well.
However, the RIA is based upon a competitive format rather than
the two-site immunometric format employed in the non-isotopic
assays, so its decreased sensitivity and working range may
reflect the different assay format rather than the detection
system.

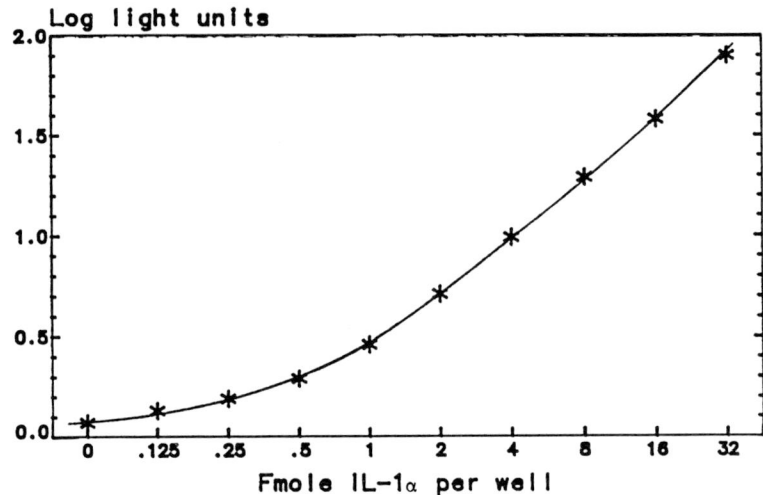

Fig. 2. IL-1α ELIA standard curve

In conclusion, these results show that it is possible to develop sensitive and convenient competitive assays for haptens and two-site immunometric assays for macromolecules based upon enhanced luminescence.

REFERENCES

1. R.F.Schall, Jr., and H.J.Tenoso, Alternatives to radio-immunoassay: labels and methods. Clin. Chem., 27, 1157-1164 (1981).

2. M.J.O'Sullivan, J.W.Bridges and V.Marks, Enzyme immunoassay: a review. Ann.Clin.Biochem., 16, 221-240 (1979).

3. L.J.Kricka and G.H.G.Thorpe, Chemiluminescent and bioluminescent methods in analytical chemistry. Analyst, 108, 1274-1296 (1983).

4. T.P.Whitehead, G.H.C.Thorpe, T.J.N.Carter, C.Groucett and L.J.Kricka. Enhanced luminescence procedure for sensitive determination of peroxidase-labelled conjugates in immunoassay. Nature, 305, 158-159 (1983).

5. B.F.Erlanger, Principles and methods for the preparation of drug protein conjugates for immunological studies. Pharmacological Reviews, 25, 271-280 (1973).

NOVEL-AND RAPID IMMUNOFILTRATION ASSAYS
WITH ENHANCED CHEMILUMINESCENCE (ECL) DETECTION

J.K. Horton, J.E. James, S. Swinburne and M.J. O'Sullivan.

*Amersham International Plc, Forest Farm, Whitchurch,
Cardiff, CF4 7YT, Wales, U.K.*

INTRODUCTION
The development of solid-phase immunoassays using nitrocellulose
as a support for either antigen or antibody have provided rapid
alternative methods to both ELISA and RIA (1). Here we describe
rapid immunofiltration assays using ECL (2), for the detection
of antibodies raised in mouse and rabbit and for human IgG,IgM
and IgA. The assays were carried out on membrane-bottomed
96-well microtitre plates (Amersham International plc, Amersham,
Bucks, U.K.) which have a typical 96-well format featuring
open-bottomed wells individually sealed with a nitrocellulose
filter. The membrane-bottomed plates are designed to be used
with a single or multiple (10 plate) vacuum manifold (Amersham).
The vacuum can be generated with a simple water pump or low cost
electrical pump.

METHODS
Detection of specific mouse and rabbit antibodies.
Membrane-bottomed plates were coated with antigen
(200μl/well;see figure legends) and incubated for 15 mins. at
room temperature. After drawing the solution through the
membrane-bottomed plate by vacuum, the filter was washed (x3) by
adding 200μl of phosphate-buffered saline (PBS) pH 7.4,
containing 0.5% (v/v) Tween-20 (PBS-T) per well under vacuum.
Any remaining active sites on the membrane were blocked by
incubating with 1%(w/v) dried milk powder dissolved in PBS
(200μl/well) for 15 mins. at room temperature. The wells were
washed as before and samples (100μl/well) containing the
appropriate antibody added and incubated for 15 mins. at room
temperature. The wells were washed to remove unbound material
as above and horseradish peroxidase-labelled detecting antibody
(100μl/well) in PBS was then added. Following an incubation for
15 mins. the plates were washed as before, and activity of the
bound enzyme conjugate revealed by the addition (50μl/well) of
ECL signal reagent (Amersham). Chemiluminescence was measured
immediately in a microtitre plate luminometer (Amersham).

Quantitative detection of human immunoglobulins.
Membrane-bottomed plates were coated with polyvalent goat
anti-human immunoglobulin (IgG fraction; 10μg/ml in PBS;
200μl/well; Sigma), and incubated for 30 mins. at room
temperature.The filter was washed with PBS-T and blocked as
described previously. The membrane was washed as before, human
immunoglobulin standards (0-100ng/well) (IgG, IgA: Sigma;
IgM:Calbiochem) added and incubated for 60 mins. at room
temperature. The wells were washed with PBS-T and 100μl of
diluted (in PBS) peroxidase-labelled goat anti-human IgG
(1:250;Sigma), peroxidase-labelled goat anti-human IgM (1:500;
Sigma) and peroxidase-labelled goat anti-human IgA (1:500;
Sigma) added, as appropriate. The labelled antibodies were
incubated for 60mins. at room temperature, the wells washed with
PBS-T and enzyme activity measured by chemiluminescence as
before.

RESULTS AND DISCUSSION
Representative dilution curves for the detection of monoclonal
antibodies to human chorionic gonadotrophin (hCG) (3) and rabbit
polyclonal antibodies to cyclic AMP (from a commercial cAMP
immunoassay kit; Amersham) are shown in figs. 1 & 2. The
monoclonal antibodies to hCG diluted 1/2000 and the rabbit
polyclonal anti-cAMP antibodies diluted 1/200000 still exhibited
activities greater than background values. Control experiments
involved the use of monoclonal antibodies and rabbit hyperimmune
antisera to irrelevant proteins.

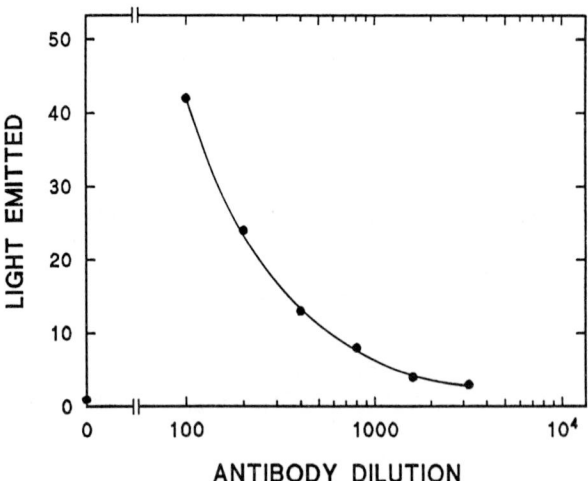

Fig.1. Enhanced chemiluminescence detection of antibodies to
hCG. Membrane-bottomed plates were coated with hCG (10μg/ml in
PBS), washed and blocked before application of anti-hCG
(1:100-1:3200 in PBS). The plates were washed and diluted

(1:500) peroxidase-labelled sheep anti-mouse Ig (Amersham) added. The plates were washed and activity of the bound enzyme measured by chemiluminescence.

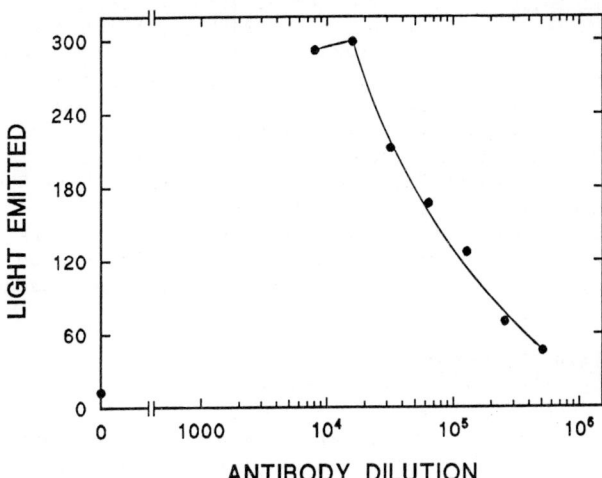

ANTIBODY DILUTION

Fig. 2. Enhanced chemiluminescence detection of antibodies to cyclic-AMP. Cyclic AMP (0.25μmoles; 10μmoles/ml) was conjugated to mouse IgG (0.025-μmoles) through an activated N-hydroxysuccinimide ester (4). The ester was prepared by reaction of the carboxylic acid group of 2'-0-monosuccinyladenosine-3'-5'-cyclic monophosphate (Sigma) with N-hydroxy-succinimide (Sigma) in the presence of 1-ethyl-3 (3-dimethylaminopropyl) carbodiimide hydrochloride (Sigma). Membrane-bottomed plates were coated with the cyclic AMP conjugate (0.025μmoles/ml, 200μl/well) in PBS and excess removed by applying a vacuum. The membrane was washed and blocked as described in the text before application of diluted anti-cyclic-AMP (1:800-1:500000 in PBS). The plates were incubated for 15 mins., washed and diluted (1:500 in PBS) donkey anti-rabbit Ig (Amersham) added. Following an incubation for 15 mins., the plates were washed and activity of the bound enzyme measured by chemiluminescence.

Dose-response curves for the detection of human IgG, IgM and IgA are shown in fig. 3. From these curves, as little as 5ng of human immunoglobulins can be detected.

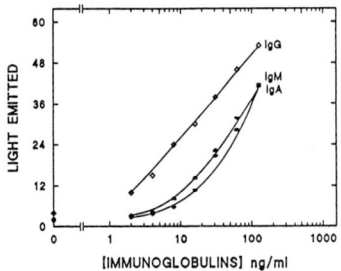

<u>Fig.3.</u> Enhanced chemiluminescence detection of human
immunoglobulins. Standard curves are shown for human IgG(◇);
human IgA(●); and human IgM(◯). The data shows the mean of
quadruplicate determinations. Experimental detail is outlined
in the text.

The immunofiltration assay procedure described here is simple
and rapid to perform and has several key advantages over
conventional techniques such as ELISA or RIA. In contrast to
microtitre plate based ELISA, using the single or ten station
vacuum manifolds, any number of murine or rabbit antibody
samples, between 1 and 960, can be processed in about 90
minutes, reducing both workload and cost per test. Furthermore
the technique has all of the advantages of ECL assays. For
example, the rapid signal generation compares very favourably
with the 30 to 60 mins. needed for many enzyme immunoassays. In
addition, the technique has none of the problems of radioactive
waste disposal and does not suffer from short isotopic
half-life. Finally, the procedure can be readily adapted for
the quantitative detection of molecules of biological interest,
such as human immunoglobulins.

REFERENCES:
1.   O.E. IJsselmuiden, P. Herbrink, M.J.M. Meddens, B. Tank,
      E. Stolz. and R.V.W. Van Eijk, Optimizing the
      solid-phase immunofiltration assay. <u>J. Immunol.Methods.</u>
      119, 35-43. (1989).

2.   T.P. Whitehead, G.H.G. Thorpe, T.J.N. Carter, C. Groucatt,
      and L.J. Kricka, Enhanced luminescence procedure for
      sensitive determination of peroxidase-labelled conjugates
      in immunoassay. <u>Nature.</u> 305, 158-159. (1983).

3.   J.K. Horton, O.M. Evans, K. Swann, and S.A. Swinburne,
      New and rapid method for the selection and cloning of
      antigen-specific hybridomas with magnetic microspheres. <u>J.
      Immunol.Methods.</u> 124, 225-230. (1989).

4.   B.F. Erlanger, Principles and methods for the preparation
      of drug protein conjugates for immunological studies.
      <u>Pharmacological Reviews.</u> 25, 271-280. (1973).

PAPILLOMAVIRUS QUANTIFICATION USING ASYMMETRIC AMPLIFICATION AND A RAPID BIOLUMINESCENT ASSAY.

P. Balaguer, B. Térouanne,A.M. Boussioux and J.C. Nicolas

INSERM Unité 58, 60, Rue de Navacelles 34090 Montpellier France

INTRODUCTION

Polymerase chain reaction is widely used to detect specific DNA sequences and the amplified DNA sequences are commonly identified using oligonucleotide probes by conventional dot blot hybridization or by hybridization in solution, the preferred format for clinical analysis (1,2). However, accuracy of the detection and the ability to quantity the amounts of viral sequence were found to be significantly influenced by the method employed for hybridization.

When hybridization is done in solution, it appears that reannealing of the target reduced the hybridization yield with oligonucleotide probes. We thus modified the PCR reaction protocol in order to produce single strand DNA (3), this enabled us to use oligonucleotide probes in solution. In these conditions, using a channeling enzymatic reaction previously described for immunoassays or biotin assays (4) and a small amount of G6PDH-labelled oligonucleotide we were able to perform a rapid and sensitive assay of viral DNA without having to carry out a separation step. Streptavidin or specific oligonucleotides were used to capture the DNA hybrid on a bioluminescent support (luciferase and oxidoreductase from marine bacteria coimmobilized on Sepharose)

MATERIALS AND METHODS

Reagents. FMN, NADH, G6PDH and streptavidin were obtained from Boehringer, BrCN activated Sepharose was from Pharmacia. Oligonucleotides were made on an Applied Biosystem synthesizer and an amino group was added to the 5' end using Aminolink 2. Oligonucleotides were labelled with biotin or G6PDH as previously described (4). Immobilization of streptavidin, FMN-oxidoreductase and luciferase on Sepharose was performed as described in (4), oligonucleotide adsorbent was synthesized using 100 μg of amino oligonucleotide per g of BrCN activated Sepharose.

DNA amplification. The amplified sequence was HPV18 specific and the following oligonucleotides were used as amplification primers labelled or not with biotin N Hydrosuccinimide ester :

  5' Aminolink GCTGGTAAATGTTGATGATTAACT3'
  5'Aminolink CGACAGGAACGACTCCAACG3'

The symmetric amplification was run as described by Saiki (5) in 100 μl solution containing 0.5 μmol/l of primers, 1 μg of genomic DNA, 0.2 mmol/l dATP, dCTP, dGTP, dTTP in reaction buffer: 16.6 mmol/l $(NH4)_2SO_4$, 67 mmol/l Tris HCl pH 8.8, 6.7 mmol/l $MgCl_2$ 10 mmol/l β mercaptoethanol, 6.7 mmol/l EDTA, 0.1% Triton X100 and 0.17 g/l gelatine. Samples were heated for 2 min at 95°C to denature the DNA. Two units of Taq polymerase (Promega) were added to each sample and incubated at 55°C for 1 min (annealing) and 2 min at 72°C for primer elongation. Further rounds, including1 min denaturation, 1 min annealing and 2 min synthesis, were continued for 30 cycles.

For asymmetric PCR, oligonucleotide concentrations were 500 nmol/l for the primer present in excess and 10 to 100 nmol/l for the second primer.

Bioluminescent detection The following oligonucleotide 5'NH2-TAAGGCAACATTGCAAGACA was labelled with G6PDH. Hybridizations were performed in solution by incubating the biotinylated target with 0.1 mIU (5 femtomoles) of an oligonucleotide-G6PDH in 0.25 mol/l NaCl, 0.1 mol/l phosphate buffer pH 7, 1g/l BSA and 10 mg/l salmon sperm DNA. The hybrid was incubated at 37°C with 0.5 mg of streptavidin bioluminescent-Sepharose.

Amplified sequences were also quantified by hybridization with two different oligonucleotides, one labelled with G6PDH and the other one immobilized on a bioluminescent Sepharose. The sequence of immobilized oligonucleotide was : 5'Aminolink ATGAAATTCCGGTTGACCT.

In both cases, the bound enzyme was detected by a channeling reaction leading to light emission (Fig1). The bioluminescent reaction was triggered by the NADH produced by G6PDH linked to oligonucleotide. NADH produced by the label in excess was oxidized in solution by pyruvate in the presence of lactate dehydrogenase. This process did not require any separation step since the unbound enzyme does not lead to light emission. The luminescent reaction was started by adding a solution containing: 1mmol/l NAD, 3 mmol/l G6P, 3 mmol/l sodium pyruvate, 10^{-2} mmol/l FMN, 6×10^{-2} mmol/l decanal and 30 mIU/ml LDH in 0.1 mol/l phosphate buffer, 0.25 mol/l NaCl, and 1g/l BSA.

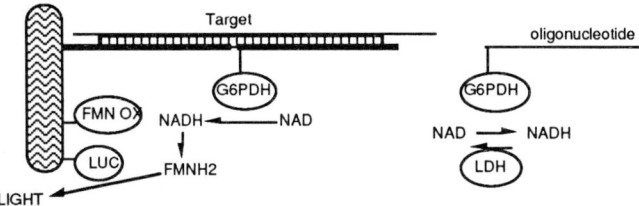

Fig 1 : Principle of the bioluminescent DNA assay.

RESULTS AND DISCUSSION

Detection with streptavidin adsorbent. Total DNA (1μg) extracted from HELA cells was amplified and various volumes of the PCR amplification medium were assayed. The hybridization reaction between target DNA and oligonucleotide-G6PDH was maximal after 2 h incubation at 37°C and 10 pg of oligonucleotide probe (2 femtomoles) gave the best signal/background ratio.

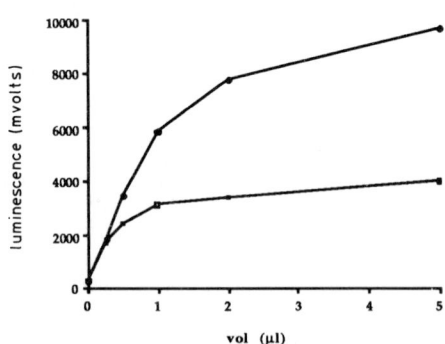

Fig 2 : Detection by hybridization in solution of amplified biotinylated DNA.

1 μg of DNA from HELA cells was amplified for 30 cycles with a biotinylated primer (lower curve symmetric amplification, upper curve asymmetric amplification). Various amounts of amplification medium were hybridized with 0.1 mU of oligonucleotide-G6PDH conjugate in 50 μl of hybridization buffer at 37°C for 2h.

Fig 2 shows results obtained with various amounts of amplification medium.These results confirm that hybridization in solution of an oligonucleotide with a double strand DNA target lead to lower hybrid yield. The amount of DNA produced by symmetric amplification was 10-fold higher than that produced in asymmetric conditions (calculated with ^{32}P dCTP being used as tracer) and we obtained higher luminescence values for asymmetric amplification. The plateau was due to the low concentration of probe used for the hybridization reaction.

Detection with oligonucleotide adsorbent. Fig 3 shows results obtained with various amounts of HPV sequences added to human DNA. For low amounts of target, symmetric PCR produced more DNA than asymmetric amplification, giving a signal which rapidly reached a plateau for target quantities greater than 10^4 molecules. With asymmetric PCR the initial target detection limit was 100 molecules and the signal increased up to 10^7 molecules. In asymmetric conditions, the hybridization can be done directly on the medium and without denaturation of the DNA target.

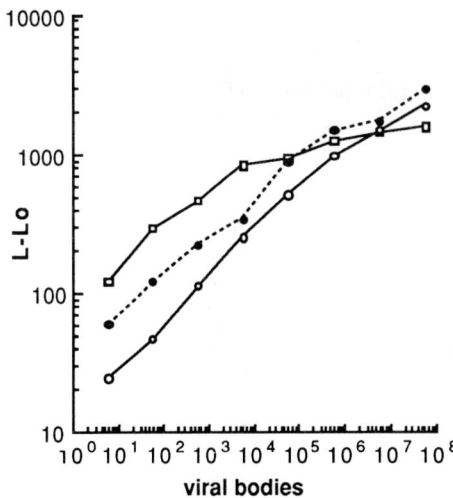

Fig 3 : Bioluminescent assay of HPV DNA added to cellular DNA using a oligonucleotide bioluminescent adsorbent. □---□ symmetric amplification ●--● asymmetric amplification and denaturation O--O asymmetric amplification.

The bioluminescent assay can be used to detect and quantify DNA sequences produced by the PCR. Streptavidin bioluminescent adsorbent performed as well as the oligonucleotide bioluminescent adsorbent. Even at 37°C the two methods were very specific, the labelled oligonucleotide did not bind to non-specific biotinylated DNA. However, we found that asymmetric amplification gave very different amplified product yields for different DNA sequences and some of them did not produce single strands even in the presence of an excess of one oligonucleotide primer. Works to improve the production of single strand nucleic acid (DNA or RNA) are in progress in our laboratory.

ACKNOWLEDGEMENTS
This work was supported by the French Ministry of Education (Grant 88 M 0886).

REFERENCES

1   L. Harju, P. Jänne, A. Kallio, M.L. Laukkanen, I. lautenschlager, S. Mattinen, A. Ranki, M. Ranki, V.R.X. Sorez, H. Söderlund and A.C. Syvänen, Affinity-based collection of amplified viral DNA : application to the detection of human immunodeficiency virus type 1, human cytomegalovirus and human papillomavirus type 16. Mol. Cell Probes.4 223-235 (1990).

2   S. Sauvaigo, B. Fouqué, A.Roget, T. Livache, H. Bazin, C. Chypre and R. Téoule, Fast solid support detection of PCR amplified viral DNA sequences using radioiodinated or hapten labelled primers. Nucleic Acids Res. 18 3175-3183 (1990).

3. Gillensten, U.B., and Erlich, H.A. Generation of single-stranded DNA by the polymerase chain reaction and its application to direct sequencing of the HLA-DQA locus. Proc. Natl. acad. Sci. USA. 85, 7652-7656. (1988).

4   B.Térouanne, M. Bencheick, P. Balaguer, A.M. Boussioux and J.C. Nicolas, Bioluminescent assays using glucose 6 phosphate dehydrogenase: Application to biotin and streptavidin detection. Anal. Biochem. 180, 40-49.(1989)

5   R.K. Saiki, D.H.Gelfand, S. Stofffel, S.J. Scharf, R. Higuchi, G.T.Horn, K.B.Mullis, and H.A Erlich. Primer-directed enzymatic amplification of DNA with a thermostable DNA polymerase. Science 239, 487-494, (1988).

BIOLUMINESCENT COMPETITIVE DNA ASSAY :
A NEW METHOD FOR HLA TYPING

B. Térouanne, P. Balaguer, J.F. Eliaou[1], A. M. Boussioux and J.C. Nicolas.

Inserm Unité 58, 60, rue de Navacelles 34090 Montpellier, France and [1]Inserm Unité 291, Laboratoire d'Immunologie, Hôpital Saint Eloi, 34000 Montpellier, France.

INTRODUCTION.

DNA amplification using Polymerase Chain Reaction (PCR) has allowed a new approach of existing methods for the characterization of DNA sequences. Sequences can be labelled during the PCR using radioactive or biotinylated nucleotides or oligonucleotides bearing biotin, fluorophore or haptens (1). These amplified products were detected according to the method described by Syvanen (2). A competitive hybridization with fluorescent probes has also been described for the detection of PCR-amplified products (3).

PCR, followed by denaturating gradient gel electrophoresis allowed the study of genome polymorphism and the detection of point mutations (4). And recently, the single-strand conformation polymorphism analysis of PCR products (PCR-SSCP) was developed and used for the detection of gene mutations (5).

We describe here a Bioluminescent Competitive DNA Assay to identify mutations in PCR amplified DNA sequences. This method examined the potentiality of hybridization after denaturation between a double-labelled double strand DNA and samples of PCR products having homologous sequences to the probe or several mismatches.

Using four lines cells having a polymorphic part of the HLA DQB gene, we constructed double-labelled sequences obtained by PCR using biotinylated and FITC oligonucleotide primers. Each probe was denatured and re-hybridized either in the presence of the homologous amplified and non-labelled DNA sequence or in the presence of the four others having some mismatches with the probe. The amount of double-labelled molecule remaining after hybridization was captured by a bioluminescent immunosorbent, (anti-FITC γ globulin, luciferase and oxidoreductase, from Beneckea harveyi, coimmobilized on Sepharose) and detected by a streptavidin-G6PDH conjugate (6). We showed that a total homology sequence between probes and targets was required to obtain a complete disappearance of the luminous signal.

MATERIALS AND METHODS

Reagents Enzymes (G6PDH-LDH) and substrates for bioluminescent assays were obtained from Boehringer.CNBr- activated Sepharose 4B and Sephadex G25 columns were purchased from Pharmacia. DNA amplifications and competitive DNA assays were performed on a programmable thermal controller (PTC 100 - Prolabo.) Taq polymerase and 10X Taq polymerase buffer were obtained from Promega.(10X = 500 mmol/l KCl, 100 mmol/l Tris HCl pH 9 at 25°C, 15 mmol/l MgCl$_2$, 0,1% gelatin and 1% triton X 100).

Oligonucleotide primers were synthesized on an Applied Biosystems oligonucleotide synthesizer (391 DNA synthesizer). A 5'amino-group was added to the PCR primers P3 and P4 with the amino-link II reagent (Appled Biosystems).The amino-group were biotinylated using biotinyl ε amino caproic acid N hydroxy succinimide ester (Boehringer) or fluoresceinated using fluorescein isothiocyanate (Sigma).

Bioluminescent measurements were done with an automated luminometer (LKB 1251) coupled to an Apple IIe microcomputer. Southern blots were performed on Immobilon N transfer membrane (Millipore) using a semi-dry transfer cell (Biometra).
Ascitic fluid containing mouse anti-FITC monoclonal antibodies were submitted to 50% ammonium sulphate precipitation and immunoglobulin solution was chromatographed on DEAE trisacryl M.The luciferase oxidoreductase enzyme system was coimmobilized on Sepharose 4B with 2 mg anti-FITC γ globulin antibody according to the previously described protocol (6).
Synthesis of the double-labelled DNA hybrid. The sequences of the oligonucleotide primers used for the two PCR steps are given in Table 1.

Primer	Codon position	Sequence	Size of amplified products
P1	6-13	5'GATTTCGTGTACCAGTTTAAGGGC3'	
P2	79-86	5'ACGTCTGTGTTGTTGATGCTCCACC3'	240
P3	39-45	5'CGCTTCGACAGCGACGTGGG3'	
P4	60-66	5'TTCTTCTGGCTGTTCGAGTA3'	87

Table 1 : Sequences of oligonucleotides used as primers in amplification reactions.

a) A first amplification was performed on 1 µg of different genomic HLA DNAs in 50 µl containing 200 µM of each dTNP, 0.2 µmol/l of P1 and P2 , 1X Taq polymerase buffer and 5 units of Taq polymerase from Promega The DNA sample was finally added to each tube before 100 µl of mineral oil to avoid evaporation. The reaction was carried out for a total of 30 cycles for 1 min at 94° C, 1 min at 55° C and 2 min at 72°C.
b) Then, after a 50-fold dilution, 5 µl of the first amplifications were submitted to a second amplification with the internal biotinylated and fluoresceinated primers (P3 and P4). The amplified sequences had 87 bp and presented several between-sequence mismatches. PCR was carried out in 200 µl with the reagents described above.
Purification of the double-labelled sequence.The DNA fragments amplified with labelled primers (biotin and FITC primers) were recovered from a 10% polyacrylamide gel as described in Maniatis (7) with the following modification : the gel was ground with 0.2 ml buffer in a 1.5 ml microfuge tube with a pellet piston, the acrylamide gel was discarded by centrifugation and the DNA was recovered by ethanol precipitation. The purity of this probe and the presence of the two labels were tested on an aliquot by Southern blot after gel acrylamide electrophoresis. Biotinylated-DNA was detected by streptavidin coupled to G6PDH and FITC-DNA by an anti-FITC monoclonal antibody.
Determination of the concentration of the double-labelled sequence DNAs. All dilutions and incubations were carried out in buffer A (0.1 mol/l phosphate buffer pH 7.2 with 0.2 mol/l NaCl, 0.1% gelatin and denatured sonicated salmon sperm DNA (10 µg/ml). The concentrations of these probes were determined by a sandwich assay. A standard curve was obtained using a synthetic double strand (a 20 mer biotinylated oligonucleotide hybridized with its 20 mer complementary oligonucleotide labelled with FITC) : 50 µl of biotinylated-fluoresceinated dimer dilutions (10^{-9} to 10^{-12} mol/l) were incubated with 50 µl of bioluminescent adsorbent (1 g of wet adsorbent suspended in 20 ml of buffer A) and 50 µl of streptavidin-G6PDH (2 mUI/ml). Incubation was carried out at room temperature for 2 h under constant shaking. The luminescent reagent was added : 1mmol/l NAD, 3.mmol/l G6P, 3 mmol/l sodium pyruvate, 10^{-2} mmol/l FMN, 6 10^{-2} mmol/l decanal and 30 mIU/ml LDH in buffer A. After 10 min, photon counts were integrated over 10 sec.
Competitive Bioluminescent DNA Assay procedure. We constructed 4 different probes from 4 genomic HLA DNAs. We also amplified the same sequences with non-labelled primers. The sequences are given below, P3 was the sense primer, P'4 was the complementary antisense primer.
P3 G GTG TAC CGG GCA GTG ACG CCG CAG GGG CGG CCT **GTT** GCC GAG P'4
P3 G GTG TAC CGC GCG GTG ACG CCG CAG GGG CGG CCT **GAT** GCC GAG P'4
P3 G GTG TAT CGG GCG GTG ACT CCG CAG GGG CGG CCT **GAC** GCC GAG P'4
P3 G GTG TAC CGG GCG GTG ACG CCG CAG GGG CGG CCT **AGC** GCC GAG P'4

Different amounts of amplification product (1 to 10 µl) were incubated in 100 µl of buffer A with 10 femtomoles of the homologous sequence labelled by biotin and FITC and purified as described above. The incubation was performed for 5 min at 95°C and then the temperature was decreased (1° per min). When the incubation temperature was 40°C, the samples were microfuged and 50 µl of the incubation medium was added to luminometer tubes with 50 µl of streptavidin-G6PDH and 50 µl of bioluminescent adsorbent as described above.

RESULTS AND DISCUSSION

<u>Preparation of the labelled probes</u> : Two steps of PCR amplification were performed to obtain the production of DNA hybrids carrying a biotin at one 5' end and a fluorescein at the other 5' end. This procedure has been used by numerous authors (1-2). Indeed, when the amplification is performed on genomic DNA or crude preparations, non-specific products can be synthesized during the first amplification. During the second step, these products are higly diluted and cannot produce any interactions.

It was necessary to perform a further purification on these probes, because in some cases we have observed the formation of dimer-primers. Several purifications were performed : Centricon 30, Mermaid kit, filtration on ultrafree-MC 30000, electrophoresis on acrylamide gel. This last procedure gave us the best and the quickest purification. It was also necessary to quantify these probes. Fig 1 shows a standard dose reponse curve for a double-labelled 20 mer hybrid. The detection limit was 20 attomoles but to have a good signal for the competitive DNA assay, we used between 1 and 20 femtomoles of the probes.

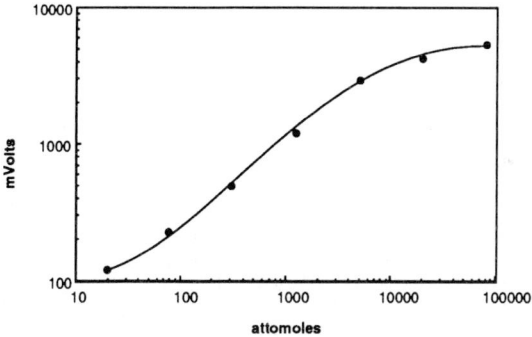

Fig 1 : Calibration curve of a double-labelled DNA

<u>HLA typing by bioluminescent competitive DNA Assay</u> : Amplified sequences and double-labelled probes were denatured and submitted to reannealing in conditions which prevent the formation of heteroduplex. The gradient temperature (1° per min) was chosen to obtain a slow annealing rate. In these conditions the reactions were very specific as shown in Table 2.

no probe	probe	probe incubated without target	probe 1 and target 1	probe 1 and target 2
100	6 500	6 000	155	2 900

Table 2 : Luminescence values (mvolts) obtained with a double-labelled probe

When the probe was incubated with the homologous sequence (not labelled and in 50-fold excess) the competition was complete and the signal was equal to the background value. On the other hand, when there were some mismatches between the DNA sequences, the double-labelled DNA was reformed during the temperature gradient. However reannealing of the probe in the presence of non-

homologous target was not complete, the hyridization conditions could be improved to obtain a
better hybridization yield.
Table 3 shows the percentage of reannealing of different probes in the presence of different DNA
targets. The reannealing percent was less than 5 when the sequence was homologous and between
30 and 60 when sequences were different. The DNA X was a heterozygote (1-4) and competed with
the reannealing of the two probes.

Target -> probe	1	2	3	4	X
1	**0**	51	34	48	3
2	30	**2**	63	44	39
3	33	40	**0**	35	41
4	35	41	50	**0**	5

Table 3 : % of reannealing of the probes in the presence of homologous or heterologous sequences

Numerous strategies have been developed for rapid detection of DNA mutations. Same sized
amplified products but with mismatches are more often detected with different electrophoresis
methods. This new method could permit a rapid and non-radioactive screening. Purified labelled
probes are very stable at -20°C for several months and the bioluminescent competitive DNA assay,
which does not require any separation step, could be performed in three hours.
Many improvements of this method remain to be done. We are studying the incubation conditions
(temperature, and probe or target concentrations), and the importance of the position and the
number of mismatches on the sequence.

ACKNOWLEDGEMENTS
This work was supported by the French Ministry of Education (Grant 88 M 0886). We thank Paul
Tchen and J.F. Vautherot for the gift of ascitic fluid containing mouse anti-FITC monoclonal
antibodies.

REFERENCES
1   S. Sauvaigo, B. Fouqué, A.Roget, T. Livache, H. Bazin, C. Chypre and R. Téoule, Fast
      solid support detection of PCR amplified viral DNA sequences using radioiodinated or
      hapten labelled primers. Nucleic Acids Res. 18, 3175-3183 (1990).
2   A.C. Syvanen, K. Aalto-Setälä, K. Kontula and H. Söderlund, Direct sequencing of
      affinity-captured amplified human DNA application to the detection of apolipoprotein E
      polymorphism. FEBS Letters 258, 71-74 (1989).
3   L.E. Morrison, T.C. Halder and L.M. Stols, Solution-phase detection of polynucleotides
      interacting fluorescent labels and competitive hybridization. Anal. Biochem. 183, 231-
      244 (1989).
4   O. Attree, D.Vidaud, M.Vidaud, S. Amselem, J.M. Lavergne and M. Goossens,
      Mutations in the catalytic domain of human coagulation factor IX: rapid characterization
      by directgenomic sequencing of DNA fragments displaying an altered melting behavior.
      Genomics 4, 266-272 (1989)
5   M. Orita, Y. Suzuki, T. Sekiya and K.Hayashi, Rapid and sensitive detection of points
      mutations and DNA polymorphism using PCR Genomics 5,874-879 (1989).
6   B.Térouanne, M. Bencheick, P. Balaguer, A.M. Boussioux and J.C. Nicolas,
      Bioluminescent assays using glucose 6 phosphate dehydrogenase: Application to biotin
      and streptavidin detection. Anal. Biochem. 180, 40-49.(1989)
7   J. Sambrook, E.F. Fritsch and T. Maniatis, Gel Electrophoresis of DNA in Molecular
      Cloning I. C. Nolan (Ed), Cold Spring Harbor Laboratory Press, 6, 46-48.(1989)

FLASH™ CHEMILUMINESCENT SYSTEM FOR SENSITIVE, NONRADIOACTIVE DETECTION OF NUCLEIC ACIDS

A. St. Louis, K. Considine and J.C. Braman

Stratagene, 11099 North Torrey Pines Road, La Jolla, CA 92037 USA

INTRODUCTION

The advent of chemiluminescent substrates for alkaline phosphatase has resulted in the development of sensitive nonradioactive protocols for the detection of nucleic acids (1,2). Most of the detection systems rely on the hybridization of biotinylated probes (produced by nick translation) to target nucleic acids and subsequent hybridization with avidin-alkaline phosphatase.

We have produced biotinylated DNA probes within ten minutes employing the Flash™-Prime-It™ random primer labeling kit (3). Probes made with this kit reveal single copy genomic sequences when used in conjunction with the Flash chemiluminescent detection system and Flash nylon membranes. The level of sensitivity rivals that of radioactivity and avoids the inconvenience of radionucleotide handling and disposal. The Flash detection system is also ideal for the detection of mRNA by Northern blot hybridization.

MATERIALS AND METHODS

Flash™ Prime-It™ and Flash™ Detection kits were obtained from Stratagene (La Jolla, CA).

RESULTS AND DISCUSSION

Random Primers Labeling of Probes   Kinetic experiments have shown that T7 DNA Polymerase (T7 Pol) and random nonamer primers (9-mers), components in Stratagene's Prime-It™ random primers labeling kit, produce radioactive DNA probes with specific activities greater that 1×10^9 dpm/μg in two minutes (4). These results, in conjunction with evidence suggesting that this enzyme efficiently uses modified nucleotides as substrates (5,6), led to the hypothesis that T7 Pol may also be used to generate biotinylated probes using nucleotide analogs such as bio-11 dUTP, an analog of dTTP.

To test this hypothesis, the incorporation rate of bio-11 dUTP was estimated using 50 μCi ^{32}P-dCTP as a tracer, essentially as described by Langer (7). Fig. 1 demonstrates that the initial incorporation rate for bio-11 dUTP was slightly lower than the rate of dTTP alone and for bio-11 dUTP/dTTP mixtures. As the ratio of bio-11 dUTP to dTTP decreased in the reaction mixture, the incorporation rate tended to increase. This same effect was described for DNA polymerase I holoenzyme (7). Despite this slightly lower initial rate, bio-11 dUTP incorporation after 10 minutes was estimated to be 38% of the total radioactivity added to the reaction mixture. This corresponds to a specific radioactivity of 1.2×10^9 dpm/μg when using 50 μCi of ^{32}P-dCTP (3000 Ci/mmol) (8). Fifty-eight percent incorporation was achieved after 10 minutes of incubation when bio-11 dUTP was one third of the total dTTP concentration in the reaction mixture. This incorporation level corresponds to a specific radioactivity of 1.6×10^9 dpm/μg.

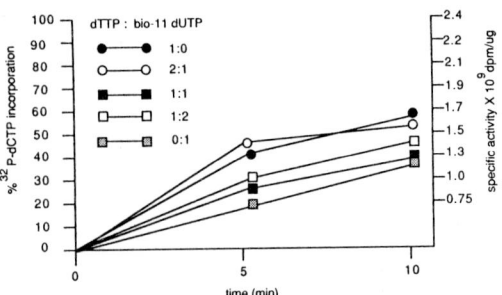

Figure #1. <u>Incorporation of bio-11 dUTP by T7 DNA Polymerase</u> 25 ng of a 3 kb denatured DNA template, random 9-mer primers, reaction buffer (containing dATP and dGTP), 50 µCi ^{32}P-dCTP (3000 Ci/mmole), and 2 units of T7 DNA pol were added to various ratios of dTTP:bio-11 dUTP in separate reaction mixtures. Incubations were performed at 37°C. One µl aliquots from each reaction mixture were removed at 5, 10, 20, and 30 min. and added to 19 µl of 0.2 M EDTA to stop the reactions. Four µl of the diluted samples were spotted onto DE 81 filter paper discs. The discs were washed 3 times for 5 min. each in 2 X SSC and once in EtOH, dried under a heat lamp and counted. The counts obtained for each timepoint were divided by the total counts added to each reaction mixture to obtain the percent incorporation. Specific activity was calculated as described previously (8).

A qualitative measure of bio-11 dUTP incorporation was made by spotting aliquots of the probes described in Fig. 1 onto Duralon-UV™ nylon membranes. The biotinylated DNA was detected using an avidin-alkaline phosphatase conjugate and the phosphatase substrate 5-bromo-4-chloro-indolyl phosphate (BCIP) and nitroblue tetrazolium (NBT). A blue signal was observed in all of the biotinylated DNA spots with the exception of the control spot containing probe DNA prepared without bio-11 dUTP (data not shown). The signal strength increased as the concentration of bio-11 dUTP in the reaction mixture increased. These results confirm the ability of T7 DNA Pol to incorporate this biotinylated nucleotide using the Prime-It random primers kit protocol.

<u>Chemiluminescent Detection</u> Nonradioactive detection of target nucleic acids immobilized on a solid support, such as a nitrocellulose membrane, has commonly been accomplished by hybridization with a nick-translated biotinylated nucleic acid probe. Following hybridization to the target, the probe is coupled to either and avidin-enzyme complex or an antibody-enzyme conjugate (7,9). The target nucleic acids are then revealed by incubating the membrane with an appropriate substrate for the enzyme. Alkaline phosphatase has proven to be a popular enzyme for the generation of signal, and the phosphatase substrate 5-bromo-4-chloro-3-indolylphosphate (BCIP), in combination with nitro-blue tetrazolium (NBT), has been used extensively. The major drawback to the BCIP/NBT substrate is lack of sensitivity in revealing single-copy or low abundance target sequences. This is mainly due to the inhibition of the enzyme as the product dye deposits on the membrane surface.

Recently, chemiluminescent substrates for alkaline phosphatase have been synthesized which allow fewer enzyme molecules to be detected (1,2). Product inhibition of the enzyme does not occur, translating into increased sensitivity in detection of nucleic acid sequences. In addition, improved hybridization and washing protocols have been developed for nylon membranes which minimize the background usually associated with the nonradioactive method. Nylon characteristically has a higher nucleic acid binding capacity than nitrocellulose and this also improves the sensitivity of detection. We have therefore employed the use of a chemiluminescent substrate (Lumigen™ PPD) to develop a system useful for sensitive and rapid nonradioactive detection of nucleic acid target sequences. The substrates are 1, 2-dioxetanes which are dephosphorylated by the action of alkaline phosphatase and subsequently decompose with the generation of light. Light emanating from a target-probe-alkaline phosphatase complex can be detected by exposure to radiographic film.

Flash™-Prime-It™ Labeling and Detection System The DNA fragment 7C22, which identifies a single copy RFLP near the cystic fibrosis gene, was labeled with the FLASH-Prime-It random primer labeling kit. The biotinylated DNA was then used as a hybridization probe in a Southern blot experiment. Detection was accomplished with the FLASH Detection system. A FLASH nylon membrane containing 10 μg of fractionated human genomic DNA was hybridized with 5 ng/ml of biotinylated 7C22. The membrane was washed, blocked, and then hybridized with an avidin-alkaline phosphatase conjugate. The blot was treated with the chemiluminescent substrate for alkaline phosphatase, and exposure to radiographic film for 30 minutes produced the result shown in Fig. 2. The fragment sizes detected were 7.2 and 5.1 kb respectively, corresponding to 24 and 17 pg of these single copy target sequences.

Figure #2. Southern Blot Analysis Using Chemiluminescent Detection Ten μg of human DNA were digested with 50 units of EcoR I and then loaded onto a 0.8% agarose gel. Following electrophoresis, the DNA was transferred to a FLASH™ nylon membrane and the blot was hybridized with a randomly primed biotinylated probe (7C22) at a concentration of 5 ng/ml. The blot was washed, blocked, and then incubated with an avidin-alkaline phosphatase conjugate. The membrane was washed again, incubated Assay Buffer (0.1 M diethanolamine, 1 mM $MgCl_2$), and then in Assay Buffer containing 0.25 mM chemiluminescent substrate (Lumigen™ PPD) as outlined in the FLASH™ detection system manual. The blot was exposed to film for 30 min at room temperature.

Northern Blot Analysis A further demonstration of the utility of the FLASH labeling and detection system is shown in Fig. 3. A human alpha-1 antitrypsin probe was used to detect abundant human mRNA transcripts in a transgenic mouse. Four μg each of transgenic mouse liver total RNA, and normal mouse liver total RNA, and 250 ng each of poly(A)-selected RNA from both transgenic and normal mouse liver were fractionated, in duplicate, through a 1% formaldehyde-agarose gel and transferred to a FLASH nylon membrane. The Northern blot was hybridized with a FLASH-Prime-It labeled alpha-1 antitrypsin probe and target transcripts detected as described in Fig. 2. Human-specific transgenic transcripts from both total RNA (lanes 1 and 2) and poly(A)-selected RNA (lanes 5 and 6) were detected with a 30 minute exposure to radiographic film. Nonspecific hybridization did not occur in the lanes containing mouse transcripts exclusively (lanes 3 and 4 for total normal mouse RNA and lanes 7 and 8 for poly(A)-selected normal mouse RNA).

1 2 3 4 5 6 7 8

Figure #3. <u>Northern Blot Analysis Using Chemiluminescent Detection</u> Four µg of total RNA (lanes 1 and 2: human alpha-1 antitrypsin transgenic mouse liver, lanes 3 and 4: normal mouse liver) and 250 ng of poly(A)-selected RNA (lanes 5 and 6:human alpha-1 antitrypsin transgenic mouse liver, lanes 7 and 8: normal mouse liver) were fractionated in a 1% formaldehyde-agarose gel and transferred to a FLASH™ nylon membrane. The blot was hybridized with a human alpha-1 antitrypsin probe labeled with the FLASH™-Prime-It™ labeling kit. The blot was washed, blocked, incubated with avidin alkaline phosphatase, washed again and then treated with the chemiluminescent substrate (Lumigen™ PPD) as outlined in the FLASH™ detection system manual. The blot was exposed to film for 30 min at room temperature.

The results described in this paper demonstrate that sensitive, nonradioactive DNA probes can be made rapidly using Stratagene's FLASH-Prime-It random primer labeling and FLASH detection system. These products offer a viable alternative to radioactive detection methods. Single copy genomic sequences and mRNA transcripts can be detected within 30 minutes in most instances, without the hazards associated with radioactive handling.

REFERENCES

1.   A.P. Schaap, Chemical and Enzymatic Triggering of 1, 2-dioxetanes. <u>Photochem. and Photobiol. 47</u>, 505 (1990).
2.   A.P. Schaap, H. Akhavan, and L.J. Romano, Chemiluminescent Substrates for Alkaline Phosphatase: Application to Ultrasensitive Enzyme-Linked Immunoassays and DNA Probes. <u>Clin. Chem. 35</u>, 1863-1864 (1990).
3.   A. Pitta, K. Considine, and J. Braman, Flash™ Chemiluminescent System for Sensitive, Nonradioactive Detection of Nucleic Acids. <u>Strategies, 3</u>, 33-35 (1990).
4.   A. Pitta and J. Braman, DNA Labeling Using the Prime-It™ Random Primer Kit. <u>Strategies, 3</u>, 25 (1990).
5.   J.M. Prober, G.L. Trainor. R.J. Dam, et al., A System for Rapid DNA Sequencing With Florescent Chain-Terminating Dideoxynucleotides. <u>Science, 238</u>, 336-341 (1987).
6.   S. Tabor and C.C. Richardson, DNA Sequence Analysis With a Modified Bacteriophage T7 DNA Polymerase. <u>Proc. Natl. Acad. Sci. USA, 84</u>, 4764-4771 (1987).
7.   P.R. Langer, A.A. Waldrop, and D.C. Ward, Enzymatic Synthesis of Biotin-Labeled Polynucleotides: Novel; Nucleic Acid Affinity Probes. <u>Proc. Natl. Acad. Sci. USA, 78</u>, 6633-6637 (1981).
8.   J. Braman, Prime-It™ Random Primer Kit: Allows Production of High Specific Activity DNA Probes in Two Minutes. <u>Strategies, 2</u>, 56 (1989).
9.   J.J. Leary, D.J. Brigati, and D.C. Ward, Rapid and Sensitive Colorimetric Method for Visualizing Biotin-Labeled DNA Probes Hybridized to DNA or RNA Immobilized on Nitrocellulose: Bio-Blots. <u>Proc. Natl. Acad. Sci. USA</u>. 80, 4045-4049 (1983).

INCREASED SENSITIVITY OF INTERLEUKIN EIA QUANTITATION BY CHEMILUMINESCENT AMPLIFICATION

J. Friberg, J. Glindre and J. Kangasmetsä, 1)
F. Blomberg, A. Loftenius 2) and B. Andersson 2)
Kabi Peptide Hormones R&D S-112 87 Stockholm, Sweden
1) Reserca AB, Box 17007, S-104 62 Stockholm, Sweden
2) National Bacteriological Laboratory, S-105 21
Stockholm, Sweden

INTRODUCTION

The objective of the study was to evaluate to what
extent SALIA (signal amplification of chemilumenescence
in enzyme immunoassay), (Ref.1) could be used to
increase the sensitivity of interleukin quantitation.
The principle of the assay is depicted in Fig.1. The
first step in the test is performed identically to a
conventional enzyme immunoassay (EIA.), (Ref.2). The
colorimetric reading of the conventional method is
however replaced by reading of a light flash that is
released within the first second after the addition of
luminol.
Our method differs from enhanced chemilumenescence
(Ref.3) where the luminol is present from the beginning
of the enzymatic reaction and a slowly emerging light
emission is recorded (see Fig.2).

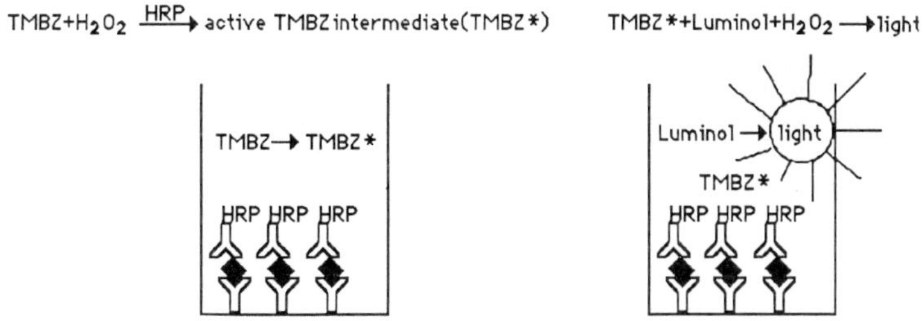

$TMBZ + H_2O_2 \xrightarrow{HRP}$ active TMBZ intermediate (TMBZ*) $TMBZ* + Luminol + H_2O_2 \longrightarrow light$

$TMBZ \rightarrow TMBZ*$

HRP HRP HRP

Luminol $\rightarrow$ light

TMBZ*

HRP HRP HRP

Fig. 1   The SALIA principle: A first step where HRP generates
(15 min, pH 5.5) an intermediate product from TMBZ,
and a second step where NaLuminol is added (at pH 8.5)
A flash-like luminescent signal appeares within 50 ms.

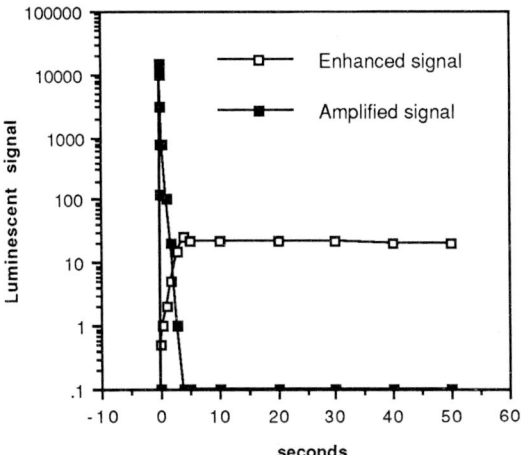

Fig. 2   Comparision of the luminescent
signal for the SALIA system and
the "enhanced system"

MATERIALS AND METHODS

Reagents. EIA reagents were conventional commercial
kits. For Interleukin-6 (IL-6) Quantikine, (Research
and Diagnostic Systems, Minneapolis MN) and for soluble
CD8 (sCD8) Cellfree T8, (T Cell Sciences Inc. Cambridge
MA) was used. Substrate for the HRP conjugates was
tetramethylbenzidine and NaLuminol was prepared from
Luminol (Sigma Chemical Co St Louis MO). For reading,
the microtiter plates were placed in a
chemiluminescence photometer (Luminoscan, Labsystems
Inc., Tyresö, Sweden). Emitted light from each cell was
measured after the addition of 150 μl of 0,5 mM
NaLuminol. The photomultiplier has a spectral response
of 350-680 nm, and the signal was measured at peak mode
during 1.0 sec total time.

RESULTS AND DISCUSSION

We compared the conventional colorimetric EIA with the
SALIA, in two tests used in current clinical immunology
at our laboratory. In Fig.3 it can be seen that
measurement of IL-6 can be made at lower concentrations
with SALIA than with EIA using the respective
commercial test kit for the assays. In the colorimetric
EIA the standard curve for IL-6 leveled off at 30
pg/ml, whereas the SALIA measurements were possible
down to 1 pg/ml.

The standard curve for sCD8 leveled off at 50U/ml when
measured colorimetrically whereas the measuring range
improved down to 1U/ml with the SALIA.
For the quantitation of minute amounts of biological
molecules assays with increased sensitivity are
presently needed. Many of the cytokines, interleukins
and other soluble lymphocyte products are not
detectable in serum and other body fluids with
conventional methods. The method described here allows
the establishment of normal values in serum and other
body fluids for some of these substances by adapting
commercially already available test kits to our
chemiluminescence procedure. Furthermore, now the
possibility also exists to study whether there are
clinical conditions with decreased concentrations of
soluble lymphocyte products previously not detectable.

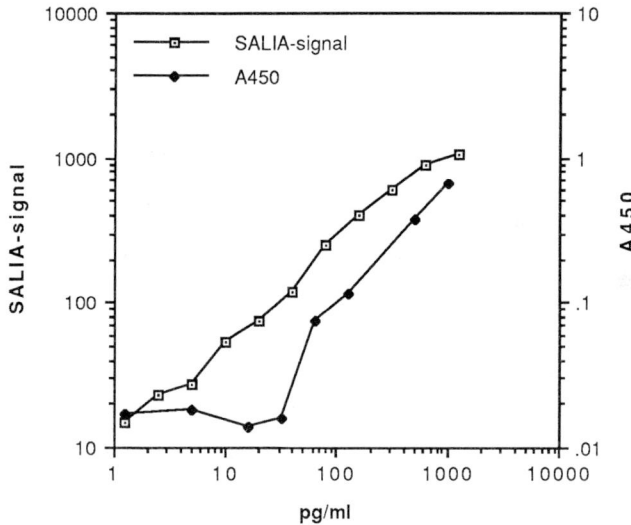

Fig.: 3   Comparison of IL-6 standard curves from
                EIA and SALIA systems.

Table 1 shows values for sCD8 (a T-lymphocyte
activation marker) and for IL-6 (a mediator of the
acute phase reaction) in normal individuals as well as
in people with acute viral and bacterial infections.
The establishment of normal serum values for IL-6 would
otherwise have required a bioassay and the normal
values for sCD8 in cerebrospinal fluid (CSF) would not
have been possible to determine with conventional EIA.
We have compared SALIA adapted assays with the original
commercial and in-house EIA kits. The detection limit,
for a number of cytokines including IL-1 alfa, IL-1
beta, IFN-gamma and sIL-2R is readily increased with a
factor of 10-50.
Table 1.

Clinical condition	Specimen*	n	SCD8**	IL-6**
Healthy	S	18	268±102	21±19
Acute EBV Infection	S	29	7225±4848	45±45
Acute Bacterial Sepsis	S	10	576±393	3036±7997
Healthy	CSF	16	25±18	Not done

*   S = serum
    CSF = cerebrospinal fluid

** detection limit with conventional colorimetric
    EIA is 50 U/ml for SCD8 and 31 pg/ml for IL-6

REFERENCES

1.   Friberg, J. et al. A new principle for
     chemiluminescent signal amplification used in an
     ELISA for determination of human growth hormone.
     In Third International Symposium on Quantitative
     Luminescence Spectrometry in Biomedical Sciences.
     State University of Ghent, Belgium.
2.   S. Avrameas, et al. (Eds) Immunoenzymatic
     Techniques. Elsevier Publ. Comp. (1983).
3.   L.J. Kricka, A.M. O Toole, G.H.G.H. Thorpe and
     T.P. Whitehead. UK Patent Application GB 2162946A
     (1986).

A CHEMILUMINESCENT IMMUNOASSAY FOR 17α 19-NORTESTOSTERONE

L.J. Van Look, E.H.J.M. Jansen[1], R.H. van den Berg[1] and C.H. Van Peteghem

Laboratory of Food Analysis, State University of Ghent
Harelbekestraat 72, 9000 Gent, Belgium
[1]*National Institute of Public Health and Environmental protection, P.O. Box 1, 3720 BA Bilthoven ,The Netherlands*

INTRODUCTION

In a lot of countries of Western Europe , it is well known that 17β 19-nortestosterone (17β-19-NT) and its esters are frequently used products in cattle production. The most important metabolite found in urine is its 17α epimer. An enzyme immunoassay (EIA) on microtitre plates with spectrofotometric detection was already developed (1) to control the misuse of those products. In this paper these enzyme immunoassay procedure was extended using the second-antibody technique and a detection based on enzyme enhanced chemiluminescence (2). A comparison was made between the two detection methods.

MATERIALS AND METHODS

Reagents

17α 19-nortestosterone was kindly provided by Prof. Martens (LUC, Diepenbeek, Belgium). All organic solvents were of analytical grade. Helix Pomatia juice was obtained from Boehringer (Mannheim,F.R.G.). Gelatine (Type AAA) was purchased from N.V. Delft (Delft, The Netherlands). The antiserum against 17α 19-nortestosterone (17α 19-NT antibody) was raised in a rabbit by immunization with a 17α 19-NT-3-carboxymethyloxime conjugate (17α 19-NT-3-CMO) with bovine serum albumin. The enzyme label of 17α 19-NT (17α 19-NT-3-CMO-HRP) was prepared by coupling the 3-CMO derivative of 17α 19-NT to horseradish peroxidase (HRP) via the N-hydroxysuccinimide approach. 17α 19-NT-3-CMO-HRP was purified on a PD-10 column and characterized by colorimetric assay on tetramethylbenzidine. More details about the preparation of the antibody and synthesis of the label and their properties have been published elsewhere (1).

Second antibody, purified swine anti-rabbit IgG, code N° Z 196 was purchased from Dakopatts (Glostrup, Denmark). Flat-bottomed 8 well strips (Type Maxisorp), transparent and black, used for spectrofotometric and chemiluminescence detection respectively were obtained

from Life Technologies (Ghent, Belgium). Phosphate
buffered saline solution {(0.01 mol/L, pH 7.25) in 0.9 %
NaCl} was used for coating the antibody to the microtitre
plate wells. Another phoshate buffered saline solution
(0.02 mol/L, pH 7.2) to which 0.05 % Tween 20 was added
was used for the washing steps. A more concentrated
phosphate buffered saline solution (0.01 mol/L, pH 7.0)
with 0.1 % gelatine was used as incubation buffer.
The reagent for chemiluminescence detection (Amersham,
Brussels, Belgium) consisted of signal reagent buffer (pH
8.4) in which Amerlite signal reagent tablets A (luminol
and p-iodophenol) and B (perborate) were dissolved as
recommended by the manufacturer. The spectrofotometric
reagent consisted of 0.150 mL tetramethylbenzidine (TMB,
stock solution: 6 mg/100mL dimethylsulfoxide) in 10 mL
0.11 mol/L sodiumacetate buffer (pH 5.5) to which 2 μL
hydrogen peroxide (30%) was added. A Luminoskan (Lab
Systems, Helsinki, Finland) was used for
chemiluminescence detection in microtitre plates. A 8-
channel microtitre plate reader of Eurogenetics
(Tessenderlo, Belgium) was used for the spectrofotometric
detection.

Assay protocol

Microtitre plates were coated with second antibody in 0.1
mL (dilution 1/8000) coating buffer per well overnight at
37°C. The plates were washed three times with 0.3 mL of
washing buffer (washing procedure). After drying the
wells the plate was incubated with 17α 19-nortestosterone
rabbit antibody in 0.1 mL incubation buffer (dilution
1/16000) at 37°C. After the washing procedure and drying
the wells on absorbing paper, the plate was incubated
with standards ranging from 0.98 pg to 2000 pg per 0.050
mL and HRP label (0.05 mL, dilution 1/100 000) in
incubation buffer for 90 min at 37°C. Again the washing
procedure was repeated, the plate was dried followed by
the determination of the enzyme activity by addition of
0.2 mL substrate solution. The chemiluminescence was
integrated for 1 s each min immediately after the start
of the enzymatic reaction during 30 min. The
spectrofotometric reaction was allowed to proceed for 20
min and stopped by the addition of 0.050 mL 1 mol/L H_2SO_4
after which the sample solution turned from blue to
yellow.

RESULTS AND DISCUSSION

* Dilution curve of enzyme label and antibody

To determine the optimal concentration of enzyme label
and 17α 19-NT antibody, a two-dimensional titration curve
was performed. Enzyme dilutions from 1/25 000 to
1/150 000 and 17α 19-NT antibody dilutions from 1/2000 to
1/256 000 were used. For all further experiments an
enzyme dilution of 1/100 000, a second antibody dilution
of 1/8000 and an NT-antibody dilution of 1/16 000 were
chosen.

* Standard curve
        **Second antibody solid phase immunoassay
From the comparison between enzyme immunoassay with
direct coating of the first antibody (1) and with the
second-antibody technique it could be concluded that with
respect to the slope of the curve, the sensitivity was
rather the same. Because the dose at 50%B_0 is better with
the last method *i.e.* 11 pg/well (n=7, %CV=11) instead of
67 pg/well (n=4, %CV=9.5) , a better limit of
determination could be expected. Those improvements
caused by the
introduction of the second antibody technique could be
explained by the smaller amount of 17α 19-NT antibodies
present on the solid phase (dilution 1/16000 instead of
1/4000).
        **Enzyme immunoassay with chemiluminescence and
            spectrofotometric detection
The dose response curve of the EIA with spectrofotometric
detection covered the range 0-2 ng and the EIA with
chemiluminescence detection covered the range 0-1 ng as
shown in Fig. 1 together with the   standard curve
obtained after direct coating of the 17α 19-NT antibody.

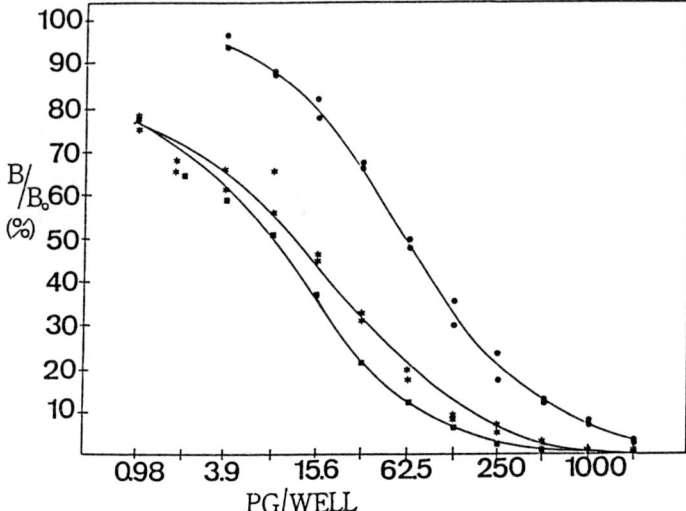

Fig. 1 Standard curves of ● EIA with direct coating,
✱EIA with spectrofotometric detection,
■EIA with chemiluminescence detection

The chemiluminescent light signal integrated for 1 s was
measured after each min during 30 min. The assay
parameters of the standard curves worked out with these
values were summarized in Table 1. After 5 min these
results were almost similar.

Table 1

ASSAY PARAMETERS	REACTION TIME (MIN)				
	1	5	10	20	30
Slope	1.01	0.99	0.96	0.96	0.98
r	0.996	0.996	0.996	0.996	0.996
Dose,20% B_0	31.5	35.9	37.7	38.3	36.9
Dose,50% B_0	6.8	7.5	7.5	7.5	7.0
Dose,80% B_0	0.76	0.78	0.78	0.75	0.70

Legend : r : correlation coefficient
doses are expressed in pg/well
Calculations are done after 4 Parameters Logistic of the results.

In comparison, no important differences in the standard curves with respect to the slope of the curve and the dose at 50% B_0 were observed between the two detection methods as concluded earlier (3). The enzyme immunoassay with spectrofotometric detection required a longer time for the development of the color (20 min) where as the enzyme immunoassay with chemiluminescence detection could be measured already after 1 min.

REFERENCES
1. L.J. Van Look, E.H.J.M. Jansen, R.H. van den Berg, K.E. Vanoosthuyze and C.H. Van Peteghem. Development of a competitive enzyme immunoassay for 17α 19-nortestosterone. J. Chromatogr., Biomed. Appl., in press
2. T.P. Whitehead, G.H.G. Thorpe, T.J.N. Carter, C. Groucutt and L.J. Kricka. Enhanced chemiluminescence procedure for sensitive determination of peroxidase labeled conjugates in immunoassays. Nature 305 , 158-159 (1983).
3. E.H.J.M. Jansen, J.J. Bergman, R.H. van den Berg and G. Zomer. Comparison between spectrophotometric and chemiluminescence detection in enzyme immunoassays for nortestosterone. Anal.Chim.Acta 227 109-117 (1989).

ENHANCED CHEMILUMINESCENT POLYPHENOL REACTIONS CATALYZED BY PEROXIDASE

L A Williams, G H Thorpe and L J Kricka.*
(Wolfson Research Laboratories, QEMC, Birmingham, B15 2TH, UK and *Department of Pathology and Laboratory Medicine, University of Pennsylvania, 3400 Spruce Street, Philadelphia PA 19104, USA).

INTRODUCTION

A range of compounds have previously been shown to increase light emission from the peroxidase catalysed oxidation of diacyl hydrazides such as luminol (1). The objective of this study was to establish if similar enhancers operated with polyphenol chemiluminescent substrates. Enhancement of the horseradish peroxidase (Type VI) catalysed chemiluminescent oxidation of polyphenols including pyrogallol and purpurogallin has been characterised. The degree of enhancement and kinetics of light emission, were determined for various phenols (p-iodophenol, p-hydroxycinnamic acid), benzothiazoles (6-hydroxybenzothiazole), naphthols (1,6-dibromo-2-naphthol) and amine enhancers (eg, 3,5,3',5'-tetramethylbenzidine, N,N,N',N'-tetramethylbenzidine). Both the peak and total light emission could be enhanced.

MATERIALS AND METHODS

Reagents

Purpurogallin, pyrogallol, firefly luciferin, 6-ethyl firefly luciferin, 3,3',5,5'-tetramethylbenzidine, hydrogen peroxide (30% w/v), dimethyl sulphoxide (Grade 1), horseradish peroxidase type VI were purchased from Sigma Chemical Co. (Dorset, UK). p-Iodophenol, p-phenylphenol and the enhancers listed in Table 1 were purchased from Aldrich Chemical Co (Kent, UK). 2,4-Dichlorophenol was obtained from BDH Chemical Co (Dorset, UK). Benzothiazole derivatives were a gift from Dr N Baggett (Birmingham University, UK).

Luminometry

Light emission was measured using a Berthold Auto-Biolumat LB 950 or a luminometer measuring photocurrent, constructed in the Wolfson Research Laboratories (2).

Enhancement Studies

Solutions of pyrogallol (1.82 mmol/1), hydrogen peroxide (1.28 mmol/1) and HRP Type VI (2.5×10^{-7} mol/1) were prepared in potassium phosphate buffer (0.18 mmol/1, pH 6.5). Purpurogallin (2 mg/ml) was dissolved in methanol and then diluted in the phosphate buffer to 1.73 mmol/l and stored in the dark. Enhancers were prepared in DMSO, halophenols in DMSO/buffer (1:100 v/v). Purpurogallin or pyrogallol (200 μl), peroxide (200 μl), enhancer solution (10 μl) and HRP (50 μl) were mixed together and the light emission monitored for up to 15 minutes. Diluent replaced the enhancer in the unenhanced control. Degree of enhancement was determined from the ratio of the light emission from the enhanced and unenhanced reactions.

RESULTS
The influence of the different compounds on the light emission from the HRP catalysed oxidation of purpurogallin and pyrogallol is summarised in Table 1.

6-Hydroxybenzothiazoles Using purpurogallin, pronounced enhancement of total and peak light emission was produced by 6-hydroxybenzothiazole, 2-cyano-6-hydroxybenzothiazole and L-firefly luciferin. Replacement of the hydroxyl group at the 6 position with a methoxy or an ethoxy group eliminated the enhancement of light emission. The kinetics of light emitted in the presence of 6-hydroxybenzothiazole derivatives were complex. A rapid, intense flash was produced using 6-hydroxybenzothiazole, however with 2-cyano-6-hydroxybenzothiazole and L-firefly luciferin a complex, delayed peak of high intensity was observed, with maximal intensity occurring approximately 250 to 600 seconds after initiation.
With pyrogallol, under the conditions employed, none of the 6-hydroxybenzothiazoles produced significant increases in total light during the 900 seconds monitored. However, several increased the intensity of the rapidly decaying initial peak eg 6-hydroxybenzothiazole produced a 40-fold enhancement, 2-cyano-6-hydroxybenzothiazole produced a less intense, much longer lived, biphasic peak.

Phenols Pronounced enhancement was achieved with several halophenols in the purpurogallin reaction, these included p-chloro, p-bromo, p-iodophenol, and 2,4-dichlorophenol. Increases in the total light emitted over the 700 second period monitored, were up to 32-fold, with increases in peak intensity of up to 97-fold. The light emission in the enhanced reactions showed a slow rise to a peak which occurred several minutes after initiation. The time to reach peak intensity varied with the halophenol used. No enhancement was seen with meta-substituted halophenols eg (m-chlorophenol) and methylation of para hydroxyl groups (eg p-chloroanisole and p-iodoanisole) eliminated the halophenols ability to act as enhancers. Although no enhancement was initially seen with o-chlorophenol after 25 minutes, an unexpected 12-fold increase in total light emission was observed. Under the conditions employed using pyrogallol as the chemiluminescent substrate, light was emitted as a rapid flash and little enhancement was achieved using the halophenols tested (Table 1).

Several p-substituted phenol derivatives produced good enhancement of the purpurogallin system. Six compounds, p-hydroxycinnamic acid, p-methylphenol, 2-chloro-4-phenylphenol, p-chloro-3-methylphenol and 4-hydroxy-3-methoxybenzaldehyde (vanillin) produced increases of 16 to 82-

Table 1. **Influence of benzothiazoles, phenols, naphthols and amines on the peroxidase catalysed chemiluminescent oxidation of pyrogallol and purpurogallin**

Compound	Pyrogallol		·Purpurogallin	
	Peak	Integral	Peak	Integral
	(-fold enhancement of light)			
Benzothiazoles[1]				
6-hydroxybenzothiazole	40	1.6	47	8
2-cyano-6-hydroxybenzothiazole	3	2.1	143	30
L-firefly luciferin	-	-	104	20
Phenols[2]				
p-chlorophenol*	1.5	1.2	86	31
p-bromophenol*	1.3	1.3	83	32
p-iodophenol*	1.6	1.3	91	19
2,4-dichlorophenol*	1.7	1.4	97	26
p-hydroxycinnamic acid	4	1.5	44	15
p-methylphenol	-	-	16	13
p-phenylphenol	4	1.3	58	11
2-chloro-4-phenylphenol	5	3	55	30
4-chloro-3-methylphenol	2	1.6	82	35
Naphthols[3]				
1-bromo-2-naphthol	18	4	25	5
1,6-dibromo-2-naphthol	1.6	1.6	33	4
Amines[4]				
3,5,3',5'-tetramethylbenzidine	9	2.4	11	7
N,N,N'N'-tetramethylbenzidine	2	1.2	7	3

Final concentrations 1) 0.18 mg/ml, 2) 116 umol/l, *) 140 umol/l, 3) 60 umol/l, 4) 100 umol/l

fold in peak emission, and 11 and 35-fold increases in total light emitted during the 600 seconds monitored. p-Hydroxycinnamic acid, p-phenylphenol and 2-chloro-4-phenylphenol produced intense rapidly decaying initial light emission. p-Methoxybenzaldehyde produced high intensity, slower peaks with maximal intensity occurring approximately 200 seconds after initiation. Similar kinetics were also observed with other compounds which produced much lower levels of enhancement. Replacement of the hydroxyl group in phenolic enhancers such as p-hydroxycinnamic acid and phenyl-phenol by a methoxy group, abolished the ability to increase light levels. Lower intensity light emission and less significant enhancement was achieved in the pyrogallol reaction. p-Hydroxycinnamic acid, p-phenylphe-nol and 2-chloro-4-phenylphenol produced increases in peak intensity of up to 5-fold, and total light emitted up to 3-fold. All three compounds pro-duced a rapidity decaying flash of light, with 2-chloro-4-phenylphenol and p-hydroxycinnamic acid exhibiting unusual, rapid secondary peaks.

Naphthols   Enhancement of light emission from purpurogallin and pyro-gallol was achieved with 1-bromo-2-naphthol and 1,6-dibromo-2-naphthol.

Amines   In addition to those substituted amines, such as o-, m-, and p-phenylenediamine, previously reported to enhance light from purpurogallin and pyrogallol two additional compounds 3,5,3',5'- and N,N,N',N'-tetra-methylbenzidine were shown to enhance light emission. Most enhance-ment was observed with 3,5,3',5,-tetramethylbenzidine (Table 1).

CONCLUSIONS
Several compounds previously shown to enhance HRP catalysed light emission from luminol have also been shown to enhance light emission from polyphenols including pyrogallol and purpurogallin. Although the complex emission kinetics with polyphenols may limit their applications, these observations may be of use in mechanistic studies on enhancement of chemiluminescent reactions catalysed by horseradish peroxidase.

Acknowledgements
The financial support of the Department of Health is gratefully acknowl-edged.

References
1. G.H.G. Thorpe and L.J. Kricka, Enhanced chemiluminescent reactions catalysed by horseradish peroxidase. Methods in Enzymology 133, 331-353 (1986).
2. R.A. Bunce, T.J.N. Carter, L.J.Kricka et al., Determination of concentra-tion of an analyte in a sample. UK Patent Appl 2025609 (1990).

IMMUNOCHEMILUMINOMETRIC SCREENING ASSAYS BASED ON ACRIDINIUM LABELS AND MICROTITRE PLATE LUMINOMETERS

J.S. Woodhead, S.A. Herbert and I. Weeks

Department of Medical Biochemistry
University of Wales College of Medicine
Heath Park, Cardiff CF4 4XN, Wales, UK

INTRODUCTION

The use of microtitre plate formats for immunoassays has become popular in recent years particularly for assay systems based on enzyme labelled reagents. Microtitre plates are particularly well-suited to high-throughput assays such as screening tests because of their small size, the requirement for small sample and reagent volumes and the availability of equipment such as plate washers to facilitate automation.

Whilst enzyme labelled systems have been used in this format, the end-point detection is not well-suited to high throughput systems due to complexity and lack of robustness.

Certain chemiluminescent acridinium salts exhibit rapid reaction kinetics and have been used as labels in the development of highly sensitive chemiluminescence immunoassays (1). Further, initiation of acridinium chemiluminescence is very simple and yields rapid and robust end-points. Despite these advantages, immunoassays based on acridinium chemiluminescence have not been used in a microtitre plate format due to lack of appropriate instrumentation.

Here we describe the preliminary development of immunoassays which combine the advantages of acridinium labels with the convenience of microtitre plates. Since this system is ideally suited for screening tests we describe assays for maternal serum alphafetoprotein (AFP) and neonatal blood spot thyrotrophin (nTSH) as tests for foetal neural tube defects and congenital hypothyroidism respectively.

MATERIALS AND METHODS

Acridinium labelled monoclonal antibodies raised against AFP and TSH were a generous gift from Ciba Corning Diagnostics (AFP Lite and TSH Lite reagents respectively). Antibodies for plate coating were monoclonal anti-AFP antibody, produced in-house by

conventional methods, or sheep (anti-TSH) antiserum
purchased from the Scottish Antibody Production Unit,
Carluke, UK.

Serum AFP standards and samples, and blood spot disc TSH
standards were obtained from the Supra Regional Assay
Service Laboratory, University Hospital of Wales,
Cardiff, UK.

Coating buffer was carbonate/bicarbonate solution
(0.1mol/l, pH 9.6), blocking buffer was phosphate
buffered saline (0.05mol/l, 0.15mol/l NaCl, pH7.4)
containing 0.5% (w/v) bovine serum albumin and 0.05%
(w/v) sodium azide. Wash buffer was PBS containing 0.05%
Tween 20.

Microplates were samples kindly provided by Dynatech
Laboratories, ICN-Flow and Labsystems. Luminometers
were, for tubes, Berthold Clinilumat (marketed by Ciba
Corning as the Magic Lite Analyser) and, for
microplates, Labsystems Luminoskan (marketed in UK by
ICN-Flow). Chemiluminescence initiator reagents were
Magic Lite Reagents 1 and 2 (Ciba Corning Diagnostics)

Coating antibodies were obtained by sodium sulphate
precipitation from ascites or antiserum. Plates were
coated overnight at room temperature using a solution of
antibodies (20ug/ml) in coating buffer (100ul), washed
three times with wash buffer and blocked with blocking
buffer (200ul) for 2 hours at 37°C. The plates were
washed a further three times with wash buffer prior to
use.

Assays were performed as follows. Serum (50ul) or 3mm
blood spot samples/standards were placed into microwells
and 100ul of the appropriate labelled antibody solution
added. The plates were briefly agitated and then
incubated for one hour at 37°C. The plates were washed
three times with wash buffer and placed in the
luminometer. The instrument added to each well 100ul of
Magic Lite Reagent 1, immediately followed by 100ul of
Magic Lite Reagent 2 and integrated photon counts over
two seconds.

Signal/noise (S/N) data were generated by measuring
chemiluminescence emission from 10ul of the appropriate
dilution of labelled antibody. Such data were used for
comparisons between different microtitre plates and also
to compare the characteristics of the plate instrument
with the tube instrument.

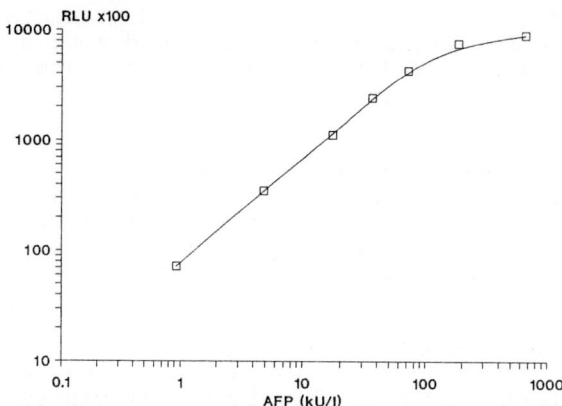

FIGURE 1. Dose-Response for AFP

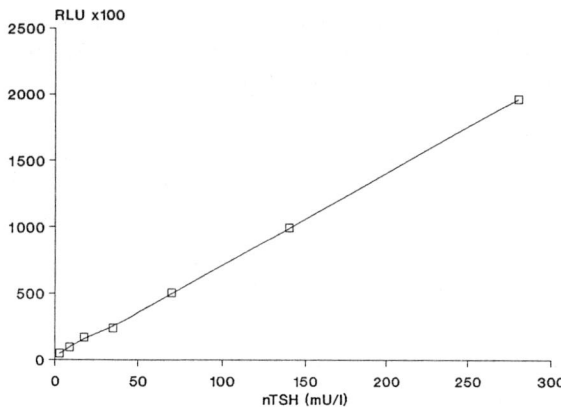

FIGURE 2. Dose-Response for nTSH

Table 1. Intra-assay precision

Mean AFP (kU/l)	%CV	Mean TSH (mU/l)	%CV
1.9	3.5	5.6	12.5
41.4	5.8	26.3	11.3
421	5.2	210	6.1

NB Mean of triplicates over three assays

RESULTS AND DISCUSSION
The dose-response curves for the immunochemiluminometric
assays are shown in Figs.1 and 2. Intra-assay precision
for both assays is shown in Table 1 and predictably is
less good in the TSH assay for 3mm blood spot samples
than for the serum AFP assay owing to the nature of the
punched-out filter paper discs.

Different microtitre plates showed considerable
differences between plates both in terms of signal
(S)(range 0.36 - 5.23 RLU) and background (N)(range 0.01
- 0.22 RLU). This is not unexpected between white, clear
and black plates but it was also seen that there exists
variation between different white plates (S range 3.54 -
5.23 RLU, N range 0.09 - 0.22 RLU). It is apparent that
under given conditions S/N is slightly worse (30%
lower) for the plate luminometer as compared to the tube
luminometer. These data were generated using the
Microfluor plate (S=4.71, N=0.18) but it was seen that
the Labsystems white strips (S=4.44, N=0.11) give
slightly better S/N which would bring the plate
instrument more in line with the tube instrument. Such
improvement is due more to lower plate background rather
than to higher signal intensity. This fact would also
benefit the immunoassays since their sensitivity is
limited by plate background rather than by assay non-
specific binding.

The data shown here clearly demonstrate that the
advantages of acridinium labelled reagents can be
exploited in a microtitre plate format to yield rapid,
simple and sensitive immunoassays which are well-suited
to high throughput screening purposes. We have recently
found that performance can be further enhanced by
optimisation of plate characteristics and oxidant
injection.

ACKNOWLEDGEMENTS
The authors wish to express their thanks to the
following for plates, instruments and reagents:
Dynatech, ICN Flow, Labsystems, Ciba Corning
Diagnostics. This work was supported by Bioanalysis Ltd,
Cardiff,Wales, UK.

REFERENCES
1.   I Weeks, M Sturgess, RC Brown and JS Woodhead,
Immunoassays using acridinium esters. Methods in
Enzymology 133, 366-387 (1986).

MECHANISM OF DYE- AND SELF-SENSITIZED
CHEMILUMINESCENCE OF LUMINOL

A.D.Klimov, S.R.Marulidy, S.F.Lebedkin and V.N.Emokhonov

*Institute of Energy Problems of Chemical Physics,
USSR Academy of Sciences, 117829, Moscow, USSR*

INTRODUCTION

Certain sensitizers (e.g. dyes) are able to trigger off luminol chemiluminescence (CL) upon photoexcitation (1,2). This photosensitized CL of luminol may prove eligible for various applications: a)Chemiluminescent assays of labeled compounds; b) Estimation of the total photooxidative activity of natural water photosensitizers (e.g.humic substances),which are believed to be of the major significance for the photochemical degradation of aquatic pollutants (3); c) Elucidation of the role of the side reactions, due to $^1O_2$, which may be essential for the luminol CL used to probe production of the reactive oxygen derivatives by phagocytic cells.

Previous attempts to study the photosensitized CL mechanism were focused mainly on the role of $^1O_2$. CL was believed to be the result of $^1O_2$ reaction with luminol (1) or with some species derived indirectly from $^1O_2$ (2), but later the reaction of $^1O_2$ with luminol was proved to be non-chemiluminescent (4). To promote the use of the photosensitized CL for the above mentioned applications, detailed investigation of the model dye-sensitized and self-sensitized reactions has been undertaken. Specific features of the photosensitized CL include initiation and side reactions, while the general picture of luminol CL mechanism has been firmly established (5).

MATERIALS AND METHODS

Reagents. Luminol (Koch-Light Lab.) was recrystallized first from 1 mol/l KOH, then from 20% HCl. Isoluminol and 4-dimethylaminophthalhydrazide were synthesized in our laboratory. Eosin Y and methylene blue (Reachim) were chromatographed on Al_2O_3 column eluted with water or EtOH. Other chemicals were of reagent grade and used without purification. Alkaline stock solutions of luminol, isoluminol, 4-dimethylaminophthalhydrazide and neutral pH solutions of dyes were stable in a darkness for months. Reaction mixture was prepared in a dark place from component stock solutions just before use.

Apparatus. Absorption, fluorescence, CL emission and CL excitation spectra were recorded using spectrophotometer (Heath-700) and spectrofluorimeter (Elumin-2M). Conventional flash photolysis was employed for transient kinetics measurements.

Continuous photolysis installation was used to study CL time-course and reagent consumption. Scattered light was cut off using glass or interference filters. The light fraction absorbed by a sensitizer was measured with a calibrated thermopile (Hilger-Schwars). Single-beam spectrophotometer scheme was used to measure the time-course of luminol consumption.

Stopped-flow combined with flash photolysis ('flash-stop') was used to study self-sensitized CL. Prior to reagent mixing luminol or/and H_2O_2 solutions in either/both channels were photoexcited to generate luminol radical and superoxide. Photoexcitation prior to mixing excludes possible undesirable interference with the primary photoprocesses caused by the presence of the second reagent. 'Flash-stop'method enables to vary easily luminol radical and superoxide concentrations and to follow light emission or transient absorption kinetics. 'Flash-stop' is convenient to study superoxide reactivity provided H_2O_2 does not interfere on the reaction timescale.

RESULTS AND DISCUSSION

Spectra. CL excitation spectrum corresponds to the dye absorption spectrum in the visible region. Photoexcitation of a dye-sensitizer in the UV-region, where luminol absorbs too, also causes CL, which is masked by bright fluorescence of luminol (fluorescence spectrum is very similar to CL spectrum). Moreover, absorption of UV-quanta by luminol in dye-sensitizer free solution results not only in fluorescence, but in CL as well. Excitation spectrum of this self-sensitized CL matches absorption spectrum of luminol. Dye- and self-sensitized CL emission spectra match fluorescence spectrum of 3-aminophthalate dianion.

Initiation. Triplet sensitizer oxidizes luminol to produce sensitizer radical and luminol radical. The rate constants for the reductive quenching of triplet methylene blue and triplet eosin by luminol are respectively $4.2 \cdot 10^9$ and $1.6 \cdot 10^9$ 1/mol·s (the rate constants for isoluminol and 4-dimethylaminophthalhydrazide are close to luminol ones).

The yields of formation of radical products escaping cage recombination (6) were found to be pH-dependent. Cage escape yield is constant from pH 8 up to pH 11 (0.13 for methylene blue and 0.11 for eosin), increases in the range from pH 11.2 up to pH 13.2 (with apparent pK_a 12.2), and becomes constant again at pH > 13.2 (0.35 for methylene blue and 0.25 for eosin). pH-dependence of cage escape yield is not a simple acid-base equilibrium, because none of the reaction components have pK's in the pH interval from 11 to 14. To account for this phenomenon we assumed that deprotonation of the primary radical pair (consists of the sensitizer radical and protonated luminol radical; the latter is in thermodynamically unfavourable ionic form in alkaline solution (pK_a 7.7 (5)) and tends to lose proton) due to reaction with OH^- competes with radical pair decay (at pH > 12 deprotonation time $1/(K_{diff} \cdot [OH^-])$ becomes comparable with typical radical pair lifetime in water (ca.10^{-9} s)) resulting in the formation of the deprotonated radical pair with reduced value of the rate constant of back electron transfer (7).

Methylene blue and eosin form the complex with luminol and related compounds. The equilibrium constants ($2.4 \cdot 10^{-3}$ mol/l for methylene blue and $3.4 \cdot 10^{-2}$ mol/l for eosin with luminol) were derived from complexation-induced changes of fluorescence intensity of the dyes.

Photoexcitation of the ground state complex does not result in the formation of free radicals because of static quenching; bound sensitizer is absolutely inefficient.

The sensitizer catalytic cycle includes: a) Photoexcitation to give triplet sensitizer; b) Electron transfer from luminol to the triplet sensitizer to yield sensitizer radical; c) Oxidation of sensitizer radical by O_2 to recover the ground state sensitizer and to generate superoxide. Eosin radical is known to react rapidly with O_2 (rate constant is $2 \cdot 10^8$ 1/mol·s (8)). The rate constant for methylene blue radical oxidation by O_2 was found to be substantially lower: $5 \cdot 10^5$ 1/mol·s.

Photoionization of luminol occurs upon absorption of UV-quantum in the self-sensitized CL. Photoionization was proved to be the single-photon process. Ejected electrons are scavenged by O_2.

Key reaction. Both the dye-sensitized and self-sensitized CL are quenched by addition of superoxide dismutase (10^{-7}-10^{-9} mol/l) and are not influenced by addition of up to 10^{-3} mol/l H_2O_2. That proves that major CL arises from luminol radical - superoxide pathway, while the contribution of diazaquinone - H_2O_2 pathway is imperceptible (5).

The rate constant for the reaction of luminol radical with superoxide was determined using 'flash-stop' method. Superoxide concentrations were calibrated with tetranitromethane (reaction stoichiometry was supposed to be 1 : 1). The second-order rate constant for the reaction of luminol radical with superoxide was derived from the dependence of the pseudo-first-order rate constant of the CL decay on superoxide concentration and was found to be $7 \cdot 10^7$ 1/mol·s. This value is lower than measured by pulse radiolysis method: $2.3 \cdot 10^8$ 1/mol·s (5).

Inhibitor. The dye-sensitized CL time-courses are different for the studied phthalhydrazides (Fig.1 a, b). Luminol gives a single splash of light. Consumption of the reagents is negligible during the splash time, but aliquot of this irradiated solution, being added to the fresh one, strongly quenches its CL indicating the inhibitor

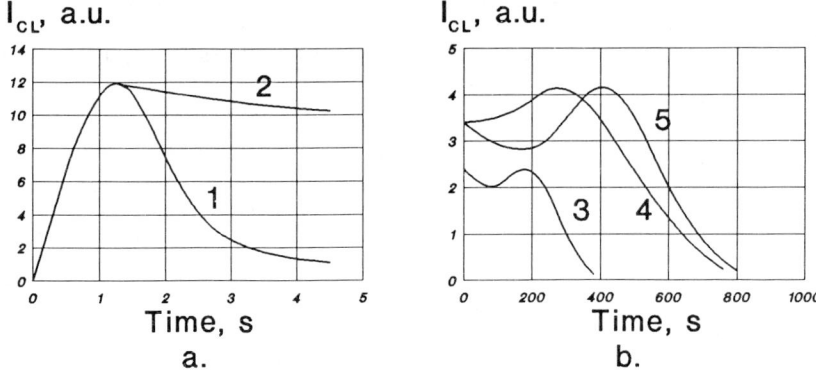

Fig.1a, b. Dye-sensitized CL time-courses. Luminol (1), isoluminol (3), 4-dimethylaminophthalhydrazide (5)-$2.5 \cdot 10^{-4}$ mol/l; methylene blue - $2 \cdot 10^{-6}$ mol/l; pH 11.9; light absorption rate - $4.8 \cdot 10^{-5}$ E/l·s. (2, 4) - added 10^{-2} mol/l sodium azide.

formation. Initial decrease of the related compounds CL is also caused by the inhibition, though not so strong as for luminol. The following prolonged CL comes to termination on account of complete consumption of O_2 or phthalhydrazide. We found that inhibitor is formed in reaction of phthalhydrazide with $^1O_2$: a)Addition of azide ($>10^{-3}$ mol/l) supresses inhibitor formation (Fig.1); b)Using tetra-(4-sulfonatophenyl)porphine as a photosensitizer (it was shown to produce only $^1O_2$, but no free radical species) instead of methylene blue or eosin, the product of $^1O_2$ reaction with luminol (or related compounds) was generated and proved to exhibit inhibitory properties. The inhibitor produced from luminol turns into a more potent inhibitor within several minutes, while the inhibitors produced from isoluminol and 4-dimethylaminophthalhydrazide decay on the same timescale. That causes time-course differences of the CL (Fig.1a and 1b). The rate constant for luminol reaction with $^1O_2$ is $3 \cdot 10^7$ l/mol·s(4). Using this value along with the other measured ones, we calculated the rate of luminol radical and inhibitor generation. The rate of luminol radical generation is much higher than for inhibitor in the reasonable luminol concentration range. That means that each inhibitor molecule acts repeatedly, presumably, in a chain process. To account for the measured rates of luminol and O_2 consumption, reduction of luminol radical and oxidation of superoxide were postulated to take place in the chain inhibitor process.

REFERENCES
1.   K.Kuschnir and T.Kuwana, Photosensitized chemiluminescence of luminol, 6-aminophthalazine-1,4(2H,3H)-dione. J.Chem.Soc., Chem.Commun., 193 (1969).
2.   I.B.C.Matheson and J.Lee, The dye-sensitized photooxidation chemiluminescence of luminol. Photochem. Photobiol. 12 , 9-16 (1970).
3.   W.J.Cooper and F.L.Herr, Introduction and overview. ACS Symp. Ser.,V327, Photochemistry of environmental aquatic systems, R.G.Zika and W.J.Cooper (Eds), ACS, Washington D.C., p.1-8(1987).
4.   I.B.C.Matheson and J.Lee, The non-chemiluminescent reaction of luminol with singlet oxygen. Photochem. Photobiol. 24, 605-607 (1976).
5.   G.Merenyi, J.Lind and T.E.Eriksen, Luminol chemiluminescence: chemistry, excitation, emitter. J.Biolum. Chemilumin. 5, 53-56 (1990).
6.   U.Steiner, G.Winter and H.E.A.Kramer, Investigation of physical triplet quenching by electron donor. J.Phys. Chem. 81, 1104-1110 (1977).
7.   S.F.Lebedkin, A.D.Klimov and V.N.Emokhonov, Prototropic reactions in radical pairs: influence on cage escape yield. Izv.Acad.Nauk SSSR, Ser.Khim., 1291-1295 (1990).
8.   V.Kasche, Radical intermediates in the fluorescein and eosin-photosensitized autooxidation of l-tyrosine. Photochem. Photobiol. 6, 643-650 (1967).

QUANTIFICATION OF LEVELS OF MUTANT RAS PROTEIN IN KERATINOCYTES TRANSFECTED WITH ACTIVATED RAS ONCOGENES

Peter Lloyd Jones, Isobel M. Greenfield, Claire E. Hooper[1], Richard Ansorge[2] and Margaret A. Stanley

Department of Pathology, University of Cambridge, Tennis Court Road, Cambridge, U.K., [1]Robens Institute, University of Surrey, Guildford, U.K., [2]Department of Physics, University of Cambridge and Image Research Limited, St Johns Innovation Centre, Cambridge, U.K.

INTRODUCTION

In vitro transformation of murine and human keratinocytes by activated oncogenes, (i.e. genes which are capable of inducing a transformed phenotype), is a useful system in which to investigate the biological mechanisms of tumour promotion and multistage carcinogenesis (1,2). The ras proto-oncogene family which encode highly conserved proteins of approximately Mr, 21,000, (termed p21), seem to play a central role in the processes of epithelial proliferation.

We have created a number of murine keratinocyte cell lines containing activated ras oncogenes (either cellular or viral) and have characterised their self-renewal capacity and differentiation potential in vitro. The aims of this study are to measure the relative levels of p21 in each cell line using an enhanced chemiluminescence (ECL) Western blotting system in conjunction with intensified CCD image analysis and relate these findings to their in vitro characteristics.

MATERIALS AND METHODS

Cell Culture Neonatal mouse epidermal keratinocytes (NEKs), RZ1-4 and SW5A were cultured on a lethally irradiated 3T3 feeder layer (3) in GMEM containing 10% foetal calf serum supplemented with 10^{-10}M cholera toxin, 0.5µg/ml hydrocortisone and 0.1mg/ml kanamycin. Cells were grown at 31°C on a tissue culture treated plastics in an humidified atmosphere with 5% CO_2 and 95% air.

Cell Lines pHO6T1 is a plasmid construct containing the cellular Ha-ras-1 oncogene, two viral enhancers and the aph gene from the bacterial transposon, Tn5. 5 µg of oncogene-plasmid DNA was transfected into subconfluent, primary cultures of NEKs using the Ca^{2+}-phosphate precipitation technique (4) and this resulted in the SW series of cell lines.

The RZ series of cell lines were derived by infection of primary cultures of NEKs with raszip6, a helper-free retrovirus containing the viral Ha-ras gene.

Characterisation of Cell Lines

i Colony Forming Efficiency (CFE), a measure of in vitro proliferation. Keratinocytes sufficient to form 100 colonies were seeded onto 60mm tissue culture dishes containing 7×10^5 irradiated 3T3 feeders and complete medium. Cultures were incubated for 14 days and the keratinocyte colonies were then scored.

ii Suspension-induced terminal differentiation.

This was carried out according to the method of Green (5). Exponentially growing keratinocytes were cultured in complete medium on a 1 % agar bed. Cells were harvested, counted and the proportion of cells possessing cornified envelopes determined.

Western Blotting and ECL Detection of p21 was performed using an anti-ras monoclonal antibody in a standard Western Blot technique (6). The protein was visualised on X-ray film after reaction with Amersham ECL reagents.

Quantitaitive Luminescent Imaging of p21 Quantitaitive analysis of emitted light was achieved using a room temperature, intensified CCD imaging system (Biomedical Quantifier, Image

Research Ltd., Cambridge, UK). See Hooper et al., in this volume.

RESULTS AND DISCUSSION

In Vitro Self-Renewal Normal, murine NEKs have a limited tissue culture lifespan and a CFE of
approximately 3% (Fig. 1). RZ1, RZ2, RZ3, RZ4 and SW5A all have an increased self-renewal
capacity, as measured by an enhanced tissue culture lifespan and an increase in CFE (Fig. 1) RZ2
had the highest CFE (26%) and this decreased in the following order, RZ4(25%), RZ1(19%), RZ3
(18%) and SW5A (8%).

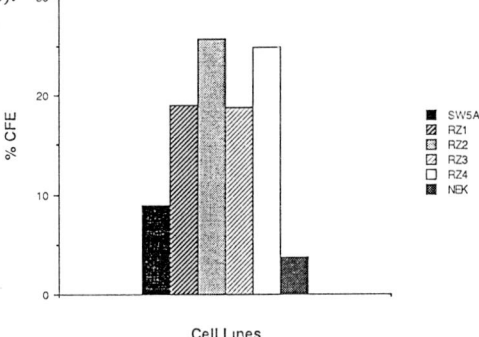

Figure 1 Colony forming efficiencies of NEKs and cell lines. The CFE was determined
as described in Materials and Methods.

InVitro Alterations in Differentiation Potential In order to assess the differentiation potential of
keratinocytes in vitro, the ability to form cornified envelopes when relieved of substrate adhesion
was determined. After eight days in suspension culture 86% of NEKs had formed cornified
envelopes (Fig. 2). RZ2 shows the greatest resistance to terminal differentiation at 2% and this
decreased as % cornification increased in the following order, RZ4(4%), RZ1(8%), RZ3(15%) and
SW5A(28%).

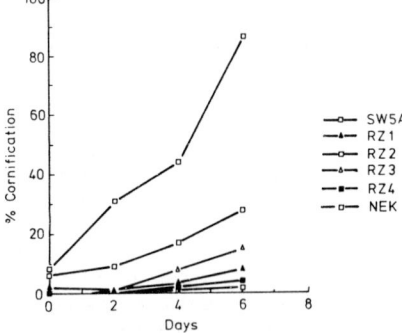

Figure 2 Cornified envelope production of NEKs and cell lines. The
production of cornified envelopes was determined as described in
Materials and Methods.

Visualisation and Quantification of p21 To correlate p21 expression levels with the in vitro
characteristics of the cell lines, the 21, 000 Mr protein was detected using an anti-ras monoclonal
antibody in an ECL Western Blotting system. The ras protein was visualised and quantified by
CCD image analysis (Fig. 3). The one dimensional histograms derived from the CCD image
analyses show NEKs to be negative for activated p21(Fig. 3a) and that increasing levels of p21

can be visualised and quantified in the following order, RZ3 (Fig. 3b), RZ1(Fig. 3c), RZ4 (Fig. 3d), RZ2(Fig. 3e) and SW5A(Fig. 3f).

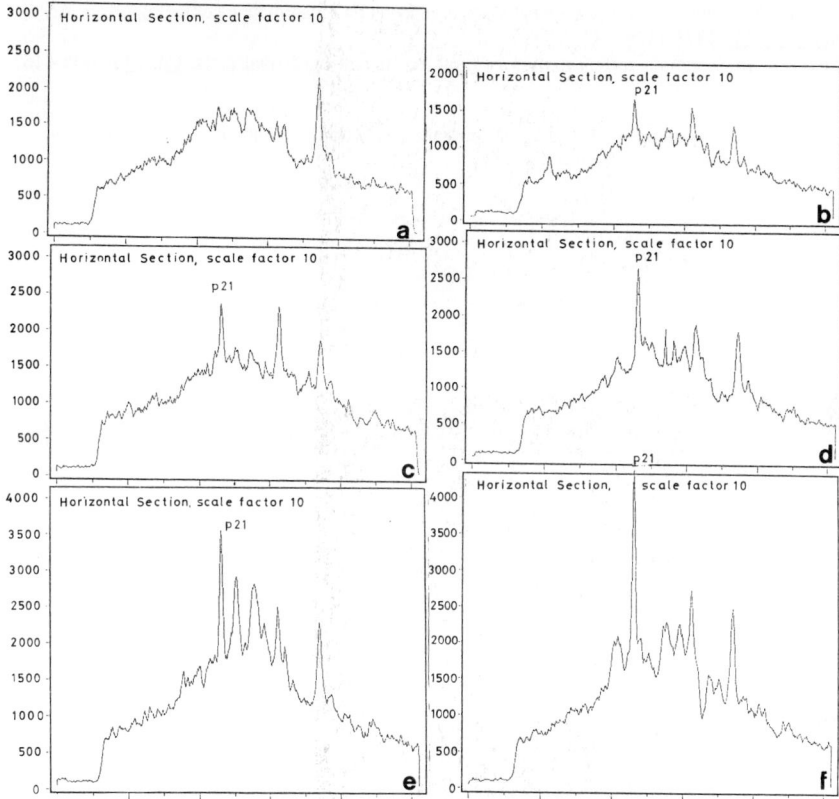

Figure 3   Quantification of emitted light using a room temperature, intensified CCD imaging system (BIQ). NEKs(a). RZ3(b), RZ1(c), RZ4(d), RZ2(e) and SW5A(f).

The combined techniques of Western-ECL and CCD image analysis used in this study provide a rapid, safe and highly sensitive method for the detection of a minor cellular protein. The data presented here argue strongly that increasing levels of v-Ha-ras p21 correlate with an increased self-renewal capacity and increased resistance to terminal differentiation. SW5A expresses most p21, but its self-renewal capacity and differentiation potential are lower than those recorded in the RZ series of cell lines. Alterations in the genomic structure of v-Ha-ras when compared to c-Ha-ras may account for these differences.

REFERENCES
1. A. Balmain, M. Quintanilla and M. Ramsden, An approach to the molecular mechanisms of cancer induction. J. Pathol., 149, 3-8 (1986).
2. J.S. Rhim, G. Jay, P. Arnstein, F.M. Price, K.K. Sanford and S.A. Aaronson, Neoplastic Transformation of Human Epidermal Keratinocytes by AD12-SV40 and Kirsten Sarcoma Viruses. Science, 227, 1250-1252 (1985).

3. M.A. Stanley and E.K. Parkinson, Growth requirements of human cervical epithelial cells in tissue culture. Int. J. Cancer, 24, 407-414 (1979).

4. M. Wigler, A. Pellicer, S. Silverstein, R. Axel, G. Urlaub and L. Chasin, DNA mediated transfer of the adenine phosphoribosyltransferase locus into mammalian cells. Proc. Natl. Acad. Sci. USA. 76, 1373-1376 (1979).

5. H. Green, Terminal differentiation of cultured human epidermal cells. Cell, 11, 405-416 (1977).

6. H-W. Birk and H. Koepsell, Reaction of Monoclonal Antibodies on Nitrocellulose:Renaturation of Antigenic Sites and Reduction of Nonspecific Antibody Binding, Analytical Biochemistry, 164, 12-22 (1987).

SENSITIVE CHEMILUMINESCENT DETECTION OF DIGOXIGENIN (DIG) LABELED NUCLEIC ACIDS. A FAST AND SIMPLE PROTOCOL AND ITS APPLICATIONS

H.J. Höltke, I. Ettl, I. Obermaier and G. Schmitz

Boehringer Mannheim GmbH, Biochemical Research Center, Nonnenwald 2, D-8122 Penzberg

INTRODUCTION

Nonradioactive labeled nucleic acids are in most cases detected by binding of a reporter molecule (e.g. streptavidin or antibodies) and a coupled enzyme reaction. The preferred marker enzyme is alkaline phosphatase (AP) with 5-bromo-4-chloro-3-indolylphosphate (BCIP) and nitroblue tetrazolium (NBT) as colour substrates. These substrates give rise to a dark blue precipitate by an AP-catalyzed redox reaction. Disadvantages of the colour reaction are 1) speed: to reach high sensitivity, the colour reaction must run for > 16 h and 2) possibility to reprobe: the colour substrates form a very strong precipitate, which is hard to remove. These drawbacks can be overcome by using a luminescent substrate instead of colour substrates.
We have developed a fast and simple protocol for highly sensitive luminescent detection of DIG-labeled nucleic acids with AMPPD as substrate for alkaline phosphatase conjugated to anti-DIG antibody Fab-fragments.

MATERIALS AND METHODS

Principle: We have optimized methods for the nonradioactive labeling of nucleic acids resulting in high yield, high sensitivity and low background: The hapten digoxigenin (DIG) is bound via a spacer arm to uridin-nucleotides (Fig. 1) and incorporated enzymatically at a predetermined density into nucleic acid probes by random primed DNA labeling, PCR, *in vitro* RNA transcription or 3'-endlabeling/tailing. DIG is an artificial hapten and thus avoids unspecific signals.

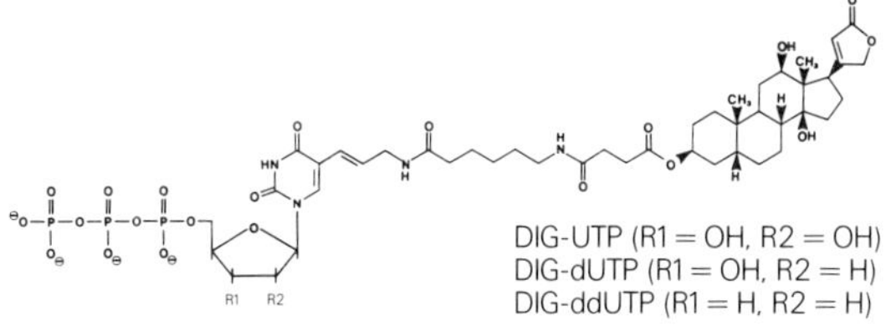

DIG-UTP (R1 = OH, R2 = OH)
DIG-dUTP (R1 = OH, R2 = H)
DIG-ddUTP (R1 = H, R2 = H)

Fig. 1: Structure of digoxigenin (DIG)-modified nucleotides.

For blotting charged nylon membranes can be used. After hybridization and blocking bound probes are detected by high affinity antibody Fab-fragments coupled to alkaline phosphatase. We have optimized the application of AMPPD for the detection of DIG-labeled nucleic acids.

Buffers: **blocking stock solution:** 10 % blocking reagent (purified casein fraction) in buffer 1, autoclaved and stored at 4°C.
hybridization buffer: 50 % formamide, 5 x SSC, 2 % blocking reagent, 0.1 % N-lauroylsarcosine, 0.02 % SDS.
buffer 1: 0.1 M maleic acid, 0.15 M NaCl, pH 7.5, autoclaved.
washing buffer: buffer 1 + 0.3 % Tween-20.
buffer 2: blocking stock sol. diluted 1:10 in buffer 1.
buffer 3: 0.1 M Tris-HCl, 0.1 M NaCl, 50 mM $MgCl_2$, pH 9.5.
AMPPD stock solution: 10 mg/ml; 23.5 mM.

Flow diagram of DIG luminescent detection	
wash	2 min
block	30 min
antibody binding	30 min
wash	2 x 15 min
equilibrate	2 min
AMPPD incubation	5 min
preincubation 37°C	10 min
film exposure	20 min
total	130 min

Procedure:
1. Blot DNA or RNA to a nylon membrane, the type of membrane influences sensitivity and background; we recommend the positively charged nylon membrane distributed by Boehringer Mannheim.
2. Hybridize with 10 - 20 ng/ml DIG-labeled DNA or 50 ng/ml DIG-labeled RNA.
3. Perform stringency washes as usual.
4. Wash membrane shortly in washing buffer.
5. Incubate for 30 min in buffer 2.
6. Dilute <DIG>AP-conjugate to 75 mU/ml (1:10000) in buffer 2.
7. Incubate membrane 30 min in diluted <DIG>AP.
8. Wash 2 x 15 min with washing buffer.
9. Equilibrate in buffer 3.
10. Dilute stock sol. of AMPPD (10 mg/ml) 1:100 in buffer 3.
11. Incubate membrane for 5 min in substrate solution (diluted substrate solution can be reused).
12. Let excess liquid drip off the membrane; blot for a few seconds on a sheet of Whatman 3MM paper but not to complete drieness.
13. Seal the damp membrane in a hybridization bag.
14. Preincubate the sealed membrane for 5-15 min at 37°C.
15. Expose for 15-25 min to X-ray or Polaroid b/w film. The time of exposure depends on strength of signal and on background. Luminescence continues for at least 24 hours, signal intensity seems even to accumulate with time.
16. For reprobing the membranes must be kept wet.
17. Reprobing: Wash membrane 2 x 15 min with 0.2 N NaOH, 0.1 % SDS at 37°C followed by a short wash in 2 x SSC. Then hybridize with the second probe etc.

RESULTS AND DISCUSSION
The following examples demonstrate sensitivity and low background of detection when DIG-labeled nucleic acids are visualized by AMPPD.

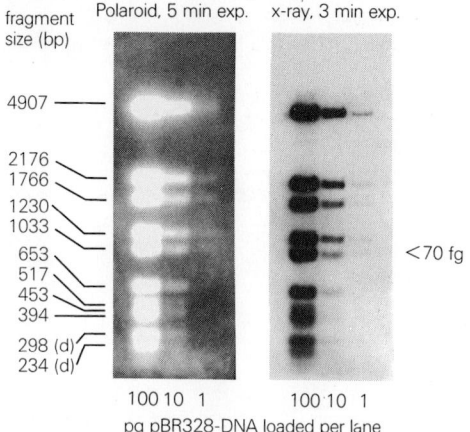

Fig. 2: Sensitivity of chemiluminescent DIG-detection in homologous Southern-blots.
pBR328 was digested separately with *BamHI, BglI and HinfI* and mixed in a ratio of 2:3:3. Probe: 20 ng/ml random primed DIG-labeled pBR328 DNA.

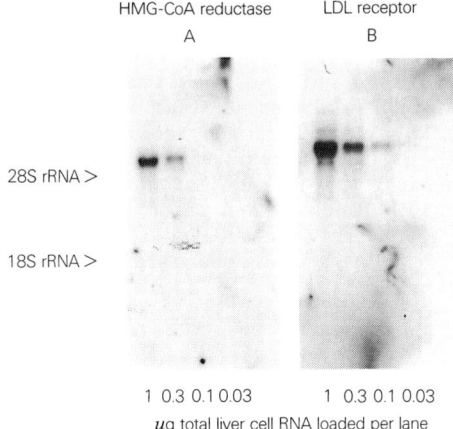

Fig. 3: Chemiluminescent detection of rare mRNAs on Northern blots by DIG-labeled RNA probes.
Total human liver cell RNA was separated on a denaturing formaldehyde gel and blotted onto charged nylon membrane (BM). Probe A: HMG-CoA reductase antisense RNA, DIG-labeled by *in vitro* transcription, Probe B: LDL-receptor antisense RNA, DIG-labeled by *in vitro* transcription.

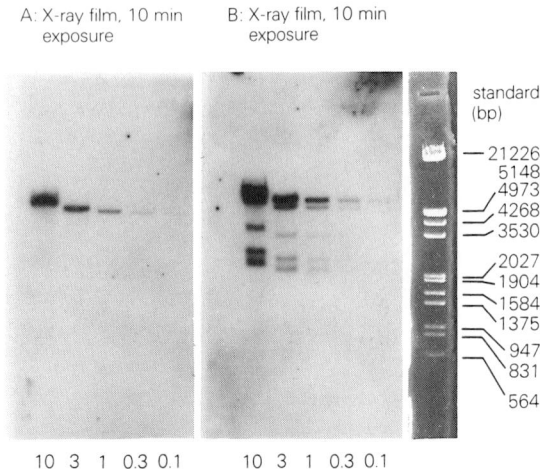

A: X-ray film, 10 min exposure   B: X-ray film, 10 min exposure

standard (bp)

—21226
5148
4973
4268
3530

2027
1904
1584
1375

947
831

564

10 3 1 0.3 0.1   10 3 1 0.3 0.1

μg human placenta DNA, digested with *Eco* RI (A) or *Bgl* II (B)

Fig. 4: Chemiluminescent detection of the single-copy human tPA gene on genomic Southern blots with DIG-labeled DNA-probes.
Human placenta DNA was digested with EcoRI (A) or BglII (B) and blotted onto nylon membrane, positively charged (BM). Probe A: 20 ng/ml gel-purified random primed DIG-labeled genomic tPA DNA fragment. Probe B: 20 ng/ml gel-purified random primed DIG-labeled tPA cDNA fragment.

We have optimized the use of the chemiluminescent substrate AMPPD for the detection of DIG-labeled nucleic acids. We have deviced a simple protocol which allows fast and sensitive detection. Our protocol has the following advantages:
1) The protocol is fast, simple and reproducible. Only three buffers are needed for washing, blocking and detection.
2) The blocking reagent can be dissolved in maleic acid buffer to a concentration of 10 %, sterilized and stored. It blocks nylon membranes very efficiently.
3) The anti-digoxigenin Fab-fragments conjugated to alkaline phosphatase exhibit significantly less unspecific binding to solid supports like charged nylon membranes than intact antibodies (containing Fc) or (strept)avidin. This reduces background and allows longer exposure times resulting in higher sensitivities.

Exposure to X-ray or Polaroid films for up to 30 minutes is sufficient for highly sensitive detection of DIG-labeled nucleic acids down to 10 fg homologous DNA. Human single-copy genes are detected in a Southern-blot of down to 0.1 μg total placenta DNA. In addition we have developed a simple and fast protocol for the reprobing of blots. The major advantages of luminescence are the speed of detection and the ease of reprobing.

THE USE OF A NEW MEASURING EQUIPMENT (MTP READER) FOR BIOLUMINESCENCE-ENHANCED TEST SYSTEMS

Werner Miska and Reinhard Geiger[+]

Dept. of Dermatology and Andrology,
University of Giessen, D-6300 Giessen, FRG and
+ MEDOR GmbH, D-8036 Herrsching, FRG

INTRODUCTION

Because of its high sensitivity, firefly (Photinus pyralis) bioluminescence has been used for many years for the sensitive determination of ATP and ATP-dependent enzyme reactions. The developement of a new type of bioluminogenic enzyme substrates based on D-luciferin derivatives opens new dimensions for the application of bioluminescence in scientific use (1-4). The test principle of these new substrates is the release of D-luciferin from D-luciferin derivatives by the action of hydrolytic enzymes. Liberated D-luciferin can be easily quantified in a luminometric assay. Because of non-efficient measuring equipment the application of bioluminogenic substrates was limited. Very recently MEDOR GmbH is offering now a MTP reader which can be used for this purpose. In our paper we describe the application of the MTP reader for bio-luminescence-enhanced enzyme immunoassays.

MATERIALS AND METHODS

Luciferin-O-phosphate, D-luciferin and luciferase were purchased from MEDOR GmbH, D-8036 Herrsching, FRG.

Buffer A: 15 mmol/l Na_2CO_3, 0.35 mol/l $NaHCO_3$, 0.2 g/l NaN_3, pH 9.6

Buffer B: 10 mmol/l KH_2PO_4, 15 mmol/l NaCl, 0.05 g/l, Tween 20 pH 7.4

Buffer C: 1.5 mmol/l $KH_2PO_4xH_2O$, 0.14 mol/l NaCl, 0.5 g/l Tween 20, pH 7.4

Buffer D: 10 mmol/l DEA, 0.5 mmol/l $MgCl_2$, pH 9.8

IgE sandwich antigen assay
Microtiter plates were coated with rabbit anti-human
IgE-IgG (0.1 ug/ml buffer A; 0.2 ml per well) at 4°C
overnight and washed (5 times) with buffer B. Human IgE
standard samples and test samples were diluted with
buffer C. Of these samples 0.2 ml were added to the
wells and the plates were incubated at 37°C for 3 h and
washed (5 times) with buffer B. 0.2 ml of goat anti-
human-IgE-alkal. phosphatase conjugate solution (dilution: 1:10,000) was added to each well and incubated at
37°C for 1 h. Thereafter, the plates were washed 5
times, 0.2 ml of D-luciferin-O-phosphate solution (10^{-6}
mol/l in buffer D) were added to the wells and incubated for 10 min at 37°C, 0.1 ml of this solution were
transferred to a black microtiterplate containing 0.1
ml detection solution (41 mmol/l HEPES, 5 mmol/l $MgCl_2$,
0.5 mmol/l EDTA, 3.5 mmol/l DTT, 2.6 mmol/l ATP, 8 ug/l
Luciferase, pH 7.75). The microtiterplate was placed
under light exclusion into the MTP reader (MEDOR GmbH,
D-8036 Herrsching, FRG)(Fig. 1).

Fig. 1: The MTP reader.

Emitted photons were counted and integrated for 10 min.
Standard curves and sample values were calculated by
the inbuilt software.

RESULTS AND DISCUSSION
Total human IgE was measured by sandwich antibody
assay. The assay system was optimized with respect to
enzyme immunoassay conditions (binding of antibody to
microtiterplates, antigen-antibody binding, etc.) and
to optimal detection system conditions (concentration
of substrates, etc.). Figure 2 shows the standard curve
of the bioluminescence-enhanced sandwich enzyme
immunoassay for total human IgE. Total human IgE can be
determined in the range between 0.2 and 100 IU/ml IgE.

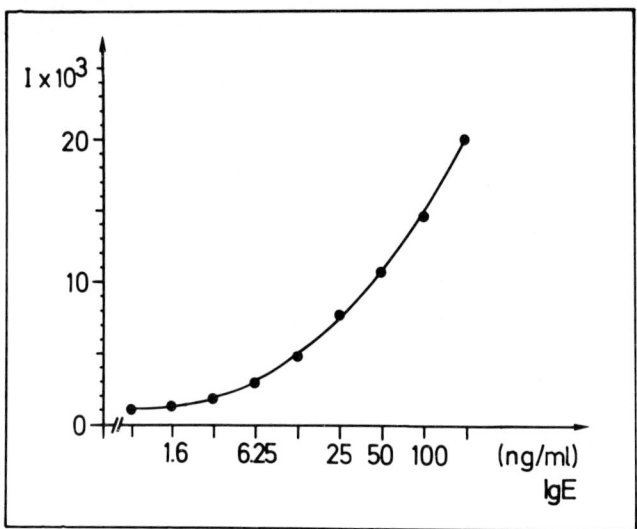

Fig. 2: Standard curve of the bioluminescence-enhanced
        sandwich enzyme immunoassay for total human IgE

Very recently a highly sensitive bioluminescence-
enhanced immunoassay has been developed for the determination of total human IgE in serum. The sensitivity
of the assay is 0.02 ng/ml of human IgE corresponding
to less than 10 mIU IgE (5). Furthermore an enzyme
immunoassay for allergen-specific IgE determination has

also been developed (5). Using microtiter plates coated
with allergens (e.g. animal hairs, plant probes, a.o.)
elevated IgE levels of allergic patients could be
determined.

ACKNOWLEDGEMENTS

We thank Mrs. P. Langbehn for excellent technical
assistance.

REFERENCES

1. **Miska, W. and Geiger, R.,** New, Ultrasensitve
   Detection Systems for Enzyme Immunoassays, I.
   Synthesis and Characterization of Luciferin Deri-
   vatives for Use in Bioluminescence Enhanced Enzyme
   Immunoassays, J. Clin. Chem. Clin. Biochem., 25, 23-
   30 (1987).

2. **Geiger, R. and Miska, W.,** New, Ultrasensitve
   Detection Systems for Enzyme Immunoassays, II. Bio-
   luminescence Enhanced Enzyme Immunoassays, J. Clin.
   Chem. Clin. Biochem., 25, 31-38 (1987).

3. **Miska, W. and Geiger, R.,** A New Type of Ultrasensitive Bioluminescence Enzyme Substrates.
   I. Enzyme Substrates with D-Luciferin as Leaving
   Group, Biol. Chem. Hoppe-Seyler, 369, 407-411,
   (1988).

4. **Geiger, R., Hauber, R. and Miska, W.,** New, bioluminescence-enhanced detection systems for use in
   enzyme activity tests, enzyme immunoassays, protein
   blotting and nucleic acid hybridization, Molecular
   and Cellular Probes (1989) 3, 309-328.

5. **Schröter, A., Langbehn, P., Miska, W. and Geiger,
   R.,** Bioluminescence-enhanced tests in allergy
   diagnosis. J.Clin.Chem.Clin.Biochem. (1990)
   (submitted for publication).

LUMINESCENT DETECTION OF DNA PROBES :
A COMPARATIVE STUDY OF FOUR LUMINESCENT DETECTION SYSTEMS

P. Balaguer, N. Bonafé, B. Térouanne,A.M. Boussioux, A. Baret* and J.C. Nicolas

*INSERM Unité 58, 60, Rue de Navacelles 34090 Montpellier France and *Laboratoire d'analyses médicales 11 allée Duquesne 44000 Nantes*

INTRODUCTION

Nucleic acid assays are widely used in research and clinical chemistry and short oligonucleotides provide the possibility for obtaining specific probes. Enzyme labelling of these probes has been developed and is a good alternative to radioactive labels. These enzymes can be detected using chromogenic substrates but also by using luminescent substrates (1-5). Luminescent substrates offer both rapid results and a better detection limit. Using a video imaging system it is possible to obtain quantitative analysis of dot blots and we describe some applications using several probes which can be detected by different enzymes.

Glucose 6 Phosphate dehydrogenase, xanthine oxidase, peroxidase, and alkaline phosphatase were coupled to streptavidin, antibodies or oligonucleotides. This detection system permitted us to simultaneously perform hybridization of several probes for rapid identification of pathogens such as human papilloma virus (HPV). Quantitative analysis of light emission and calculation of the luminescence ratio performed on the same sample enabled identification of of allelic variations or mutations to be made, using specific oligonucleotides.

MATERIALS AND METHODS

Reagents. Glucose 6 Phosphate dehydrogenase, xanthine oxidase, peroxidase, and alkaline phosphatase were from Boehringer.Bioluminescent measurements were performed on an automated luminometer (LKB 1251). Oligonucleotides were made on a Applied Biosystem synthesizer and amino group was added to the 5' end using aminolink 2.

Labelling of DNA. DNA plasmids or probes were labelled by nick translation or using the PCR in the presence of biotin-dUTP or digoxigenin dUTP. Oligonucleotides with an amino group at the 5' end were coupled to biotin N hydroxysuccinimide ester or labelled at the 3' end using dUTP digoxigenin and terminal transferase. The labelled DNA was recovered by gel filtration on a Sephadex G 50 column equilibrated with 0.1 % SDS..

Hybridizations. The following buffers were used : 1 X SSC buffer: 150 mmol/l NaCl, 15 mmol/l Na3 Citrate. FPGe solution: 0.02 % gelatin, polyvinylpyrolidone, Ficoll. 1X hybridization buffer: 2X SSC, 5X FPGe, 25 mmol/l KH_2PO_4, pH 7, 2.5 mmol/l EDTA, salmon sperm DNA (0.2 mg/ml). Denatured DNA was serially diluted in carrier DNA (0.2 mg/ml) in 10 X SSC. Amplified DNA was denatured by heating, and diluted two-fold with 20 X SSC. Dilutions (1 µl) were spotted on nitrocellulose or Hybond C extra filters. Filters were allowed to dry at room temperature, baked at 80 °C for 1 h and stored dessicated at room temperature until used. The filters were prehybridized in 2X hybridization buffer at 64°C in a water bath. The DNA probe was heated in a boiling water bath for 10 min and added to the 1X hybridization buffer at 1 µg/ml for large DNA probes or at 10 ng/ml for oligonucleotide probes, and thoroughly mixed. Filters were hybridized overnight at 64°C for large DNAs or for 2 h at 55°C for oligonucleotides. Filters were washed twice for 10 min with 0.2 X SSC 0.2% Tween at room temperature, and twice for 10 min at 50°C.

Bioluminescent detection Enzyme labelled oligonucleotides were detected after hybridizaton using a photon counting system (Argus 100/CL Hamamatsu). The Vim camera includes a double

microchannel plate intensifier. Photon counting was performed after 3 to 5 min accumulation. The samples were wetted with the following luminescent reagents :
- Xanthine oxidase with reagent described in (4),
- Glucose 6 phosphate dehydrogenase with reagent described in (5)
- Peroxidase with the ECL reagent from Amersham
- Alkaline phosphatase with AMPPD (Photogene from Gibco BRL)
The filters were placed between two sheets of polyethylen film and luminescence was monitored using a video camera, photon emission was measured for each spot after a 5 min exposure. Between each incubation the filters were washed with the buffer corresponding to the next assay.

RESULTS AND DISCUSSION
Detection limit of luminescent detection systems. Detection limits were determined with streptavidin conjugates in order to compare the four detection systems in the same conditions. However the quality of streptavidin conjugates can have affected the performance of each system.

detection system	blanc	1 pg	10 pg	50 pg
Xanthine oxidase	5 000	14 000	210 000	867 000
Peroxidase	26 000	48 000	473 000	1 800 000
G6PDH	1 700	8 600	140 000	487 000
Alkaline Phosphatase	1 200	5 400	72 000	306 000

Table 1 : Photon count obtained for each spot after 5 min exposure with the four luminescent detection systems. 1 µl of biotinylated plasmid (pBR 322) was applied on each spot.

Table I shows results obtained with the four detection sytems and a biotinylated DNA probe, the best detection limit was achieved using alkaline phosphatase and G6PDH, xanthine oxidase was more sensitive than peroxidase even when the signal was weaker than that of peroxidase. For high target concentrations, peroxidase gave a very strong signal but the signal-background ratio decreased rapidly for low target concentrations. Alkaline phosphatase produced the lowest background using nitrocellulose as support. Nylon gave a better signal with AMPPD but the background was also increased. The four detections were performed with nitrocellulose (Amersham C extra) filters since the Argus 100 provided a highly sensitive detection..
Luminescence detection and multihybridization. Since it is possible to detect several probes using different luminescence detection, we studied the possibility of revealing different probes hybridized simultaneously with the same target. For this purpose we used biotin, and digoxigenin probes and after hybridization we revealed with phosphatase labelled antibody and with G6PDH, peroxidase, or xanthine, labelled streptavidin. The different enzymes were detected after a rapid washing step with the buffer used to perform the enzymatic reaction. To avoid interference between the various assays, it was necessary to perform the phosphatase assay after the other reactions since AMPPD is more difficult to remove from the filters. We found that this luminescent reagent led to some inhibition of the G6PDH.

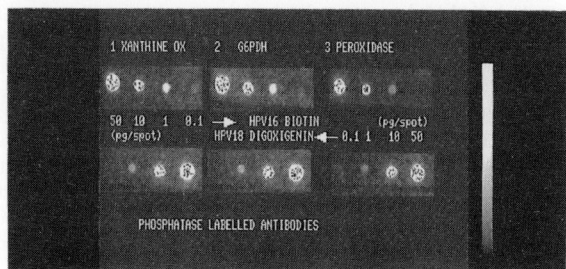

Fig:1 : Detection of biotin and digoxigenin labelled hybrid using streptavidin labelled with xanthine oxidase (1) G6PDH (2) or peroxidase (3),and phosphatase labelled antibodies.

Fig 1 shows results obtained with the different detection systems, we did not notice any interference between the different detection systems. In this experiment the detection was performed with a streptavidin conjugate and then the phosphatase was detected by AMPPD luminescence. Using general primers it was possible to amplify several homologous HPVs and after amplification the identification could be done using several labelled probes (6). The multidetection system avoided having to perform several hybridizations and moreover the competition between the probes improved the specificity of the hybridization reaction. Using oligonucleotide probes detected by phosphatase labelled antibodies and xanthine oxidase, G6PDH, or peroxidase labelled streptavidin it was possible to identify papillomavirus 16 and 18. We even determined that the two viruses were present in different quantities.

Multidetection was also used to identify allelic variations and identification of heterozygotes. We performed hybridization with 2 oligonucleotides and compared the different intensities of the signal produced by the different probes.

Fig 2 shows results obtained after hybridizations with oligonucleotides used for HLA DQB typing. These oligonucleotides were labelled using biotin and digoxigenin, one hybridized with a common part contained in all amplified sequences and the other was allele specific. Using the imaging system it was possible to calculate the ratio of the two signals (common and allele-specific), thus permitting identification of homozygous and heterozygous DNAs.

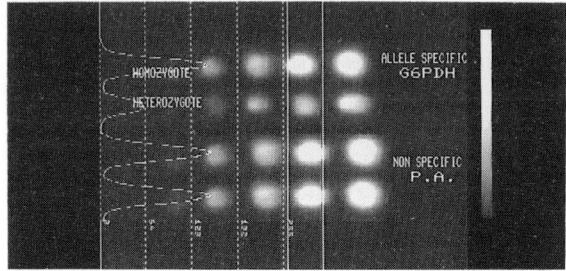

Fig 2 : Detection of hybrid with biotinylated and digoxigenin oligonucleotides (HLA DQB).

Various amounts of amplified material were applied to the filter and hybridized simultaneously with the two oligonucleotides.The hybrids were detect with G6PDH-streptavidin and phosphatase labelled antibodies. The specific oligonucleotide gave a lower signal for heterozygous DNAs in each case. For each spot it was possible to calculate the luminescence ratio (G6PDH/Phosphatase) which was close to 1 for homozygotes and 0.5 for heterozygotes. The use of a ratio eliminated errors which could be due to variations in the amounts of immobilized target, provided by a low production of target during the amplification process or by an incomplete immobilization on the filter.

We have shown that using luminescent detection systems it was possible to detect several probes on the same sample The advantages are the rapidity of the assay, and the possibility to obtain quantitative results with a video analyser such as the argus 100 from Hamamatsu. However to obtain accurate results it was necessary to plot a standard curve since, for very low concentrations, the signal is not directly proprotional to the amount of target. Quantitative analysis using luminescent probes have considerable potential for the identification of mutation or gene amplification.

This method can be extended to other labels since numerous dehydrogenases using different substrates are available and some of them are thermostable. We think that a system using oligonucleotides coupled to different enzymes and detected by luminescent reagents could be a new method for nucleic acid analysis.

ACKNOWLEDGEMENTS
This work was supported by the French Ministry of Education (Grant 88 M 0886).

REFERENCES
1   Pollard-knight, D., Read, C.A., Downes, M.J., Howard, L.A., Leadbetter, M.R., Pheby, S.A., Mcnaughton, E., Syms, A., Brady, M.A.W., Nonradioactive Nucleic Acid Detection by enhanced chemiluminescence using probes directly labelled with horseradish peroxidase, Anal. Biochem. 185, 1, 84-89, (1990).

2   Pollard-knight, D., Simmonds A.C., SchaapA.P., Akhavan H. and Brady A.W. Non radioactive DNA detection on southern blots by enzymatically triggered chemiluminescence. Anal. Biochem. 185, 1, 353-358, (1990).

3   Bronstein, I., Voyta, J.C., Lazzari, K.G., Murphy, O., Edwards, B., Kricka, L.J., Rapid and sensitive detection of DNA in Southern blots with chemiluminescence, Biotechniques, 8, 3, 310-314, (1990).

4   Baret, A., Fert, V., Aumaille, J., Application of a Long-Term Enhanced xanthine oxidase-induced luminescence in solid-phase immunoassays. Anal. Biochem. 187, 1, 20-26, 1990.

5   Balaguer, P, Térouanne, B., Eliaou, J-F., Humbert, M., Boussioux, A-M, and Nicolas, J-C. Use of glucose 6-phosphate dehydrogenase as a new label for nucleic acid hybridization reactions. Anal. Biochem. 180, 50-54.(1989)

6   Van Den Brule A.J.C., Snijders P.J.F., Gordijn R.L.J. , Bleker O.P., Meijer C.J.L.M. and Walboomers J.M.M., General primer-mediated polymerase chain reaction permits the detection of sequenced and still unsequenced human papillomavirus genotypes in cervical scrapes and carcinomas. Int. J. Cancer 45 644-649 (1990)

LONG TERM ENHANCED XANTHINE OXIDASE SYSTEM IN PROTEIN BLOTTING

A. Baret, J.C. Nicolas [1], Perot C.

*Laboratoire d'analyses médicales, 11 allée Duquesne
44000 Nantes, France.
[1]INSERM Unité 58, 60 rue de Navacelles, 34090 Montpellier,
France.*

The detection of antigens by immunodot binding and Western blotting assays using enzymatic probes is used widely in many clinical and research laboratories.(1-3). Generally, the use of chromogenic substrates for the detection of antigen-antibody complexes are reported. However,recently the advantages of luminescent detection (sensitivity, versatility) have been emphasized (5-7). We have demonstrated that xanthine oxidase (XO) dependent luminol oxidation is dramatically enhanced in the presence of iron EDTA complex and peroxide in alkaline buffer (8). The mechanism is consistent with an O_2^- driven Fenton reaction, leading to the generation of hydroxyl radical $OH^{\cdot}$, which can oxidize luminol with high efficiency.

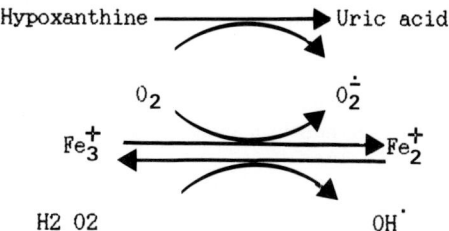

XO is detected with a high sensitivity (1 amole) and can be used as a label in immunoanalysis (9-10).
We describe here the application of XO label to the luminescent detection of immunodot binding and Western blotting.

MATERIAL AND METHODS

Purified XO (Buttermilk, sp act 1.23 units/mg) was purchased from Biozyme as an ammonium sulfate suspension. Streptavidin was purchased from Sigma; goat anti mouse IgG and goat anti human IgG + IgM antisera were from Jackson Laboratories, and mouse IgG (MAK 33) from Boehringer. Streptavidin-XO and anti Ig-XO conjugates were prepared as previously described (10).
The composition of the optimized luminescent reagent is the following : Borate 0.2 M pH 10.2, luminol 25 µM, iron EDTA complex (molar ratio EDTA/Fe : 20) 62.5 µM, Hypoxanthine 1 mM, Na Perborate 20 µM).

Immunodots : Mouse IgG is diluted in Phosphate buffered saline (PBS) containing 0.1 % BSA. One μl of each dilution containing 2.0 ng to 0.0 ng IgG, is spotted on nitrocellulose strips (Schleicher et Schüll BA 85). The strips are then air dried, blocked with PBS containing 2.5 % BSA for 1 h, and then incubated under slow agitation for 1 h with anti mouse IgG-XO conjugate diluted to 1 μg/ml in PBS, 0.1 % BSA. The strips are then washed four times soaking first for 30 s and three times for 5 min each pH 8.6 25 mM borate containing 0.05% BSA and 0.1% Tween 80

Detection of biotinylated proteins Proteins equilibrated in pH 8.6 20 mM borate buffer containing 50 mM NaCl by Sephadex G 25 column chromatography (PD 10 column, Pharmacia) are biotinylated with biotin aminocaproic acid N-hydroxysuccinimide ester (Pierce) (w/w 20 %). The unreacted reagent is eliminated by a further PD 10 chromatography. Sequential dilutions of the biotinylated proteins are separated by PAGE (0.9 % acrylamide), then electrotransferred to a nitrocellulose support. This support is then blocked in PBS containing 2.5 % BSA for 1.0 h, then incubated in the same buffer containing XO-SA conjugate (2 ug/ml) for 1.0 h at ambient temperature. The filters are then washed as described above.

Detection of human anti HIV 1 antibodies (Western blot) : The nitrocellulose support containing HIV1 viral proteins was obtained from Diagnostics Pasteur. The detection of human anti HIV1 antibodies is performed according to the manufacturer´s recommendation (colorimetric detection), or by immunoluminescence. In the latter case, after the incubation with the diluted sera the strips are washed three times with PBS containing 2.5 % BSA, then incubated with antihuman IgG + IgM - XO conjugates diluted in the same buffer for 1 h (final concentration : 1 μg/ml). The strips are then washed as described above.

Luminescence detection : After the washing step, the filters are dried, immersed in the luminescent reagent for 5 min and then placed between two sheets of clear plastic wrap. The light emission is detected by exposure to a photographic film (Hyperfilm, Amersham).

RESULTS AND DISCUSSION

As shown in Fig.1, it is possible to detect picogram quantities of mouse IgG immobilized on nitrocellulose filters, using anti mouse IgG-XO conjugate, revealed by the enhanced luminescent reaction coupled to photographic film. These results were obtained for a 15 min exposure time. Exposures were repeated without loss of sensitivity (0 - 10 h) at various times after triggering the luminescent reaction. The presence of an EDTA complex in the luminescent reagent leads to the improvement of the detection level of mouse IgG (Fig 1, 0 μM iron EDTA : 100 pg; 50 μM iron EDTA : 10 pg). Previously data obtained with XO in liquid phase or immobilized in microwells (8), showed that this enhancement is much higher (10^3 fold) and that the luminescent signal from unenhanced reaction (0 uM iron EDTA) is very low. When using a nitrocellulose support, an XO dependent luminescent signal can be readily visualized and mouse IgG can be detected with a sensitivity of 100 pg. It is thus possible that the nitrocellulose support itself leads to an enhancement of XO dependent light emission.

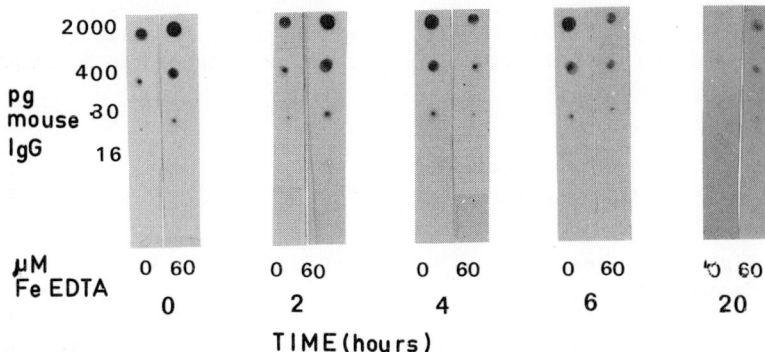

Fig.1 Immunodot binding of mouse IgG

The enhanced chemiluminescent detection of XO has been applied to
the detection of electrotransferred biotinylated BSA or urinary
proteins, using XO-SA conjugate (Fig 2) and less than 0.1 ng of
protein can be detected with this procedure. It thus appears that
this method can be used for the analysis of low protein level
biological fluids.

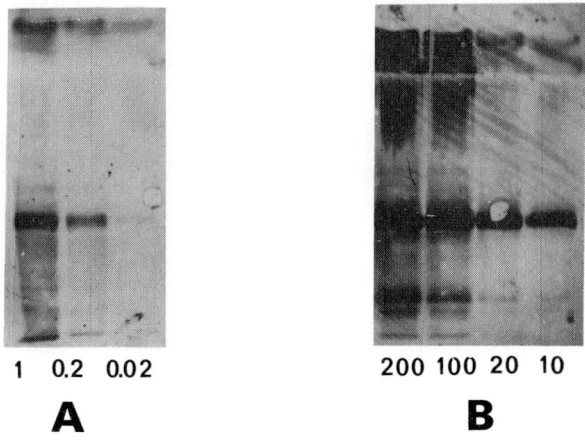

Fig. 2 Luminescent detection of electrotransferred biotinylated
proteins. Exposure time 10 min. A : ng biot. BSA, B : ng biot.
urinary proteins.

Chemiluminescent detection of anti human IgG + M-XO conjugate can
be applied to the detection of anti HIV1 antibodies after Western
blotting. Fig 3 shows that the resolution is excellent allowing a
clear identification of every specific antibody. This result is
probably related to the nature of the highly reactive oxyradical
responsible for luminol oxidation. The reaction of OH˙ radical with

luminol occurs immediately after its generation, at a very close distance from the immunological fixation site of the XO conjugate.

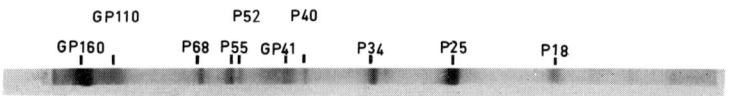

Fig. 3 Luminescent detection of human anti HIV 1 antibodies. Exposure time 10 min.

It thus appears that the detection system consisting of chemiluminescent enhanced XO detection on autoradiography film, is well suited for immunodot binding and Western blotting assays, with the following advantages (1). High sensitivity (2). Long term signal (3). High resolution (4). Low cost.

REFERENCES
1. H. Towbin and J. Gordon, Immunoblotting and dot immunoblotting current status and outlook. J. Immunol. Meth. 72,313-340 (1990).
2. M.S. Blake, K.H. Johnson, G.J. Russel-Jones and E.C. Gostschlich, A rapid, sensitive method for detection of alkaline phosphatase-conjugated antibody on Western Blots. Anal. Biochem. 136, 175-179 (1984).
3. M.S. Brower, C.L. Brakel and K. Garry, Immunodetection with streptavidin-acid phosphatase complex on Western blots, Anal. Biochem. 147, 382-386 (1985).
4 . M.M.L. Leong, C. Milstein and R. Pannel, Luminescent detection method for immunodot, Western and Southern blots, J. Histochem. Cytochem. 34, 1645-1650 (1986).
5. R. Hauber, W. Miska, L. Schleinkofer and R. Geiger, New, sensitive, radioactive-free bioluminescence-enhanced detection system in protein blotting and nucleic acid hybridization. J. Biolumin. Chemilum. 4, 367-372 (1988).
6. M.M.L. Leong and G.R. Fox, Enhancement of luminol-based immunodot and Western blotting assays by iodophenol. Anal. Biochem 172, 145-150 (1988).
7. A. Baret, V. Fert and J. Aumaille, Application of a long-term enhanced xanthine oxidase induced luminescence in solid phase immunoassays, Anal. Biochem. 187, 20-26 (1990).
8. A. Baret and V. Fert, T4 and ultrasensitive TSH immunoassays using luminescent enhanced xanthine oxidase assay. J. Biolumin. Chemilumin. 4, 149-153 (1989).
9. V. Fert and A. Baret, Preparation and characterization of xanthine oxidase-antibody and hapten conjugates for their use in sensitive chemiluminescent immunoassays, J. Immunol. Meth. 131, 237-247 (1989).

ACKNOWLEDGMENT : This work was supported by grants from Packard Instrument B. V., Groningen, Holland.

BUFFER EFFECTS ON THE TRIGGERING OF HORSERADISH PEROXIDASE ENHANCED CHEMILUMINESCENCE

Bibijana Cercek, R.J.Obremski, C.S.Oh

Beckman Instruments,Inc., 200 South Kraemer Boulevard, Brea, CA 92621, USA

INTRODUCTION

The effects of the chemical composition of buffers in the triggering of HRP enhanced chemiluminescence were investigated. Although the importance of the para-iodophenol enhancer, luminol and peroxide in this system has been demonstrated (1,2), the interaction of the buffer itself as a crucial fourth component influencing the signal reagent background, kinetics of light production, and hence, both the linearity of the HRP dose response curve and the HRP detection limits, has not been previously investigated. The significant influence of six different buffers on this system cannot be attributed to pH alone. Changing the molarity of the buffers also significantly affected both the kinetics and the steady-state light yields.

MATERIALS AND METHODS

Reagents

Horseradish peroxidase (HRP) Type VI, luminol, sodium perborate, sodium tetraborate 'borax', Trizma base, bis-tris-propane, glycine, glycylglycine free base, tricine, were all obtained from Sigma Chemical Co. Para-iodophenol was from Aldrich.

Assay Protocol

A stock solution of luminol and para-iodophenol at 2 mmol/l and 4.4 mmol/l, respectively, was prepared in the appropriate 50 mmol/l buffer, final pH 8.4. This was either diluted ten fold in the same buffer at pH 8.5 to give a solution of final pH 8.5, or in the same buffer at pH 6.0 to give a solution of final pH 7.3. To this was added freshly prepared perborate (16 mg/ml) in the appropriate buffer, pH 8.5. Thus, the final luminol, para-iodophenol and perborate concentrations in the signal reagent were 200 umol/l, 440 umol/l and 194 umol/l, respectively.

The luminol, 3x recrystallised according to protocol in (3), was dissolved at 8 mg/ml in the appropriate buffer, pH 10.5. The para-iodophenol was prepared as a 60 mg/ml stock soluion in 0.5 mmol/l sodium hydroxide and then diluted to 23 mg/ml in the appropriate buffer, pH 8.5. The HRP was prepared as a 2 mg/ml stock and then diluted in borax buffer, pH 8.5.

In the HRP assay, all measurements were carried out in the kinetics mode (20 second integration time) on the Berthold Clinilumat LB 9502, in which the chemiluminescence signal units are reported as Relative Light Units (RLU). The signal reagent background was first determined on a 400 ul aliquot of the signal reagent over 100 seconds. A 40 ul aliquot of the HRP sample was then added to the signal reagent, instantly vortexed, and the kinetic profile followed for an additional 200 seconds. The chemiluminescence signal at the 300 second time point, was used for the dose response curves.

RESULTS AND DISCUSSION

At any given HRP concentration within the range studied, from 8 to 0.016 fmol/40ul assay sample, whether at pH 7.3 or 8.5, the kinetic time course of the light generating reaction and/or the magnitude of the signal itself, are both significantly influenced by the signal reagent buffer composition (Fig. 1 and 2). This is most evident when the signal reagent is prepared with tricine, glycylglycine and glycine, where the steady-state is not yet reached at 300 seconds (200 seconds after addition of HRP) and is preceded by a distinct, buffer dependent lag phase. The HRP dose response at pH 7.3 and 8.5, showing low sensitivity, sigmoidal curves typical of this buffer group, can be seen in Fig. 3. In contrast, signal reagents prepared in the borax, bis-tris-propane and tris buffers result in rapid rise to steady-state kinetics, within the 300 second measurement time, with only an indication of a lag phase for the bis-tris-propane and tris buffers at pH 8.5. Thus, this latter buffer group gives rise to greater dose response linearity and low-end sensitivity under the two pH conditions studied. A representative diagram is shown in Fig. 4.

The only pH related difference observed is in the signal reagent background, which is generally up to two fold greater at pH 8.5 than pH 7.3.

The involvement of luminol and enhancer radicals in the mechanism of light generation has been proposed (2,4). Our data not only indicate that the HRP generated signal is completely inhibited on addition of the known free radical scavengers, ascorbate and hydroquinone, but also show a significant buffer-concentration dependent interaction in the light generating process as evidenced by the effect of change in buffer molarity on both the kinetics of the light generating process and the steady-state luminescence yields. Further experiments are in progress.

In conclusion, our studies show that buffers are not inert, and must be considered as a crucial fourth component of the HRP enhanced chemiluminescence light generating system. This further suggests the importance of keeping to a single buffer medium, as far as is possible, in determining the pH optimum for a given signal reagent system.

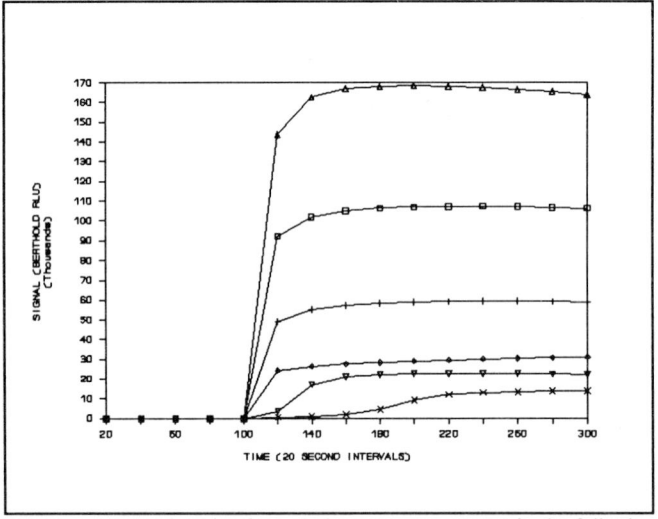

Figure 1. HRP (4 fmol/40ul assay) time course at pH 7.3 in the following buffer based signal reagents : Borax (□); Tris (╬); Bis-Tris-Propane (◇); Glycine (▲); Tricine (✕); Glycylglycine (▽).

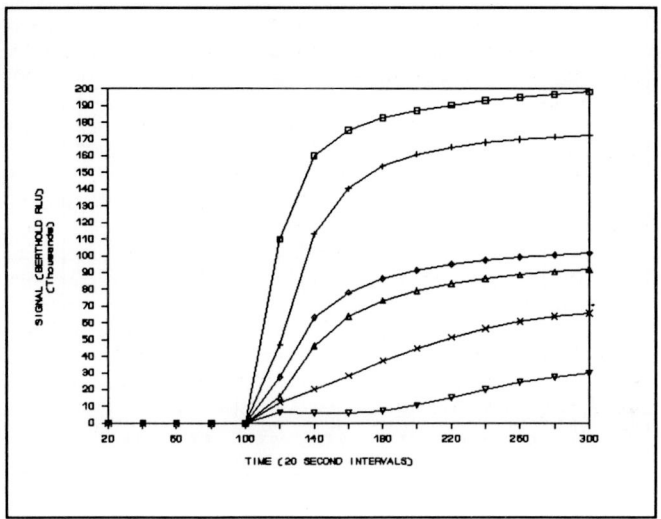

Figure 2. HRP (1.6 fmol/40ul assay) time course at pH 8.5 in the following buffer based signal reagents : Borax (▢); Tris (╈); Bis-Tris-Propane (◇); Glycine (▲); Tricine (✕); Glycylglycine (▽).

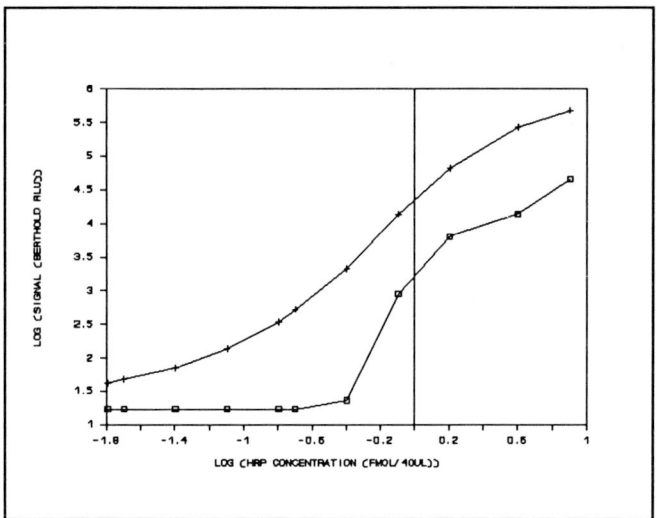

Figure 3. HRP dose response curves in Tricine signal reagent at pH 7.3 (▢) and pH 8.5 (╈).

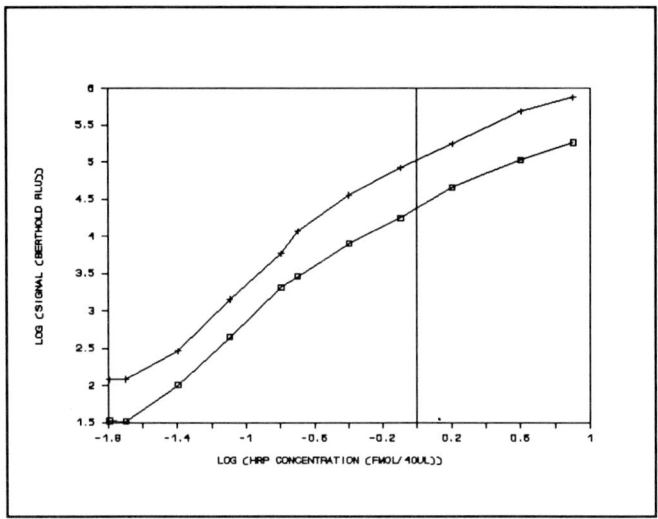

Figure 4. HRP dose response curves in Borax signal reagent at pH 7.3 (☐) and pH 8.5 (+).

ACKNOWLEDGEMENTS

We would like to thank Mr. T. Dugan for his diligent technical assistance and Mr. N. Shah for his assistance with the computer programming. We are also grateful to Dr. B. Cercek for helpful discussions.

REFERENCES

1.   G.H.G. Thrope, L.J.Kricka, S.B.Moseley and T.P. Whitehead, Phenols as enhancers of the chemiluminescent horseradish peroxidase-luminol-hydrogen peroxide reaction: Application in luminescence-monitored enzyme immunoassays. Clin. Chem. 31, 1335-1341 (1985).

2.   G.H.G. Thorpe and L.J. Kricka, Enhanced chemiluminescent assays for horseradish peroxidase: characterisation and applications. In Bioluminescence and Chemiluminescence: New Perspectives, J. Scholmerich, R. Andreesen, A. Kapp, M. Ernst and W.G. Woods(Eds), Wiley, Chichester, p.199-208 (1987).

3.   G. Ham, R. Belcher, L.J. Kricka and T.J.N. Carter, Stability of trace iodine solutions. Anal. Lett. 12, 535-541 (1979).

4.   M. Hodgson and P. Jones, Enhanced chemiluminescence in the peroxidase-luminol-peroxide system: Anomalous reactivity of enhancer phenols with enzyme intermediates. J. Biolumin. Chemilumin. 3, 21-25 (1989).

DEVELOPMENT OF IMMUNOCHEMILUMINOMETRIC ASSAY (ICMA) FOR THE MEASUREMENT OF AFP AND MAPPING OF THE ANTIGENIC EPITOPES OF AFP

S.H.Kim[1], C.K.Kim, O.J.Kwon, K.S.Chung and J.B.Kim

Department of Animal Products Sciences, Kon-Kuk University and
[1]Research Lab. of Korea Green Cross Corp., Seoul, Korea

INTRODUCTION

Alpha-fetoprotein(AFP) is a single polypeptide chain glycoprotein with a Mw of 69,000 daltons(1) and its measuring in biological fluids is clinically important in diagnosis of tumor and pregnancy-related disorders.

This study was performed to develop a non-isotopic immunoassay system for measuring AFP using immobilized antibody to polystyrene tube and ABEI-H labelled antibody based on namely two-site sandwich immunochemilumino-metric assay(ICMA). For this, AFP was purified by immunoaffinity chromato-graphy and used to produce monoclonal antibodies to it. Monoclonal antibodies with different specifities were characterized, used to develop ICMA and investigate the antigenic epitopes of AFP. This method was evaluated in terms of sensitivity, accuracy, and correlation with RIA and EIA.

MATERIALS AND METHODS

Reagents Hypoxanthine, aminopterine, thymidine, PEG 1500, microperoxidase (MP-11), hydroxysuccinimide, dimethylformamide, dicyclohexylcarbodiimide were purchased from Sigma Co., DMEM, complete and incomplete Freund's adjuvant, gentamycine were from Gibco Lab. Bovine calf serum and fetal bovine serum were from Hyclone, ABEI-H was from LKB Co., Protein A-agarose gel and its binding & elution buffer were from Pierce Co., mouse hybridoma subtyping kit was from Boehringer Mannheim Co., polystyrene tube (12x75mm) was from NUNC Co., and AFP RIA and EIA kits were from Korea Green Cross Corp.

Preparation of monoclonal antibody(McAb) to AFP

AFP was purified by immunoaffinity chromatography from cord serum. McAb to AFP was produced according to a procedure previously deseribed(2). Antibodies from ascitic fluids were purified by protein A-agarose column for IgG and AFP coupled immunoaffinity column for IgG.

ICMA procedure Twenty μl of AFP standard or serum sample and 200μl of incubation buffer containing bovine calf serum were added to the antibody-coated tubes and incubated for 2 hs at room temperature. The solution was aspirated and washed 3 times with distilled water. ABEI-H labeled McAb(200μl) prepared according to the method of Barnard et al(3) was them added further incubated for 2 hs at room temperature and followed aspiration and washing. After reaction, 200μl of 5N NaOH was added to each tube and incubated for 1 h at 60℃. The tubes after cooling were placed in the luminometer(Berthold CliniLumat LB 9520). The chemiluminescence reaction was initiated by the rapid injection of microperoxidase(100μg/ml) and 200μl hydrogen peroxide(0.3%) at the same time. The light emitted was integrated for 4 s.

Epitope analysis by ICMA

Accordingly, the interrelationship of antibody binding sites on the surface of the AFP molecule has been investigated using the ICMA principle similar to the approach described previously for hGH(4). For this, all possible combinations of the anti-AFP McAb were used to construct a matrix of 10 capture antibodies x 10 labeled antibodies.

These combinations were studied for their ability to bind to AFP simultaneously.

RESULTS AND DISCUSSION

Epitope analysis of AFP

The quantitative data of all 10 x 10 combinations have been expressed as positive or negative shown in Fig.2 The black squares symbolize compatibility of AFP binding sites(i.e. apparently nonoverlapping epitopes), while open squares represent combinations of antibodies that cannot bind to AFP simultaneously. Distinct patterns of positive and negative binding could be distinguished.

		LABELED ANTIBODY									
		2	4	9	11	22	32	75	94	176	239
IMMOBILIZED ANTIBODY	2		■		■	■		■		■	■
	4	■		■	■	■	■	■	■	■	■
	9		■		■	■		■		■	■
	11	■	■	■		■	■	■	■	■	■
	22	■	■	■	■		■		■		
	32		■		■	■		■	■	■	■
	75	■	■	■	■		■		■		
	94		■		■	■	■	■		■	■
	176	■	■	■	■		■		■		
	239	■	■	■	■		■		■		

Fig.1. Cheas-board two-site ICMA of a 10 x 10 matrix of McAbs combinations. The numbers refer to the individual McAb code. Reactions classified as positive are represented here as ■, indicating that two McAbs recornize different antigenic determinants.

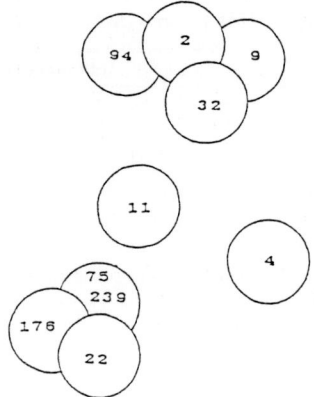

Fig. 2. Diagram of AFP epitopes mapped by 10 x 10 matrix
         of ICMA data. Each number denotes McAb.

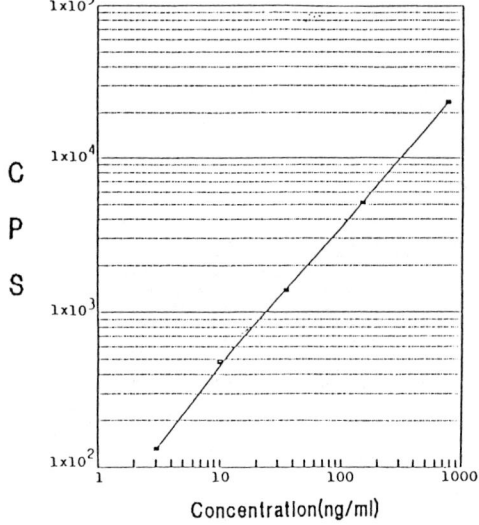

Fig. 3. Standard curve of ICMA system for measuring AFP.

The results shown in Fig. 1 indicate similar reaction patterns for McAb NO.
22, 75, 239, 176, and also for McAbs 2, 9 and 32.   Accordingly, from the
two-site epitope analysis, it may be concluded that 10 McAbs express nine
distinct reaction patterns namely, McAbs No 2, 4, 9, 11, 22, 32,
75/239, 94, and 176, which can form the basis for the construction of a n
epitope map represented in Fig. 3 in which the spatial relationships
between the different antigenic regions can be expressed.

Table 1. Intra- and inter-assay result.

		Intra-assay		Inter-assay	
Serum	Number of Determination	Mean± SD (ng/mℓ)	CV(%)	Mean± SD (ng/mℓ)	CV(%)
A	10	26.3± 3.2	12.0	25.0± 3.3	11.7
B	10	94.5± 6.6	6.9	91.4± 7.5	8.1
C	10	191.1± 15.4	8.0	222.2± 23.9	10.6

Table 2. The results of recovery test.

Added AFP (ng/mℓ)	Observed AFP (ng/mℓ)	Expected AFP (ng/mℓ)	Recovery (%)
35	138	153	90.0
150	256	268	95.6
500	674	618	109.1

* Serum sample contained 118ng/mℓ of AFP

Evalnation of ICMA for the measurement of AFP in serum

One combination, McAb 22-McAb 94 of 10 combinations was chosen for the ICMA system for the assay of AFP in serum.   Its dose-response curve are show in Fig.2.   The minimum concentration of AFP that could be significantly distinguished from zero was 1.0ng/ml.

The intra- and inter-assay variences was shown in Table 1.   Increasiny amounts of AFP were added to serum which had been analysed previously and -were subjected to the analysis.   The results are shown in Table 2.   The result of ICMA corretaled with those obtained with RIA ; Y=0.91X + 11.8 where X was RIA(r=0.98) and ELISA ; Y=0.82X + 13.3, where X was ELISA(r=0.97).

We have found that the ICMA developed for the measurement of AFP are at least as sensitive, precise and specific as conventional RIA and ELISA.

REFERENCES

1. M. Masseyeff, C. Aussel and P. Soubiran.   Recent data on alpha-feto-protein. Blood transfusion and Immunohaematology. 27, 6, 767-783, (1984).
2. S.W. Kimm, B.S. Chai, H.K. Chung, P.G. Suh and N.K. Kim. Production of monoclonal antibody to human α-fetoprotein. Korean J. Biochem 16, 2, 41-47,(1984).
3. G.J. Barnard, J.B. Kim, J.L. Brockelbank, W.P. Collins, B. Gaier and F. Kohen., Measurement of choriogonadotropin by Chemiluminescence Immuno-assay and Immunochemiluminometric Assay : 1. Use of Isoluminol Derivatives., Clin. Chem., 30. 538-541 (1984)
4. C.J. Strasburger, J. Kostyo, T. Vogel, G.J. Banard and F. Kohen.   The antigenic epitopes of human growth hormone as mapped by monoclonal antibodies. Endocrinology, 124, 3, 1548-1557, (1989).

MEASUREMENT OF ENTEROTOXIN A AND OCHRATOXIN A BY SOLID-PHASE CHEMILUMINESCENCE IMMUNOASSAY

C.K. Kim[1], J.S. Han, J.H. Lee, H.K. Shin and J.B. Kim

Department of Animal Products Science, Kon-Kuk university, and
[1]Yang Ji Chemical Co. LTD, Seoul, Korea

INTRODUCTION
The detection of staphylococcal enterotoxin A(SEA) in foods is very important because staphylococcal food poising is one of the most common food-borne diseases(1). Ohratoxin A(OcA) is one of the most toxic isocoumarin containing mycotoxins produced by various species of *Aspergillus* and *Penicillium*(2). Because of their association with certain mycotoxicoses in animals and possibly in humans, they have received considerable attention in recent years.
One of the most important tasks for monitoring these toxins is to establish the assay system. Solid-phase chemiluminescence immunoassay(CIA) systems for the measurement of SEA and OcA were therefore developed and evaluated.

MATERIALS and METHODS
Reagents The labels used in this study were horseradishperoxidase(HRP) for ELISA and N-(4-aminobutyl-N-ethyl)-isoluminol(ABEI) and N-(4-aminobutyl -N-Ethyl)-isoluminol hemisuccinamide (ABEL-H) for chemiluminescent assays. HRP and chemiluminescent markers were conjugated to the antigens to the assayed according to the methods of Kim et al(3) and Barnard et al(4). SEA purified by the method of Bergdoll et al(5) and OcA purchased from Sigma(USA)were used as standards and to raise antibodies. The antisera to enterotoxin A and ochratoxin A-BSA were raised in rabbits(6) and IgG fractions of antisera used for coating were prepared by protein A-sepharose column(Sigma;USA).
Assay Protocol One hundred μl of sample or standard solution was added in duplicate to the antibody-coated tube. Subsequently, we added 100μl of tracers, SEA-ABEI-H and OcA-ABEI for CIA, SEA-^{125}I for RIA, and OcA-HRP for ELISA. The mixture was incubated at 37℃ for 2 hs. and the contents of the tube were removed by aspiration. The antibody bound fractions were washed three times with phosphate buffer. For the CIA we added 200μl of 5mol/L NaOH to each tube and incubated the contents at 60℃ for 1h. Microperoxidase solution(100μl) and 100μl of diluted H_2O_2 was autometically injected to the assay tube in the luminometer(Berthold:West Germany) and the signal integrated for 4s. Minireader Ⅱ(Dynatch:USA) for ELISA and MiniGamma 1257(LKB:Finland) for RIA were used.

RESULTS AND DISCUSSION
Typical calibration curves for the measurements of SEA and OcA in solid-phase CIA are shown in Fig.1. The least amounts of SEA and OcA that could be distinguished from zero(mean-2 SD) was calculated from 10 calibration curves.
The values were 0.5ng and 20pg per tubes, respecitively, which are similar with other results of RIA and ELISA(1,2).
Different known amounts of authentic SEA and OcA were added to sausage and corn, respecively, which had been analysed previonsly, and were analyzed by solid-phase CIA. The analytical recoveries of SEA were more than 95% but those of OcA were less than 70%.
The low recoveries in the assay of OcA may be due to the loss caused during the extraction procedure which has been indicated as a common problem encounted with the assay of OcA(2).

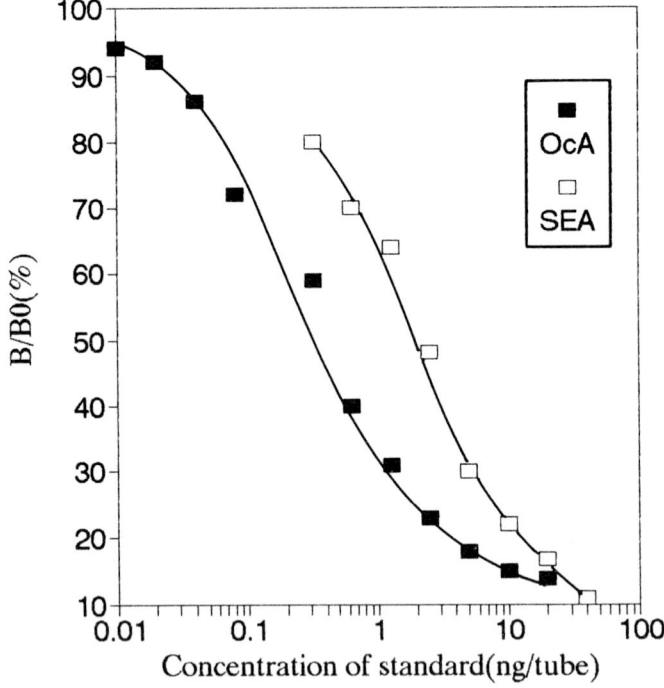

Fig. 1.   Standard curves for SEA and OcA in solid-phase CIA

An estimate of the intra-assay variations was obtained by analysing replicate samples within a single assay. The corresponding value for inter-assay variations was also obtained from the measurement of OcA in 3 corn samples SEA in one sausage sample used for internal quality control over a period of 3 months. The results are shown in Table 1.

Table 1. Intra- and Inter-assay variances of SEA and OcA as
          determined by a solid-phase CIA

		Intra-assay variance[a]			Inter-assay variance[b]		
		Mean ± S.D		C.V	Mean ± S.D		C.V
OcA	High	925 ± 27.4		2.5	903.8 ± 49.1		5.4
	Midium	576 ± 13.1		2.3	561.1 ± 76.8		13.6
	Low	210 ± 21.1		10.0	218 ± 8.5		3.9
SEA		6.6± 0.5		7.4	4.2 ± 0.17		4.0

[a]Data are means of twenty analyses
[b]Data are means of ten analyses
Coefficient variation (%) = Standard Deviation / Mean × 100

The concentration of SEA in 29 samples was determined by both CIA and RIA
and that of OcA in 20 samples was also determined by both CIA and ELISA to
calculate correlation with each other. The results are shown in FIG. 2 and
Fig. 3. The results of solid-phase CIA developed in this study were in
good agreement with the values derived by ELISA and RIA used for the
reference methods.

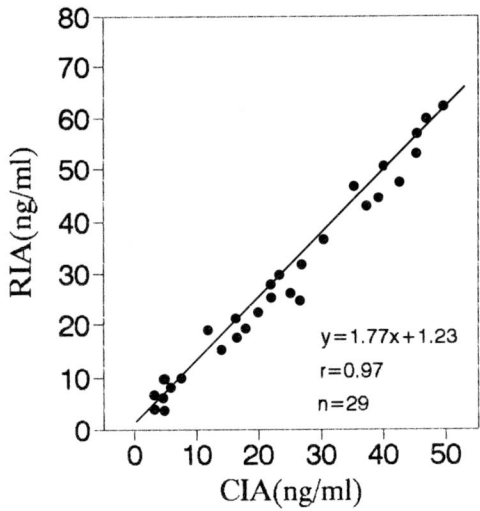

Fig. 2. Correlation between CIA and RIA for SEA

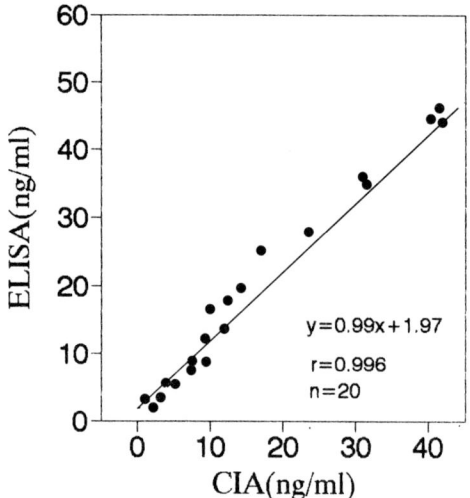

Fig. 3.   Correlation between CIA and ELISA for OcA

Although several analytical methods have been developed for the assay of
SEA and OcA in agricultural commodities(1,2), there has been much
controversy over the use of alternative, non-isotopic labels in
immunoassay(7).
The resalts of the present study have shown that it is possible to develop
a simple non-isotopic immunoassay to monitor various mycotoxins and
entrotoxins, although more studies should be done to be accepted as a
routine method specially for the assay of OcA.

REFERENCES
1. D.S. Orth, Statistical analysis and   quality control in radioimmunoassays for staphylococcal enterotoxins A, B and   C. Appli.
   Environ. Microbiol. 34, 710-714 (1977)
2. S.S. Lee and F.S. Chu, Enzyme-Linked Immunosorbent Assay of ochratoxin A in   wheat.   J. Assoc. Off. Anal. Chem. 67, 45-49 (1984)
3. J.B. Kim, G.J. Barnard, W.P. Collins,   F. Kohen,   H.R. Linder and
   Z. Eshhar, Measurementof plasma estradiol-17β by solid-phase
   chemiluminescence immunoassay. Clin. Chem. 28, 34-38 (1982)
4. G.J. Barnard, J.B. Kim, J.L. Brockelbank, W.P. Collins, B. Gaier and
   F. Kohen,   Measurement of choriogonadotropin by chemiluminescence
   immunoassay and immunochemiluminometric assay : 1. Use of isoluminol
   derivatives. Clin. Chem. 30, 538-541 (1984)
5. M.S. Bergdoll and R. Reiser,   Application of radioimmunoassay   for
   detection of staphylococcal enterotoxin in foods. J. Food. Prot.
   43, 68-72 (1980)
6. F.S. Chu, F.C.C. Chang and R.D. Hinsdill,   Production of antibody
   against ochratoxin A.   Appli. Environ. Microbiol. 31, 831-835 (1976)
7. R.F. Schall, H.J. Teuoso,   Alternatives to radioimmunoassay : labels
   and methods.   Clin. Chem. 27, 1157-1164 (1981)

DEVELOPMENT OF A NOVEL METHOD FOR EMBRYO SEXING BASED ON THE UTILIZATION OF A CAMERA LUMINOMETER

J.M.Paik, J.H.Kim, K.H.Jang, K.S.Chung and J.B.Kim

Department of Animal Products Science, Kon-kuk University, Seoul, Korea

INTRODUCTION

Production of offspring of a predetermined sex has become a goal of the embryo transfer industry since its early days. For this, various techniques for sexing embryo have been developed(1). The most widely used method for this has been the indirect immunofluorescence method utilizing antibody to histocompatability Y-chromosomal(H-Y) antigen present in only XY embryos, but not in XX embryos(2). We have already shown that when 8-cell embryos of mouse and bovine pre-treated with H-Y Ab are incubated with FITC-labeled 2nd Ab and observed with fluorescence microscope, only XY embryos are fluorescent(5). It was therefore possible to select XY embryos using this method. This technique however requires an expensive fluorescence microscope which is not convenient to transport from the laboratory to the farm. We investigated the possibility of Camera Luminometer(4) designed for use with enhanced luminescence assays as an alternative method to overcome some drawbacks associates with the indirect immunofluorescence method for the sexing of embryos.

MATERIALS and METHODS

Reagents 3-(2'-spiroadamantane)-4-methoxy-4-(3"-phosphoryloxy)phenyl-1,2-dioxetane disodium salt(AMPPD), Emerald enhancement solution, Polaroid type 612 film and Camera Luminometer(ICL 901) were purchased from Tropix,Inc.,USA. Goat anti-rabbit IgG-AP(alkaline phosphatase) and goat anti-rabbit IgG-FITC conjugate used for the indirect immunofluorescence test were purchased from Bio-Yeda,Israel.

Assay Protocols Polyclonal antibody to H-Y antigen was produced from the rabbit immunized with H-Y Ag purified by immunoaffinity column(5). 8 to 16-cell embryos were flushed from the oviducts of ICR mice on the second day after mating. Embryos were washed and cultured for 1 h in 100ul of Whitten's Medium(WM), which contained polyclonal H-Y antibody. After rinsing with WM, the embryos were further incubated for 1 h in 100 μl of WM containing goat anti-rabbit IgG-AP(1:1000) or goat anti-rabbit IgG-FITC conjugates. The embryos were washed and subsequently placed one by one in the wells of 96 well microtiter plate and 100 μl of Emerald enhancement solution prepared according to the mannual of Tropix Inc.. After the reaction, the plate was properly positioned into Camera Luminometer and exposed for 60s.

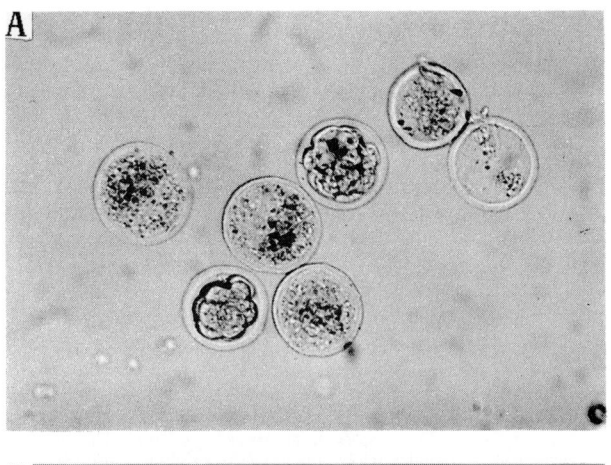

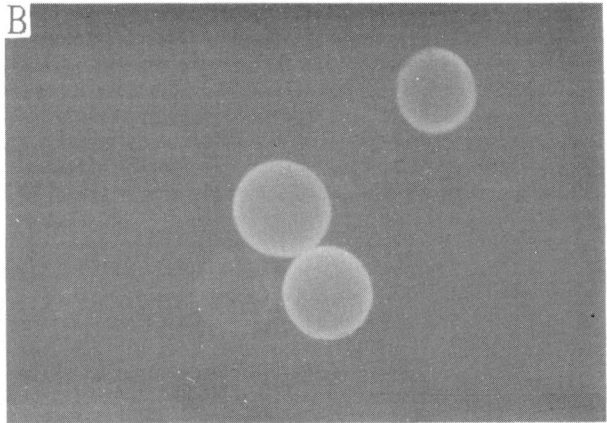

Fig.1.   Embryo sexing with polyclonal H-Y antibody by
         indirect immunofluorescence test.
         A: Observed under the phase contrast microscope.
         B: Observed under the fluorescence microscope.

RESULTS AND DISCUSSION
When embryos treated with FITC-labeled second antibody were observed under
inverted fluorescence microscope equipped for FITC excitation, fluorescing
embryos(Fig.1) were distinguished from non-fluorescing embryos areas of
fluorescence associated with one or more blastomeres. Embryos with no
fluorescence, which appeared as a faint haze rather than bright
fluorescence were classified H-Y negative. As shown in Table 1, 62
embryos(51.7%) from 120 embryos displayed cell-specific fluorescence which
might be regarded as XY embryos and the remaining 58 embryos(48.3%) with
no or faint fluorescence.

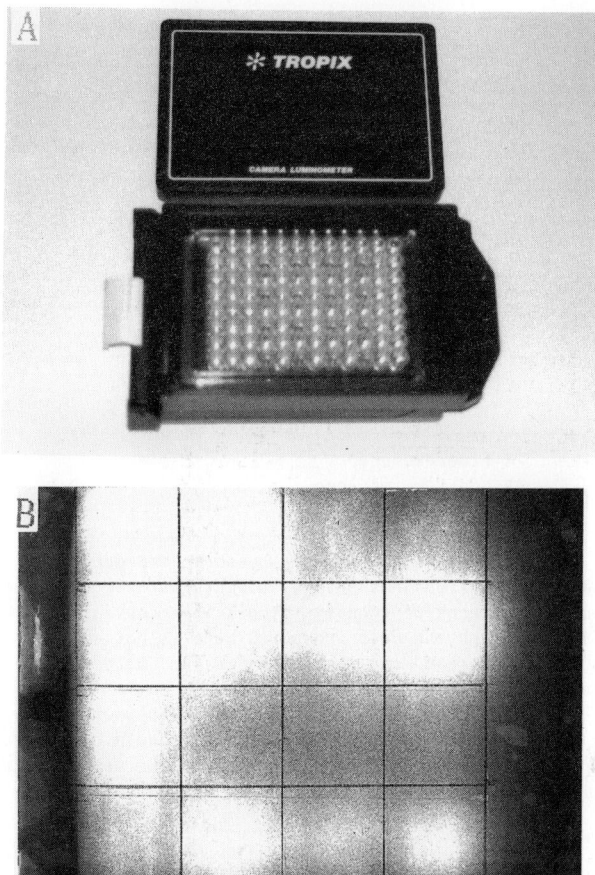

Fig. 2.   The result of indirect immunochemiluminescence test
          with camera luminometer.
          A: The sites of reaction on the microtiter plate.
          B: The result of the reaction.

In the same way as indirect immunofluorescence method, Camera Luminometer
was applied for indirect immunochemiluminescence test. As shown in Fig. 2,
one group emitted light which can be regarded as embryos(H-Y positive)
could be distinguished from others with no or faint lights which can be
scored H-Y negative; 9 from 16 embryos tested shown a strong
chemiluminescence and the rest 7 embryos which might be XX embryos did not
emit light. It was therefore possible to discriminate XX embryos from the
XY embryos which produced light.

Table 1. Presumptive H-Y phenotype in mouse embryos treated
with rabbit polyclonal Ab and rabbit IgG-FITC

Number of experiments	Number of embryos	H-Y+ (%)*	H-Y- (%)**
1	20	12	8
2	24	13	11
3	11	4	7
4	24	9	15
5	28	16	12
6	12	8	5
Total	120 (100)	62 (51.7)	58 (48.3)

* Fluorescence
** Non-fluorescence

The use of a single Camera Luminometer based on the detection of the light
from luminescent reactions has several advantages over the immunofluorescence technique in terms of economy, simplicity, convenience of handling
and carrying. Although the data presented here is not enough to conclude
that this novel method can be used as an alternative method to
conventional immunofluorescence method, we strongly believe that this
technique using simple portable Camera Luminometer would allow comparative
and acceptable results. For this, more studies should be done:the effects
of light, pH, AMPPD and enhancement solution on the development of embryos
are under investigations.

REFERENCES
1. S.S.Wachtel, H-Y antigen in the study of sex determination and
   control of sex ratio. Theriogenology. 21, 18-28 (1984)
2. K.L.White, G.M.Lindner, G.B.Anderson and R.H.BonDurant, Cytolytic
   and fluorescent detection of H-Y antigen on preimplantation mouse
   embryos. Theriogenology. 19, 5, 701-705 (1983)
3. I.Bronstein, J.C.Voyta, Chemiluminescent detection of Herpes
   Simplex Virus I DNA in blot and in-situ hybridization assays.
   Clinical chemistry. 35, 9, 1856-1857(1989)
4. Tropix, Guide chemiluminescence detection.
5. J.M.Paik, Studies on the purification and characterization of H-Y
   antigen. Master's thesis. Konkuk univ, Seoul, Korea (1990).

AUTOMATED CHEMILUMINESCENCE MEASUREMENTS OF TRANSIENT AND ENZYME LABELS IN MICROTITRE PLATES.

R.H. van den Berg, E.H.J.M. Jansen, E.J.M. Reinerink and G. Zomer

*National Institute of Public Health and Environmental Protection,
P.O. Box 1, 3720 BA Bilthoven, The Netherlands.*

INTRODUCTION

With microtitre plate technology large number of samples can be handled simultaneously and prepared for endpoint detection in (immuno)assays. Chemiluminescence or bioluminescence detection devices for microtitre plates are available now for enzyme enhanced chemiluminescence. The measurement of transient labels, however, requires an advanced injection system for each well. Recently a microtitre plate reader for chemiluminescence with the possibility for injections became available (Labsystems).

In the present report the chemiluminescence measurements in microtitre platesare described of both transient labels (luminol and acridinium compounds), enzyme labels (horseradish peroxidase and xanthine oxidase) and other chemiluminescence reactions used in biochemical toxicology (like luminous bacteria, NADH and cytochrome P-450 isoenzymes). Both kinetic and integrated signals were recorded and evaluated with respect to sensitivity and reproducibility.

MATERIALS AND METHODS

Reagents Luminol was obtained from Aldrich and used after additional purification. The synthesis of the N-functionalized acridinium compound will be described elsewhere. Xanthine oxidase was obtained from Biozyme, Blaenavon, UK. The signal reagent for xanthine oxidase was composed as described earlier [1]. Horseradish peroxidase (Type VI) was obtained from Boehringer. The signal reagent was obtained from Amersham Benelux. The rat liver microsomal preparation of cytochrome P-450 and the composition of the signal reagent has been described elsewhere [2]. Microtitre plates and strips were obtained from Amersham (white strips), Dynatech (white plates), Labsystems (black plates) and Greiner (white and transparant plates).

Measuring protocols The chemiluminescence measurements in microtitre plates were performed in the Luminoskan (manufactured by Labsystems, Helsinki, Finland) in the continuous or integrating mode as described in the text.

RESULTS AND DISCUSSION

Characteristics of the luminometer. The Luminoskan can measure microtitre plates or strips in a holder of standard size. The measurements are performed by one photomultiplier. In principle three measuring modes can be used: 1. an integral mode which measures the chemiluminescence signal for a defined period of time, 2. a continuous signal acquisition mode in which the signal is measured 50 times within a defined limit of time with a integral time of 10 msec and 3. a repeat-integral mode in which the signal is measured 50 times within a defined time integrated during a defined period of time. The precision of the repeat-integral mode is much better which is illustrated by the following experiment. A number of 50 measurements have been performed for three levels of xanthine oxidase (100 pg, 20 ng and 20 µg/well) with both measuring modes. The coefficients of variation (%) in the continuous mode are 33.9, 4.1 and 0.58%, respectively and in the repeat-integral mode 1.8, 1.1 and 0.23%, respectively.

For the assessment of the repeatability and reproducibility the following experiments were performed with the xanthine oxidase system [1]. The effect of longer integration times (0.2, 0.9, 1.9, 3.9, and 7.9 sec) on the repeatability was studied for

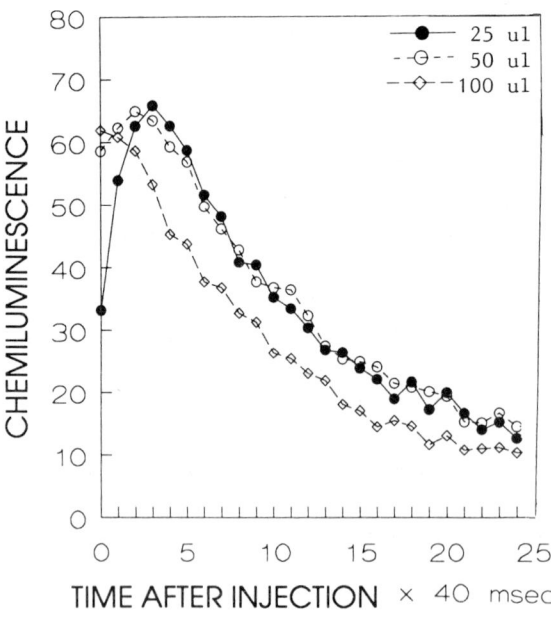

FIGURE 1: Kinetics of the chemiluminescence of an acridinium ester in microtitre plate as function of the volume of the starting reagent.

50 measurements of 20 ng xanthine oxidase in one well. The coefficients of variation were 0.99, 0.51, 0.36, 0.30 and 0.34%, respectively. The reproducibility was investigated for three levels of xanthine oxidase (100 pg, 20 ng and 20 µg) in 12 different wells. Coefficients of variation were 25, 6.3 and 2.7 % as measured with an integration time of 200 msec.

The dynamic range of the luminometer is 6 orders of magnitude (from 0.01 to 9999 counts). In addition, the microtitre plate can be thermostatted at temperatures above room temperature.

Measurements of transient labels. Both luminol and acridinium esters can be measured very conveniently with the microtitre plate luminometer. Before each measurement three independent addition can be done by dispensors in the apparatus. The injection volumina can be varied between 5 and 350 µl. In addition, the whole plate can be vibrated in order to achieve a optimal mixing of reagents. For the measurement of luminol, microperoxidase (25 µl) was added with the first injector followed by the addition of hydrogen peroxide (25 µl) by the second injector. In case of acridinium compounds only one injector is used for the hydrogen peroxide. If the continuous mode is used, the signal measurement starts at the end of the addition of the reagent. As a result, very fast transient signals can only be measured kinetically when small volumes are added. In Fig.1 the effect of different volumes on the kinetics of the chemiluminescence signal of an acridinium ester is shown. The measurements in the integral mode, however, can be started at any given time.

The coefficients of variation (N=4) were 2-4% (for luminol) and 6-7% (for acridinium compounds). The time required for the measurement of one plate (96 samples) is 15 min for luminol with the automatic addition of microperoxidase and hydrogen peroxide (after 2 sec) and an integration time of 10 sec. For acridinium compounds, with the automatic injection of hydrogen peroxide and an integration time of 2 sec, the total measuring time of one plate was 4 min.

Measurements of enzyme labels. For the chemiluminescence measurements of enzyme labels no injector system is needed, because of the long-term steady state signals during 30 min (for peroxidase) or during 30 hrs (for xanthine oxidase). Dilution experiment of both chemiluminescent enzyme systems lead to the conclusion that xanthine oxidase shows a linear response even at very high dilutions (11 attomol = 38 pg), whereas horseradish peroxidase shows a non-linear relationship at amounts below 200 attomol (= 70 pg) as was observed earlier [3]. The measuring time for one plate (96 wells) is 35 sec or 2 min for integration times of 0.1 or 1.0 sec, respectively.

Biochemical reactions. The microtitre plate reader was used also for the chemiluminescence determination of cytochrome P-450 isoenzymes in liver microsomes. In Fig. 2 the kinetic signals are shown of rat liver microsomes from an

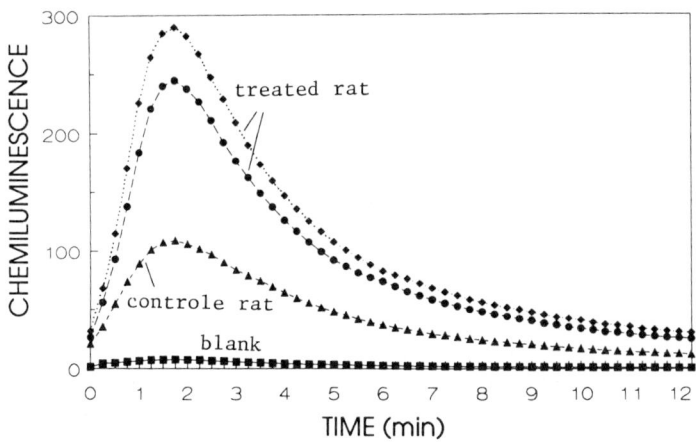

FIGURE 2: Chemiluminescence detection of cytochrome P-450 in rat liver microsomes from a controle rat and from two male rats treated with β-naphtoflavone.

induction experiment with β-naphtoflavone. If the kinetics are known a single-point measurement can be used for the determination of large number of samples in microtitre plates.

Also tests were performed with commercial kits for toxicity testing of luminous bacteria (Microtox) at 20 C and the chemiluminescence the determination of NADH with a luciferin/luciferase kit (Boehringer).

ACKNOWLEDGEMENT

Mrs. T. Jernstrom of Labsystems Oy (Helsinki, Finland) is acknowledged for technical advice.

REFERENCES

1. E.H.J.M. Jansen, R.H. van den Berg and G. Zomer. Characteristics and detection principles of a new label producing a long-term chemiluminescent signal. J. Biolumin. Chemilumin. 4, 129-135 (1989).
2. E.H.J.M. Jansen, E.J.M. Reinerink and R.H. van den Berg. Chemiluminescence detection of haem-containing microsomal proteins. Anal. Chim. Acta, 227, 49-55 (1989).
3. E.H.J.M. Jansen, C.A.F. Buskens and R.H. van den Berg. A sensitive CCD image system for detection of chemiluminescent reactions. J. Biolumin. Chemilumin., 3, 53-57 (1989).

PHOTON EMISSION FROM BRAIN CELLS: SUBCELLULAR LOCALIZATION, NON-ENZYMATIC REACTANTS AND KINETICS

H. Reiber and M. Uhr

Neurochemisches Labor, Universität Göttingen, D-3400 Göttingen, FRG

INTRODUCTION

Vital cells from post-natal rat brain homogenate as well as isolated oligodendrocytes could be kept at 37°C for hours without any detectable emission of photons (1). Only after disintegration of the cells a long lasting (hours) photon emission could be observed. Data have been presented (1, 2) for a non-enzymatic source of photon emission independent of external oxygen or oxygen free radicals.

A first approach to subcellular localization and the characterization of a low molecular weight substance from the 100 000 g supernatant of brain tissue homogenate (2) led meanwhile to convincing results: Post-vital photon emission from brain cells needs the plasma membrane fraction, endogenous ascorbic acid (3) together with high ionic strength. The details of the membrane associated part of the reaction together with influences on the reaction kinetics are reported in this paper. The change from an antioxydant function of ascorbate in the biological system to a prooxydant reaction system with lipid peroxidation in the post-vital cells will be discussed.

MATERIAL AND METHODS

Subcellular fractions: Grey matter of pig brain (1) was washed two times in reaction buffer and suspended in reaction buffer with a syringe (without needle) followed by homogenization with a Potter homogenizer. The homogenate was diluted to a total protein concentration of 4 g/l. During all steps the material was kept at 0 - 4°C.

The centrifugation of cell homogenate resulted in the following subfractions (resuspended in reaction buffer): Pellet from 200 g, 5 min, 4°C, (4 mg protein/ml). Pellet from 900 g, 10 min, 4°C, (2 mg protein/ml). Pellet from 20 000 g, 20 min, 4°C, (2 mg protein/ml). Pellet from 100 000 g, 30 min, 4°C, (2mg protein/ml). The 100000g supernatant contained 0.8 mg protein/l.

Membrane fraction: a) The 200 g pellet was washed three times in water and resuspended to 2 mg protein/ml with reaction buffer or water. b) To get a metal ion-free membrane fraction the procedure in a) was followed by an incubation with $1 \cdot 10^{-2}$ mol/l EDTA at pH 7.2 and ultrasonication. The EDTA-metal ion complex was washed out of the membrane fraction (3 times washed with water and resuspended in reaction buffer by ultrasonication).

Reaction Buffer: Hanks balanced salt solution with 0.02 mol/l HEPES, pH 7.2, 0.13mol/l glutamine. For long term incubation the reaction buffer contained penicillin 50 units/ml + streptomycin 50 ug/ml.

Lipid peroxydation: Lipid peroxidation was measured by the thiobarbituric acid method (4): To 1 ml membrane suspension, 1 ml 10% cold trichloroacetic acid was added, 10 min precipitated on ice and centrifuged (10 000 g, 10 min, 4°C). To 1 ml supernatant 1.3 ml of 0.5% thiobarbituric acid were added and incubated 20 min in the boiling water bath. The concentration has been determined from the absorbance at 532 nm by comparison with a malondialdehyde standard curve.

<u>Single photon counting assay:</u> The photon counting apparatus (Biolumat, Berthold) was kept at 4°C. The reaction temperature in the counting chamber was 37°C. The sample (1 ml) was stirred continously. Cell or membrane samples (1 mg protein/ml), kept at 4°C, were ultrasonicated with a microtip (Branson sonifier, energy level 2, 20% pulsed, 20 strokes) in a microvessel (Eppendorf) and than transferred into the counting vessel. Photon emission intensity was calculated as impulses/s (I/s).

RESULTS AND DISCUSSION

Table 1 represents the luminescence intensities of the single subfractions from total brain tissue homogenate. The comparison of maximal photon emission intensity showed a dominant contribution from the 200 g fraction (plasma membranes) with 80 % of the intensity of the total homogenate. Only minor contribution were observed from nuclei, microsomes or mitochondrial fraction.

If total photon counts (area under the curves, e.g. in Fig. 1) were compared instead of emission intensity in the maximum of the curve, more than 90 % of photon counts from total homogenate originated from the 200 g pellet.

Table 1

Subcellular distribution of photon emission from brain cells. Intensities (I/s) were compared at maximum of emission (average from two preparations).

	Yield of intensity (%)	Protein related intensity (I/s/mg)
total homogenate	100	154
Cell debris, (200 g, 5 min.)	80 *)	60
nuclear fraction (900 g, 10 min.)	4 *)	80
mitochondrial fraction (20 000 g, 20 min.)	8 *)	53
microsomal fraction (100 000 g, 30 min.)	2 *)	210
supernatant (100 000 g)	0	0

*) resuspended in 100 000 g supernatant;

Contributions of the single reactants

Pure membrane fractions in the reaction buffer did not show any luminescence signal at 37°C, pH 7 (Fig. 1). Only the addition of 100 000 g supernatant or ascorbic acid (3) resulted in the photon emission.

As shown in Table 2 the membrane fraction lost its luminescence activity completely if preincubated at 37°C for 14 hours. This is in contrast to the effect of a short heating (30 min, 95°C) of the membrane fraction in buffer which increased the maximal photon emission intensity due to a faster overall reaction. The total photon counts and the extent of lipid peroxidation remained rather constant (Fig. 1 and Table 2). The efficiency of this heat treatment proves the non-enzymatic origin of chemiluminescence in this membrane fraction.

In all our experiments the appearance of a photon emission was concomitted by a lipid peroxidation (Table 2). The lipid peroxidation was finished within 30 minutes in contrast to the photon emission which reached maximal intensity after about one hour (depending on conditions) and was lasting for additional hours. The function of the lipid peroxidation for the photon emission process remains unsolved so far.

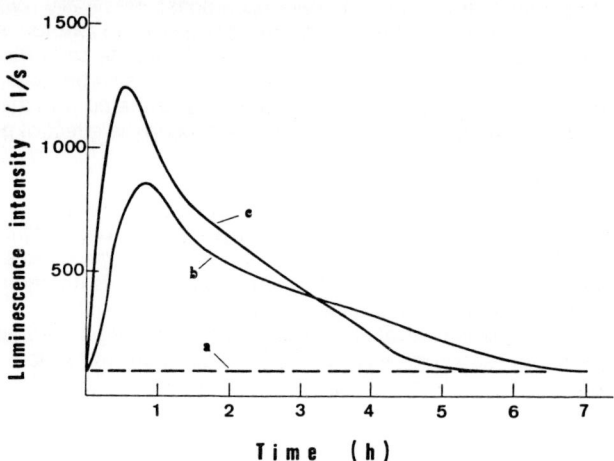

Fig. 1.: Photon emission from brain cell membranes: a) Membrane fraction (1mg protein/ml) in reaction buffer). b) Membrane fraction + 100 000 g supernatant in reaction buffer. c) Pretreated membrane fraction (30 min at 95°C preincubated, cooled to 4°C) + 100 000 g supernatant in reaction buffer.

Table 2
Comparison of untreated and heated membrane fractions. The maximal luminescence intensity is given in impulses/s (I/s), total photon counts represent the integral of the curves in Fig. 1 and the peroxidation is characterized by the malondialdehyde formation (nmol/assay). Assay as described for Fig. 1

	untreated membranes	preincubated membranes 90°C, 30 min	37°C, 14h
CL-Intensity (I/s)	730	1140	0
Total photon counts (I)	$7.0 \cdot 10^6$	$8.3 \cdot 10^6$	0
Lipid peroxydation (nmol)	4.2	4.7	0.2

The emission spectrum showed a maximum of intensity at a wavelength larger than 700 nm. Due to the low sensitivity of the photomultiplier in the range above 600 nm the complete spectrum could not be analysed. But obviously there was no maximum at 640 nm where an oxygen radical emission would be located.

If the membrane fraction has been pretreated with EDTA to extract endogenous metal ions, there was no chemiluminescence obtained inspite of optimal reaction conditions (ascorbic acid, buffer). The photon emission could be obtained again by adding Fe^{2+} to this system. From these data we conclude that the ascorbic acid which is identified as an endogenous reactant from the supernatant of brain tissue homogenate (3) together with the endogenous metal ion (e.g. Fe^{2+}, Cu^{2+} or Mn^{2+}) in the membrane fraction forms a prooxydant system reacting with the membrane components.

Buffer components influence the photon emission process drastically. Without any additional ions in the HEPES buffered reaction there was no photon emission induced. The addition of 0.05 mol/l glycin or 0.5 mol/l $CaCl_2$ or 3 mol/l NaCl induced a comparable photon emission intensity. With these different ions there is a different time necessary to reach the maximal intensity. Compared with the effect of NaCl the influences of bivalent cations or zwitter ions exceed the effect of pure ionic strength.

Interpretation and biological relevance:
1. The photon emission of brain cells could be explained as a prooxidant effect of the endogenous components, ascorbate and Fe^{2+} in the membrane fraction together with suitable buffer ions. The photon emitting process is not identical with that part of the lipid peroxidation process which forms malondialdehyde.
2. The observation of an inhibition of chemiluminescence and lipid peroxidation at high concentrations of ascorbic acid (3) is of biological relevance. In our chemiluminescence assay with membrane fraction (1 mg protein) 0.05 mmol/l ascorbic acid are optimal for maximal photon emission. At higher ascorbate concentrations the chemiluminescence intensity was decreased down to zero.
In human cerebrospinal fluid mean ascorbate concentrations of 0.2 mmol/l have been reported (5). Extracellular concentrations of ascorbic acid in rat brain have been reported to be about 0.5 mmol/l and of 1 - 3 mmol/l in the whole tissue of rat brain cortex, i.e. ascorbate is 2 to 10 fold higher concentrated in the cells (6). The change from the antioxidant effect of ascorbate in the vital system to a prooxidant effect in the disintegrated cells with a consequence of lipid peroxidation and photon emission could be a consequence of the dilution of ascorbic acid under in vitro conditions.
Together with data from earlier reports (1, 2) we can summarize:
- Photon emission is seen only in freshly isolated cells (as long as kept at 4°C).
- The photon emission occures only after breakdown of the membrane potential.
- Physical effects on the membrane accelerate the photon emitting process.
These observations could be interpreted as a relaxation of metastable states in membranes (7) which were induced and stabilized only in the vital, complete biological system (e.g. intact neuron glia interaction).

        REFERENCES
1.   H. Reiber, Discrimination between different types of low-level luminescence in mammalian cells: The biophysical radiation. J. Biolumin. Chemilumin. 4,245-248 (1989)
2.   H. Reiber, Low-level luminescence from disintegrated brain cells, subcellular distribution and kinetic of the biophysical radiation. In Biological Luminescence B. Jezowska-Trzebiatowska, B. Kochel, J. Slawinski, W. Strek (Eds.) World Scientific Publ., Inc., Singapore, (1990).
3.   M. Uhr and H. Reiber, Characterization of ascorbic acid as an endogenous reactant of non-enzymatic chemiluminescence from brain cells. This volume.
4.   L. Flohé, G. Niebch and H. Reiber, Zur Wirkung von Divicin in menschlichen Erythrozyten. J. Clin. Chem. and Clin. Biochem. 9, 431-437 (1971).
5.   R. Brau, S. García Castineiras and N. Riffkinson, Cerebrospinal fluid ascorbic acid levels in neurological disorders. Neurosurgery 14, 142-146 (1984).
6.   R. Todd and P. Bauer, Ascorbate modulates 5-hydroxytryptamin binding to central 5-HT3 sites in bovine frontal cortex. J. Neurochem. 50,1505-12 (1988).
7.   H. Fröhlich, The biological effects of microwaves and related questions. Electronics and Electron Physics 53, 85-152 (1980).

THE EFFECT OF EXPERIMENTAL CONDITIONS ON
BIS(2,4,6-TRICHLOROPHENYL)OXALATE CHEMILUMINESCENCE

A.C. Capomacchia, S.M. Hemingway and N.H. Do

*Department of Pharmaceutics, College of Pharmacy,
University of Georgia, Athens, GA 30602, U.S.A.*

INTRODUCTION

The reactions of bis(2,4,6-trichlorophenyl)oxalate (TCPO) with hydrogen peroxide (HOOH), a fluor like diphenylanthracene (DPA) in the presence of a catalyst yield chemiluminescence (CL) by means of a complicated reaction mechanism (1-3). The latter is still unproven, however, recent studies have provided much needed information on the extent and complexity of the mechanism (3,4). The TCPO reaction is strongly dependent on experimental reaction conditions. If the reaction is run in an aqueous organic solvent system, the CL burst demonstrates a first-order rise and decay (3). However, in primarily organic media the burst occurs in two sequential events (5). The effect of TCPO, HOOH, DPA and imidazole (I_m) concentrations on the first-order CL rise and decay has been examined in aqueous acetonitrile at pH 7 (3). The effect of pH on TCPO CL in a flow system, and stability in a stopped-flow system has been reported (6). The present study was designed to examined the effect of H_2O on the TCPO reaction. In addition, the effect of TCPO, HOOH and I_m concentrations were also examined, but at concentrations greater than those reported (3). Finally, the effects of pH and ionic strength were studied and compared to an earlier report (6).

MATERIALS AND METHODS

Apparatus All experiments were conducted with the novel stopped-flow chemiluminometer RML-10 developed by Hi-Tech Scientific Ltd (Salisbury, England) and our laboratory. CL was monitored at the emission maximum of DPA (430 nm) using a narrow band pass filter.

Reagents Freshly prepared stock solution of TCPO and DPA in HPLC grade acetonitrile, HOOH in aqueous acetonitrile and aqueous imidazole buffer solutions were reacted at the final concentrations stated in the Tables. The test analyte for all experiments was 40 μl of DPA (1×10^{-4}M).

Data Analysis Figure 1 shows a typical pseudo-first-order CL curve for the reaction of TCPO, HOOH and DPA in the presence of imidazole catalyst. The time-profile parameters k_r, k_f (rate constants for CL rise and fall, respectively), AUC (area), I_{max} (peak intensity) and t_{max} were calculated by nonlinear least squares analysis using 60-120 data points. Each value is the mean (standard deviation) from 3 to 6 experiments.

RESULTS AND DISCUSSION

The data in Table 1 shows the effect of varying H_2O concentration on the time-profile parameters. An increase in H_2O concentration causes a decrease in k_r but not k_f, and k_r is 100-fold greater than k_f as shown by Alvarez, et al. (5). The most interesting feature in the table, however, is that as H_2O is increased, I_{max} also increases to a plateau region at a HO concentration near 13 M. This is opposite the effect seen with DNPO (bis(2,4-dinitrophenyl)oxalate) which shows a decline in I_{max} as H_2O is increased (7). TCPO is not only not hydrolyzed as readily as DNPO, but the presence of H_2O enhances I_{max} while not appreciably changing either the rise or decay kinetics. Table 2 shows the effect of changing pH on the time profile parameters. The values for k_r and k_f are greatest at pH 8.5 and 8.0, respectively, while

TABLE 1. The Effect of H_2O Concentration on the I_{max}, T_{max}, AUC, Rise (K_r) and Fall (K_f) Rate Constants in the TCPO-HOOH CL Reaction, [TCPO] = 1.4 mM, [HOOH] = 0.23 M, $[I_m]$ = 1.4 mM, pH = 7.0, [DPA] = $1x10^{-4}$M.

$[H_2O]$ M	K_f s^{-1}	K_r s^{-1}	T_{max} s	I_{max} (au)	AUC
0.9	.0345(.0023)	5.403(0.440)	.9452(.0705)	2.254(0.089)	67.20(6.191)
5.0	.0328(.0013)	6.055(0.484)	.8688(.0528)	3.619(0.011)	112.8(3.896)
10.0	.0297(.0003)	3.540(0.128)	1.363(0.038)	5.600(0.084)	195.7(4.754)
15.0	.0333(.0059)	4.065(0.999)	1.227(0.231)	6.295(0.044)	199.9(28.91)
20.0	.0387(.0094)	3.264(0.996)	1.437(0.329)	6.720(0.044)	189.0(36.92)

TABLE 2. The Effect of pH on the I_{max}, T_{max}, AUC, Rise (K_r) and Fall (K_f) Rate Constants in the TCPO-HOOH CL Reaction, [TCPO] = 1.4 mM, [HOOH] = 0.23 M, $[H_2O]$ = 0.48 M, $[I_m]$ = 1.4 mM, [DPA] = $1x10^{-4}$M.

pH	K_f s^{-1}	K_r s^{-1}	T_{max} s	I_{max} (au)	AUC
6.8	.0248(.0020)	3.523(0.124)	1.419(0.0563)	1.477(0.060)	61.77(7.546)
7.2	.0452(.0016)	6.194(0.176)	0.801(0.022)	2.571(0.030)	58.56(2.480)
7.6	.0723(.0065)	8.562(0.185)	0.545(0.046)	3.131(0.054)	45.08(4.603)
8.0	.0833(.0028)	11.39(0.292)	0.435(0.011)	3.337(0.077)	41.35(2.128)
8.3	.0834(.0083)	11.97(0.673)	0.418(0.013)	3.309(0.064)	41.14(3.772)
8.5	.0805(.0027)	12.74(0.786)	0.401(0.019)	3.245(0.036)	41.40(1.174)
8.7	.0789(.0040)	11.90(0.315)	0.410(0.027)	3.105(0.074)	40.51(2.515)
9.0	.0741(.0027)	11.51(0.596)	0.442(0.020)	2.985(0.109)	41.42(1.904)

TABLE 3. The Effect of HOOH Concentration on the I_{max}, T_{max}, AUC, Rise (K_r) and Fall (K_f) Rate Constants in the TCPO-HOOH CL Reaction, [TCPO] = 1.4 mM, $[H_2O]$ = 4.3 M, $[I_m]$ = 1.4 mM, pH 7.0, [DPA] = $1x10^{-4}$M.

[HOOH] M	K_f s^{-1}	K_r s^{-1}	T_{max} s	I_{max} (au)	AUC
0.5	0.039(0.002)	5.300(0.114)	0.938(0.034)	4.340(0.148)	94.09(3.761)
1.0	0.053(0.003)	8.220(0.015)	0.620(0.005)	4.600(0.040)	89.50(2.50)
1.5	0.073(0.004)	9.790(0.026)	0.510(0.005)	5.430(0.09)	76.50(4.500)
2.0	0.094(0.002)	10.95(0.545)	0.441(0.011)	6.350(0.25)	69.50(1.50)
2.5	0.118(0.002)	11.74(0.195)	0.395(0.005)	6.770(0.528)	60.00(5.00)

TABLE 4. The Effect of TCPO Concentration on theI_{max}, T_{max}, AUC, Rise (K_r) and Fall (K_f) Rate Constants in the TCPO-HOOH CL Reaction, [TCPO] = 1.4 mM, [HOOH] = 0.23 M, [I_m] = 1.4 mM, pH = 7.0, [DPA] = 1.0×10^{-4} M.

[TCPO] mM	K_f s^{-1}	K_r s^{-1}	T_{max} s	I_{max} (au)	AUC
0.5	0.038(0.001)	5.164(0.236)	0.958(0.039)	0.427(0.001)	11.06(0.316)
1.0	0.035(0.003)	4.959(0.392)	1.005(0.086)	0.882(0.013)	25.06(1.75)
1.5	0.036(0.004)	5.833(0.411)	0.880(0.086)	1.187(0.005)	33.33(0.215)
2.0	0.035(0.007)	5.898(0.184)	0.872(0.021)	1.499(0.031)	42.99(0.937)
2.5	0.034(0.005)	6.128(0.510)	0.855(0.075)	1.874(0.009)	55.68(1.004)
3.0	0.034(0.001)	6.062(0.506)	0.864(0.079)	2.176(0.025)	65.25(2.853)

TABLE 5. The Effect of I_m Concentration on the I_{max}, T_{max}, AUC, Rise (K_r) and Fall (K_f) Rate Constants in the TCPO-HOOH CL REaction, [TCPO] = 1.4 mM, [HOOH] = .14 M, [H_2O] = 0.68 M, [DPA] = 1.0×10^{-4}M at pH 6.6 and 7.0.

[Im] M	K_f s^{-1}	K_r s^{-1}	T_{max} s	I_{max} (au)	AUC
			pH 6.6		
5×10^{-4}	1.08	1.9×10^{-3}	5.90	14.7×10^{-2}	69.1
1×10^{-3}	1.45	3.6×10^{-3}	4.14	18.5×10^{-2}	46.0
5×10^{-3}	0.235	6.6×10^{-3}	74.7	8.1×10^{-2}	15.0
			pH 7.0		
5×10^{-4}	2.13	7.7×10^{-3}	2.64	2.52	329
1×10^{-3}	6.57	11.3×10^{-3}	0.97	2.53	224

maximum intensity was attained at pH 8.0. Solution ionic strength was shown not to be a factor. The above pH values are greater than those reported by Honda, et al., under similar experimental conditions (6). Tables 3 and 4 present the effects of changing the HOOH and TCPO concentrations, respectively, on the time-profile parameters. As HOOH is increased, k_r, k_f and I_{max} increase whereas t_{max} and AUC decrease. These results are in partial agreement with those reported by Orlovic, et al., since they report that k_f remains essentially constant (3). It should be pointed out that our concentration range of HOOH (0.5 M - 2.5 M) is much greater than theirs (0.005 M and 0.03 M), and they examined only the two HOOH concentrations. As with DNPO k_r, k_f and I_{max} all increase with increasing HOOH and $k_r >> k_f$ (7). Changing the TCPO concentration produces results similar to those reported earlier (3). Table 5 shows the effect of I_m concentration on the time-profile parameters at two pH values. At pH 6.6, kr seems to be variably dependent on I_m concentration whereas k_f and I_{max} decrease and increase, respectively, as I_m is decreased. At pH 7.0 the trend for k_r and k_f seems to continue, but I_{max} does not change. The results appear to substantiate the idea that TCPO undergoes general base catalysis as has been suggested (3). The above results indicate that optimum reactant concentrations for the TCPO system, in a stopped flow apparatus used for analysis is: TCPO, limiting solubility; I_m, 1×10^{-3}M, HOOH, 2M and perhaps greater; H_2O, 13M; pH 8.0 and disregard ionic strength effects.

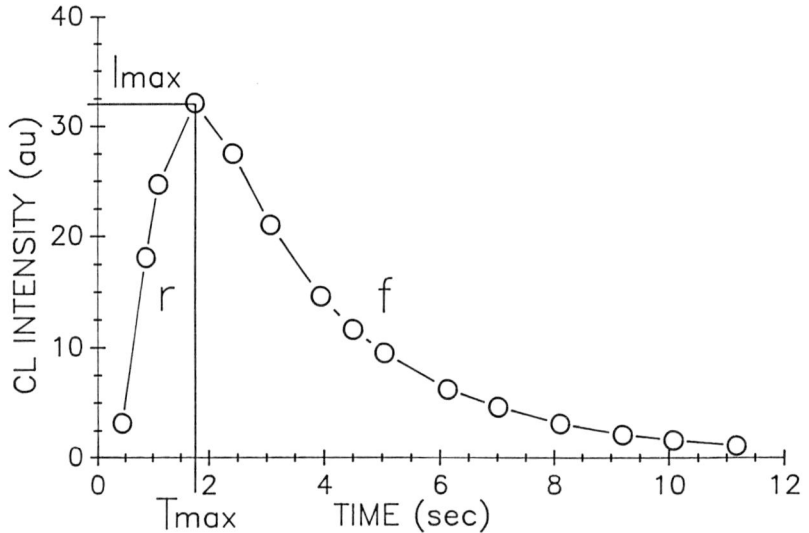

FIGURE 1: CL TIME PROFILE FOR THE REACTION OF 7.4 mM TCPO, 0.14 M HOOH, 0.68 M
H_2O, 1 mM, Im, 1×10^{-4} M DPA AT pH 7.0.

REFERENCES

1. Rauhut, M.M., Bollyky, L.J., Roberts, B.G., Loy, M., Whitman, R.H., Ianotta, A.V., Semsel, A.M., and Clarke, R.A., (1967), Chemiluminescence from reactions of electronegatively substituted aryl oxalates with hydrogen peroxide and fluorescent compounds, J. Am. Chem. Soc., 89, 6515-6522.

2. Gunderman, K.-D. and McCapra, F. (1987), Chemiluminescence in Organic Chemistry, Springer-Verlag, Berlin Heidelberg, Chap. VI.

3. Orlovic, M. Schowen, R.L., Givens, R.S., Alvarez, F., Matuszewski, B. and Parekh, A., (1989), Simplified model for the dynamics of chemiluminescence in the oxalate-hydrogen peroxide system: Toward a reaction mechanism, J. Org. Chem., 54, 3606-3610.

4. Orosz, G., (1989), The role of diaryl oxalates in peroxyoxalate chemiluminescence, Tetrahedron, 45, 3493-3506.

5. Alvarez, F.J., Parekh, N.J., Matuszewski, B., Givens, R.S., Higuchi, T., and Schowen, R.L. (1986), Multiple intermediates generate fluorophore-derived light in the oxalate/peroxide chemiluminescence system, J. Am. Chem. Soc., 108, 6435-6437.

6. Honda, K., Miyaguchi, K., Imai, K. (1985), Evaluation of aryl oxalates for chemiluminescence detection in high performance liquid chromatography, Anal. Chim. Acta, 177, 103-110.

7. Capomacchia, A.C., Hemingway, S.M. and Do, N.H. (1990), The effect of reactant concentration on bis(2,4-dinitrophenyl)oxalate chemiluminescence, this journal.

THE EFFECT OF REACTANT CONCENTRATION ON BIS(2,4-DINITROPHENYL)OXALATE CHEMILUMINESCENCE

A.C. Capomacchia, S.M. Hemingway and N.H. Do

Department of Pharmaceutics, College of Pharmacy,
University of Georgia, Athens, GA 30602, U.S.A.

INTRODUCTION
Aryl oxalate esters like bis(2,4-dinitrophenyl)oxalate (DNPO), hydrogen peroxide (HOOH) and a fluorophor like diphenylanthracene (DPA) are believed to produce chemiluminescence (CL) by the following general reaction scheme (1,3):

oxalate ester + HOOH → RI-F or RI⁻ F⁺

$$RI\text{-}F \text{ or } RI^- F^+ \rightarrow {}^1F + non\text{-}CL \text{ products}$$
$$^1F \rightarrow Hv \text{ (CL)} + F$$

The effect of changing the ester or HOOH concentration on the reaction has been reported (1). However, the effect of H_2O on DNPO CL has not been studied as both the ester and HOOH concentrations are varied in the absence of a catalyst. DNPO is known to degrade in protic solvents, especially H_2O. For this reason, the effect of H_2O and its concentration limits in analytical use is very important. In this regard and as an extension of earlier work on DNPO hydrolysis and its oxidation by HOOH, this study was intended to examine the effect of H_2O, HOOH and DNPO concentrations on DNPO CL (3,4).

MATERIALS AND METHODS
Apparatus   All experiments were conducted with the novel stopped-flow chemiluminometer RML-10 developed by Hi-Tech Scientific Ltd (Salisbury, England) and our laboratory. CL was monitored at the emission maximum of DPA (430 nm) using a narrow band pass filter.

Reagents   Freshly prepared stock solutions of DNPO and DPA in HPLC grade acetonitrile, and HOOH in aqueous acetonitrile were reacted at the final concentrations stated in Tables 1-3. The test analyte for all experiments was 40 μl of DPA (4×10^{-6}M).

Data Analysis   Figure 1 shows a typical pseudo-first-order CL curve for the reaction of DNPO, HOOH and DPA in aqueous acetonitrile. The time-profile parameters k_r, k_f (rate constants for CL rise and fall, respectively), AUC (area), I_{max} (peak intensity) and t_{max} were calculated by nonlinear least squares analysis using 60-120 data points. Each value is the mean (standard deviation) from 3 to 6 experiments.

RESULTS AND DISCUSSION
The data in Table 1 shows the effect of varying HOOH concentration on the time-profile parameters at a constant ratio H_2O:HOOH = 1.88:1.0 (both HOOH and H_2O increase as HOOH is increased). Under these conditions both k_r and k_f increase with increasing HOOH (or H_2O) but values for k_r are much larger than those for k_f. These results agree with those reported by Orlovic, et al. for the reaction of TCPO, HOOH, DPA in acetonitrile-imidazole buffer solutions with a constant water concentration (4). Also in agreement is that I_{max} increases linearly with increasing HOOH. However, both AUC and t_{max} decrease. The reason for this is that at low HOOH concentrations the CL burst is slow which shifts t_{max} to much longer times. As HOOH is increased the reaction

TABLE 1: The Effect of HOOH and H_2O Concentration on the I_{max}, T_{max}, AUC, Rise (K_r) and Fall (K_f) Rate Constants in the DNPO-HOOH CL Reaction at a Ratio H_2O:HOOH of 1.88:1 in Acetonitrile.

[HOOH] M	[H_2O] M	K_r s^{-1}	K_f s^{-1}	T_{max} s	I_{max} (au)	AUC
		[DNPO] = 1.11 mM				
0.019	0.035	0.046(.003)	0.034(.002)	27.3	14.0(.30)	73.7(9.4)
0.056	0.105	0.106(.008)	0.096(.01)	9.6	16.5(1.2)	38.9(1.9)
0.074	0.14	0.150(.01)	0.109(.01)	7.3	15.8(1.4)	30.2(1.3)
0.112	0.21	0.366(.04)	0.161(.02)	4.3	15.2(1.5)	18.6(.75)
0.149	0.28	0.632(.02)	0.175(.004)	2.8	15.1(.25)	13.8(.34)
0.224	0.42	1.051(.02)	0.281(.02)	1.7	22.3(.21)	12.0(.36)
0.298	0.56	0.155(.05)	0.362(.02)	1.5	25.7(.20)	11.1(.26)
0.390	0.73	1.222(.07)	0.471(.01)	1.4	31.7(.40)	10.0(.79)
0.702	1.32	1.340(.04)	0.763(.03)	1.0	38.9(2.3)	9.8(.72)
1.080	2.03	1.592(.11)	1.02(.08)	0.8	49.1(3.8)	9.7(1.2)
1.520	2.86	2.470(.27)	1.24(.21)	0.6	65.7(7.4)	9.6(1.0)
		[DNPO] = 0.74 mM				
0.019	0.035	0.044(.003)	0.028(.003)	26.2	8.0(.41)	64.2(5.7)
0.056	0.105	0.093(.003)	0.071(.001)	11.6	10.1(.66)	29.1(1.3)
0.074	0.14	0.118(.01)	0.103(.005)	8.9	8.3(.94)	18.6(.78)
0.112	0.21	0.246(.02)	0.128(.02)	5.5	9.2(.75)	17.2(.92)
0.149	0.28	0.588(.02)	0.149(.02)	2.9	11.2(1.0)	11.3(.44)
0.224	0.42	0.938(.03)	0.220(.01)	2.0	13.4(.96)	11.0(.56)
0.298	0.56	1.11(.02)	0.272(.02)	1.8	16.0(.83)	8.7(.36)
0.390	0.73	1.16(.04)	0.362(.02)	1.5	18.5(.66)	8.6(.18)
0.702	1.32	1.40(.03)	0.634(.03)	1.0	25.4(.79)	7.5(.33)

TABLE 2: The Effect of HOOH Concentration and H_2O:HOOH Ratio (R) on the I_{max}, T_{max}, AUC, Rise (K_r) and Fall (K_f) Rate Constants in the DNPO-HOOH CL Reaction at 1.11M H_2O in Acetonitrile.

[HOOH] M	[R]	K_r s^{-1}	K_f s^{-1}	T_{max}	I_{max} (au)	AUC
		[DNPO] = 1.11 mM				
0.019	58:1	0.231(.04)	0.017(.002)	9.9	2.98(.03)	21.2(2.3)
0.056	20:1	0.362(.02)	0.038(.003)	6.5	5.9(.04)	20.2(1.7)
0.074	15:1	0.430(.02)	0.046(.01)	5.3	7.58(.27)	19.2(2.1)
0.112	10:1	0.542(.025)	0.087(.007)	4.0	11.7(1.1)	18.8(.87)
0.149	7.4:1	0.592(.04)	0.165(.009)	3.6	14.9(1.0)	15.5(.77)
0.224	4.9:1	0.635(.06)	0.207(.021)	2.7	18.6(1.5)	15.8(1.5)
0.298	3.7:1	0.736(.05)	0.265(.04)	2.2	23.4(.92)	13.6(.33)
0.390	2.8:1	1.02(.06)	0.364(.03)	1.6	27.7(1.3)	13.4(1.6)
0.477	2.3:1	1.16(.05)	0.421(.04)	1.3	29.9(1.1)	11.4(.96)
		[DNPO] = 0.74 mM				
0.019	58:1	0.262(.05)	0.036(.001)	8.0	1.72(.06)	13.5(4.2
0.056	20:1	0.436(.04)	0.048(.002)	5.7	5.23(.42)	18.8(1.3)
0.074	15:1	0.487(.04)	0.058(.001)	5.1	6.5(.19)	16.6(1.1)
0.112	10:1	0.560(.03)	0.083(.003)	4.3	9.2(.27)	15.3(1.3)
0.149	7.4:1	0.660(.04)	0.144(.006)	3.3	11.1(.93)	12.9(3.4)
0.224	4.9:1	0.787(.05)	0.165(.013)	2.6	13.9(1.1)	12.7(1.2)
0.298	3.7:1	0.814(.03)	0.223(.01)	2.3	17.9(.76)	11.8(.96)
0.390	2.8:1	0.930(.05)	0.269(.008)	1.9	20.6(2.1)	10.6(.88)
0.477	2.3:1	1.12(.06)	0.295(.011)	1.6	23.4(1.1)	12.2(1.4)

TABLE 3: Effect of H_2O Concentration and H_2O:HOOH Ratio (R) on the I_{max}, T_{max}, AUC, Rise (K_r) and Fall (K_f) Rate Constants in the DNPO-HOOH CL Reaction at 0.39M HOOH and 1.11M DNPO in Acetonitrile.

$[H_2O]$ M	[R]	K_r s^{-1}	K_f s^{-1}	T_{max}	I_{max}	AUC
0.617	1.6:1	1.22(.07)	0.471(.01)	1.40	31.7(.40)	10.0(.79)
1.11	2.8:1	1.02(.06)	0.364(.03)	1.56	27.7(1.3)	13.4(1.6)
2.19	5.6:1	0.976(.09)	0.493(.01)	1.7	16.9(.90)	7.3(.83)
3.07	7.8:1	1.47(.01)	0.220(.01)	1.58	7.8(.33)	4.3(.18)
3.94	10.1:1	1.67(.01)	0.130(.006)	1.65	4.0(.09)	3.5(.28)

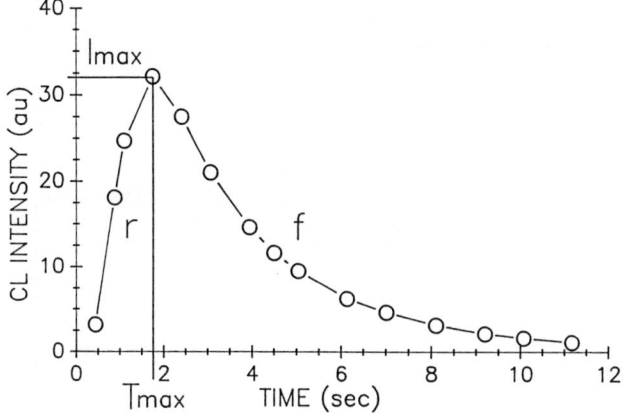

FIGURE 1: CL TIME PROFILE FOR THE REACTION OF 0.39M HOOH, 1.11M DNPO, 0.6M H_2O AND DPA 5.3 X 10^{-7}M IN ACETONITRILE.

proceeds faster and both t_{max} and AUC decrease. Table 2 shows the effect on the time-profile parameters of varying HOOH but at a constant H_2O concentration of 1.11 M. Increasing HOOH increases both k_r and k_f. However, the initial values for k_r are much higher than those contained in Table 1. The AUC, I_{max} and t_{max} follow the same pattern as those in Table 1. The difference between the data in Tables 1 and 2 is that in the latter the H_2O concentration is constant and, therefore, is very high compared to HOOH in the beginning of the table (H_2O/HOOH=1.11 M/0.019M = 58 in the first entry and decreases as HOOH is increased). The relatively high H_2O concentration apparently and unexpectedly affects the rise phase of the CL burst since the initial k_r values in the first few entries in Table 2 are larger than comparable entries in Table 1. This is more clearly seen from the data in Table 3 which shows the effect on the time-profile parameters as H_2O concentration is varied at a constant HOOH concentration of 0.39 M. The values for k_r increase dramatically at H_2O concentrations greater than 2.0 M. This result was unexpected since previous work on DNPO hydrolysis showed increased hydrolysis rate as the H_2O was increased (3). Although H_2O is seen to increase k_r, k_f values were constant until the H_2O concentration reached 2 M at which point they decreased. I_{max} and AUC values decrease as the H_2O is increased, however, t_{max} increases slightly. The decrease in I_{max} correlates with the DNPO hydrolysis work. Increasing the H_2O concentration of aqueous acetonitrile solutions decreases the effective DNPO concentration due to hydrolysis. A possible explanation for the increased k_f values is that the pre-CL reaction, HOOH + DNPO → CL intermediate occurs much more rapidly than the hydrolysis reaction, H_2O + DNPO → non-CL intermediate, and is assisted by H_2O forming a strong solvent cage around the carbonyl carbons. The decrease in k_f as H_2O is increased may be related to the loss of DNPO due to hydrolysis. Changing the

DNPO concentration from $1.11 \times 10^{-3}M$ to $0.740 \times 10^{-3}M$ causes an almost two-fold decrease in I_{max} while little change is seen in the k_r, k_f or t_{max} values (Table 1 versus 2). These results resemble those published for TCPO (5). Although H_2O affects the CL rise phase it has a deleterious effect on I_{max} unless the $H_2O/HOOH$ is held constant. This means that for analytical applications of DNPO that require a H_2O concentration greater than 0.22 M (4%; 2) the HOOH concentration must also be increased as the H_2O is increased in order to maintain the same or slightly greater I_{max}.

REFERENCES

1.   Rauhut, M.M., Bollyky, L.J., Roberts, B.G., Loy, M., Whitman, R.H., Ianotta, A.V., Semsel, A.M., and Clarke, R.A., (1967), Chemiluminescence from reactions of electronegatively substituted aryl oxalates with hydrogen peroxide and fluorescent compounds, J. Am. Chem. Soc., 89, 6515-6522.

2.   Gunderman, K.-D. and McCapra, F. (1987), Chemiluminescence in Organic Chemistry, Springer-Verlag, Berlin Heidelberg, Chap. VI.

3.   Jennings, R.N., Capomacchia, A.C. (1988), Solution kinetics of the degradation of the chemiluminescent reagent bis(2,4-dinitrophenyl)oxalate in water/acetonitrile, Anal. Chim. Acta, 205, 207-213.

4.   Jennings, R.N. and Capomacchia, A.C,. (1989), Prechemiluminescent biexponential degradation of bis(2,4-dinitrophenyl)oxalate in hydrogen peroxide/acetonitrile, Anal. Chim. Acta, 227, 37-48.

5.   Orlovic, M. Schowen, R.L., Givens, R.S., Alvarez, F., Matuszewski, B. and Parekh, A., (1989), Simplified model for the dynamics of chemiluminescence in the oxalate-hydrogen peroxide system: Toward a reaction mechanism, J. Org. Chem., 54, 3606-3610.

HPLC CHEMILUMINESCENCE DETECTION BASED ON PHOTOCHEMICALLY ON-LINE GENERATED DIOXETANES

H.A.G. Niederländer, F.W. Engelaer, C. Gooijer[*] and N.H. Velthorst

Department of General and Analytical Chemistry, Free University, De Boelelaan 1083, 1081 HV Amsterdam, The Netherlands

INTRODUCTION

It is well known (1) that phosphorescence detection in high performance liquid chromatography (HPLC) necessitates thorough oxygen removal from the eluents applied. This is due to the fact that the longlived triplet excited state of the phosphorescent compound is quenched by molecular oxygen in solution generally at a diffusion controlled rate (1).

In the present paper the attention is also directed on the electronic triplet state of analytes, but now use is made of a reaction with oxygen. Shellum an Birks (2) have shown that the triplet quenching ability of molecular oxygen can be utilized in a detection method based on photochemical oxidation. Upon quenching of an analyte molecule in the triplet excited state, triplet ground state molecular oxygen $O_2(^3\Sigma_g^-)$ is promoted to the $O_2(^1\Delta_g)$ singlet excited state, either directly or indirectly via the $O_2(^1\Sigma_g^+)$ singlet excited state. In the approach of Shellum and Birks the amount of singlet oxygen produced is determined indirectly, based on its addition to either 2,5-dimethylfuran or 2,5-diphenylfuran. The decrease in UV absorption of the furan under consideration is related to the concentration of the analyte producing singlet oxygen.

Scheme 1:

$$Analyte \xrightarrow{\text{hv}} {}^1Analyte^* \qquad\qquad 1$$

$$^1Analyte^* \xrightarrow{k_{isc}} {}^3Analyte^* \qquad\qquad 2$$

$$^3Analyte^* + O_2(^3\Sigma_g) \xrightarrow{k_q} Analyte + O_2(^1\Delta_g) \qquad 3$$

$$O_2(^1\Delta_g) + Reagent \xrightarrow{k_r} Dioxetane \qquad\qquad 4$$

$$Dioxetane \xrightarrow[\Delta]{k_d} Carbonyl + {}^3Carbonyl^* \qquad 5$$

$$^3Carbonyl^* + Fuorophore \xrightarrow{k_q'} Carbonyl + {}^1Fluorophore^* \qquad 6$$

$$^1Fluorophore^* \xrightarrow{k_{fl}} Fluorophore + hv' \qquad\qquad 7$$

In our approach singlet oxygen reacts with a substituted ethylene compound and the product of this reaction, a substituted dioxetane, is detected by thermally induced chemiluminescence (the sequence of reactions is schematically given in Scheme1). Herein the choice of the substituted ethylene reagent is crucial. The following points must be considered:
- The rate constant for addition to the reagent has to be as high as possible, because of the short lifetime of singlet oxygen in most reversed phase HPLC solvents (typicaly 1 to 100 µs (3)).
- The reagent should not absorb any light at the wavelengths at which the photochemical reactor is operated. Absorption will always give some triplet excited state molecules and thus generation of singlet oxygen and some background oxidation.
- The dioxetane should decompose quickly upon chemical or physical triggering inside the detector and should give a high yield of electronically excited decomposition products.
- Energy transfer to the fluorophore should be efficient and the fluorescence quantum yield of the fluorophore should be as high as possible.

Most of these conditions could be met by using 1,2-diethoxyethylene (DEE) (a mixture of the cis and trans isomers) as the reagent for singlet oxygen. DEE has a high reactivity towards singlet oxygen (k_r = 4.7 . 10^7 mol.l^{-1}.s^{-1})(4) as compared to various reagents that form dioxetane-type products with $O_2(^1\Delta_g)$. Utilizing DEE concentrations in the order of 10^{-3} mol.l^{-1} the rate constant is high enough to guarantee an almost complete scavenging of singlet oxygen. Furthermore DEE has hardly any absorptivity at wavelengths longer than 254 nm, so that auto-oxidation can easily be circumvented.

The dioxetane product that is formed (diethoxy-1,2-dioxetane, cis and trans isomers (DEDO)) upon addition of singlet oxygen is stable at room temperature (5) with a halflife time of several hours, but can easily be triggerd by raising temperature. At 70°C the halflife time for decomposition is less then two minutes.

Upon decomposition of DEDO triplet excited ethyl formate (EF) is formed with an efficiency near 100 percent (5), but no detectable chemiluminescence will result from these excited triplets in the abscence of a fluorophore. One of the few fluorophores able to quench excited triplets efficiently producing singlet excited states is 9,10-dibromoanthracene (DBA) (6). Unfortunately DBA has a relatively low fluorescence efficiency, which additionally is strongly temperature- and solvent dependent.

In this paper some preliminary results are presented. It will be obvious that the approach is worth to be further explored: in the most simple set-up, applied till now, the detection limits achieved are already in the same range as achieved with absorption detection.

MATERIALS AND METHODS

Experimental set-up: For convenience, three aspects of the experimental set-up are described seperately i.e. the chromatographic system, the photochemical reaction system and the chemiluminescence detection system.

The chromatographic system consists of a Gilson 302 solvent pump, an analytical column (14 cm x 3.1 mm) packed with 5 µm RoSil C18HL particles and a home-made injector with a 20 µl injection loop. Chromatograms are recorded on a Kipp & Zonen BD 12 recorder.

The photochemical reactor system consists of a 90 Watt medium pressure mercury arc (Philips, 93110, arc length is 25 mm), mounted with a cylindrical quartz filter cuvet (9 mm path length). A 1.20 % KI solution in 0.5 mol.l^{-1} NaOH is pumped through the filter cuvet, serving as a cut off filter (50 % transmittance at 268 nm). The solution is also used to cool the lamp and to thermostate the reactor coil (ca. 24 °C). The PTFE reactor tubing, i.d. 0.33 mm, o.d. 0.73 mm, is coiled around the filter cuvet (diameter 5.7 cm). The tubing is 2.5 - 12.5 m in length, providing reaction times ranging from 42 s. to 7 min 45 s. at flow rates of 0.15 - 0.34 ml.min^{-1}. Before entering the reactor, a flow of DEE in acetonitrile is added to the eluent by means of a home-made syringe pump. Directly after the photochemical reactor, DBA in tetrahydrofuran is added (by means of a home-made syringe pump) and the mixture enters the chemiluminescence detector consisting of an ABI-Spectroflow-980 fluorescence detector with the lamp turned off and a 25 µl detector-cell installed. The stainless steel bar containing the cell-compartment is adapted with a thermocouple and a thermocoax electrical heater. The temperature is controlled with a home-made electronic temperature programmer at 73.0 ± 0.1 °C. A 6.5 mm 1.14 mol.l^{-1} CuSO$_4$ filter is used in front of the photomultiplier tube, to cut off the red emission of the thermocoax heater. Boiling and creation of gas bubbles inside the detector cell were avoided by operating the detector at about 4.5 bar backpressure.

For comparison with UV absorption detection the chemiluminescence detector is exchanged for a Kratos spectroflow 757 variable wavelength detector.

Reagents and chemicals: 1,2-Diethoxyethene was prepared as described by Baganz et. al. (7) and purified by preparative gas chromatography in small portions prior to use, since high background oxidation signals resulted with unpurified DEE. 9,10-Dibromoanthracene was obtained from Aldrich. The solvents were purchased from Baker. Tetrahydrofuran (THF) was used without purification, methanol (MeOH) and acetonitrile (ACN) were HPLC grade and toluene was Z.A. quality. Water was destilled once.

RESULTS AND DISCUSSION

The crucial step in the reaction sequence scheme 1 is the production of DEDO from DEE in solvents appropriate for HPLC. For this reason we have studied the reaction in ACN and MeOH in batch experiments, at varying lamp intensities , utilizing   anthracene as a model analyte. The DEDO concentration was monitored by means of its chemiluminescence upon decomposition at 70 °C in the presence of DBA.

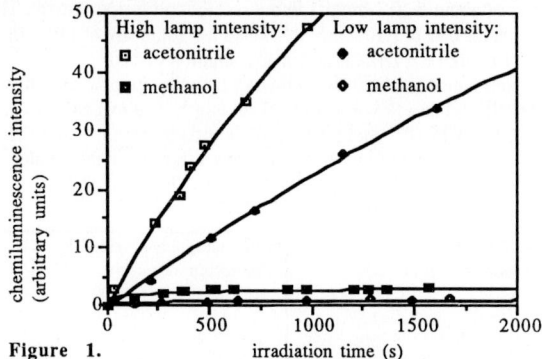

Figure 1.   irradiation time (s)

DEDO concentration, monitored by chemiluminescence intensity as a function of irradiation time.

The results are collected in Fig 1. Two qualitative conclusions can be drawn:

1- At the same lamp intensities the rate of build-up of DEDO in ACN is much higher than in MeOH, which is in-line with the difference in singlet oxygen life times.

2- For MeOH a constant, steady-state concentration of DEDO is reached after a short irradiation time. This indicates that in this solvent (and presumably also in other hydroxylic solvents) chemical decomposition is not negligible. Dioxetanes and other cyclic peroxides have been proven to be unstable in MeOH (8), hence it is important to use an intense lamp.

In the analytical flow system, variation in irradiation time is realized by varying the length of the reactor coil. For anthracene as a model compound at a concentration of 2.10^{-6} mol.l^{-1} the result is depicted in Fig 2.

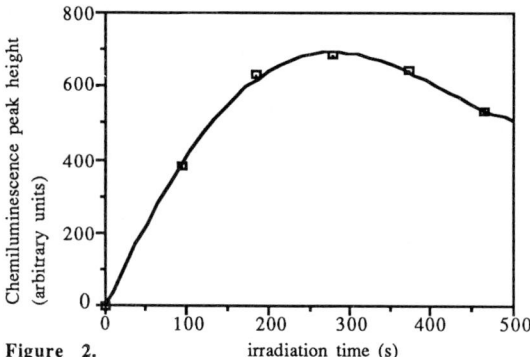

Figure 2.   irradiation time (s)

Chemiluminescence peak height as a function of irradiation time, for DEE oxidation sensitized by 2.exp-6 mol/l anthracene injected.

From a chromatographic point of view the reactor length should be as short as possible in order to reduce band broadening. Therefore we conclude from Fig 2 that the optimum irradiation time is about 3 min, corresponding to a coil of 10 m (ca. 0.93 ml) at a reactor flow rate of 0.34 ml.min^{-1}. Although not fully explored yet it seems not unrealistic to utilize this irradiation time also in other solvent compositions in which hydroxylic solvents are expected to induce chemical decomposition of DEDO.

In the flow system also the influences of the concentrations of DEE and DBA have been examined. The general conclusion is that for both compounds the concentration should be as high as possible. In practice a DEE concentration of about 5.10^{-3} mol.l^{-1} in the reactor coil is appropriate. The optimum DBA concentration in the detector cell is 2.10^{-3} mol.l^{-1}. Since the solubility of DBA in ACN/water mixtures is lower than about 10^{-4} mol.l^{-1}, DBA is added as a solution in THF.

On the basis of the above optimization studies, the following chromatographic system was applied. The eluent is ACN/water (90/10 v/v) at a flow rate of 0.25 ml.min^{-1}. The reagent flow rates of DEE (2.10^{-2} mol.l^{-1} in ACN) and DBA (5.10^{-3} mol.l^{-1} in THF) are 0.09 ml.min^{-1} and 0.20 ml.min^{-1}, respectively. In Fig 3 a typical chromatogram abtained for a mixture of the model compounds anthracene (AN), biphenyl (BP) and phenanthrenequinone (PQ) in eluent is depicted.

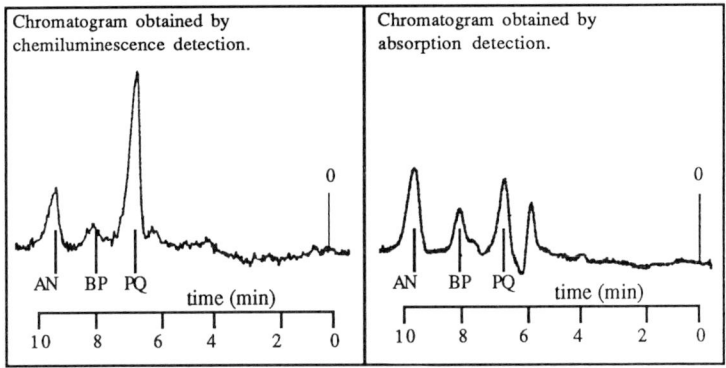

Figure 3.

Chromatograms obtained by chemiluminescence- and absorption detection.
[AN] = 2.52 exp-7, [BP] = 4.18 exp-7 and [PQ] = 4.30 exp-7 mol/l.

Comparison with the UV-detected chromatogram (at 254 nm) reveals that the detection limits are of the same order of magnitude. Studies directed on the reduction of the chromatographic band broadening (photochemical reactor performance), increase of chemiluminescence intensities and reduction of experimental complexity are currently under way.

REFERENCES

*   To whom correspondence should be addressed.
1   C.Gooijer, R.A. Baumann and N.H. Velthorst, The potential of liquid state molecular phosphorescence in analytical chemistry. Prog. Anal. Spectrosc. 10, 573-599 (1987).
2   C.L. Shellum and J.W. Birks, Photochemical amplifier for liquid chromatography based on singlet oxygen sensitization. Anal.Chem. 59, 1834-1841 (1987).
3   B.M. Monroe, Singlet oxygen in solution: lifetimes and reaction rate constants. Singlet O₂ vol.I, CRC Press, Boca Raton, Florida, 181-187 (1985).
4   A.P. Schaap and K.A. Zaklika, 1,2-Cycloaddition reactions of singlet oxygen. Singlet oxygen, Academic Press, New York, 231-237 (1979).
5   T. Wilson and A.P. Schaap, The chemiluminescence from cis-diethoxy-1,2-dioxetane. An unexpected effect of oxygen. J. Am. Chem. Soc. 93, 4126-4136 (1971).
6   W. Adam, Determination of chemiexcitation yields in the thermal generation of electronic excitation from 1,2-dioxetanes. Chemical and biological generation of excited states, Academic Press, New York, 122-125 (1982).
7   H. Baganz, K. Praefcke and J. Rost, Synthese von 1,2-dialkoxy-äthenen aus 1,2-dichlor-1,2-dialkoxy-äthanen. Chem Ber. 96, 2657-2660 (1963).
8   C.W. Jefford and C.G. Rimbault,Reaction of singlet oxygen with norbornenyl ethers. Characterization of dioxetanes and evidence for zwitterionic peroxide precursors. J. Am. Chem. Soc. 100, 6437-6445 (1978).

ELECTROCHEMILUMINESCENT DETECTION FOR HPLC AND FIA

M.Sato and T. Yamada

Faculty of Textile Science, Kyoto Institute of Technology
Matsugasaki, Sakyo-ku, Kyoto 606, JAPAN

INTRODUCTION

The high sensitivity and the selectivity are always essential for
the detection system of the high performance liquid chromatography
(HPLC) as well as for the flow injection analysis (FIA).
We would like to report here a new kind of detection system based
on the electrochemiluminescence(ECL). ECL is the phenomenon of the
light emission following the electrode oxidation or reduction of
target molecules. The reason for the light emission is generally
due to the formation of excited state molecule after the electron
transfer at the electrode interface. Therefore, it is known that
the ECL spectra of luminescent molecule coincides with the
fluorescence spectra of the same molecule.
One of the typical examples is the ECL due to the electrode
oxidation of luminol(5-amino-2,3-dihydro-1,4-phthalazinedione)(1).
Though the mechanism of ECL of luminol has been studied since long
ago(2-4), the analytical application of ECL of luminol for the HPLC
or FIA has not been fully developed until recently.
In the course of studies on the ECL of luminol and its analogous
compounds, we have found several advantages of ECL detection system
for HPLC and FIA(5-8).
Firstly, the electrolysis current and the ECL intensity can be
obtained simultaneously by single experiment. The two physical
quantities give us different kind of information for the target
molecules.
Secondly, ECL detection system generally needs much simpler
experimental device than the chemiluminescence system, because the
light source is not necessary and the effect of impurities which
causes the fluorescence is much more reduced.
A useful application of ECL detection is the determination of the
electro-inactive and non-luminescent molecules after the
derivatization of such target molecules with luminol through the
chemical modification of amino group of luminol. The sensitive
determination of oligo-peptides such as triglycine, hystidylglycine
has become possible(6). Another application has been found by
using the enhancement of ECL intensity of luminol in the presence
of hydrogen peroxide. For example, the trace amount of glucose was
successfully determined after the passage of glucose through the
column of immobilized glucose oxidase. As the quantities of

hydrogen peroxide produced was equivalent to the amount of glucose, the increase of ECL intensity was proportional to the amount of glucose included in the sample solution.

MATERIALS AND METHODS
Reagents The reagents used were of analytical-reagent grade. Commercial luminol was purified according to the usual procedure and stored in a light-tight desiccator. The electrochemically inactive substances such as oligo peptides or proteins were derivatized with diazoluminol according to the reported procedures (8). Diazotized luminol was coupled with the target molecules having amino or imidazol group.
Apparatus The observation of ECL was performed using the thin layer electrochemical cell of flow-through type, assembled in our laboratory. The three-electrode system was used for the potentiostatic control of working Pt foil electrode(5 X 5 mm). The counter electrode was a stainless steel pipe and the reference electrode was Ag/AgCl. The cylindrical quartz window was set in front of the working electrode to observe the total emission intensity using photomultiplier.
FIA-system A low pressure plunger pump supplied a carrier solution at the constant flow rate. The sample injector made of Teflon (0.1 ml) was used. The peak-shaped current as well as the luminescence intensity was recorded by two-pen type Y-T recorder.
HPLC-system Two-line system was adopted; one was the ordinary HPLC line equiped with the reversed phase column for the separation of mixed sample solution, another was the post-column flow line to carry 0.1 mol/l sodium carbonate solution which is indispensable to make the solution alkaline.

RESULTS AND DISCUSSION
Determination of oligo peptide and protein Luminol and its analogue compounds were determined by using 0.1 mol/l sodium carbonate as carrier solution. The most favorable potential for the oxidation of luminol was +0.7 V. Oligo peptide such as L-Gly-Gly-Gly, L-His-Gly were determined after derivatization with diazoluminol. Bovin serum albumin(BSA) was also determined by the same procedure. Although the coupled number of luminol per one molecule of BSA is not exactly known, the luminescence intensity was linearly dependent on the concentration of BSA down to micro g/ml.
Determination of glucose In the course of the ECL observation of luminol using FIA, it was found that emission intensity increased in the presence of hydrogen peroxide. The reason for this increase must be the reaction of oxygen produced by the oxidation of hydrogen peroxide with the electrochemically oxidized luminol. As an analytical application of this phenomenon, the flow system to determin glucose was constructed. A system was composed of two flow lines: One line carries buffered solution (pH 5.0) and has the injector for glucose sample and the column of immobilized glucose oxidase; another line carries alkaline solution necessary for the light emission of luminol. Two lines are connected just before the ECL cell. Before injection of sample solution to the flow system, a constant amount of luminol was added.

The difference of luminescence intensity, in the presence and the absence of hydrogen peroxide, was proportional to the amount of glucose in a sample solution. An analytical application of this phenomenon was the determination of glucose included in wine or soft drinks. There is no need for the preliminary treatment of samples, and the thousand times dilution is sufficient for the determination.

Separation of luminol and its analogues by HPLC For example, the separation of mixed compounds including luminol, isoluminol, aminophthalic acid and phthalhydrazinedione was successfully performed by using C18 reversed phase column with buffered carrier solution(pH 6.0) containing 10 % acetonitrile. Post-column solution was mixed with a flow line of alkaline solution to obtain ECL. Simultaneous detection of absorbance(280 nm), oxidation current (+0.7 V), and luminescence intensity showed a different kind of peak intensity pattern.

Formation of luminol derivatives of diamines It is known that several kind of diamines are physiologically important substances. To develop an analytical procedure of diamines by ECL, a reaction mechanism for the derivatization of ethylendiamine and its analogue, as a model compound, with diazoluminol was followed up by using HPLC. According to the passage of time after the mixing, the peak corresponding to diazoluminol gradually disappeared and a new peak of derivatized compound increased instead.

Advantages of ECL detection system In conclusion, several advantages are found for ECL detection. Though the sensitivity is comparable with fluorescence, ECL system does not require a light source. Moreover, ECL system is usually simple compared with the chemiluminescence system, because the mixing of reagent is not necessary. The observation of ECL is dependent on the oxidation potential which is easily controlled. ECL detection is applicable even to the electrochemically inactive compounds after the derivatization with electrochemically active compounds.

The enhancement of luminescence intensity of luminol in the presence of hydrogen peroxide will find a wide range of application for the determination of biologically important substances combined with a reaction of immobilized glucose oxidase.

REFERENCES
1.   O.H.Hercules,Technique of Chemistry,Vol.4,Wiley,New York, p.275(1971).
2.   T.Kuwana, Electro-oxidation followed by light emission,J. Electroanal. Chem.,6, 164-167(1963).
3   B.Epstein and T.Kuwana,Electrooxidation of phthalhydrazides, J.Electroanal.Chem.,15, 389-397(1967).
4   B.Epstein and T.Kuwana,Electrochemical Generation of Solution Luminescence III Photochem.Photobiol.,6,605-611(1967).
5   M.Sato and T.Yamada,Chap.4,Electrochemiluminescence Analysis, Trace analysis (in Japanese),Kagaku-dojin,p.63-77(1978).
6   M.Sato, T.Yamada and M.Fujino,Selective Microanalysis of Substituted Anthracenes by Electrogenerated Chemilumi- nescence, Nippon Kagaku Kaishi,1981, 74-78(1981).

7   M.Sato,T.Yamada and M.Horikawa,Determination of Luminol-
     Labeled Amino Acids by Electrogenerated Chemilumi-
     nescence,Denkikagaku,51, 111-112(1983).
8   J.S.A.Simpson,A.K.Campbell,M.E.T.Ryall and J.S.Woodhead,
     A stable chemiluminescent-labelled antibody for immuno-
     logical assays, Nature,279, 646-647(1979).

Measurement of Free Fatty Acid, Phospholipid, Triglyceride, and Cholesterol Ester Hydroperoxides by a CL-HPLC Method

E. Wieland, F. Diedrich, P.D. Niedmann and D. Seidel*
Departments of Clinical Chemistry, University Hospital, Göttingen, FRG and *University Hospital Großhadern, Munich, FRG.

INTRODUCTION

The reaction of oxygen with unsaturated fatty acids has important pathophysiological consequences. It has been speculated that lipid hydroperoxides and their reaction products are highly potent biological agents which are deleterious to cell components. Based on this knowledge lipid peroxidation has been implicated as a factor in a variety of diseases such as cancer and atherosclerosis (1,2). Many assays have been developed to measure the various lipid peroxidation products . Recently highly sensitive chemiluminescence methods for the detection of lipid hydroperoxides have been published (3-8). These assays are based on the breakdown of hydroperoxides by heme containing proteins in the presence of luminol or isoluminol. Since this reaction is not specific for hydroperoxides and is sensitive towards antioxidants it has been combined with HPLC separation. However, none of the published methods allows the simultaneous determination of hydroperoxides in different lipid classes without interference from lipophilic antioxidants or reduced coenzyme Q 10 (6). In order to overcome these difficulties we have modified previously described CL-HPLC methods (5-8)

MATERIALS AND METHODS

Analytical protocol. Samples were extracted with butanol and 100 μl of the organic phase were subjected to HPLC separation using a combination of a 1/4 normal phase slica gel (Bischoff, Leonberg, FRG) and a 1/1 C18 reversed phase column (Macherey and Nagel, Düren, FRG). Methanol/butanol/water (54/38/8) was used as the mobile phase at a flow rate of 1 ml/min. The eluted fractions were simultaneously monitored by a UV detector (Uvicon 720 LC, Contron, München, FRG) set at 233 nm and a LKB 1250 luminometer (Turku, SF) equipped with a flow cell (Helma, Mühlheim, FRG). The chemiluminescence cocktail was composed of 100 ml borate buffer (0.1 mol/l, pH 11), 80 mg/l cytochrome C (Boehringer Mannheim, FRG) and 40 μmol/l luminol (Sigma, München, FRG). Its flow rate was adjusted to 0.9 ml/min. Light production was initiated by continously mixing this CL-cocktail with the eluted column fractions in a mixing coil.

Standard Lipid Hydroperoxides. Lipids were from Sigma (München, FRG) and Serva (Heidelberg, FRG) and were used without further oxidation.

Preparation of low density lipoproteins (LDL). LDL were either isolated by precipitation or by ultracentrifugation as previously described (8).

Blood Collection. Blood was collected into syringes prepared with butylated hydroxytoluene (100 μmol/l) and EDTA (2 mmol/l).

RESULTS AND DISCUSSION

The flow diagram of our CL-HPLC system is shown in Fig. 1. In order to improve the HPLC separation we combined a normal phase silica gel and a C 18 reversed phase HPLC column.

Fig. 1   Flow Diagram of the CL-HPLC Assay

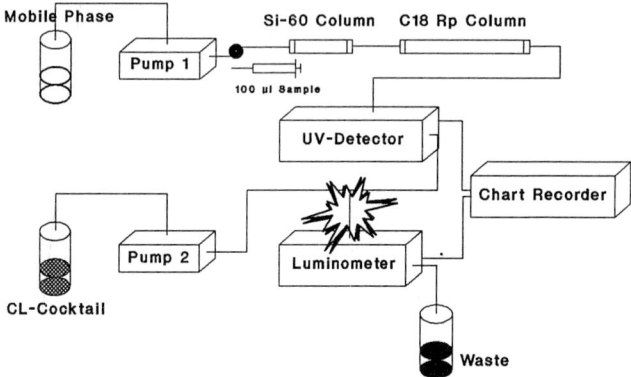

The retention times of lipids and antioxidants tested are summarized in Table 1

Table 1.   Retention Times

Analyte	Minutes	Peak	UV
Oleic acid hydroperoxide	4.6	+	y
Linoleic acid hydroperoxide	4.2	+	y
Beta-carotene	4.2/12.5	+/+	y/n
7-alpha-hydroxycholesterol	5.8	+	y
BHT	6.0	-	y
Vitamin E	9.9	-	y
Phophatidylcholine hydrop.	11.8/17.4	+/+	y/y
Triolein hydroperoxide	12.8	+	n
Trilinolein hydroperoxide	21.2	+	y
Trilinolenin hydroperoxide	14.8	+	y
Chol. linoleate hydrop.	28.5	+	y
Chol. linolenate hydrop.	26.0	+	y
Chol. arachidonate hydrop.	26.9	+	y
Coenzyme Q 10	35.0/56.0	-/-	y/n
Tripalmitine	no signal	o	n

Experimental conditions as described in "Materials and Methods". + = positive CL-peak; - = negative CL-peak; o = no CL-peak; y = peak at 233 nm; n = no peak at 233 nm.

Using luminol and cytochrome C to detect lipid hydroperoxides (6) instead of isoluminol plus microperoxidase (4,6) reduced and oxidized coenzyme Q 10 caused quenching of the light signal after 35.0 and 56.0 minutes respectively. However, coenzyme Q 10 eluted after cholesterol ester hydroperoxides and does not therefore intefere with the assay. Detection of conjugated dienes (UV, 233 nm) correlated in most cases with hydroperoxide induced CL. However, oleic acid which cannot form conjugated dienes also absorbed at 233 nm.

Light production was linearly related to fatty acid hydroperoxide concentrations up to 10 μmol/l. The lower detection limits of the various hydroperoxides depended on the lipid class. As little as 50 nmol/l triglyceride, cholesterol ester and fatty acid hydroperoxides could be detected compared to 200 nmol/l phosphatidylcholine. Application of this improved assay to plasma or low density lipoproteins (LDL) of healthy volunteers failed to detect hydroperoxides in vivo. This is in agreement with Frei et al. (7) who evaluated a similar assay originally published by Yamamoto et al (5) on human plasma samples. However, these findings are in contrast to our own previously reported results (8) and to Myiazawa (6) who has measured between 10 μmol/l and 500 μmol/l phospholipid hydroperoxides in plasma of healthy donors. In our previous publication reduced coenzyme Q 10 had been erroneously considered as cholesterol ester hydroperoxide.

Since we were interested in the oxidative modification of LDL we used this improved method to follow the formation of hydroperoxides in the different lipid classes during oxidation of LDL by copper ions in vitro. An original recorder graph of LDL oxidised for 24 hours with copper ions is shown Fig. 2. It can be seen that more than 80% of the hydroperoxides were formed in the cholesterol esters. Detection of conjugated dienes at 233 nm compared well with that of hydroperoxides.

Fig. 2   Recorder Graph of LDL oxidised by CuSO$_4$

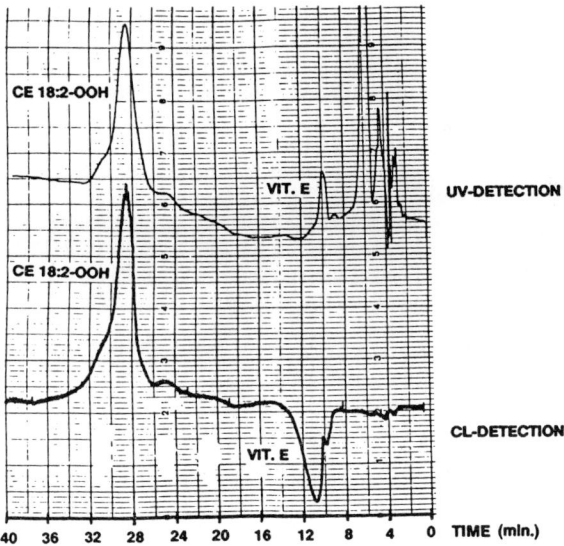

Fig. 3 compares the formation of hydroperoxides and aldehydes (9) during copper-mediated LDL oxidation. Hydroperoxide formation preceded the formation of aldehydes (Fig.3). This is in accord with the different stages of lipid peroxidation where hydroperoxides being precursor of aldehyde formation.

Fig. 3   Time Course of LDL Oxidation by CuSO$_4$

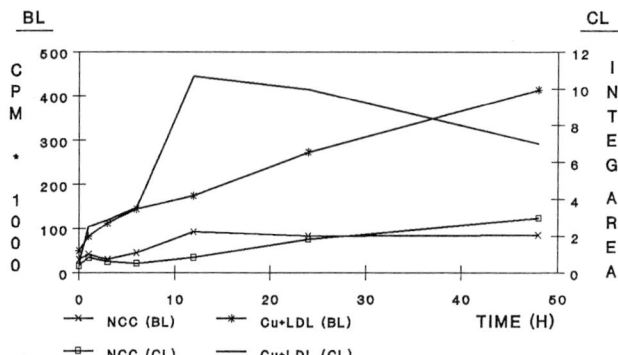

In summary an improved CL-HPLC method for the sensitive detection of lipid hydroperoxides has been applied to plasma and LDL samples in vivo and in vitro. Whereas no hydroperoxides were detectable in vivo the method offered valuable information about hydroperoxide formation during LDL oxidation in vitro. In combination with a bioluminometric aldehyde assay (9) the close relation between hydroperoxide formation and aldehyde generation was confirmed.

REFERENCES

1.   A.K. Basu and L.J.Marnett. Uneqivocal demonstration that malondialdehyde is a mutagen.
*   Carcinogenesis 4, 331-33 (1983).
2.   D. Steinberg, S. Parthasarathy, T. Carew, J.C. Khoo, and J.L Witztum. Beyond Cholesterol. Modifications of Low-Density Lipoprotein that increase its atherogenicity.New Engl. J. Med. 320, 915-924(1989).
3.   E. Wieland, S. Parthasarathy, and D. Steinberg.Chemiluminescence assay for lipid hydroperoxides: application to monitoring low density lipoprotein (LDL) oxidation in vitro. In Bioluminescence and Chemiluminescence: New Perspectives. J. Schölmerich, R. Andreesen, A. Kapp, M. Ernst, and W.G. Wood (Eds.), Wiley Chichester, p 321-324 (1987).
4.   K. Belghmi, J.-C. Nicolas, and A.C. Crastes de Paulet. Chemiluminescent assay of lipid hydroperoxides. J. Biolumin. Chemilumin. 2, 113-119 (1988).
5.   Y. Yamamoto, M.H. Brodsky, J.C. Baker, and B. Ames, Detection and characterization of lipid hydroperoxides at picomole levels by high-performance liquid chromatography. Anal. Biochem 160, 7-12 (1987).
6.   T. Myiazawa. Determination of phospholipid hydroperoxides in human blood plasma by a chemiluminescence-HPLC assay. Free Rad. Biol. Med. 7, 209-207 (1989).
7.   B. Frei, Y. Yamamoto, D. Niclas and B. Ames. Evaluation of an isoluminol chemiluminescence assay for the detection of hydroperoxides in human plasma. Anal. Biochem. 175, 120-130 (1988).
8.   E. Wieland, H. Kather, D. Niedmann, F. Diedrich, and D. Seidel, Luminescence in the study of lipid metabolism. J. Biolumin. Chemilumin. 4, 436-445 (1989).
9.   E. Wieland, H. Kather, A. David, and V.W. Armstrong. Bioluminometric determination of aldehyde lipid peroxidation products during the oxidation of low density lipoproteins (LDL). Abstract VIth Internatonal Symposium on Bioluminescence and Chemiluminescence, Cambridge (1990).

EFFECT OF SULFITE ON OXIDATIVE METABOLISM
OF NEUTROPHILS

ANIL[1] MISHRA AND INGRID BECK-SPEIER
Project Inhalation, GSF-Munchen,
8041-Neuherberg, FRG.
1. Present address : Inhalation Toxicology
Div. ITRC, P.O.Box 80,
Lucknow-226 001, India.

INTRODUCTION

Sulfur dioxide and sulfates are well known air pollutants and the role of sulfite forms of sulfur oxides
in the ambient air is gaining increasing attention(6 & 12)
Substantial exposure to sulfites is envisages as they
are frequently used as preservatives in foods, beverages
and many drug preparations. Clinical fluids for intraveno-
us infusion and peritoneal dialysis also contain sulfite
in concentrations upto 15mM (7 & 9). We have earlier
reported that sulfite can adversely affected energy metabolism and that target organ sensitivity is related to
the activity of sulfite oxidase(3) which detoxifies sulfite
by oxidation to sulfates(8). Beck Speier, et al, (5) reported
that sulfite in low concentrations (0.01 to 0.1mM) increases the lucigenin-dependent chemiluminisence (CL) of
resting and stimulated neutrophils indicating an increased
production of superoxide anions. The activation effect of
low sulfite concentration on the oxidative metabolism
with sulfite induced changes in the plasma membrane
of neutrophils.

In the present study, we examined the effect of
sulfite on the oxidative metabolism of resting and stimulated neutrophils, using the criteria of their NADPH-
oxidase activity, lucigenin and leuminol-dependent CL
and myeloperoxidase (MPO) activity **in vitro.**

MATERIALS AND METHODS

Materials : Leuminol, lucigenin and phoropol myristate
acetate (PMA) were purchased from Sigma (Deisenhofen,
FRG), human myeloperoxidase enzyme from Calibiochem
(Frankfurt, FRG) and all other chemicals were from
Merck (FRG).

Isolation of Neutrophils : Human neutrophils were isolated
from citrate treated whole blood, viability and purity
of prepeıation were varified with trypan blue according
to Beck-Speier **et al.**,(4).
Sulfite solution : A stock sulution (10mM) sodium sulfite
was prepared by direct in phosphate buffered saline(PBS)

pH 7.0. Suitable dilution of this stock were made just prior to use in various measurements.

Measurement of lucigenin and leuminol dependent CL : The lucigenin dependent and leuminol dependent CL of human neutrophils were measured at 37°C in a six channel Biolumat LB 9505. Neutrophils (5×10^6 Cells) were taken in 0.5 ml PBS containing 0.1% glucose, 0.2mM leuminol or lucigenin. The cells were maintained at 37°C during the CL-measurement of resting and PMA stimulated cells. Total time of measurement was 30 min in both cases.

In vitro myeloperoxidase activity with leuminol dependent CL : After 10 min pre-incubation of PBS with 0.2 mM leuminol at 30°C in the Biolumat, 10μl of (1:800 diluted) myeloperoxidase and sulfite (0.05 to 2.0 mM) using 0.3mM H_2O_2 as a substrate, the MPO— activity was measured for 20 min and expressed in terms of counts integrated over 20 min/mg protein.

Assay for NADPH- oxidase activity with Lucigenin dependent CL : Membranes of freshly isolated human neutrophils (10^7 cells) were prepared and NADPH-oxidase activity of membranes was determined by the lucigenin-dependent CL according to Minkenberg and Ferberg (11). The assay mixture contained of membrane suspension (30 to 40 μg protein) per 0.5 ml reaction mixture. The reaction was started by adding 1mM NADPH as a substrate and its oxidation was observed in terms of counts integrated over 25 mim/mg protein. Protein content was estimated by the Lowery (10).

RESULTS AND DISCUSSION

The results of sulfite on stimulated human neutrophils is summarized in Table 1. Sulfite in concentration from 0.1 to 1.0 mM resulted 6 to 7 fold increase in lucigenin-dependent CL of stimulated neutrophils. Higher concentrations (above 1.0mM) of sulfite did not further increase their CL instead it was gradually decreased with increa-

Table 1 : Chemiluminiscence of Neutrophils±Sulfite

Sulfite Conc. (mM)	Lucigenin-dependent CL		Luminol-dependent CL	
	Resting Cells	Stimulated Cells	Resting Cells	Stimulated Cells
0	10	276±23 (n=7)	100	
0.1	3.5±11(n=10)	626±41 (n=4)	101±18(n=6)	94±24(n=10)
0.5	495±2(n=16)	630±114(n=5)	105±16(n=6)	101±16(n=9)
1.0	595±22(n=12)	745±110(n=4)	87±25(n=6)	72±17(n=9)
2.0	430±24(n=11)	726±115(n=4)	67±14(n=6)	30±18(n=10)

Mean ± SE ;
n=number of experiment.

sing dose of sulfite. The CL of stimulated neutrophils, measured in the presence of lucigenin, is due to the reaction of O_2^- with lucigenin and this was confirmed in experiments with exogenous addtion of superoxide dismutase, which is a catalyst dismutation of superoxide dismutase inhibited the CL resting neutrophils. Beck-Speier, **et al.**,(3) reported that sulfite in higher concentrations significantly decreased the ATP content and oxygen consumption of tissues. Therefore, higher concentrations not only affect energy metabolism, but also inhibit processes associated with phagocytosis. In our experiment inhibition of CL by superssion of superoxide anions supports the above contention since superoxide burst in phagocytosis is a well known phenomenon.

Activity of NADPH-oxidase determined by lucigenin dependent CL was to reflect superoxide anion production in membranes. Sulfite (1 to 8 mM) enhances the CL 4 to 5 folds over the controls (Fig.1). This enchancement indicates that sulfite is a stimulating agent for human

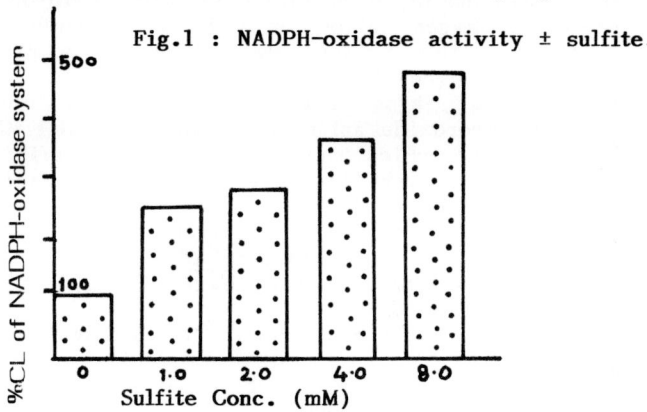

Fig.1 : NADPH-oxidase activity ± sulfite.

neutrophils by activating the oxygen metabolism to produce superoxide anions possibly through an activation of NADPH-oxidase. This effect is apparently brought about by the inhernt ablity of sulfites to act as electron donor, thus particepate in electron transport system and readily reactive with numerous biomolecules(11). Reduction of NADPH-oxidase by dithionate supports these observations (2).

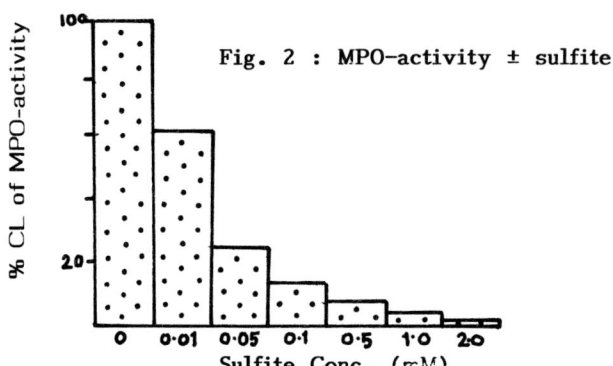

Fig. 2 : MPO-activity ± sulfite.

The leuminol dependent CL in human neutrophils was significantly decreased on exposure to sulfite. Approximate seventy percent decrease in CL was recorded with 2.0mM sulfite (Table 1). Pure myeloperoxide (MPO) activity, which is isolated from human neutrophils was seen inhibited by different concentrations of sulfite (Fig.2). The intracellular MPO-activity in stimulated human neutrophils is also inhibited up to 35 percent. (data not shown). Earlier Allen (1) reported that myeloperoxidase is directly involved in highly efficient leuminol CL response, which is discharged into phagolysosome by the PMN'S during phagocytosis. Therefore, our observation consistent to existing literature and confirm to the inhibition of phagocytosis by sulfites through interference in oxygen metabolism of human neutrophils.

In conclusion present observations have demonstrated that sulfite even in low concentrations is able to activate neutrophils. These activated neutrophils increasingly produce superoxide anion, H_2O_2 and inhibited MPO conected reactions which may lead to harmful effect because cells contain a very low level of sulfite oxidase. Therefore, the use of sulfite as preservatives, in drug preparations and infusion fluides should be carefully considered.

REFERENCES

1. Allen,R.C.(1982).Eds.Adam,W. & Cilenlo,G. pp.310-344 Acadmic. Press. New York.
2. Aviram,I. and Sharabani,M.(1985).Biochem.Biopsy.Acta. 841, 81-89.
3. BeckSpeier,I.,Henze,H. and Holzer,H.(1985).Biochem. Biophys.Acta. 841, 81-89.
4. BeckSpeier,I.,Leuschal,L.,Luppold,G.B. and Maier,K.L. (1988). FEBS Lett. 277, 1-4.
5. Beck-Speier,I.,Mishra Anil,Luippold,G.B.and Godleski,J (1989). Biochem. Biophy. Acta.(communicated).
6. Chang,S.G., Littlejhon,D. and Hu,K.Y.(1987). Science 237, 756-58.
7. Gunnison,A.F. and Jacobsen,D.W.(1987). CRC Crit. Rev. Toxicol. 17, 185-214.
8. Heimberg,M.,Fridovich,I. and Handler,P.(1953). J.Biol. Chem. 204, 913-926.
9 . Kleninhans,D.(1982). Dtsch. Med. Wschr. 107, 1409-1411.
10 .Lowery,O.H.,Rosebroughm,N.J.,Farr,A. and Randell,R.J. (1951). J. Biol. Chem. 193, 165-275.
11 . Mikenberg,I. and Ferberg,E(1984) J. Immunol. Method. 71, 61-67.
12 .Ware,J.H.,Ferris,B.G. , Dockery,D.W.,Sprengler,J.D., Stram,D.O. and Spiezer,F.E.(1986). Am. Rev. Respir. Dis. 133, 834-842.

Session III
CELLULAR LUMINESCENCE

RECENT DEVELOPMENTS IN CLINICAL APPLICATIONS OF CELLULAR LUMINESCENCE

P. De Sole

Clinical Chemistry Laboratory, Catholic University (Policlinico Gemelli)
00168 Roma, Italy

INTRODUCTION

Polymorphonuclear leukocytes (PMNs) form one of the first line of host defense against invading organisms. Their interaction with phagocytosable particles or soluble stimuli is mediated by different kinds of receptors (1,2) and triggers a well known biochemical pathway that ultimately leads to the destruction of the phagocytized microorganism (3).

Because of their oxidative potentiality, phagocytic cells can be very harmful not only to external agents but also to host tissues and organ structures (4).

PMN metabolic activation induced by phagocytosable or soluble stimuli can be easily measured among other things by means of the photon emission that accompanies their metabolic activity.

The phenomenon of chemiluminescence (CL), discovered by Allen (5), has been widely used to study many metabolic and functional aspects of PMNs(6,7,8). In particular, the use of very small amounts of blood has made it possible to perform the CL measurements directly on whole blood without any PMN purification procedure (8,9). Clinical applications of CL measurements have been performed with less than a microliter of whole blood so that the measurement of PMN CL is one of the most sensitive analytical techniques in laboratory medicine.

However, in spite of its great sensitivity PMN CL has a low specificity, because of the interaction of many different factors. Nevertheless, specificity can be increased by measuring the CL at zero order for all the parameters except the one under study (8).

In my experience a wide overlapping exists between the CL results of control people and those of patients with various diseases that can be related to neutrophil dysfunction. However, although the lack of specificity greatly hampers the use of CL in clinical practice, its use as a screening test or as a means for following up individual patients with selected diseases can be found of diagnostic utility.

In this paper, I discuss the correlation between CL and blood lipids in the first part while in the second part I correlate the zymosan induced CL to the binding of monoclonal antibody specific for the Fc receptors.

MATERIALS AND METHODS

Reagents: Zymosan and modified Krebs-Ringer phosphate medium were prepared as reported elsewhere (9). Luminol and phorbol myristate acetate (PMA) were obtained from Sigma (St. Louis, MO, USA) and dissolved respectively 50 and 5 mmoles/l in dimethylsulfoxide. Further dilutions were in modified Krebs-Ringer phosphate. CD16 monoclonal antibody FITC coniugated was obtained from Becton Dickinson. MAXEPA (1 g capsules) was from Vivax (Lt, Italy).

Opsonification: fresh plasma opsonification (OZt) was performed as reported elsewhere (9); opsonification by complement depleted plasma (OZc) was performed as OZt with the difference that the plasma was incubated 30 min at $56\,^{\circ}C$.

For the CL measurements, 0.5 ul of whole blood were added to 100 nmoles of luminol, 0.5 mg of opsonified zymosan or 0.15 nmoles of PMA in 1.0 ml final volume of modified Krebs-Ringer phosphate.

In the case of measurement on purified PMNs, 50,000 cells and 2.5 mg opsonified zymosan (OZt or OZi) were used.

CL was measured at $25\,^{\circ}C$ at 3-5 minute intervals for 1 or 2 (whole blood CL) hours.

CL AND BLOOD LIPIDS

Recent experimental evidence clearly indicates the fundamental role of membrane lipids in the cell activation processes. The activation of membrane NADPH- oxidase is preceded by and mediated through a modification of a membrane-bound phospholipase C or the translocation-activation of a protein kinase C from cytosol to plasma membrane. Different stimuli can exert their action on the phospholipase C step through interaction with a GTP-binding protein or can directly induce the translocation-activation of the protein kinase C (3).

This simplified scheme highlights the role of plasma membrane lipids in PMN function. In fact, both ligand binding to the extracellular side of a receptor and the following transmission of information through the bilayer into the cytoplasmic side are greatly affected by membrane lipid composition and the related membrane fluidity. It is well known that lipids on the membrane bilayer are subject to various types of molecular motions: rotation along the molecular axis, lateral diffusion in the bilayer plane or transverse translocation (flip-flop) across the two layers of membrane. Time constants for these different processes differ in many orders of magnitude from 10^{-10} to 10^5 sec.

Quite interestingly, membrane fluidity is greatly affected by any modification of the unsaturation degree and/or chain length of membrane fatty acids.

Both opsonified zymosan and PMA stimulate the PMNs by involving a plasma membrane lipid rearrangement. In the case of opsonified zymosan (phospholipase C activation), G-protein is cleaved inside the membrane into a dimeric (Gbeta - Ggamma) and a monomeric form (Galpha) that are split apart during the activation process. In the case of PMA, the activation proceeds through the translocation-activation of protein kinase C from cytosol to plasma membrane. In both cases, however, a dependence of PMN activation on blood lipid content is higly probable.

In order to verify this hypothesis, whole blood CL in response to both opsonified zymosan and PMA has been correlated to blood lipids in a group of 22 subjects under medical observation for hypertension (Table1).

TABLE 1

Correlation between CL parameters and blood lipids

	Correlation coefficient		
	Cholesterol		Triglycerides
Stimulus	total	HDL	
Zymosan	0.70***	-0.55**	0.43*
PMA	0.83***	-0.54**	0.50*

$(^{*})\, p < 0.05 \quad (^{**})\, p < 0.01 \quad (^{***})\, p < 0.001$

The results obtained with both opsonified zymosan and PMA show a clear correlation between blood lipids and whole blood CL. In particular, CL stimulated by zymosan and PMA is positively correlated with total cholesterol and triglycerides, while it is negatively correlated to HDL-cholesterol.

Because a similar correlation is also present between lipids and blood pressure in the same patients, the relationship between blood pressure and PMN CL has also been analyzed (Table 2).

TABLE 2

Correlation between CL parameters and blood pressure

| | Correlation coefficient | |
| | Pressure | |
Stimulus	systolic	diastolic
Zymosan	0.70***	0.83***
PMA	0.69***	0.88***

(***) $p < 0.001$

Because both systolic and diastolic pressure are positively correlated to PMN CL, a possible relationship between blood lipids, blood pressure and PMN CL is highly probable.

Moreover, because membrane fluidity is highly sensitive to the degree of unsaturation of fatty acids, any modification of membrane lipid composition by in vivo administration of unsaturated fatty acids is expected to induce a significant variation of CL response.

With this aim in view, whole blood CL has been analyzed in a group of five volunteers before and after a three day treatment with MAX-EPA (pharmaceutical preparation containing eicosapentaenoic and docosahexaenoic acid).

As expected a reduction of CL response to both opsonified zymosan and PMA has been induced by the treatment (Table 3).

TABLE 3

CL values (cpsc) before and after MAX-EPA treatment

MAX-EPA treatment Stimulus	before	after	P level
Zymosan	5.8 ± 2.1	4.4 ± 1.5	< 0.05
PMA	8.0 ± 2.4	6.0 ± 2.4	< 0.05

cpsc = counts per second per cell

CL RESPONSE TO ZYMOSAN OPSONIFIED BY
FRESH OR COMPLEMENT DEPLETED PLASMA

Zymosan opsonification is mediated by complement derived factors (mainly C3b, iC3b) and immunoglobulins. Therefore, cell response to zymosan opsonified by fresh plasma issues from particle interaction with receptors for C3 (CRI and CRIII) as well as receptors for Fc fragment of immunoglobulins (FcR). Human PMNs can express three distinct classes of Fc receptors: FcRI, FcRII, and FCRIII. Resting neutrophils do not express FcRI but they do express 10,000 - 40,000 copies of FcRII and 100,000 - 300,000 copies of FcRIII (2).

Receptor expression is changed by mediators of inflammation . FcRI is induced by gamma- but not by alpha- or beta-interferon as well as by the granulocyte/macrophage colony stimulating factor. The expression of FcRIII depends on upregulation and extracellular release of this receptor.

With the aim of analyzing the function of receptors for Fc (FcR) and C3 derived factors (C3R), the CL response of purified neutrophils was measured in response to zymosan opsonified by fresh plasma (OZt) or complement depleted plasma (OZi). The difference between OZt and OZi is roughly a measure of zymosan opsonification by complement factors (OZc) (equation 1):

$$OZc = OZt - OZi \qquad (1)$$

CL response of neutrophils to OZt can thus be split into two parts: the first toward OZc and the second toward OZi (equation 2)

$$CL(OZt) = CL(OZc) + CL(OZi) \qquad (2)$$

Furthermore, the ratio $CL(OZc)/CL(OZi)$ is a measure of the ratio between the number and/or activity of C3b receptors over Fc receptors.

CL response to OZt, OZc and OZi has been anlyzed in a group of psoriatic patients. It is well known, in fact, that, among the inflammatory diseases, psoriasis is closely linked to neutrophil alterations. Although the data in the literature are not uniform, the results of different Authors are indicative of cell and/or serum modification in psoriasis. In particular, the alteration of neutrophil function seems to be related to the disease intensity (10).

In table 4, the CL data of two groups of patients (one with active spreading form of psoriasis and the other with a chronic stable phase psoriasis) and a control group are shown.

Quite interestingly, the active form of psoriasis shows a statistically significant increase of CL in response to OZi and a reduction of the ratio $CL(OZc)/CL(OZi)$, clearly indicating an imbalance between C3b and Fc receptors in this group of patients.

TABLE 4

CL of PMNs from psoriatic patients and controls

	cpsc (mean + SD)			
Stimulus	spreading PSO (A)	chronic PSO (B)	control (C)	P level
OZt	7.0 ± 2.0	6.0 ± 1.0	6.0 ± 1.7	NS
OZc	4.0 ± 1.2	3.9 ± 0.8	4.1 ± 1.2	NS
OZi	3.0 ± 0.9	2.1 ± 0.4	1.9 ± 0.6	A/B*
				A/C*
				B/C NS
OZc/OZi	1.4 ± 0.2	1.9 ± 0.4	2.2 ± 0.5	A/B*
				A/C*
				B/C NS

(*) $p < 0.001$
cpsc = counts per second per cell
PSO = psoriasis

With the aim of comparing the zymosan induced CL with the expression of the Fc receptors, a monoclonal antibody (CD16) that specifically reacts with FcRIII has been used. The expression of this receptor has been measured by means of flow cytofluorimetry on purified neutrophils from the same groups (Table 5).

TABLE 5

PMN fluorescence intensity dependent on CD16 monoclonal antibody density

Group	Fluorescence intensity mean + SD	t-test vs. Control
Spreading PSO	344 ± 69	NS
Chronic PSO	445 ± 156	< 0.05
Control	333 ± 70	

PSO = psoriasis

It is interesting to note that there is no difference between group A (active spreading psoriasis) and control group, while FcRIII expression of group B (chronic stable phase psoriasis) seems to be higher than that of both groups although with a low statistical significance.

Cytofluorimetric analysis indicates a normal density of FcRIII on neutrophils of patients with spreading psoriasis, while CL data seem to be indicative of an activation of the signal transduction mechanism or of the oxidase complex. On the contrary in stable phase psoriasis the number of FcRIII seems to be probably increased. Because in this group CL activity does not differ significantly from that of the control group, some down regulation mechanism for ligand-receptor binding or cell metabolic activity can be hypothesized.

REFERENCES

1. G.D. Ross, Structure and function of membrane complement receptors, Fed. Proc., 211, 3089 (1982).
2. T.W.J. Huizinga, D. Roos, A.E.G.Kr von dem Borne, Neutrophil Fc-gamma Receptors: a two way bridge in the immune system, Blood, 75, 1211-1214 (1990).
3. F. Rossi, The O2- -forming NADPH oxidase of the phagocytes: nature, mechanisms of activation and function, Biochim. Biophys. Acta, 853, 65-89, 1986.
4. J.E. Weiland, W. B. Davis, J.F. Holter, J.R. Mohammed, P.M. Dorinsky, J.F. Gadek, Lung neutrophils in the adult respiratory distress syndrome - Clinical and pathophysiologic significance, Am. Rev. Respir. Dis., 133, 218- 225 (1986).
5. R.C. Allen, R.L. Stjernholm, R.H. Steele, Evidence for the generation of an electronic excitation state(s) in human polymorphonuclear leukocytes and its partecipation in bactericidal acivity, Biochem. Biophys. Res. Comm., 47, 679-684, 1972
6. R.C. Allen, M.M. Lieberman, Kinetic analysis of microbe opsonification based on stimulated polymorphonuclear leukocyte oxygenation activity, Infect. Immun., 45, 475-482 (1984).
7. C. Dahlgren, G. Briheim, Comparison between the luminol dependent chemiluminescence of polymorphonuclear leukocytes and the myeloperoxidase-HOOH system: influence of pH, cations nd proteins, Photochem. Photobiol., 41, 605-610, 1985.
8. R.C. Allen, Phagocyte leukocyte oxygenation activities and chemiluminescence: a kinetic approach to analysis. In Methods in Enzymology. M.A. De Luca and W.D. McElroy (eds), Academic Press, New York, vol.133, p. 449-493, 1986.

9.   P. De Sole, S. Lippa, G.P. Littarru, Whole blood chemiluminescence: a new technical approach to assess oxygen dependent microbicidal activity of granulocytes, J. Clin. Lab. Autom., 3, 391-400, 1983.

10.   C. De Simone, P. De Sole, G. Di Mario, A. Venier, D. Cerimele, F. Serri, Reactive oxygen species production in circulatingpolymorphonuclear leukocytes in psoriasis, Acta Derm. Venereol. (Suppl.) 146, 50-52, 1989.

ACKNOWLEDGMENTS:

I thank the following colleagues for their cooperation in this research: G. Di Mario, R. Fresu, C. De Simone, S. Buffa.
I also thank Mrs P. Massari for the help in reviewing the manuscript.
This work has been supported in part by a grant of the "Ministero della Pubblica Istruzione".

FACTORS AFFECTING A CHEMILUMINESCENCE ASSAY FOR ERYTHROPHAGOCYTOSIS

I Downing, JG Templeton, R Mitchell, RH Fraser

Glasgow and West of Scotland Blood Transfusion Service
Law Hospital, Carluke, Lanarkshire ML8 5ES, UK

INTRODUCTION

Mononuclear phagocytes have been shown to bind to erythrocytes (RBC) coated with non-complement binding IgG via Fc receptors. This ensures the destruction of the RBC. However non-complement binding IgG alloantibodies to some high incidence RBC antigens can be unpredictable in their clinical effect. Normal test procedures depend on the agglutination of sensitised RBCs by a second antibody and will only indicate the presence of an antibody and not prove whether it is clinically significant. Thus provision of blood for patients with antibodies against relatively common antigens can be difficult, especially when antibodies of mixed specificity are involved.

We have developed a chemiluminescence assay in order to address the potential clinical problems caused by the above anti-RBC antibodies.[1] The effector cells are the monocyte portion of a mononuclear leukocyte cell fraction, and the target cells RBCs sensitised with anti-D IgG. RBCs thus sensitised are known to be clinically significant and were used to optimise the assay before comparing it to the standard microscopic assay.

METHODS

Preparation of Mononuclear Leukocytes

A mononuclear leukocyte cell fraction was obtained from healthy blood donors by apheresis using a Haemonetics V-50 plasmapheresis unit. The mononuclear leukocytes were purified and frozen using essentially the methods of Bøyum (1983)[2] and Weiner (1976).[3] The number of monocytes in a preparation was assessed by the method of Hogg *et al* (1984).[4] Viability was established by phase contrast microscopy.

Chemiluminescence (CL) Assay

The mononuclear leukocytes were taken from liquid nitrogen and thawed rapidly at 37°C. The cells were transferred to tubes containing RPMI 1640 + 10% foetal calf serum and then centrifuged at 500g for 5 minutes. The cells were resuspended in RPMI 1640 (Searle Modification) without phenol red (RPMI-5M), pH8.0. Mononuclear leukocytes (3×10^6 cells) were placed in cuvettes and incubated for 15 minutes at 37°C. $100\mu l$ luminol (8×10^{-5} mol/l final concentration) was added. This was followed by $75\mu l$ RPMI-SM containing 1×10^6 RBCs (rhesus positive) sensitised (apart from controls) with either 5.25IU anti-Rh(D) immunoglobulin (Protein Fractionation Centre, Edinburgh, UK, lot 100470120) or monoclonal anti-Rh(D) IgG_1 (a gift from Dr M McCann, South East of Scotland Blood Transfusion Service and Dr K James, University of Edinburgh) or monoclonal anti-Rh(D) IgG_3 (a gift from Dr B Kumpel, UKTS, Bristol, UK). The cuvettes were placed in the carousel of an LKB 1251 luminometer linked to an IBM PS2 running the LKB phagoprogram.

Light emission was monitored for 120 minutes with mixing for the first cycle only. The area under the curve of light output for sensitised RBCs was divided by that for unsensitised RBCs to give the opsonic index where appropriate.

Microscopic Monocyte Assay (MMA)

This phagocytic assay was a modification of that of Schanfield *et al* (1980)[5], the principal differences being the use of Labtek tissue culture slides (Miles Laboratories, Slough, UK) and their incubation in a 5% CO_2 atmosphere. The total association index (TAI) was obtained by counting all the RBCs associated with 100 monocytes. The percentage phagocytosis (PRBC) was determined by counting the RBCs associated with 100 monocytes after treatment with ammonium oxalate. Two hundred monocytes were evaluated for each parameter.

RESULTS AND DISCUSSION

The batch of mononuclear cells used consisted of 48.9 $\pm$ 1.5% monocytes, with better than 80% of recovered cells viable. As can be seen from Fig 1 the amount of IgG used to sensitise the RBC, and thus by implication the amount of IgG actually bound to the RBC has a sigmoidal relationship to the chemiluminescent response.

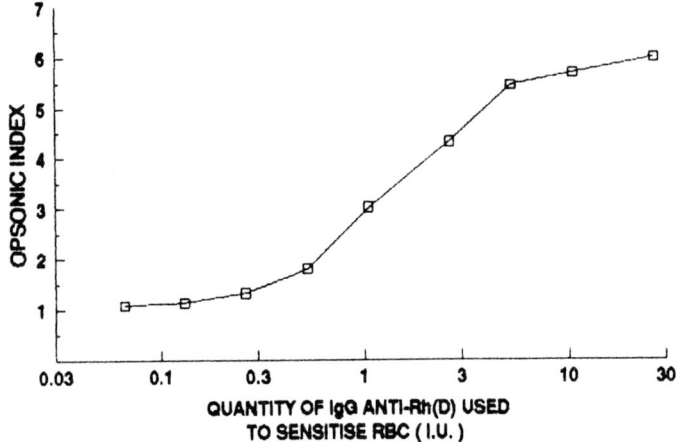

Fig 1. The effect of the concentration of anti-Rh(D) used to sensitise the RBCs on the CL assay. Each point is the mean of three observations.

It was found that the pH of the reaction medium rises during the course of an experiment. It was further found that this was an effect of the RPMI-SM medium alone and did not depend on any of the cellular constituents. The rise in pH can be attributed to the loss of CO_2 by the medium altering the HCO_3^- /H_2CO_3 balance which maintains the pH. The use of more effective buffers was tried but found to quench the luminol reactions. Fig 2 shows the effect of initial pH of the medium on light emission. It was found that an initial pH of 8.0 was most efficient and also limited further pH rises to 0.5 pH units while providing maximal luminescence. The effect of initial pH on light emission probably reflects the effects of an alkaline pH on the integrity and health of the monocytes on one hand and the efficiency of the luminol-radical reaction at a lower than optimal pH on the other. The monocytes normal environment has a pH of 7.4 while the optimal for the luminol reaction is pH 12(6). Thus pH 8.0 is the best compromise on the monocytes synthesis of radical species and their reaction with luminol. Obviously the radical synthesis is the rate limiting and more sensitive part of the system.

RBC concentration was quite critical with errors either side of the optimum reducing the luminescence signal as previously demonstrated (1), by either not optimally stimulating the monocytes or the increased concentration of haemoglobin absorbing the emitted light and possibly interfering with the luminol-radical reaction. The interference by haemoglobin was not unexpected due to its broad absorption spectra. It was possible to eliminate the haemoglobin by preparation of RBC membrane ghosts. However it is difficult to relate the amount of membranes to a given number of RBCs, and there is the probability that some will be "inside out", not exposing the Rh complex. The response obtained in a trial was poor compared to intact RBCs, giving a low opsonic index and a poor light emission curve lacking a well defined peak with large amounts of sensitised membranes.

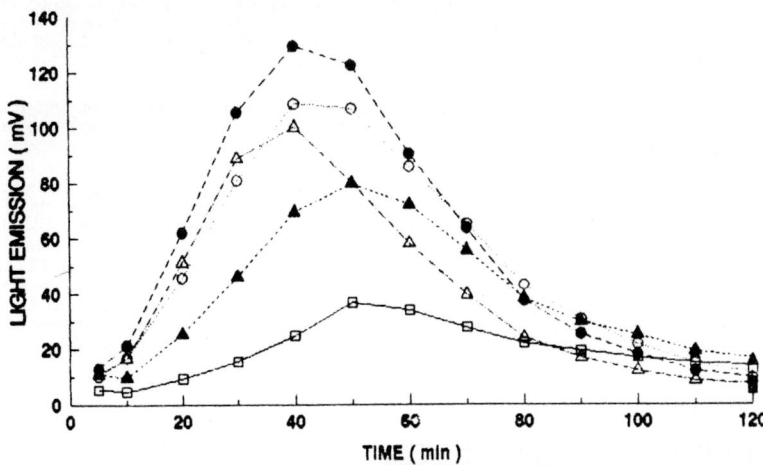

Fig 2. The effect of initial pH of the RPMI-SM medium on the light emission by the CL assay, pH 6.5 (—□—), pH 7.0 (···▲···), pH 7.5 (··○··), pH 8.0 (-●-), pH (-△-). Each point is the mean of three observations.

The above work led on to an examination of the target cell:effector cell ratio using a polyclonal IgG and monoclonal IgG_1 and IgG_3. Table 1 shows that the polyclonal IgG and monoclonal IgG reagents had identical optimum target cell:effector cell ratios of 1:1.

TABLE 1

EFFECT OF TARGET CELL:EFFECTOR CELL RATIO FOR CL ASSAY

RBC SENSITISING REAGENT	RATIO OF TARGET CELL:EFFECTOR CELL			
	1:1	3:1	10:1	30:1
Polyclonal IgG_{1+3}	10.73	4.96	2.32	1.27
Monoclonal IgG_1	5.44	2.79	1.96	1.20
Monoclonal IgG_3	6.03	3.78	2.28	1.37

Results Shown as Opsonic Index

The results obtained are at variance with Hadley et al (1988)(7) who found optimal target:effector ratios of 100:1 and 10:1 for IgG_1 and IgG_3 respectively. The simplest explanation is that the difference is caused by the use of different monoclonal antibodies and differing experimental techniques. However, while it may be possible for the two monoclonal antibodies to have different adherence properties they still ultimately trigger the same Fc receptor and initiate the same intracellular events. Hence when measuring the interaction of monocytes and RBCs at a step removed from the attachment phase, if the monocyte has an equivalent number of receptors stimulated a similar chemiluminescence response would be expected. This reponse is finally determined by the number of IgG molecules on the RBC. So that if the monocyte was maximally stimulated the occupation of further Fc receptors, by for example further IgG_3 coated RBCs, would cause no increase in chemiluminescence.

Finally we compared the ability of the MMA and the CL assay to detect naturally occurring anti-Rh(D) antibodies. The antibodies ranged from very high to very low concentrations with the latter predominating. Of the 29 patient samples investigated the CL assay identified 79% as containing "clinically significant" antibodies while the MMA identified 69% as containing "clinically significant" antibodies. The two assays returned results in agreement for 25 of the 29 samples. For the remaining four samples the CL assay gave a positive result for three, the MMA a positive for the other. In all the CL assay gave six negative results, the MMA nine in samples which were known to have very low levels of Anti-Rh(D). The CL assay compared well with the MMA, which has already been used to predict the clinical significance of patient anti-RBC antibodies[8]. It is intended that in practice the CL assay should be used for other than Anti-Rh(D) antibodies and has proven at least as sensitive as the MMA, faster to perform, less labour intensive free from subjective errors and more reproducible (CL assay, intra-assay CV 5%, MMA CV 10%). Experiments are continuing to establish criteria to interpret results from clinically uncertain Anti-RBC antibodies.

REFERENCES

1   I Downing, JG Templeton, R Mitchell & RH Fraser. A chemiluminescence assay for erythrophagocytosis. **J Biolumin Chemilumin**. In Press.

2   A Bøyum. Isolation of human blood monocytes with Nycodenz - a new non-ionic iodinated gradient medium. **Scand J Immunol** **17**; 429-436 (1983).

3   RS Weiner. Cryopreservation of lymphocytes for use in *in vitro* assays of cellular immunity. **J Immunol Meth** **10**; 49-60 (1976).

4   N Hogg, S Macdonald, M Slusdrenko and PCL Beverley. Monoclonal antibodies specific for human monocytes, granulocytes and endothelium. **Immunology** **53**; 753-767 (1984).

5   MS Schanfield, SL Schoeppner and JO Stevens. New approaches to detecting clinically significant antibodies in the laboratory. In **Immunobiology of the Erythrocyte** SG Sandler, J Nusbacher and MS Schanfield (Eds) Alan R Liss New York p305-323 (1980).

6   J Lee, HH Seliger. Quantum yields of the luminol chemiluminescence reaction in aqueous and aprotic solvents. **Photochem Photobiol** **15**; 227-237 (1972).

7   AG Hadley, BM Kumpel and AH Merry. The chemiluminescent response of human monocytes to red cells sensitised with monoclonal anti-Rh(D) antibodies. **Clin Lab Haemat** **10**; 377-384 (1988).

8   SJ Nance, P Arndt and G Garratty. Predicting the clinical significance of red cell alloantibodies using a monocyte monolayer assay. **Transfusion** **27**; 449-452 (1987).

REAPPRAISAL OF WHOLE BLOOD CHEMILUMINESCENCE FOR THE DETECTION OF PHAGOCYTE SELECTIVE INHERITED DISORDERS IN PEDIATRIC CLINICAL RESEARCH

Béatrice DESCAMPS-LATSCHA

INSERM U 25, CNRS UA 122, Hôpital Necker, 161 Rue de Sèvres, 75015 Paris

INTRODUCTION

Increasing evidence has accumulated that determination of chemiluminigenic probe amplified chemiluminescence (CL) in whole blood is a highly sensitive means for diagnosing major inherited phagocyte disorders such as chronic granulomatous disease (CGD) and lymphocyte function associated antigen (LFA-1) deficiency. During the past years this methodology has been extensively developed in our laboratory to investigating phagocyte oxidative dysfunctions, both in experimental animal models and in adult patients with various immune disorders (recently reviewed in 1). In the latter situation, control subjects for defining the normal range of whole blood CL responses were easily obtained among volunteer blood donors. On the contrary the ethical obstacle with respect to the use of normal children as controls for clinical studies, as well as the lack of precise knowledge on the influence of age on phagocyte oxidative functions , make it critical to develop applications of whole blood CL determination in pediatric clinical research.

Taking advantage of the regular health check ups which are performed in normal children in the context of the French preventive medicine system, and of the fact that whole blood CL determination requires a minimal volume of blood (<0.5ml), we recently investigated a wide population of normal children with the aim of defining the normal range of whole blood CL in childhood and of having a control population for further pediatric clinical research. The present report will summarize the results obtained in this comprehensive study (2) which allowed a reappraisal of the diagnosis and the predictive value of whole blood CL in children with inherited disorders associated with phagocyte abnormalities.

METHODS

Subjects

Normal children having a regular health check up by Dr. Labadie at the centre des Bilans de Santé de la Caisse Centrale d'assurance Maladie directed by Pr Rossignol were studied at 10 months (32 cases), 2 years (48 cases) and 4 years (40 cases). Normal adults (67 cases) were recruited among volunteer blood donors of the Centre de Transfusion Sanguine of Necker Hospital.

Children with CGD (31 cases), LFA-1 deficiency (7 cases) or Chediak Higashi (CHS) (5cases) were followed in the department of Pr. Griscelli and children with cystic fibrosis (CF) (23 cases) were followed by Pr. Lenoir (Hôpital des Enfants Malades).

CL determination

Materials and methods for CL measurements were as earlier described (3) with the exceptions that both luminol and lucigenin were used as chemiluminigenic probes and that CL probes were added to 10^{-2} diluted blood samples together with the various stimulating agents which included particulate (latex beads and serum AB opsonized zymosan) and soluble (phorbol myristate acetate (PMA), ConcanavalinA (ConA), and formyl-Methionyl-leucyl-phenylalanine (fMLP)) agents at the concentrations given in (3).

RESULTS AND DISCUSSION

Normal children

No major differences were observed among the three groups of children except for 4 year old children who demonstrated significantly decreased CL responses to zymosan as compared to 10 month and to 2 year old children ($P < 0.01$). In contrast, as compared to the 67 adults, the overall population of the 120 children demonstrated significantly increased CL responses to all stimuli in the presence of luminol ($P < 0.01$) but not lucigenin.

To our knowledge, this important difference in whole blood CL responses among adults and children depending on the type of CL amplifier has not been previously reported. Among these CL probes it is now well established that luminol mainly measures myeloperoxidase (MPO) associated reduced oygen intermediates whereas lucigenin almost exclusively involves the superoxide anion and thus more selectively reflects the NADPH oxidase activity. It is also generally accepted that at a 10^{-2} dilution whole blood CL production mainly reflects that of circulating PMN, and in our study a close correlation existed between CL intensity and the number of PMN present in whole blood. One may therefore suggest that the major difference observed between children and adults is related to differences in MPO content and or catalytic activity of circulating PMN. It must be stressed, however, that even if it is verified this MPO defect remains clinically asymptomatic. Based on our observations we concluded that precise evaluation of phagocyte oxidative metabolism abnormalities in childhood requires normal children as controls.

Chronic granulomatous disease (CGD)

In CGD, the well characterized genetic defect in phagocyte oxidative responsiveness (4), we confirmed the accuracy of whole blood CL determination for diagnosing the disease. Regardless of the stimulus used and the CL probe, nil CL responses were found in the 20 children with X-linked CGD and the 11 with autosomal recessive (AR-CGD)tested. Family studies further demonstrated that, as compared to normal adult subjects, mothers of X-linked CGD children have 30 to 50% reduced CL responses ($P < 0.01$), whereas AR-CGD mothers have CL responses within the normal range. These latter observations are in agrement with other reports in the literature indicating that O_2^- production measured by spectrophotometry and/or NBT reduction is significantly decreased in X-linked CGD mothers (5).

Lymphocyte function associated antigen (LFA-1) deficiency

Several reports have established that children with LFA-1 deficiency suffer from widespread recurrent bacterial infections that lead to early death in the absence of a bone marrow transplant (BMT) and the gene mutation responsible for this autosomal recessive disease is now well characterized (recently reviewed in 6). Over the past years our laboratory has contributed to the diagnosis of LFA-1 deficiency in 7 children on the basis of a selective CL unresponsiveness of whole blood to opsonized zymosan contrasting with normal responses to other stimulating agents. In six patients with successful BMT, partial or almost complete recovery of CL responsiveness to opsonized zymosan was observed in coincidence with the improvement of other phagocyte adhesion dependent functions (7,8). These findings support the conclusions that the magnitude of defective phagocyte responses to opsonized particles is related to the level of expression of adhesion proteins (6).

Chediak-Higashi (CHS)

The exact biochemical and genetic abnormality leading to the formation of giant granules in neutrophils of patients with CHS has not yet been characterized. Some authors have reported that neutrophils of CHS patients have an accelerated respiratory burst in response to phagocytized particles (9).
Five children with CHS were tested in our laboratory. A significant decrease (of 25 to 30%) of whole blood CL responses to latex, ConA and fMLP ($P < 0.02$), and of more than 50% to zymosan and PMA ($P < 0.001$) was observed. These findings are in contrast with this previous report but are more compatible with the defective bactericidal activity and recurrent infections frequently observed in these patients.

Cystic fibrosis (CF)

Considerable advances in the knowledge of the gene defect responsible for CF have been made during the past few years (10). In contrast, the underlying mechanisms of the great susceptibility of CF patients to infections are still unclear. Most studies have concentrated on the search of serum opsonizing dysfunctions and only a few on phagocyte associated disorders.
Study of 23 children with CF showed that, as compared to normal children, basal whole blood CL production was significantly decreased in the presence of both CL probes ($P < 0.001$); CL responses to stimulating agents were dramatically decreased in the presence of luminol ($P < 0.001$) whereas with lucigenin they were within the normal range except for zymosan-induced CL responses which were also significantly decreased ($P < 0.001$). The pattern of CL responses of isolated PMN and MN in these patients is currently under study in order to verify the hypothesis of a selective abnormality in phagocyte oxidative metabolic activation in CF.

In conclusion, our findings, derived from large scale development of whole blood CL analysis in pediatric clinical studies, demonstrate that phagocyte genetic disorders associated with major oxidative metabolism abnormalities can easily be detected but that the search for more mild abnormalities requires reference of the results to a normal range defined among normal children. In view of the numerous factors capable of interfering with CL measurements, it appears necessary that each laboratory should define its own CL reference control population. By doing so we were recently able to detect selective whole blood CL response abnormalities in children with genetic disorders susceptible of being associated with phagocyte dysfunctions. However the origin of these oxidative abnormalities and their role in the pathophysiology of these diseases still remains to be further elucidated.

REFERENCES

1.   B. Descamps-Latscha, New insights into allograft immunity and immunopathology afforded by chemiluminescence evaluation of phagocyte oxidative metabolism. In Bioluminescence and chemi-luminescence: New perspectives, J. Schölmerich et al (Eds), Wiley & Sons, Chichester, New York, Brisbane, Toronto, Singapore, p. 23-32 (1987).

2.   B. Descamps-Latscha, A.T. Nguyen, G. Brun-Cottan, J-C. Brun- Cottan, D. Descamps, O. Spatzierer, M.D. Labadie, C. Rossignol, F. Veber and G. Lenoir, Comprehensive investigation of whole blood chemiluminescence in normal children: its diagnostic value in chronic granulomatous disease and cystic fibrosis. Submitted (1990).

3.   B.Descamps Latscha, A.T. Nguyen, R.M. Golub and M.N. Feuillet-Fieux, Chemiluminescence in microamounts of whole blood for   investigation of the human phagocyte oxidative metabolism function. Ann. Immunol. 133C, 349-364 (1982).

4.   B. Royer-Pokora, L.M. Kunkel, A.P. Monaco, S.C. Goff, P.E. Newburger, R.L. Baehner, F.S. Cole J.T. Curnutte, S.H. Orkin, Cloning the gene for an inherited human disorder - chronic granulomatous disease - on the basis of its chromosomal location. Nature (Lond) 322: 32 (1986).

5.   P.E. Newburger, H.J. Cohen, S.B. Rothschild, J.C.Hobbins, S.E. Malawista and M.J. Mahoney, Prenatal diagnosis of chronic granulomatous disease. N. Engl. J. Med. 300, 178-181 (1979).

6.   A. Fisher, B. Lisowska-Grospierre, Leukocyte adhesion deficiency: Molecular basis and fuctional consequences. Immunodeficiency Rev, 1, 39-54 (1988).

7.   A. Fisher, P.H. Trung, B. Descamps-Latscha, B. Lisowska-Grospierre I. Gerota , N.P. Perez, C. Scheinmetzler, A. Durandy, J.L. Virelizier, and C. Griscelli, Bone marrow transplantation for inborn error of phagocytic cells associated with defective adherence, chemotaxis, and oxidative response during opsonised particle phagocytosis. Lancet 2, 473-476 (1983).

8.   F. Ledeist, S. Blanche, H. Keable and al., Successful HLA non identical bone marrow transplantation in three patients with the leukocyte adhesion deficiency. Blood 74, 512-516 (1989).

9.   R.K.Root , A.S. Rosenthal and D.J. Balestra, Abnormal bactericidal metabolic and lysosomal functions of Chediak-Higashi syndrome leukocytes. J. Clin. Invest. 51, 649 (1972).

10.   J.R. Riordan, J.M. Rommers, B.S. Kerem and al., Identification of the cystic fibrosis gene: Cloning and characterization of complementary DNA. Science 245, 1067-1072 (1989).

ASSAYS OF HUMAN PHAGOCYTE FUNCTION USING

MICROTITRE PLATE LUMINOMETERS

I.A. Cree

Department of Pathology, University of Dundee,
Ninewells Hospital and Medical School, Dundee, DD1 9SY, Scotland

INTRODUCTION

Phagocytes are integral to the immune response in man. Those in
whom phagocytic function is deficient are predisposed to a large
number of diseases, particularly infections. This is most
evident in genetic disorders such as Chronic Granulomatous
Disease, but recent studies of phagocyte function in a range of
conditions suggest that acquired impairment is more common than
has been thought in the past.

Paradoxically, enhanced phagocyte function is also a feature of
several diseases and may be an important factor in the
pathogenesis of several common conditions, such as asthma and
psoriasis. The application of chemiluminescence to these
clinical problems has been ably reviewed by De Sole (5, vide
infra). Microtitre plate luminometers offer particular advantages
for the study of clinical defects of phagocyte function and use a
format which is familiar to most immunologists.

Since the production of reactive oxygen species (ROS) is
intimately linked to microbial killing, methods for the
measurement of ROS production have been widely used to assess
phagocyte function. One of these, chemiluminescence (CL) relies
upon the interaction of ROS with chemical indicators to produce
light and has proved to be very sensitive.

Previous CL techniques have been limited by the inability of
luminometers to measure more than a few samples at one time and
have often required large quantities of blood. However, as
microtitre plate luminometers are able to measure CL from one
plate in 90 seconds, up to four plates (384 wells) can be
measured concurrently in a single experiment - if plates are
changed over between each reading. CL responses can be obtained
from as few as 5,000 PMNL per well, so that little blood is
required. Using appropriate experimental design, CL assays can be
used to examine each step in the phagocytic process.

MATERIALS AND METHODS

The methods (1-4) used for phagocyte CL with microtitre plate
luminometers are scaled down versions of those used with other
luminometers.

The basic method is shown in Fig. 1. We have successfully used
two commercial media and our own to separate polymorphonuclear
leukocytes (PMNL) and mononuclear cells from peripheral blood.
The amount of blood required is small: 5ml from adults is usual,
but it is possible to obtain enough PMNL for six wells from just
200μl of capillary blood from preterm infants. Our standard assay
system uses 50 μl of 1 x 10^6 PMNL/ml in each well.

Fig. 1. Basic assay method for phagocyte CL.

Add 50μl 1 x 10^6/ml PMNL to wells (triplicate).

Add 50 μl of 5 x 10^{-4} mol/l lucigenin to wells.

Allow to equilibrate at 37 C for 15 min.

Take background CL readings for a further 15 min.
in an Amerlite microtitre plate luminometer.

Add Stimulant - Opsonised Zymosan, PMA, etc.

Continue to take CL readings at 5 min intervals
for up to 120 min.

Lucigenin and luminol are relatively easy to use, although the
latter requires myeloperoxidase to be present to work
efficiently. This can be added if the cells present do not
possess their own. Monocyte myeloperoxidase decreases as the
cells migrate into tissues and become activated: lucigenin is
certainly the enhancing agent of choice for macrophage work. In
our hands, the optimal concentration of lucigenin is 10^{-4}
mol/l for microtitre plate assays.

The debate about whether phagocyte CL assays should be shaken or
stirred continues! However, in all microtitre plate luminometers
except the Luminoskan (Labsystems), one cannot use either. Either
way, the baseline CL in unstimulated wells should remain very low
- if it begins to climb after 30 min. or refuses to fall, this
tends to indicate cell death or aggregation and the experiment
should be abandoned.

The two latest microtitre plate luminometers, the Dynalite
(Dynatech) and the Luminoskan (Labsystems) both incorporate plate
incubation. It is obviously preferable to do phagocyte assays at
37°C, although more sluggish responses can be obtained at the
30°C internal operating temperature of the Amerlite (Amersham).

The facility to cycle, i.e. to perform measurements every few minutes automatically, is useful since it allows the operator to set up the next assay, have lunch, or perform other assays.

The choice of stimulant depends upon what one wishes to measure: there is a wide choice. Opsonised zymosan particles rely upon ligation of complement and Fc receptors on the cell surface to trigger CL via both DAG generation and calcium influx. FMLP also acts via a cell surface receptor. Such receptors can be up-regulated very quickly on the surface of PMNL, since they are stored in their a-granules. This can happen during cell separation and may be one reason behind the erratic results obtained from sequential studies of clinical specimens. Non-receptor-mediated stimulation can be achieved with stimulants such as phorbol esters, which act on Protein Kinase C, and calcium ionophores.

The use of microtitre plates poses some problems of data analysis, since one plate will generate about 3000 measurements during an assay. We transfer our results to a microcomputer and use the macro facility of a large spreadsheet, Supercalc 5 (Computer Associates) to analyse our results. This produces the mean, standard deviation, and coefficient of variation for triplicate wells, together with the integrated total CL and the slope from the last point measured. Graph generation is also simplified.

RESULTS AND DISCUSSION

Opsonisation   Measurement of opsonisation defects in sera and differences in microbial susceptibility to opsonisation, phagocytosis and ROS production can be studied using these methods. One useful methodological variation was presented by Hastings (4) in which bacteria are opsonised within the microtitre plates, and washed using a centrifuge fitted with a plate holder, before the PMNL are added to start the CL response. This has advantages for analysis of strain differences or large numbers of sera.

Cellular defects   The major genetic defects of human phagocyte ROS production are Chronic Granulomatous Disease and Myeloperoxidase Deficiency. Both are rare, and it as simple a matter to screen patients for these problems using microtitre plate luminometers as it is with any other luminometer.

Acquired disorders of phagocyte function are much more common, but tend not to be the all-or-nothing phenomena that one sees in genetic diseases. In certain conditions, such as renal failure, it has been suggested that serial CL measurements could be used to follow the response of patients to treatment and warn of impending exacerbations of disease. Cost-effective provision of such facilities requires many samples to be handled at one time, preferably automatically, as can be done with microtitre plate luminometers.

Immunopharmacology One of the problems which has caused phagocyte CL to be mistrusted in the past was the variability seen in the results from phagocytes from different donors, or from the same donor at different times. Such variation made direct comparison of the effects of drugs on phagocyte function in vitro very difficult. Using microtitre plates, it is possible to look at three drugs with 8 dilutions simultaneously or to try various other combinations. This makes it relatively simple to use CL as a screening method for drug effects on phagocytes.

Phagocyte Activation Assays   Since monocytes can be purified from the MNC fraction of peripheral blood directly into the microtitre wells used by the Amerlite and cultured with lymphokines, CL can be used as the basis for a monocyte activation assay (4).

The possibility that disorders of phagocyte activation by cytokines (priming) exist and have clinical significance have not been investigated. Monocytes form a circulating pool of resting macrophages which undergo a complex process of activation following their migration into sites of inflammation. The activation process relies upon mediators such as IFNy released during lymphocyte activation. It is therefore a major part of the cell-mediated immune response and produces cells capable of enhanced microbial killing, tumour cytolysis, and granuloma formation.

ACKNOWLEDGEMENTS
I wish to thank Prof. JS Beck, Miss AL Blair and Miss LA Pierce. This research which was supported by grants K/MRS/50/C983 and K/MRS/50/C1211 from the Scottish Home and Health Department.

REFERENCES
1.   A.L. Blair, I.A. Cree, J.S. Beck and J.G.M. Hastings. Measurement of phagocyte chemiluminescence in a microtitre plate format. J. Immunol. Methods 112, 163-168 (1988).
2.   A.L. Blair, I.A. Cree and J.S. Beck. Measurement of phagocyte chemiluminescence using a microtitre plate luminometer. J. Biolum. Chemilum. 3, 67-70 (1989).
3.   I.A. Cree, A.L. Blair and J.S. Beck. Use of a microtitre plate chemiluminescence reader to study surface phagocytosis by human monocytes. J. Biolum. Chemilum. 3, 71-74 (1989).
4.   J.G.M. Hastings, L.A. Jewes and K.M. Oxley. Opsono-phagocytosis of bacteria studied by chemiluminescence in microtitre plates. J. Biolum. Chemilum. 4, 267-271 (1989).
5.   P. De Sole. Polymorphonuclear chemiluminescence: some clinical applications.   J. Biolum. Chemilum. 4, 251-262 (1989).

THE ROLE OF F_C-GAMMA AND C3 RECEPTORS IN OPSONIZED AND NON-OPSONIZED BACTERIA INDUCED CHEMILUMINESCENCE OF HUMAN NEUTROPHILS

L. Leino, P. Peltonen[1] and E-M. Lilius[1]

Department of Hematology, Turku University Central Hospital,
20520 Turku, Finland, and [1]Department of Biochemistry,
University of Turku, 20500 Turku, Finland

INTRODUCTION

The main opsonins in human serum are IgG and opsonic forms of C3. By binding to the surface of micro-organism, the opsonins initiate the recognition of targets by specific receptors on the plasma membrane of host phagocytes. The activation of phagocytes results finally in ingestion and intracellular killing of pathogen or degranulation and the extracellular release of digestive enzymes and reactive oxygen forms. Three distinct classes of receptors for the Fc region of IgG have been characterized (FcRI, FcRII, FcRIII) (for review see 1). Moreover, at least two different receptors on human neutrophils have been described to recognize opsonic fragments of C3: CR1 binds C3b and CR3 C3bi, respectively (for review see 2). Recently it has been shown that FcRII on human neutrophils is essential for IgG-induced superoxide production and phagocytosis (3) whereas a phosphatidylinositol-linked form of FcRIII mediates the signal for degranulation (4). CR1 and CR3 have been suggested to participate in binding of opsonized particles, but their ability to initiate phagocytosis and internalization of targets is still controversal (5,6). In the present study, we have examined the role of Fc-gamma receptors (FcRs) and C3 receptors in opsonized and non-opsonized bacteria induced luminol-dependent chemiluminescence (CL) of human neutrophils.

MATERIALS AND METHODS

Reagents A stock solution of 10 mmol/l luminol (5-amino-2,3-dihydro-1,4-phthalazinedione) (Sigma Chemical Co.) was prepared in 0.2 mol/l sodium borate buffer, pH 9.0 and stored at -20 °C. Hanks balanced salt solution (HBSS, pH 7.4) was made without phenol red and supplemented with 0.1 % gelatine. Purified anti-FcRI (CD64), anti-FcRII (CD32), and anti-FcRIII (CD16) whole monoclonal antibodies (mAbs) and anti-FcRII monoclonal Fab and anti-FcRIII monoclonal F(ab)$_2$ fragments were purchased from Medarex Inc., USA. Anti-CR1 (CD35) and anti-CR3 (CD11b) mAbs were obtained from Immunotech S.A., France. The bacteria used were single clinical isolates of Streptococcus pneumoniae type 3. The organisms (kindly provided by Dr P. Huovinen, Dept. of Medical Microbiology, University of Turku) were grown in Todd-Hewitt broth and after the cell number determination the bacteria were stored in tryptic soy broth containing 20 % glycerol at - 40 °C. Prior use, the cells were washed with HBSS.

Human neutrophils Peripheral EDTA-anticoagulated blood samples were collected from healthy blood donors. Erythrocytes were sedimented with 6 % dextran (Dextran 500, Pharmacia) in 0.9 % NaCl (1 ml per 10 ml blood) for 45 min at RT. The leukocyte rich plasma was transferred over a Ficoll-Paque (Pharmacia) density gradient and the granulocytes were separated by centrifuging for 20 min at 400 x g at RT. Subsequently the erythrocyte contamination was lysed with 0.83 % ammonium chloride at RT. Neutrophils were washed twice with HBSS and the cell count of the final neutrophil suspension (>90% pure) was determined with an automated blood cell analyser (Coulter counter S Plus, Coulter Electronics Inc).

Assay procedure To block the FcRs and CRs on neutrophils the cells (1 x 10^6/ml) were incubated with various concentrations of mAbs at 4 °C for 60 min. After incubation the cells were washed with HBSS to remove the unbound mAbs. The bacteria were opsonized in luminometer vials (2 x 10^6/vial) for 30 min at 37 °C in 380 µl HBSS containing 4 % pooled, heat-inactivated (56 °C for 30 min) human serum. Subsequently 100 μl of neutrophil suspension and 20 μl of luminol stock solution were added to each vial. The CL was measured at 37 °C for 60 min with an automated luminometer set up allowing the simultaneous and continuous measurement of 25 samples (LKB Wallac 1251 luminometer) (7). Every assay was run in duplicate.

RESULTS AND DISCUSSION

Both anti-FcRII and anti-FcRIII whole mAbs inhibited the opsonized bacteria induced CL of human neutrophils in a dose dependent manner (Fig. 1). The highest inhibitions obtained by blocking FcRII and FcRIII were 45 % and 47 %, respectively, with the mAb concentration of 0.4 ng/1000 neutrophils. The blocking of both FcRII and FcRIII simultaneously did not increased the maximal inhibition. When the cells were incubated with anti-FcRII Fab fragments a similar, but slightly weaker inhibition (mean and SEM = 39 ± 9 %, n=3) was observed. The exposure of neutrophils to anti-FcRIII F(ab)$_2$ fragments had no effect on CL. Anti-FcRI mAbs did not inhibit CL. On the contrary, a slight increase in CL response was observed. The role of CR1 and CR3 in neutrophil CL was also examined. In opsonized bacteria induced CL the anti-C3 receptor mAbs were unable to alter the CL response. However, when the neutrophils were activated with non-opsonized bacteria, anti-CR1 and anti-CR3 mAbs decreased the CL. The maximal inhibition was 41 ± 13 % (mean and SEM, n=3) with anti-CR1 mAb concentration 0.02 ng/1000 neutrophils and 64 ± 8 % (mean and SEM, n=3) with anti-CR3 mAb concentration 5 ng/1000 neutrophils, respectively.

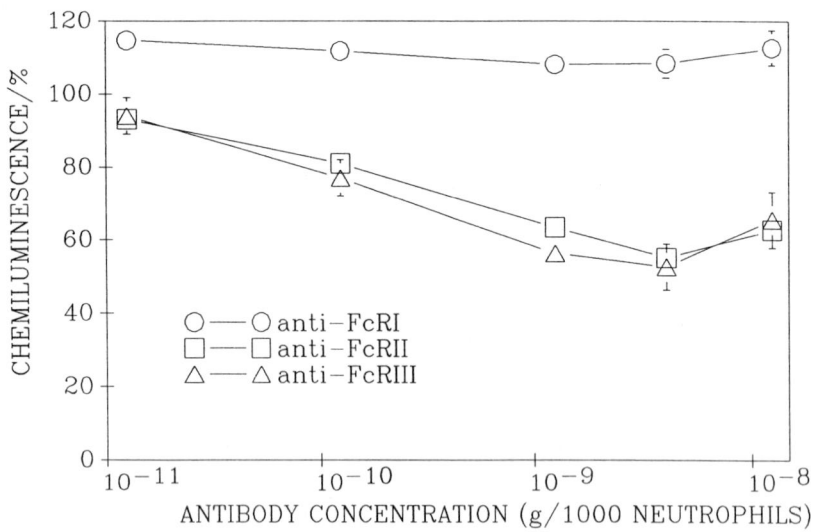

Figure 1. The effect of Fc-gamma receptor mAbs on chemiluminescence of human neutrophils. Activator: opsonized bacteria (2 x 10^6/10^5 neutrophils). Mean and SEM, n=7-8. The CL data are presented as percentage of the control CL without mAbs.

Our results suggest that opsonized bacteria induced CL is mediated via FcRII and FcRIII on human neutrophils. Since in heat-inactivated serum the main opsonins are heat-stabile IgGs, it is possible that the crosslinking of FcRIIs and FcRIIIs by antibody-antigen complexes triggers the neutrophil oxidative response generating CL. The observation that the blocking of either FcRII or FcRIII could inhibit the CL response suggests that both these two receptor types are capable of mediating the stimulus leading to CL emission. Apparently, however, the contribution of both receptors is required for complete activation. This contrasts with the conclusion of Huizinga et al. (3), which have postulated that only FcRII is responsible for IgG-induced activation of respiratory burst. Despite methodological differences - the respiratory burst activity was measured with less sensitive ferricytocrome c -assay and neutrophils were challenged with IgG-opsonized S. aureus bacteria and latex particles - their experimental results, that the blocking of FcRIIIs with anti-FcRIII F(ab)$_2$ fragments did not inhibit the respiratory burst, parallel to ours. They did not, however, test at all the whole anti-FcRIII mAb. Our interpretation to the observation that anti-FcRIII F(ab)$_2$ fragments failed to inhibit CL is that either the affinity of F(ab)$_2$ fragments is much lower than that of whole mAbs or the steric hindrance of bound F(ab)$_2$ is incomplete allowing the binding of IgG-Fc to the receptor. Another possibility is that FcRIII actually does not mediate neutrophil activation and the CL inhibition caused by anti-FcRIII whole mAbs is a consequence of the competative binding of free Fc ends of antigen-bound anti-FcRIII mAbs and opsonizing bacteria-bound IgGs to FcRIIs. However, we do not favor this explanation, for the following reasons. First, similar inhibition was not detected when neutrophils were incubated with anti-FcRI, anti-CR1 and anti-CR3 mAbs and second, one would expect that the interaction of FcRII with antigen-bound mAb-Fc would induce CL rather than inhibit it.

Experiments with anti-CR1 and anti-CR3 mAbs demonstrated that these receptors play an important role in non-opsonized bacteria induced CL. It is possible that CR1 and CR3 have lectin-like properties and are capable of interacting with surface of bacteria, what initiates the opsonin-independent activation of neutrophils. This result parallels previous reports which have shown that CR3 reacts with non-opsonized zymosan by binding the zymosan polysaccharide component glucan (8,9). As yet, it remains to be elucidated if the opsonin-independent transduction of activation is associated with a certain structure on C3 receptors or if the binding sites of non-opsonized particles are variable and ligand specific. Obviously, CR1 and CR3 on human neutrophils have a multifunctional role in binding both C3-opsonized and non-opsonized ligands.

None of the mAbs that were tested in the present study decreased the CL response more than 50 - 60 %, although saturating mAb concentrations, as determined with flow cytometry, were also used. This observation suggests that the reaction between mAbs and their antigens is not complete, and the remaining unblocked receptors are capable of mediating the activation signal. However, since the number of activating receptors in these conditions is apparently low, the neutrophil response to stimulation remains depressed and only a part of the maximal CL is generated. This hypothesis is supported by our experiments with non-opsonized zymosan: Although non-opsonized zymosan induced CL is totally CR3-dependent, as the CL response decreases more than 95 % when the Ca^{2+}-ions are removed from the reaction mixture, the blocking of CR3s with anti-CR3 mAbs inhibits CL only approximately 50 %.

ACKNOWLEDGEMENTS

This study was supported by grants from Finnish Allergy Research Foundation, Finland, and Emil Aaltonen Foundation, Finland.

REFERENCES

1. T.W. Huizinga, D. Roos and A.E. von Borne, Neutrophil Fc-gamma receptors: a two-way bridge in the immune system. Blood 75, 1211-1214 (1990).
2. C.D. Ross, Complement and complement receptors. Curr. Opinion Immunol. 2, 50-62 (1989).
3. T.W. Huizinga, F. van Kemenade, L. Koenderman, K.M. Dolman, A.E. Borne, P.A. Tetteroo and D. Roos, The 40 kDa Fc-gamma receptor (FcRII) on human neutrophils is essential for the IgG-induced respiratory burst and IgG-induced phagocytosis. J. Immunol. 142, 2365-2369 (1989).

4. T.W. Huizinga, K.M. Dolman, N.J. van der Linden, M. Kleijer, J.H. Nuijens, A.E. von der Borne and D. Roos, Phosphatidylinositol-linked FcRIII mediates exocytosis of neutrophil granule proteins, but does not mediate initiation of respiratory burst. J. Immunol. 144, 1432-1437 (1990).
5. J.A. van Strijp, K.P. van Kessel, M.E. van der Tol and J. Verhoef, Complement mediated phagocytosis of Herpes simplex virus by granulocytes. Binding or ingestion. J. Clin. Invest. 84, 107-112 (1989).
6. L.S. Schlesinger and M.A. Horwitz, Phagocytosis of Leprosy bacilli is mediated by complement receptors CR1 and CR3 on human monocytes and complement component C3 in serum. J. Clin. Invest. 85, 1304-1314 (1990).
7. E-M. Lilius, M. Waris and M. Lang, Automated luminometer set-up for continuous measurement of 25 samples. In Analytical Applications of Bioluminescence and Chemiluminescence, L.C. Kricka, P.E. Stanley, G.H. Thorpe et al. (Eds), Academic Press, Orlando, p. 461-464 (1984)
8. J. Lindena, H. Burkhardt and A. Dvenger, Mechanism of non-opsonized zymosan induced and luminol enhanced chemiluminescence in whole blood and isolated phagocytes. J. Clin. Chem. Clin. Biochem. 25, 765-778 (1987).
9. G.D. Ross, J.A. Cain and P.J. Lachmann, Membrane complement receptor type three (CR3) has lectin-like properties analogous to bovine conglutinin and functions as a receptor for zymosan and rabbit erythrocytes as well as a receptor for iC3b. J. Immunol. 134, 3307-3315 (1985).

SUPRESSION OF THE RESPIRATORY BURST OF ACTIVATED POLYMORPHONUCLEAR NEUTROPHILS (PMN) THROUGH PLATELET-DERIVED 5-HYDROXYTRYPTAMINE (5-HT)

P. Schuff-Werner[1], G. Huether[2], F. Schmidt[1] and A. Reimer[2]

Department of Clinical Chemistry[1] and Department of Psychiatry[2], University Hospital, Göttingen, Robert-Koch-Str. 40, D-3400, FRG

INTRODUCTION

Platelet aggregation and degranulation leading to local vasoconstriction and thrombus formation are closely linked to inflammation. Several platelet-derived factors have been identified as important growth factors or immunoregulatory mediators (1), but there is little knowledge on the involvement of released serotonin in the inflammatory process.

Because the indole nucleus of this biological active amine is easily oxidized we investigated the possible antioxidative capacity of 5-HT using a chemiluminescence assay to measure the release of reactive oxygen metabolites by stimulated PMN's.

MATERIALS AND METHODS

Isolation of peripheral blood cells. Polymorphonuclear cells (PMN) were isolated from heparinized blood of healthy donors and layered onto neutrophil isolation medium (NIM, Los Alanos Diagnostics, Los Alanos, USA). After centrifugation (400 x g, 30 min) the lower of the two visible interphases containing > 97 % polymorphonuclear cells was removed and washed three times, adjusted to 10^6 cells/ml phosphate buffered saline (PBS) and stored on melting ice (4 °C) until use.

Platelets were obtained by differential centrifugation from fresh human blood. The heparinized blood (5 U/ml) was centrifuged at 800 x g for 15 min at room temperature to obtain platelet-rich plasma.

Finally the platelets were pelleted by centrifugation (1000 x g, 20 min) washed twice in PBS and adjusted to 10^9 platelets/ml.

To obtain platelet supernatants, platelets were either repeatedly frozen and thawed or activated by thrombin and the membrane residues spun down at 2000 x g.

Chemiluminescence assay. Luminol- or lucigenin-amplified chemiluminescence was measured in a six-channel luminometer (Biolumat LB 9505, Berthold, Wildbad, FRG) and analyzed for maximum light emission (peak CL), time of peak CL (t_{max}) and total light generation (integral).

The PMN assay was performed in 400 μl total volume consisting of 100 μl PMN suspension (10^5 cells), 100 μl luminol (4×10^{-5} M) or lucigenin (10^{-3} M), 100 μl stimulant (opsonized staph. aureus, opsonized zymosan or phorbol myristate acetate) and 5-HT at serial dilutions in PBS.

The interaction of platelet supernatant or 5-HT with the reactive oxygen metabolites (ROM) generated during the enzymatic degradation of H_2O_2 was monitored by a luminol-amplified cell-free luminescence assay as described in detail by Ernst et al. (2) the peroxidative conversion of 5-HT was expressed as loss in peak and total chemiluminescence.

<u>Characterization of peroxidative products of 5-HT</u>. The oxidative loss of 5-HT in the reaction mixture was determined by HPLC and electrochemical detection (ECD).

Reaction products of 5-HT were seperated by thin layer chromatography on silica gel plates and by liquid chromatography using sephadex LH-20 (Pharmacia, Freiburg, FRG) as stationary phase and a 9 : 1 (v/v) water-methanol mixture adjusted to pH 2.5 by HCL.

Mass spectrometry of the major oxidation product of 5-HT isolated on the LH-20 column was performed by ^{252}Cf-plasma desorption time-of-flight mass spectrometry (PDMS) as described by Roepstorff et al. (3).

RESULTS AND DISCUSSION

Polymorphonuclear phagocytes activated by opsonized particles such as staph. aureus or zymosan or by the tumor promotor PMA immediately respond by generating ROM which can be either measured by direct methods or by the luminol- or lucigenin-amplified chemiluminescence.

In the presence of serial dilutions of the supernatant of 10^9 platelets the chemiluminescence signal is significantly suppressed in a dose-dependent manner (Fig. 1). This suppression is independent of the stimulus used in the experiment.

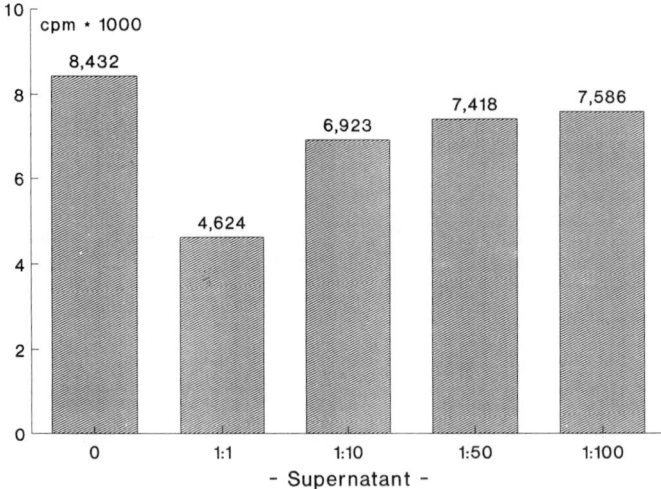

Fig. 1: Influence of serial dilutions of platelet supernatant (10^9 cells/ml) on the lumino -enhanced total chemiluminescence of PMN, stimulated by opsonized staph. aureus.

Similar results were obtained if equivalent amounts of authentic 5-HT were used in the assay instead of the complete mixture of platelet release products (Fig. 2). In the course of the reaction, 5-HT was almost completely oxidized and no longer measurable by HPLC and ECD. As shown in autoradiograms of TLC seperations, the loss of 5-HT coincides with the appearance of several oxidation products of 5-HT with smaller RF values.

The major oxidation product was separated by Sephadex LH-20 liquid chromatography. It was identified by plasma desorption mass spectrometry as a 5-HT-dimer formed under abstraction of 2 protons. The most likely structure of this dimer is that of 4,4'-bi-5-hydroxytryptamine.
The biological significance of the oxidation of serotonin by oxygen radicals formed during the respiratory burst of phagocytes remains unclear. The findings of this study emphasize serotonin as a potent antioxidant. As shown by Benedict et al. (4) platelets release serotonin in concentrations corresponding to those used in our experiments at sites of inflammation and during thrombus formation. Hence, 5-HT may play an important role as an endogenous locally releasable antioxidant.

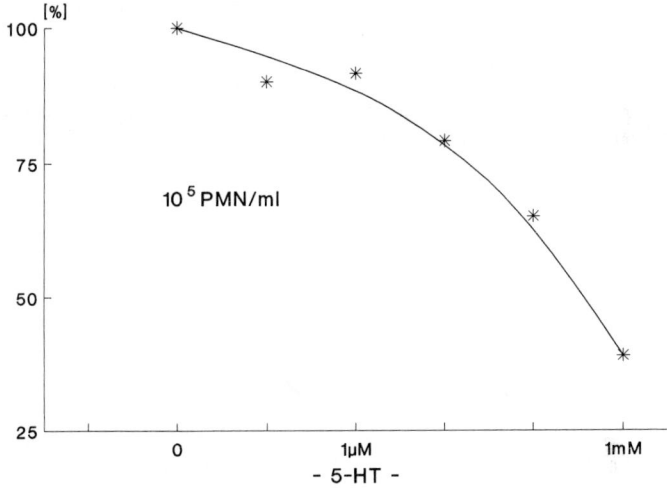

Fig. 2: Dose-dependent 5-Hydroxytryptamine-induced suppression of luminol-enhanced total chemiluminescence of PMN, stimulated by opsonized staph. aureus.

REFERENCES

1. R. Ross, A. Vogel. The platelet-derived growth factor.
   Review Cell 14, 203 (1978)

2. M. Ernst, M. Heber, and H. Fischer. Chemiluminescence measurements of immune cell - a tool in immunobiology and clinical research.
   J.Clin. Chem. Clin. Biochem. 21, 550-560 (1983)

3. P. Roepstorff. Plasma desorption mass spectrometry of peptides and proteins. Acc.Chem. Res. 22, 421-427 (1989)

4. C. R. Benedict, B. Mathew, K.A. Rex, jr. J. Cartwright, L.A. Sordahl.
   Correlation of plasma serotonin changes with platelet aggregation in
   an in vivo dog model of spontaneous occlusive coronary thrombus
   formation. Circ. Res. 58, 58-67 (1986)

ACKNOWLEDGEMENTS
We thank Dr. M.M. Brudny for the performance of plasma desorption mass spectrosmetry.

ALDEHYDE-ENHANCED PHOTON EMISSION FROM CRUDE EXTRACTS OF SOYBEAN SEEDLINGS

Haruo Watanabe[1], Masaki Kobayashi[1], Sohkichi Suzuki[1],
Masashi Usa[1], and Humio Inaba[1,2]

[1]
Biophoton Project,
Research Development Corporation of Japan,
2-1-1, Yagiyama-minami, Taihaku-ku, Sendai 982 Japan
[2]
Research Institute of Electrical Communication,
Tohoku University,
2-1-1, Katahira, Aoba-ku, Sendai 980 Japan

INTRODUCTION

It is known that imbibed soybeans or soybean seedlings show relatively higher photon emission(1,4). Especially, wounded seeds or seedlings (5,6) show an enhanced photon emission with the addition of indole acetic acid (6) or hydrogen peroxide (5). Cilento (2) has proposed that peroxidase or oxidase generates triplet carbonyl via oxidation of aldehyde or carboxylic acid through dioxetane formation by activated oxygen. Triplet carbonyl tansfers the excited energy to a fluorescent compound such as chlorophyll. Boveris et al. (1) observed photon emission from a mass of soybean seeds upon imbibition. Approximately 50-70 % of the photon emission was contributed by emissions in wavelengths greater than 600 nm. We found that acetaldehyde remarkably enhanced photon emission from intact soybean seedlings, and that it also enhanced chemiluminescence in supernatants of autoclaved crude extracts of seedlings.

MATERIALS AND METHODS

Sample: Soybean (Glycine max var. Tanrei)(4) seeds were imbibed in tapwater for 2 hrs prior to cultivation. A whole soybean seedling, grown in complete darkness for 4 days at 25° C, was used for the analysis of emission spectra in the presence of acetaldehyde.
For the preparation of crude extracts of the seedlings, parts of the cotyledon and parts of the root and hypocotyl (further referred to as root) were segmented. Cotyledons and roots were homogenized respectively in 50 mmol/l phosphate buffer, pH 7.0. Crude extracts were obtained after centrifugation. Supernatant from autoclaved crude extracts were also used. For measurements of photon emission, one ml of

samples was injected into the reaction mixtures.
Photon counting measurements: For spectral analysis of
weak photon emission, a filter-equipped photon-counting
type spectrometer was developed in our laboratory. The
rotatory filter-disc is faced on photomultiplier. In
the measurements, the gate time was one second for each
flter. A filter-disc was circulated for one minute.
Time course of the photon emission was measured by a
Lumiphotometer TD 4000 at 25°C.
Reagents: β-carotene was dissolved in benzene and
acetone, and then diluted by 50 mmol/l phosphate buffer,
pH 7.0. Reagents were dissolved in the same buffer.
For reduction of the sample (crude extracts), a small
amount of hydrosulfite powder was added.

RESULTS AND DISCUSSION

The emission spectrum and its intensity depending on
the time from a whole seedling with its root immersed in
acetaldehyde are presented in Fig. 1. This spectrum
indicates peaks at around 670 nm and 610-615 nm, with a
shoulder at 530-540 nm. In the absence of acetaldehyde,
photon emission was too weak to obtain the spectrum.

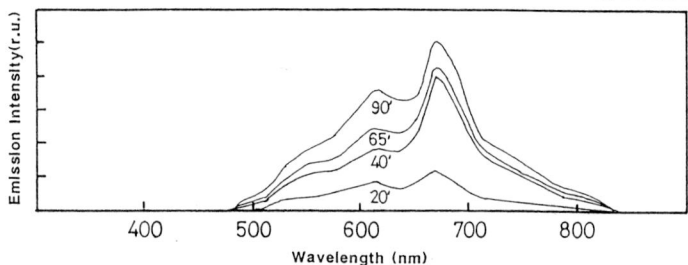

Fig. 1. Spectra of a whole soybean seedling in the
presence of acetaldehyde. The time dependency of
spectral intensity change was measured at 37°C. Each
spectrum is presented as summation of five scans.
Ordinate shows the relative emission intensity of the
spectra. Abscissa, wavelength (nm).

The effect of the aldehyde chain length on the
enhancement of photon emission was examined in crude
extracts of the root. Acetaldehyde showed the most
effective stimulation on the photon emission in 1 min;
some stimulative effect was also observed with
benzaldehyde (data not shown). On the other hand, the
emission spectra of root extracts and supernatants of
autoclaved cotyledon appeared peaks at around 670 and
610-615 nm with a shoulder at 530-540 nm, in the
presence of acetaldehyde. It resembles the spectrum of
a whole seedling immersed in acetaldehyde (see Fig. 1).
Scavengers of active oxygen species were tested with
oxidized (non-reduced) samples, in the presence of
acetaldehyde (Table 1). Obvious inhibition for photon
emission was only observed with the addition of sodium

	Cotyledon	Root
Control	100	100
Azide(50mM) (before)	59.3	66.0
Azide(50mM) (after)	117.5	90.5
Mannitol (0.1M)	173.1	95.7
Catalase (1250U/ml)	106.8	87.9
SOD (500U/ml)	85.6	88.1
β-carotene (2μg/ml)	97.7	108.7

Table 1. The effect of scavengers for active oxygen and radicles on photon emission from autoclaved crude extracts of seedlings.

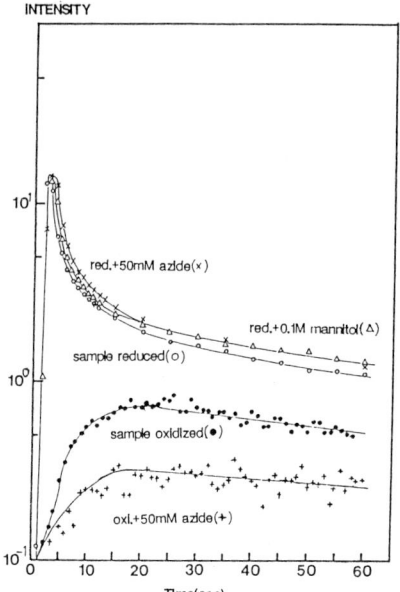

Fig. 2. Time course of photon emission from reduced or oxidized supernatants of autoclaved crude extract of roots in the presence of acetaldehyde. Ordinate shows relative intensity of photon emission. Abscissa, time (sec).

azide to acetaldehyde before injection of samples("azide before"). However, in a case which the sample was mixed first with acetaldehyde ("azide after"), the inhibition of azide disappeared. This suggests that azide and acetaldehyde are competitive for binding to the emitter in samples.

The time course of photon emission from reduced sample is presented in Fig. 2. Reduced samples from autoclaved root extracts showed a rapid increase and a quick decay of photon emission, followed by a slow decay in a second phase, whereas oxidized samples showed a slow increase followed by a similar slow decay to the reduced samples. Azide inhibited photon emission in oxidized samples (as shown in Table 1). In reduced samples, however, azide and mannitol showed no inhibition.

The emission spectra from a intact soybean seedling showed the peaks at longer wavelengths, 670 and 610-615 nm, and produced high emission intensity with aldehyde. Even in the crude extracts of seedlings, the same spectral profile appeared. Chain length of aldehydes affected the intensity and the decay rate (data not shown) of photon emissions in crude extracts. Benzaldehyde also showed

the stimulative effect.
     Reduced   crude extracts showed a rapid   increase   in
the   photon emission intensity followed by a fast   decay
upon mixing the sample with aldehyde and oxygen.   This
suggested that, at least in part, triplet oxygen (ground
state)   could react with a reduced intermediate   of   the
emitter   in the presence of aldehyde. This is   supported
further   by the fact that azide and   mannitol have   no
inhibition   on the reactivity of a reduced   intermediate
with oxygen (Fig. 2). The difference in the   inhibition
by   azide between oxidized and reduced samples,   may   be
due to a conformational change between two dihydro-   and
hydroperoxy- intermediates.   Kemal   and   Bruice   (3)
reported   that 4a-hydroperoxy FMN showed a slow increase
and a slow decay of light emission with   formaldehyde,
while   dihydro FMN in the presence of formaldehyde and
oxygen showed   faster reactivity and faster   decay of
light emission. These reactivities were similar to   the
photon emission in crude extracts of seedlings.

REFERENCES

1. A. Boveris, A.I. Varsavsky, S. Goncalves da Silva,
   and R.A. Sanchez, Chemiluminescence of soybean seeds:
   spectral analysis, temperature dependence and effect
   of inhibitors. Photochem. Photobiol.,38,99-104(1983).
2. G. Cilento, Generation of electronically excited
   triplet species in biochemical systems.   Pure & Appl.
   Chem., 56, 1179-1190(1984).
3. C. Kemal and C. Bruice, Simple synthesis of a
   4a-hydroperoxy adduct of 1,5-dihydroflavine:
   preliminary studies of a model for bacterial
   luciferase.   Proc. Natl. Acad. Sci. USA, 73, 995-999
   (1976).
4. R. Saeki, M. Kobayashi, M. Usa, B. Yoda, T. Miyazawa,
   and H. Inaba, Low-level chemiluminescence of water-
   imbibed soybean seeds. Agr. Biol. Chem., 53,
   3311-3312(1989).
5. M.L. Salin, and S.M. Bridges, Chemiluminescence in
   wounded root tissue. Evidence for peroxidase
   involvement. Plant Physiol., 67, 43-46(1981).
6. M.L. Salin, and S.M. Bridges, Chemiluminescence in
   soybean root tissue: effect of various substrates
   and inhibitors. Photobiochem. Photobiophys., 6,
   57-64(1983).

ULTRAWEAK LUMINESCENCE FROM CANCER TISSUES

F.Grasso, F.Musumeci, A.Triglia, G.Rodolico[1], F.Cammisuli[1],
C.Rinzivillo[1], G.Fragati[1], A.Santuccio[1] and M.Rodolico[1]

*Istituto di Fisica and [1]Istituto di I Clinica Chirurgica Generale,
Universita' di Catania, Viale A.Doria 6, I-95125 Catania, Italy*

INTRODUCTION

In the field of biological luminescence some interest has been
growing recently in the western countries on the ultraweak cell
radiation emitted by all biological system. The occurrence of this
radiation, suggested by several eastern scientist (1), was evidenced
in Italy fourty years ago by Facchini et al. (2), who discovered
that living systems emit light at very low rate.
The emitted radiation seems to be strictly connected to the
physiological state of the system and to biochemical processes
occurring; this has suggested the possibility of using the emitted
radiation as a non destructive tool to investigate the "status" of a
living system (3,4,5).
Application of the low level luminescence measurements have been
proposed both to agriculture and to medicine, where they could be
used as a diagnostic method for different pathologies including
cancer. Some preliminary results have been reported in this field
(6,7), but the correlation with the pathological characterization of
the examined samples is not complete.
Present paper reports measurements of radiation intensity emitted by
25 samples of human tissues coming from surgical operations. The
results refer to the emission rates of normal and tumor human
tissues.

EXPERIMENTAL

A specific experimental set up was designed and assembled for the
measuring the ultraweak luminescence from the biological tissues.
It consists of a dark chamber where the sample to be analyzed can be
maintained at constant temperature within ± 0.1°C. Photons of
wavelength ranging between 190 to 800 nm are counted by a low dark
current cooled photomultiplier, the output of which is treated by a
low noise electronic chain and stored in a multiscaler. The
experimental set up allows to perfom the spectral analysis of the
emitted radiation by means of a proper filers set. The apparatus is
controlled by a computer which allows the data treatment.
In operating conditions the background counting rate is about 8
pulses/sec as required to detect the low rate emission from samples.
Samples where human tissues coming from surgical operations; they
have size of the order of one or few cm^2.
The emitted light intensity depends on several factors due in part

to the pathological conditions of the patient and in part to the
sample handling; to reduce masking effects due to sample handling a
standard procedure has been followed in sample treatment.
Samples as soon as taken from the living body were put in a dark
thermostatic holder at 37°C, and immersed in ringer solution to
allow the metabolic processes to maintain; samples where put in the
measuring chamber 30 - 100 minutes after the take.
The intensity of light emitted decreases with the time elapsed from
surgical take away, with an half life time of the order of 1 hour,
so that the emission can be revealed for about three hours.

RESULTS AND DISCUSSION.
The examined samples where 25, among them 16 were tumor tissues,
while 9 were normal (non tumor) tissues. The sample characteristics
are listed in Tab.1, which reports the sample number, the patient
identification code, the sample type, the affecting pathology, the
intensity of light emitted (number of photons/cm^2min after
subtraction of background), the estimated error on the counting
rate, the time elapsed from the surgical take away of the sample and
the starting of emission measurements.
From Tab.1 it is evident that all the examined normal (non tumor)
samples emit with negligible intensity .
Tumor samples instead emit light with an intensity distribution
ranging from zero to 1400 photons/cm^2min.
Even if the statistics is at present too poor to perform a
complete statistical analysis of the results in terms of a
probability distribution, the two sub-sets are clearly different.
To evidence this difference one can consider the two samples sub-
sets: the first includes the 9 non tumor samples while the second
includes 16 cancer tissue samples.
All the samples belonging to the former sub-set have an emission
rate lower than 40 photons/cm^2min, with an average value of 22 ± 6
photons/cm^2min.
The samples of the latter sub-set (tumoral tissues) are
characterized by an higher emission intensity and by a wider
distribution, with an average value of 300 ± 90 photons/cm^2min.
It is clear that the two sub-sets have different populations: the
statistical significance of this statement can be evaluated by
computing the probability that the samples of the latter sub-set
belong to the former; if this were the case the measured
intensities of the tumor samples should be distributed with equal
probability below and above the average value of the former sub-set
(22 photons/cm^2min); experiments show that only 3 samples of 16
fall below the average, and 13 samples above this figure.
It is very unlikely, with a calculated probability lower than 1%,
that this result is a casual one; the calculation has been done in
the most unfavorable condition, i.e. by subtracting from the
measured value the statistical error due to the photon counting.
One can conclude then that the performed measurements show with high
statistical significance that photon emission from tumor samples is
different from that of non tumor samples. This suggests the possibility of using the Photon Emission from Biological Systems as a non
destrucive analitycal tool to determine the occurrence of cancer.

TABLE 1

MEASURED PHOTON EMISSION FROM TISSUES

SAMPLE #	PATIENT CODE	SAMPLE TYPE	EMISSION photons cm^2/min	STANDARD DEVIATION photons cm^2/min	ELAPSED TIME min
1	sa46	axillary node carcinoma	266	3	30
2a	fm56	tymome	66	50	100
2b	"	limph node	0 *	12	110
3	xx	anthrogastric carcinoma	348	12	45
4	pg60	colon tumor	1 440	80	120
5	lr69	kidney tumor	228	12	70
6	va58	breast cancer	510	24	90
7	cl74	metastasis omentum	819	15	55
8	ma57	breast cancer	24	1	30
9	g	appendix	34 *	2	30
10	fr	cholecystesis	13 *	1	45
11	na	rectum tumor	84	6	50
12	mr70	axillary node carcinoma	84	6	350
13	rs67	cunnus cancer	24	3	35
14	dc	retro peritoneum tumor	138	8	40
15	s	appendix	34 *	2	30
16	fr	rectum tumor	108	12	30
17a	cr68	breast carcinoma	24	24	95
17b	"	breast tissue	12 *	12	85
18	at	melanoma	320	24	20
19a	tn61	breast cancer	306	36	55
19b	"	breast tissue	40 *	6	85
20a	vm39	appendix	40 *	9	90
20b	"	cholecyste	0 *	9	100
21	m66	thyroid gland	26 *	3	130

The simbol * indicates non tumor tissues.

REFERENCES
1. A.A.Gurwitsch, A historical review of the problem of mitogenetic radiation, Experientia 44, 545 (1988)
2. L. Colli, U. Facchini, G. Guidotti, R. Dugnani-Lonati, M.Orsenigo and O. Sommariva, Further measurements on the bioluminescence of the seedlings, Experientia 11, 479 (1955)
3. J.W. Dobrowolski, A. Ezzahir, M. Knapik, Possibilities of chemiluminescence application in comparative studies of normal and cancer cells with special attention to leukemic blood cells, in PEBS, J. Slawinski et al (Eds), World Scientific, Singapore, 170 (1986)
4. R. Van Wijk and D.H.J. Schamhart, Regulatory aspects of low intensity photon emission, Experientia 44, 586 (1988)

5.   J. Slawinski and D. Slawinska,Low level luminescence from biological objects, in <u>Chemi</u> <u>and</u> <u>Bioluminescence</u>, J. G. Burr (Ed), Marcel Dekker Inc., New York, 495 (1985)
6.   M. I. Yanbastiev, On some applied-medical aspects of biophotonics, in <u>PEBS</u>, J. Slawinski et al (Eds), World Scientific, Singapore, 184 (1986)
7.   B. Yoda, Y. Goto, A. Saeki, H. Inaba, Chemiluminescence of smoker's blood and its possible relationship to cigarette smoke components, in <u>PEBS</u>, J. Slawinski et al (Eds), World Scientific, Singapore, 199 (1986)

CHEMILUMINESCENCE OF MACROPHAGES FROM BRONCHOALVEOLAR LAVAGE

L. Frigieri, G. Di Mario*, R. Fresu*, N. Grilli, C. Pizzoli, G. Pagliari
and P. De Sole*
*Department of Internal Medicine, *Clinical Chemistry Laboratory
Catholic University (Policlinico Gemelli) 00168 Roma, Italy*

INTRODUCTION

Interstitial lung diseases are a group of heterogeneous chronic inflammatory disorders involving the entire pulmonary parenchyma and ultimately leading to fibrotic degeneration.

Initial histological appearances show two alveolitis patterns: the first is dominated by macrophages and lymphocytes (as in sarcoidosis), the second by macrophages and neutrophils (as in idiopathic pulmonary fibrosis) (1). In any case, macrophages play a fundamental role in the disease evolution. Therefore, they have been amply studied after being harvested by means of bronchoalveolar lavage (BAL) (2).

In vivo activation or in vitro stimulation of macrophages can be studied by means of chemiluminescence (CL) induced by soluble or particulate stimuli and amplified by luminol (3).

The aim of the present paper is to analyze the CL response of alveolar macrophages (AM) from BAL in two interstitial lung diseases (sarcoidosis and idiopathic pulmonary fibrosis) and one inflammatory airway disease (chronic bronchitis).

MATERIALS AND METHODS

20 patients (aged 50 + 15 years) divided into three groups (10 sarcoidosis, 4 pulmonary fibrosis and 6 chronic bronchitis) were investigated.

Modified Krebs-Ringer phosphate and zymosan was prepared as reported elsewhere (4).

Luminol and PMA were supplied by Sigma (St. Louis, MO, U.S.A.). Both reagents were prepared in dimethylsulfoxide, 50 and 5 mmoles/l respectively. Final concentrations were prepared in modified Krebs-Ringer phosphate.

Pulmonary macrophages were harvested by means of broncho-alveolar lavage performed by injecting 60 x 4 ml of sterile saline into a wedged bronchial airway.

CL was measured after adhesion of macrophages on the bottom surface of the luminometer vials (5). To each cuvette were added 100 nmoles luminol and 0.5 mg of opsonized zymosan or 0.15 nmoles PMA, respectively for the zymosan or PMA stimulation in a final volume of 1.0 ml with modified Krebs-Ringer phosphate. CL was measured at 25 C with a Packard automatic luminometer mod. 6500 at 3 minutes intervals for 60 minutes. Specific activity (counts per second per cell = cpsc) was obtained by dividing the peak CL value by the number of adhered macrophages per cuvette.

RESULTS AND DISCUSSION

CL data of BAL macrophages for the patients under study are shown in Table 1.

TABLE 1

CL data for the different disease groups

	cpsc (mean + SD)			
	S	P	C	p < .05
Basal CL	0.08 ± 0.07	0.05 ± 0.06	0.05 ± 0.03	
Zymosan CL	0.33 ± 0.36	0.54 ± 0.37	0.15 ± 0.05	P/C
PMA CL	0.18 ± 0.28	0.14 ± 0.15	0.07 ± 0.05	
Zymosan-Basal	0.25 ± 0.29	0.49 ± 0.33	0.10 ± 0.03	P/C
PMA-Basal	0.10 ± 0.21	0.08 ± 0.13	0.03 ± 0.04	

S = Sarcoidosis P = Pulmonary fibrosis C = Chronic bronchitis
cpsc = counts per second per cell

Although the CL activities obtained are one or two orders of magnitude lower than that of peripheral granulocytes, the method used makes it possible to perform reliable measurements.

It is interesting to note that PMA CL is not significantly different from basal CL in all the three groups, while the difference between zymosan and basal CL is statistically significant. However, both sarcoidosis and pulmonary fibrosis macrophages show a higher response to zymosan compared with bronchitis macrophages.

These results are clearly indicative of a selective modification of lung macrophages, probably involving the number and/or function of receptors for complement derived factors or immunoglobulin Fc-fragment.

In table 2, CL data are correlated to the other BAL parameters studied (total and differential cell count, lysozyme, alkaline phosphatase, C3, C4, IgG, IgA, IgM, OKT8 +, OKT4 +).

TABLE 2

Correlation between Zymosan CL and other BAL parameters

BAL parameter	r	P
Alkalyne phosphatase	0.661	< .01
Neutrophils	0.555	< .05
Others		NS

r = correlation coefficient

It is very interesting to note that zymosan CL is positively correlated only with neutrophil percentage and alkaline phosphatase activity. Because the separation method we used allows an almost selective adhesion of macrophages, the CL activity can be exclusively referred to the macrophages. Moreover, in some cases with high neutrophil percentages, CL has also been determined on the total BAL cells without the adhesion step . The results obtained (unreported) induce to exclude any direct contribution of BAL neutrophils on CL activity. Therefore, the positive correlation between macrophage CL and neutrophils is probably due to some factors (colony stimulating factors ?) that can prime the macrophages (6).

In table 3, the range of values of three selected BAL parameters found in the three groups of patients under study are shown.

TABLE 3

Range of values for some BAL selected parameters

Parameter	S	P	C
Basal CL (cpsc)	< 0.22	< 0.17	< 0.11
Zymosan CL (cpsc)	< 1.00	< 1.30	< 0.30
Macrophages (%)	38-100	63-93	95-100

From the analysis of the results obtained it seems possible to delineate a useful diagnostic index that is function of macrophage percentage, basal and zymosan CL. Further research will tell us if this index can help to follow up individual patients with lung parenchymal inflammatory diseases.

REFERENCES

1. R.G. Crystal, J.E. Gadek, V.J. Ferrans, J.D. Fulmer, B.R. Line, G.W. Hunninghake, Interstitial lung disease: current concepts of pathogenesis, staging and therapy. Am.Rew.Respir.Dis. 70, 542-568(1981)
2. G.S. Davis, M.S. Giancola, M.C. Costanza, R.B. Low, Analyses of sequential bronchoalveolar lavage samples from healthy human volunteers. Am.Rev.Respir.Dis. 126, 611-616 (1982)
3. R.C. Allen, L.D. Loose, Phagocytic activation of a luminol dependent chemiluminescence in rabbit alveolar and peritoneal macrophages. Biochem. Biophys.Res.Commun. 69, 245-252(1976).
4. P. De Sole, S. Lippa, G.P. Littarru, Whole blood chemiluminescence: a new technical approach to assess oxygen-dependent microbicidal activity of granulocytes. J.Clin.Lab.Automation 3, 391-400 (1983).
5. W.J. Calhoun, S.M. Salisbury, L.W. Chosy, W.W. Busse, Increased alveolar macrophages chemiluminescence and airspace cell superoxide production in active pulmonary sarcoidosis. J.Lab.Clin.Med. 112, 147-156 (1988)
6. N.A. Nicola, D. Metcalf, Specificity of action of colony- stimulating factors in the differentiation of granulocytes and macrophages. In Biochemistry of macrophages, D. Evered, J. Nugent and M. O'Connor (Eds), Ciba Foundation Symposium 118, Pitman Publ., London, p.7-28 (1986).

WHOLE BLOOD CHEMILUMINESCENCE ASSAYS FOR CLINICAL APPLICATIONS: STANDARDISATION IMPROVEMENTS

E. NOEL, J. DECUYPER and H. A. OOMS

**Laboratory of Clinical Chemistry, Erasme Hospital,
Free University of Brussels, 1070 Brussels, Belgium**

INTRODUCTION

Oxygen activated species are generated by phagocytes during the respiratory burst. Chemiluminescence (CL) is a reliable method of assessing this phagocytic function. Nevertheless, to be used as a tool in clinical investigations, the CL technique need to be standardized.

The aim of this work is the improvement of a luminol enhanced CL assay of phorbol myristate acetate (PMA) activated leukocytes performed on diluted whole blood (1). This type of assay present some advantages. The use of whole blood avoids the induction of damages on leukocytes during their isolation and an intense activation is obtained by the use of PMA. However, the influence of some parameters on the CL remain to be determined: (i) the dilution of the whole blood which could be accompanied with modifications in scavenging and/or with quenching due to blood components; (ii) the influence of the delay between the venipuncture and the CL assay on the leukocytes activity; (iii) the incubation time of the diluted whole blood with the luminol before the stimulation by PMA; (iv) the possible competitive behaviour of leukocytes. This last point is of a peculiar importance.

MATERIALS AND METHODS

Whole blood was obtained by venipuncture with K-EDTA as anticoagulant. PMA and luminol were from SIGMA. Modified Eagle's medium without Phenol Red (MEM) was from GIBCO. Reconstituted blood leukocytes free was prepared by mixing plasma and red blood cells giving an hematocrit of 40%. CL was recorded on a luminometer (Lumicon, HAMILTON).

For the CL standart assay, 10 μL of 10x diluted whole blood in MEM were added to 890 μL of MEM containing 0.1 μmol/L luminol. After various incubation times at 37°C, leukocytes were stimulated by the addition of 100 μL of MEM containing 2 μg/mL of PMA. The CL was recorded during 15 minutes after the PMA addition. The CL intensity is expressed in counts per minute (cpm) at the maximum peak height.

RESULTS AND DISCUSSION

In order to determine the scavenging and/or quenching effects of the blood components in the dilution range used, increasing amounts of reconstituted leukocytes free blood (10x diluted in MEM) were added to the reaction mixture and the CL intensity was recorded. As shown in Fig. 1, it appears that the CL intensity is unaffected in the range of 0 to 10 μL added leukocytes free blood. We conclude that, in our working conditions, the scavenging and/or quenching effects due to blood components are negligible.

The incubation time of whole blood with luminol varied between 0 and 30 minuts. No modification of the CL intensity were observed. Thus, we performed the experiments without this preincubation. This allowed us to execute the CL assay a very short time after the blood sampling.

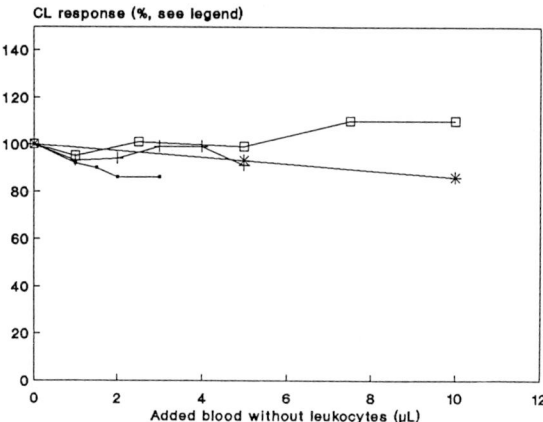

<u>Figure 1</u>: Scavenging and/or quenching effect of blood components. Y-axis: CL response is expressed as a percentage of the peak intensity (cpm) recorded without leukocytes free blood addition. X-axis: leukocytes free blood (10x diluted in MEM) volume (μL) added to the standart assay.

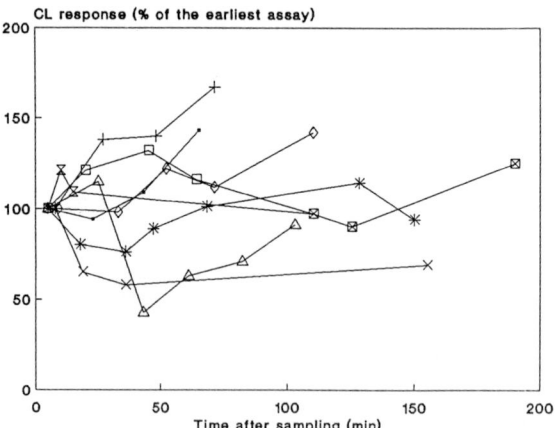

<u>Figure 2</u>: Effect of the delay between blood sampling and assay on the CL response. Y-axis: CL intensity is expressed as a percentage of the peak intensity (cpm) recorded at the earliest measurement (100%). X-axis: delay between blood sampling and PMA activation (min).

In order to determine the effect of the delay between testing and blood sampling, we performed the CL assay at various times after the venipuncture. The blood samples were left at room temperature before the assay. As shown in Fig. 2, the CL intensity varies when PMA activation is delayed after sampling. It seems better to start the assay as soon as possible after the blood sampling.

PMA stimulated phagocytes are known to generate cytotoxic species. Phagocytes are also the target of these species (2). In order to investigate the importance of this autotoxicity in our working conditions, we have determined the intensity of CL at various numbers of leukocytes in the reaction mixture. This was obtained by various dilutions of the whole blood. As shown in Fig. 3, the efficiency of PMA activated leukocytes to induce CL seems to be depending of the number of leukocytes in the reaction mixture. More diluted, more the leukocytes appear to be efficients. Thus, it is of peculiar importance to perform the CL assay within a defined scale of number of leukocytes in the reaction mixture. One has to establish reference values along this scale. Otherwise, misinterpretations of the results are possible as in the case of the study of leukocytes phagocytes from leucopenic patients, which could seem to be falsly highly efficients.

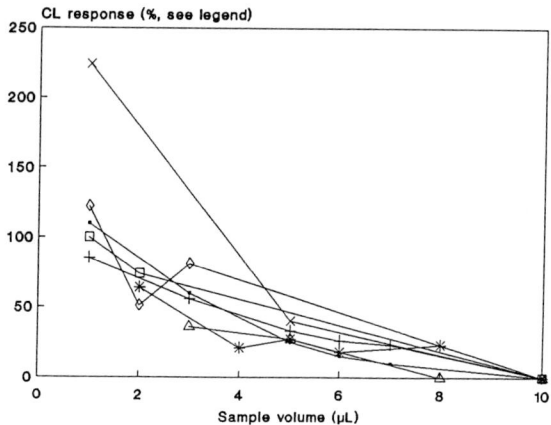

Figure 3: Effect of the number of leukocytes in the reaction mixture on the CL response. Y-axis: CL response (cpm/leukocyte) is expressed as a percentage of the CL response recorded in the standart assay conditions. X-axis: volume of the sample (μL).

In conclusion, the CL assay must be performed as soon as possible after the blood sampling. The whole blood must be highly diluted, at least 1000x, with carefull attention that the number of leukocytes in the reaction mixture is within the assigned scale.

REFERENCES

1. R.C. ALLEN, Phagocytic leukocyte oxygenetion activities and chemiluminescence: a kinetic approach to analysis. In Methods in Enzymology 133 Bioluminescence and Chemiluminescence, part B, M.A. DE LUCA and W.D. McELROY (Eds), Academic

Press, Inc., p.449-493 (1986).

2. M.-F. TSAN, Phorbol myristate acetate induced neutrophil autotoxicity. <u>J. Cell. Physiol.</u> <u>105</u>, 327-334 (1980).

CHEMILUMINESCENCE RESPONSE OF WHOLE BLOOD AND PMN LEUKOCYTES FOLLOWING EXPERIMENTALLY INDUCED NECROTISING PANCREATITIS

T. Zimmermann[1], S. Albrecht[2], T. Luther[3], T. Freidt[2], R. Schuster[1]

Department of Surgery [1], Institute of Clinical Chemistry [2] and Institute of Pathology [3] of the Medical Academy Dresden, Dresden, Germany

INTRODUCTION

Toxic oxygen metabolites seem to be involved in the development of pancreatic necroses [1]. The underlying cause seems to be an imbalance between the massive release of O_2-radicals and the reduced antioxidative capacity of the pancreatic tissue. The present study is intended to analyse the activation of granulocytes in whole blood and after separation in experimentally induced acute pancreatitis by means of chemiluminescence (CL), using luminol and lucigenin as sensitisers. For comparison we simultaneously performed flow cytometric analysis (FCA) of respiratory burst using dihydrorhodamine (DHR)123.

MATERIAL AND METHODS

Following anaesthetization with phenobarbital acute necrotising pancreatitis was induced in dogs by injection of bile. Catheters were placed postpancreatically in the portal vein and prepancreatically in the coeliac artery. Samples were taken within 15 min after placing the catheters, at the moment of bile injection and within 1, 2, 4, 6 and 24 hours after the injection of bile. At these times CL response was determined in whole blood and separated granulocytes from the portal vein and the coeliac artery. Samples of 10 µl of zymosan activated whole blood and of 10^6 separated granulocytes stimulated zymosan and suspended in 1 ml of a nutrient solution as measured in a Cliniluminat LB 9502 (Fa. Berthold, FRG) after 30 min of incubation. 100 µl of luminol or lucigenin (10^{-4} M in PBS) were added as sensitiser to each sample. For FCA of respiratory burst of PMA-stimulated and nonstimulated leucocytes we used the DHR123 method as described by others [2].

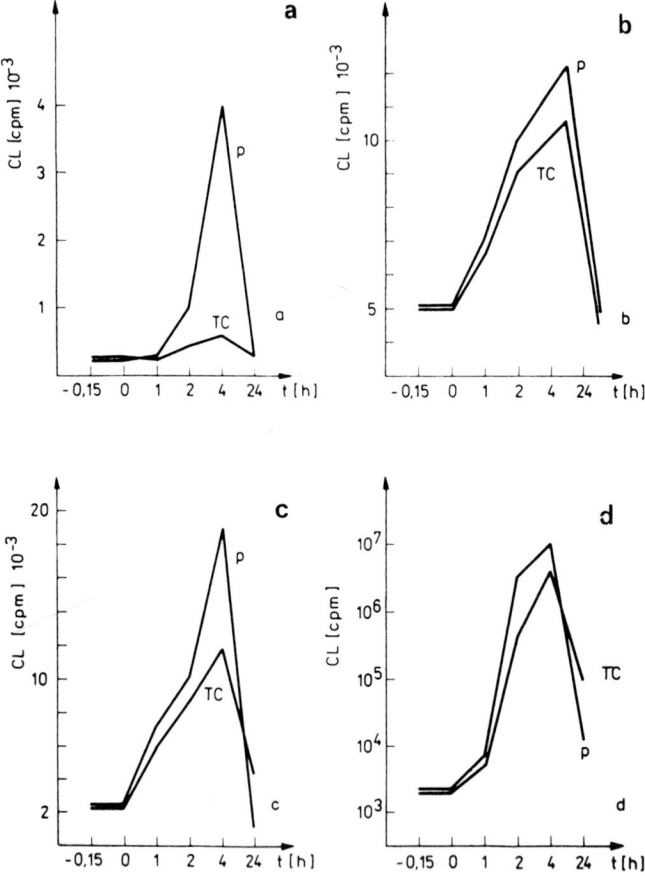

Fig. a-d. CL response of whole blood and separated granulocytes from the portal vein (P) and the coeliac artery (TC): **a** spontaneous CL of whole blood; **b** zymosan-activated whole blood; **c** spontaneous CL of granulocytes; **d** zymosan-stimulated granulocytes,

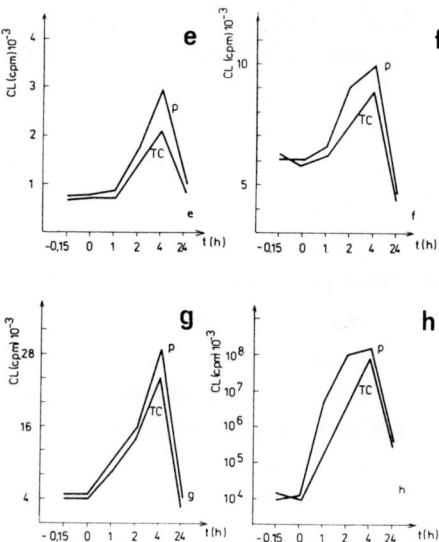

Fig. e–h   CL response of whole blood and separated granulocytes from the portal vein (p) and the coeliac artery (TC); e) spontaneous CL of whole, f) zymosan activated whole blood, g) spontaneous CL of granulocytes, h) zymosan stimulated granulocytes Lucigenin was used as sensitizer $(10^{-4}m)$.

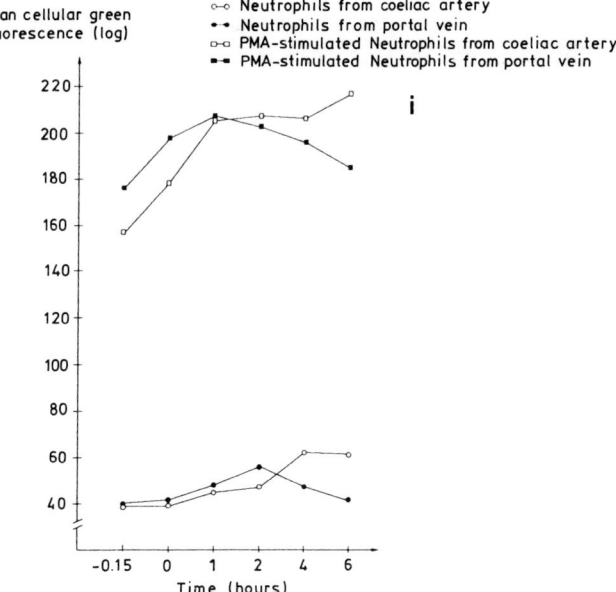

Fig. i   Flow cytometric analysis of leukocytes with DHR123: Mean cellular green fluorescence of neutrophiles.

RESULTS AND DISCUSSION

As appears from Fig. (a-d), whole blood and especially separated granulocytes from the portal vein (to much a lower degree also from the coeliac artery) show a sharp increase in spontaneous (only luminol-enhanced) CL already within 1 hour after injection of bile. The maximum is reached within 4 hours, with CL response being much higher in blood and granulocytes from the portal vein than in samples taken from the coeliac artery. Within 24 hours after injection of bile CL response of granulocytes from the coeliac artery surpasses that of granulocytes from the portal vein. Stimulation with zymosan-activated plasma results in a 10-fold increase in CL response of whole blood and a 1,000-10,000-fold increase in CL response of granulocytes. The use of lucigenin yields similar results (Fig. e-h) apart from the fact that lucigenin intensifies the CL response in separated granulocytes more and in whole blood less than luminol. According to the CL response, the maximum respiratory burst of neutrophils from the portal vein analysed by flow cytometry with DHR123 is reached within 2 hours after injection of bile (Fig. i).

The pancreas was found to be the site where the activation of granulocytes and thus the release of toxic oxygen metabolites takes place. Later on, the release of toxic oxygen metabolites activated granulocytes persists in a systemic circulation level and becomes independent of the inflammatory process in the pancreas. That is why CL response of granulocytes is much higher in systemic blood than in blood from the portal vein after 24 hours. The different sensitising power of luminol and lucigenin might be due to the possibly greater susceptibility of luminol to the catalytic activities of some enzymes (peroxidases) and the capability of lucigenin to react directly with O_2^-. The consecutive reactions of the released superoxide anions forming H_2O_2 and singulett O_2 obviously compete in whole blood with the primary reaction between O_2 and lucigenin.

REFERENCES

[1] D. Kelemen, The role of toxic oxygen radicals in the pathogenesis of acute pancreatitis. XI th Hungarian Congress of Experimental Surgery, A 42, Szeged, Hungary (1987)

[2] G. Rothe, A. Oser and G. Valet, Dihydrorhodamine 123: a new flow cytometric indicator for respiratory burst activity in neutrophil granulocytes. Naturwissenschaften 75, 354-355 (1988)

MITOGEN INDUCED LEUCOCYTE CHEMILUMINESCENCE AS A MONITOR OF CELLULAR IMMUNE STATUS IN CANCER THERAPY

S.J. Marshall and A.J. Bater

*Radiation Science Laboratories, Velindre Hospital,
Whitchurch, Cardiff, CF4 7XL, UK*

INTRODUCTION

Leukocyte chemiluminescence (CL) may be used to detect and investigate a variety of diseases (1,2). Treatment of diseases like cancer is often immunosuppressive and is routinely monitored by leukocyte counts. These, however, only reflect the overall population of the various leukocyte sub-populations, and give no indication of the activity or viability of the cells of the immune system. A sensitive assay based on the metabolic activities of the immune cells would thus be of benefit in monitoring cellular damage during the course of treatment. Leukocyte CL may fulfil this role. Previous studies have shown that leukocyte CL may be reduced in exposures to radiation at doses of less than 10 Gy (3) and during chemotherapy. (4). A crude extract of phytohaemagglutinin from *Phaseolus vulgaris* induces a characteristic three-peaked response of luminol-enhanced CL in isolated mixed leukocyte preparations (5), the peaks representing the activities of various cell types present. The *in vitro* effects of cancer therapy on peripheral blood cells was investigated using this response to monitor leukocyte activity.

MATERIALS AND METHODS

Blood samples. Blood was obtained by venipuncture from healthy donors (age range 21 to 53 years) and collected with 10ml vacutainer tubes containing lithium heparin (Becton Dickenson).

Isolation of leukocytes. Heparinized blood was diluted with an equal volume of Hank's Balanced Salts Solution (HBSS), layered onto Dextran-Metrizoate gradient (density 1.077 g/ml) and centrifuged at 1100g for 20 min at 19°C. The leukocyte interface was collected and washed twice in cold calcium and magnesium free HBSS. The cell suspension was kept on ice until required. Immediately before assay, the cell concentration was adjusted to 1.1×10^6 cells/ml with warm (37°C) HBSS containing calcium and magnesium.

Measurement of CL. 0.9 ml of cell suspension was placed into a disposable plastic cuvette and loaded into an LKB 1251 luminometer maintained at 37°C. 10 µl luminol in DMSO was added to give final concentrations of 11 µmol/l luminol, 1% (v/v) DMSO. After monitoring the background emission for 5 min, the cells were stimulated by the addition of 100 µl PHA-C (Reactif IBF, France, 1/100 dilution) and the CL was measured for 15 minutes. The CL signal was recorded at 1 second intervals by a microcomputer linked to the luminometer.

Treatment of cells. Prior to the CL assay, cell samples were irradiated in a ^{137}Cs unit to total doses of 2 Gy or 5 Gy, or were incubated with Vinblastine sulphate (Lederle, UK) at either 1 mg/ml or 10 mg/ml for 1 hour at 37°C.

RESULTS AND DISCUSSION

The cell preparations used in this study typically comprised 50% lymphocytes, 40% monocytes, and 10% granulocytes. Stimulation of untreated control samples with PHA-C produced the characteristic three peaked response as shown in the control curves of Figs. 1 and 2. The second peak is due to monocyte activity, the third peak to granulocytes, and the first peak to a cooperative action by both monocytes and lymphocytes (5). After irradiation at 2 Gy the peak response disappears, and the CL emission is reduced to a flat signal below 5mV. 5 Gy irradiation further reduces the signal nearly to baseline level at around 1.5 mV. (see Table 1). Treatment with Vinblastine has a similar effect (Fig. 2 and Table 2). If the cells treated at 2 Gy were allowed to recover on ice after irradiation, the peak response begins to reappear, with the signals for peaks 2 and 3 becoming larger as the recovery time increases (see Fig. 1); no recovery of the first peak was observed. The recovery was variable, and in several samples there was no peak activity even after 1 hour incubation. No recovery was observed after 1 hour incubation of the 5 Gy irradiated sample or of samples treated with Vinblastine. Some blood samples were also used to prepare suspensions enriched in granulocytes by dextran sedimentation, and these were tested with both PHA-C stimulation and with activation by opsonized zymosan (OZ, results not shown); the CL stimulated by either PHA-C or OZ was a single broad peak corresponding to the

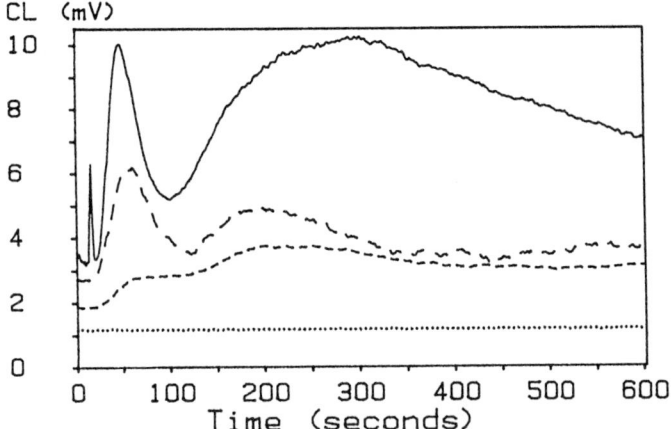

Figure 1. Effect of radiation on PHA-induced leukocyte CL.
Solid line, unirradiated control cells.   Dotted line, cells assayed immediately after 2 Gy irradiation.   Small dashed line, cells assayed after 45 min recovery.   Large dashes, cells assayed after 60 min recovery.

Sample	CL (mV)		
	Peak 1	Peak 2	Peak 3
Control	7.14 ± 1.51	12.66 ± 2.71	23.07 ± 7.84
2 Gy	4.15 ± 0.52	4.66 ± 0.76	4.63 ± 0.94
5 Gy	1.53 ± 0.04	1.76 ± 0.13	1.58 ± 0.17

Table 1. Effect of radiation on PHA-induced leukocyte CL. Values shown are in mV and are the means and standard deviations of six determinations.

third peak of the total leukocyte preparation. The granulocyte preparations gave similar results to the total leukocytes when exposed to radiation or Vinblastine; the peak response was abolished, and recovery was observed only from 2 Gy treated samples (results not shown). The reduction of CL stimulated by PHA-C or OZ may be related to radiation or drug effects on the cell membranes. In drug treated cells phagocytosis may be affected, since the Vinca alkaloids are transported across the cell membrane and bind to the microtubule system (6). Similar affects may occur with PHA activation, since reorganization of cross-linked mitogen receptors by the microtubule system may be part of the activation process. Alternatively, the absence of CL in treated samples may be due to the lack of available energy sources, as radiation in particular has been shown to deplete the internal energy pool. Comparisons of the intracellular ATP concentration before and after irradiation would be useful in elucidating the underlying mechanisms. The above results

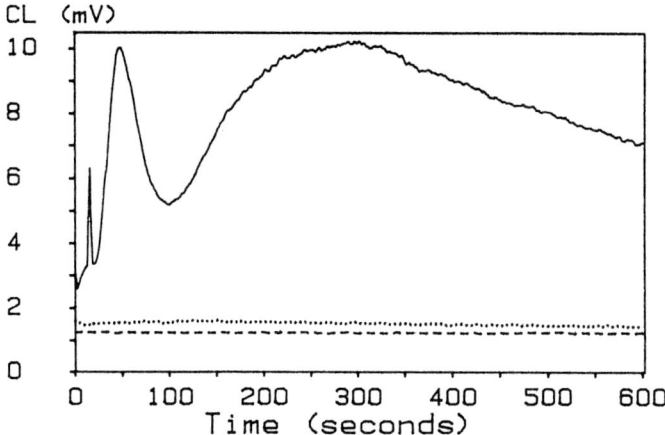

Figure 2. Effect of Vinblastine on PHA-induced leukocyte CL. Solid line, unirradiated control cells; Dotted line, cells treated with 1 mg/ml vinblastine; Dashed line, cells treated with 10 mg/ml vinblastine.

| Sample | CL (mV) | | |
	Peak 1	Peak 2	Peak 3
Control	7.14 ± 1.51	12.66 ± 2.71	23.07 ± 7.84
1 mg/ml	1.53 ± 0.19	1.53 ± 0.10	1.51 ± 0.08
10 mg/ml	1.32 ± 0.14	1.36 ± 0.17	1.51 ± 0.24

Table 2. Effect of Vinblastine on PHA-induced leukocyte CL. Values shown are in mV and are the means and standard deviations of six determinations.

demonstrate that mitogen-induced luminol-enhanced leukocyte CL is a sensitive indicator of cellular damage cause by *in vitro* exposure to agents used therapeutically in the treatment of cancer. It is possible that it could be used as a simple and rapid monitor of immunological damage resulting from *in vivo* exposure to therapeutic agents. Such a study of the PHA-induced leukocyte CL in blood samples from cancer patients undergoing therapy is now being undertaken.

ACKNOWLEDGEMENTS
The authors are grateful for the financial support of the South Wales Cancer Research Council and the South Glamorgan Health Authority. We would like to thank Mrs. Pam Cowburn and Mrs. Ann Reece for their invaluable technical assistance.

REFERENCES
1. P. DeSole, Polymorphonuclear chemiluminescence: some clinical applications. J. Biolumin. Chemilumin. 4, 251-262 (1989).

2. K. Van Dyke and C. Van Dyke, Cellular chemiluminescence associated with disease states. In Methods in Enzymology, M.A. DeLuca and W.D. McElroy (Eds), Academic Press, London, p.493-506 (1986).

3. M.L. Patchen, T.J. MacVitte and M.M. D'Alesandro, Radiation-induced hematologic and non-specific effects in the canine. In Bioluminescence and Chemilumnescence: New Perspectives, J. Schölmerich, R. Andreesen, A. Kapp, M. Ernst and W.G. Woods (Eds), Wiley, Chichester, p.61-64 (1987).

4. E.M. Lilius, A.L. Mähi, D.J. Proskin and A. Rajamki, Leukocytes as immunosensors: whole blood chemiluminescence (CL) in human leukemias. In Bioluminescence and Chemiluminescence: New Perspectives, J. Schölmerich, R. Andreesen, A. Kapp, M. Ernst and W.G. Woods (Eds), Wiley, Chichester, p.53-56 (1987).

5. N. Warren, A.E. Jones, and A.J. Bater, Three-peaked chemiluminescent response of human peripheral blood leukocytes following stimulation with phytohaemagglutinin. J. Biolumin. Chemilumin. 5, 235-241 (1990).

6. W.A. Bleyer, S.A. Frisby and V.T. Oliverio, Uptake and binding of Vincristine by murine leukemia cells, Biochem. Pharmacol. 24, 633-639 (1975).

THE FAILURE TO DETECT OH· RADICAL PRODUCTION BY PHAGOCYTIC CELL LINES WITH PHTHALHYDRAZIDE

G. Bottu

Laboratorium voor Biochemie, Vrije Universiteit Brussel,
Paardenstraat 65, 1640 Sint-Genesius-Rode, Belgium

INTRODUCTION

In order to destroy microorganisms, phagocytic leukocytes produce hydrolytic enzymes, cationic proteins and also reactive oxygen derivatives. It is now firmly established that macrophages possess a NADPH oxidase which, after triggering of the oxidative burst, insures the production of $O_2^{\div}$ (and hence of H_2O_2). Monocytes and neutrophil granulocytes have in addition a myeloperoxidase which produces ClO- and eosinophil granulocytes have an eosinophil peroxidase which makes BrO- as well. The production of OH· and $^1O_2$ remains controversial (1). Reactive oxygen derivatives can be measured easily with a photometer and a chemiluminescent substance such as luminol or lucigenin (2).

In the present work, the formation of OH· radicals in cultured phagocyte-like cell lines was investigated by means of phthalhydrazide chemiluminescence. Phthalhydrazide was suggested by Merényi and Lind (3) as a reliable detector of OH· radicals. This substance is itself not chemiluminescent, but can react with an OH· radical to form a luminol analog (with an OH instead of a NH_2 function) which produces light on further oxidation.

Several cell lines have been used. 2C11-12 and LA5-9 are hybridomas between a mouse lymphosarcoma derived fusogenic cell line and mouse macrophages (4). They show some properties of resting macrophages. Only LA5-9 produces a weak oxidative burst when triggered, but both cell lines acquire the capacity for a strong oxidative burst after having been exposed to a T cell supernatant (4). HL-60 is a human promyelocytic cell line (available from ATCC); it differentiates into a neutrophil granulocyte-like cell under the influence of retinoic acid and into a monocyte-like cell under the influence of calcitriol (5).

MATERIALS AND METHODS

Reagents Cu/Zn-superoxide dismutase (SOD) from bovine erythrocytes was purchased from Boehringer Mannheim, lipopolysaccharide from S. enteriditis from Difco, retinoic acid and phorbol-12-myristate-13-acetate (PMA) from Sigma and phthalhydrazide from Ventron-Alfa. 1α,25-dihydroxyvitamin D_3 (calcitriol) was a kind gift from Products Roche N.V.

Chemiluminescence The luminescence of luminol after addition of H_2O_2 and NaOCl, Fe^{2+} or horse radish peroxidase (HRP) was measured as before (6).

Cellular chemiluminescence 1 ml culture medium (RPMI 1640 + 10% fetal calf serum + 2 mmol/l glutamine + 100 units/ml gentamycin) containing 1.5 10^4 cells was transferred into a polystyrene cuvette and a modulator was added (100 units/ml recombinant interferon-γ, 1 μg/ml lipopolysaccharide, 1 μmol/l retinoic acid or 0.1 μmol/l calcitriol). After 1-4 days the culture medium was decanted (the cells adhere to the wall) and 1 ml Phosphate Buffered Saline with a luminescent probe was added (0.1 mmol/l lucigenin, luminol or phthalhydrazide, eventually supplemented with 1 mmol/l desferal, 1 mmol/l NaN_3 or 0.01 mg/ml HRP). After 5 min a trigger was added (either 0.1 mg/ml *Micrococcus luteus* opsonized with rabbit antiserum and guinea pig complement or 1 μmol/l PMA). The chemiluminescence was measured with a six-channel

TABLE Cellular luminescence of phagocyte-like cells, amplified with a chemiluminescent probe. Given are the counts per minute at the peak of luminescence emission and (within parentheses) the number of minutes elapsed after the addition of the trigger.

triggered with opsonized *Micrococcus* and probed with:

	2C11-12-2A grown 3 days with mouse interferon-γ	2C11-12-2A grown 3 days with lipopolysaccharide	LA5-9 grown 1 day with mouse interferon-γ	LA5-9 grown 2 days with lipopolysaccharide	HL-60 grown 4 days with retinoic acid	HL-60 grown 4 days with calcitriol
lucigenin	161300 (10)	270000 (19)	80000 (17)	31600 (40)	41300 (14)	326400 (25)
luminol	341700 (7)	509400 (9)	428300 (6)	28700 (18)	10850000 (15)	4138600 (14)
luminol + desferal	165300 (7)	346400 (9)	153600 (7)	23000 (18)	8563000 (15)	1344000 (13)
luminol + NaN3	334600 (7)	242700 (7)	200000 (6)	14700 (14)	100300 (11)	145700 (25)
luminol + NaN3 + HRP	1929600 (13)	2390400 (15)	1337000 (11)	1222600 (11)	569300 (15)	1169600 (17)
phthalhydrazide	1000	3000	0	300	2000	400

triggered with PMA and probed with:

	2C11-12-2A grown 3 days with mouse interferon-γ	2C11-12-2A grown 3 days with lipopolysaccharide	LA5-9 grown 1 day with mouse interferon-γ	LA5-9 grown 2 days with lipopolysaccharide	HL-60 grown 4 days with retinoic acid	HL-60 grown 4 days with calcitriol
lucigenin	892000 (53)	76400 (4)	53000 (4)	64000 (39)	83700 (40)	943700 (45)
luminol	636000 (4)	153300 (5)	187000 (6)	207700 (6)	645000 (32)	2369000 (39)
luminol + desferal	698000 (5)	149200 (5)	177000 (5)	148300 (6)	317300 (20)	1064000 (10)
luminol + NaN3	680500 (5)	53300 (5)	74300 (8)	104400 (6)	119300 (38)	567000 (29)
luminol + NaN3 + HRP	1825000 (52)	96300 (6)	23300 (8)	144400 (33)	8830700 (41)	16547300 (22)
phthalhydrazide	1000	300	0	0	1000	1300

Berthold LB9505 Biolumat. The presented values are averages over 3 experiments; the luminescence just before addition of the trigger has been substracted. All operations were performed at 37°C.

RESULTS AND DISCUSSION

First, it was ascertained that at a pH of 7.2 (as occurs in culture media) $OH\cdot$ radicals can be formed and detected specifically with phthalhydrazide. When a H_2O_2 solution is injected in phosphate buffer pH 7.2 with luminol and Fe^{2+} (from Mohr's salt), a brief flash of light is produced. The luminescence can be strongly quenched with the typical $OH\cdot$ quenchers ethanol, t-butanol, mannitol and sodium benzoate and also by SOD. This suggests that luminol reacts first with $OH\cdot$ and then with $O_2^{\bar{\cdot}}$, both being produced in the reaction medium:

$$Fe^{2+} + H_2O_2 \longrightarrow Fe^{3+} + OH^- + OH\cdot$$
$$H_2O_2 + OH\cdot \longrightarrow O_2^{\bar{\cdot}} + H_2O + H^+$$
$$Fe^{3+} + H_2O_2 \longrightarrow Fe^{2+} + O_2^{\bar{\cdot}} + 2 H^+$$

When phthalhydrazide is used instead of luminol it turns out that there is still a strong chemiluminescence, which is totally abolished by $OH\cdot$ quenchers. The system is strongly buffer-dependent: in veronal buffer the chemiluminescence of luminol is strongly quenched only by ethanol and mannitol (which are readily oxidizable), while phthalhydrazide produces very little light. This suggests that in this buffer the reaction of luminol does not involve $OH\cdot$ nor $O_2^{\bar{\cdot}}$. It seems that in the absence of phosphate and at near neutral pH, Fe^{2+} does not form $OH\cdot$, but another oxidant, which is perhaps the ferryl ion:

$$Fe^{2+} + H_2O_2 \longrightarrow FeO^{2+} + H_2O$$

Luminol reacts then first with FeO^{2+} and next with H_2O_2.

When a solution of NaOCl is injected into a buffered solution of luminol and H_2O_2 a brief flash of light is produced, while a solution of luminol and H_2O_2 injected into a solution of horse radish peroxidase (HRP) produces luminescence that rises fast and decays slowly (with a halflife of 85 s). SOD is not inhibitory and phthalhydrazide cannot replace luminol. This shows that although luminol chemiluminescence can be elicited by a variety of oxidators (but not by H_2O_2 or $O_2^{\bar{\cdot}}$ alone), phthalhydrazide produces only light when $OH\cdot$ and H_2O_2 are both present in appreciable amounts.

The results obtained with the cell lines are summarized in the Table. The oxidative burst was triggered either with opsonized *Micrococcus* or with PMA (which is known to bypass the natural activation pathway by binding to protein kinase C) and the production of reactive oxygen derivatives was probed with either lucigenin or luminol. Lucigenin cannot penetrate the cell membrane and reflects mainly extracellular $O_2^{\bar{\cdot}}$ production (7). Luminol can react intracellularly as well as extracellularly and produces light in the simultaneous presence of H_2O_2 and a variaty of co-oxidators (but not $O_2^{\bar{\cdot}}$). To determine the intracellular contribution, desferal was added, which quenches all luminol chemiluminescence (6), but does not penetrate the cell. We can see that there are considerable differences, depending on the cell type and the trigger. When phthalhydrazide is used as probe instead of luminol, in all cases light production does not reach above the background. This argues against the production of free $OH\cdot$ radicals. The production of an amount of $OH\cdot$ too small to be detected cannot be excluded, but if a substantial part of luminol chemiluminescence were due to $OH\cdot$, some light production should have been observed with phthalhydrazide. 1 mmol/l of the myeloperoxidase-inhibitor NaN_3 reduces the light output of the HL-60 cells (grown in the presence of retinoic acid or calcitriol) with luminol to less than 5%. This is what we expect if the luminescence is mainly due to ClO^- produced by myeloperoxidase, an enzyme which is present in neutrophil granulocytes and monocytes, but not in mature macrophages. With the other cell lines, which are macrophage-like, azide inhibits much less or not at all. So, these cells must have a pathway for luminol oxidation that does not depend on myeloperoxidase or some other azide-inhibitable peroxidase. Horse radish peroxidase (HRP) is only weakly inhibited by azide, so that the increase in luminescence after addition of HRP provides an estimate of the H_2O_2 that leaks out of the cell or is formed extracellularly by dismutation of $O_2^{\bar{\cdot}}$ (8). We can see that here too there is considerable variation: there are cases where there is much luminescence, although the same cell line-trigger combination produces little luminescence with lucigenin. This shows that, although there is a production of oxidants, there is little extracellular release of O_2^-. There are also cases where there is much less luminescence with luminol + NaN_3 + HRP than with luminol alone, what is only possible if H_2O_2 is scavenged by something else than myeloperoxidase.

ACKNOWLEDGEMENTS
We greatly thank L. Brys for access to and technical assistance with the cell lines.

REFERENCES
1.   J.K. Hurst and W.C. Barrette Jr., Leukocytic oxygen activation and microbicidal oxidative toxins. CRC Crit. Rev. Biochem. Mol. Biol. 24, 271-328 (1989)
2.   R.C. Allen, Phagocyte oxygenation activities: quantitative analysis based on luminescence. In Bioluminescence and Chemiluminescence: New Perspectives, J. Schölmerich, R. Andreesen, A. Kapp, M. Ernst and W.G. Woods (Eds), Wiley, Chichester, p.13-22 (1987)
3.   G. Merényi and J. Lind, In situ generation of chemiluminescent substances through hydroxylation. In Analytical Applications of Bioluminescence and Chemiluminescence, L.J. Kricka, P.E. Stanley, G.H.G. Thorpe and T.P. Whitehead (Eds), Academic Press, London, p.569-572 (1984)
4.   P. De Baetselier, L. Brys, L. Mussche, L. Remels, E. Vercauteren and E. Schram, Generation and analytical applications of luminescent macrophage cell lines. In Cellular Chemiluminescence: Volume III, K. Van Dyke and V. Castranova (Eds), CRC Press, Boca Raton, p.19-49 (1987)
5.   H. Tanaka, E. Abe, C. Miyawra, Y. Shiina and T. Suda, $1\alpha,25$-dihydroxyvitamin D_3 induces differentiation of human promyelocytic cells (HL-60) into monocyte-macrophages, but not into granulocytes. Biochem. Biophys. Res. Commun. 117, 86-92 (1983)
6.   G. Bottu, The effect of quenchers on the chemiluminescence of luminol and lucigenin. J. Biolumin. Chemilumin. 3, 59-65 (1989)
7.   H. Gyllenhammar, Lucigenin chemiluminescence in the assessment of neutrophil superoxide production. J. Immunol. Methods 97, 209-213 (1987)
8.   M.P. Wymann, V. von Tscharner, D.A. Deranleau and M. Baggiolini, Chemiluminescence detection of H_2O_2 produced by human neutrophils during the respiratory burst. Anal. Biochem. 165, 371-378 (1987)

THERAPEUTIC EFFECTS OF THE ANTIOXIDANT MDTQ-DA ON EXPERIMENTALLY INDUCED ACUTE PANCREATITIS IN DOGS

S. Albrecht, T. Zimmermann[1], T. Freidt, B. Török[2],
H.-C. Gabsch, R. Schuster[1] and W. Jaross

Institute of Clinical Chemistry and [1]Department of Surgery of the Medical Academy Dresden, Germany, [2]Institute of Experimental Surgery of the Medical University Pecs, 'Hungary

INTRODUCTION

So far, no effective methods have been found to bring the inflammatory processes in acute pancreatitis under control. The inhibition of the classical cascade systems involved in acute pancreatits (complement, kallikrein-kinin-system and clotting system) and the arachidonic acid metabolism is no practicable therapeutic approach, because important physiological functions would be suppressed and because the inhibition of one cascade system would trigger off the over-activity of another one. However, reducing the toxic oxygen metabolites formed in abundance during the inflammatory process seems to be a practicable solution.

CL response, histological and laboratory tests (liver enzymes) were used in vitro and in vivo experiments to evaluate the therapeutical effect of the new antioxidant MDQT-DA (2,2-dimethyl-4-methanesulfonic acid-1,2-dihydrocholine), allopurinol, and the complement inhibitor Suramin (Bayer 205) on experimentally induced pancreatitis in dogs.

MATERIAL AND METHODS

Haemorrhagic pancreatitis was produced by injection of autologous bile into the pancreatic duct. Catheters were placed prepancreatically in the coelic artery and post-pancreatically in the vena portae. [1]

Blood sampling was done after placing the catheters, immediately before the injection of bile, and 1, 2, 4 and

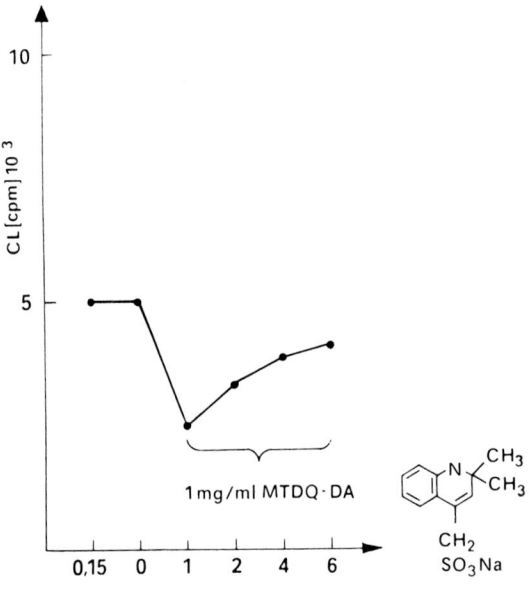

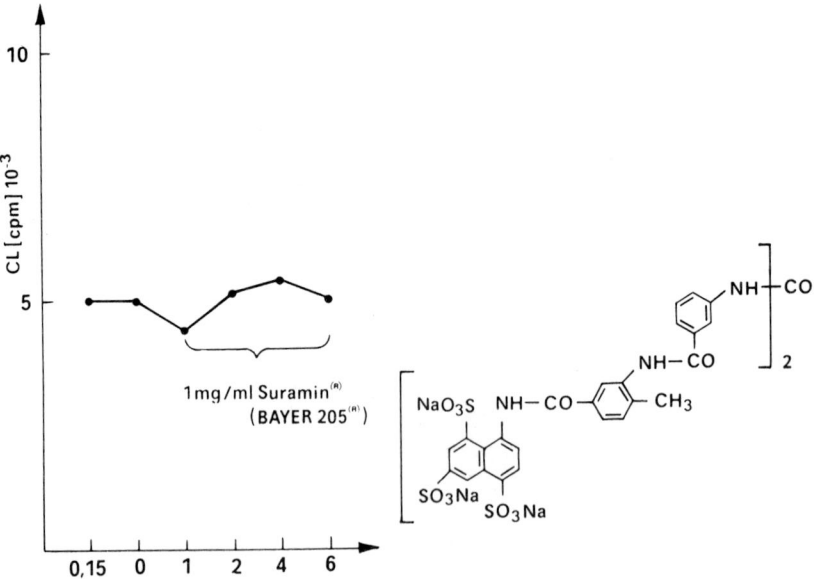

Fig.1. Therapeutic effect of MTDQ-DA and Suramin on CL-response in zymosan stimulated whole blood <u>in vitro</u>

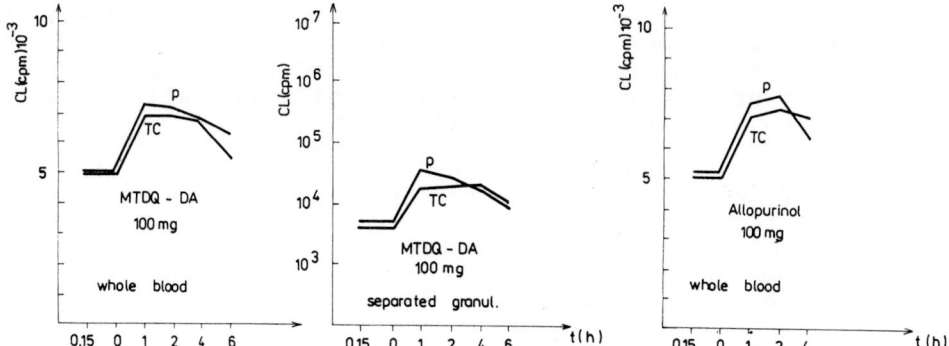

Fig.2.   Therapeutic effect of MTDQ-DA and Allopurinol
          on CL-response in zymosan-stimulated whole
          blood and separated granulocytes in vivo

LIVER   ENZYMES

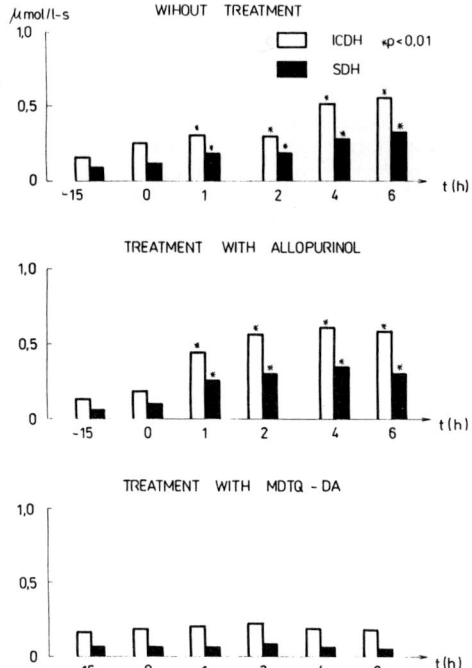

Fig.3.

6 hours after induction of acute pancreatitis. The separation of granulocytes and the measurement of chemiluminescence were done as previously described. [1]
For the in vivo tests, the animal received 100 mg/kg BW
MDTQ-DA or 100 mg/kg BW Allopurinol resp. after the
induction of acute pancreatitis.
For the in vitro tests an adaequate quantity of MDTQ-DA
(1 mg/mL) or Suramin (1 mg/mL) resp. was added to whole
blood or separated granulocytes of animals untreated with
radical traps.
In addition, we determined the liver enzymes (sorbit
dehydrogenase, lactate dehydrogenase, isoleucin
dehydrogenase) as indicator of the inflammation.

RESULTS AND DISCUSSION

Compared to the values obtained from untreated animals,
MDTQ-DA reduced the CL response in vivo by 60 p.c. and in
vitro (zymosan-stimulated whole blood) by 70-80 p.c.
(Fig. 1 and 2). A reduction of the CL response in the
same order of magnitude was found in separated granulocytes in the in vivo test. In the in vitro test, Suramin[R]
(Bayer 205) reduced the CL response by 40 p.c. whereas
allopurinol reduced it in vivo by maximally 30 p.c.
without mitigating the inflammatory processes and preventing the development of the pancreatic necroses and the
consecutive problems (pulmonary and hepatic failure). In
the animals treated with MDTQ-DA the liver enzymes remained normal.
MDTQ-DA proved to be the most effective of all the radical traps used in this study. It seems to react primilary
very quickly with compounds, such as lipidhydroperoxides,
which play an important part in the damage of live cells.
The respiratory burst can be well controlled with MTDQ-
DA, whereas allopurinol and Suramin proved less
effective. The rearkable difference in the behaviour of
the liver enzymes underlines this fact (Fig.3). Histological examinations showed that the development of ARDS or
MOF could prevented only by MDTQ-DA.

REFERENCES
[1] T. Zimmermann, S.Albrecht, R. Schuster, G. Lauschke
and W. Jaross, Chemiluminescence response of whole blood
and polymorphonuclear leukocytes following experimentally
induced haemorrhagic-necrotising pancreatitis.
Fresenius J.Anal.Chem. 337, 91-92 (1990)

CHARACTERIZATION OF ASCORBIC ACID AS AN ENDOGENOUS REACTANT OF NON-ENZYMATIC CHEMILUMINESCENCE FROM BRAIN CELLS

M. Uhr and H. Reiber

Neurochemisches Labor, Universität Göttingen, D-3400 Göttingen, FRG

INTRODUCTION

A non-enzymatic low-level luminescence from disintegrated brain cells has been reported by Reiber (1). Vital mammalian cells do not emit a measurable intensity of photons. The photon emission originates from cytoplasmatic cell membranes (2). The luminescence intensity from brain homogenates could be completely restored with three components: the plasma membrane fraction, a low molecular weight substance from 100 000 g supernatant of the brain tissue homogenate together with a suitable reaction buffer (2).

In this paper we present data from which we conclude that this soluble endogenous reactant from brain tissue is ascorbic acid.

MATERIAL AND METHODS

Tissue homogenate and subcellular fractions: Dissected grey matter of adult pig brain (3) was washed two times in reaction buffer and homogenized in reaction buffer with a syringe (without needle).For subfractionation cells were disintegrated with a Potter homogenizer. The homogenate was diluted to a protein concentration of 4 mg/ml. All preparation steps were performed at 4°C. The membrane fraction was obtained from the 200 g centrifugation (5 min, 4°C). The 200 g pellet was washed 3 times with water (centrifugation at 100 000 g, 20 min, 4°C and resuspended by ultrasonication). This membrane fraction which is finally resuspended in water or reaction buffer (2 mg protein/ml) could be stored frozen (-30°C) without loss of activity. The 100 000 g (30',4°C) supernatant was obtained after a preceeding centrifugation (12 000 g, 15 min) of the Pottered tissue homogenate (10 mg protein/ml).

Chromatography: The reactant from 100 000 g supernatant was separated by column chromatography (DEAE-Sephacel, Pharmacia) and by gel filtration (Sephadex G 25 superfine). The anion exchanger was equilibrated with 20 mmol/l triethanolamine and 65 μmol/l dithiothreitol (DTT) at pH 7.

The substance was extracted with a NaCL gradient 0 - 0.1 mol/l NaCL, (20 mmol/l triethanolamine, 65 μmol/l DTT, pH 7). The relevant fractions were identified by the UV absorption at 254 nm and the activity in the chemiluminescence assay. Sephadex G 25 was equilibrated with triethanolamine and DTT at pH 7 for further gel filtration of the enriched fractions.

HPLC: The separation was carried out on a reversed phase column LiChrosorb RP-18(7μm), Hibar 250-4 (Merck). The mobile phase consisted of 0.1 mmol/l disodium hydrogenphosphate (Na_2HPO_4), 0.07 mmol/l EDTA, O.15 mmol/l sodium-octylhydrogensulphate, adjusted to pH 3.2 with orthophosphoric acid (4). The flow rate was 2 ml/minute, resulting in a pressure of 20 MPa. The UV absorption at 255 nm was recorded as shown in Fig. 3.

<u>Single Photon Counting Assay:</u> The photon counting apparatus (Biolumat, Berthold, Wildbad, FRG) was kept at 4°C. The reaction temperature in the counting chamber was 37°C. 1 ml sample (kept at 4°C) was inserted into the cuvette housing (37°C). The stirred sample reached a temperature of 37°C in less than 4 minutes. The signal was calculated as impulses/sec (l/s).

Freshly prepared samples were ultrasonicated (4°C) with a microtip (Branson sonifier, energy level 2, 20% pulsed, 20 strokes) in a microvessel (Eppendorf) and than transferred into the counting vessel (Berthold, Wildbad, FRG).

<u>Reaction Buffer:</u> Hanks balanced salt solution with 0.02 mol/l HEPES, 0.13 mol/l glutamine, pH 7.2. For long term incubation the reaction buffer contained penicillin 50 units/ml + streptomycin 50 µg/ml.

<u>RESULTS AND DISCUSSION</u>

<u>Partial separation of the reactant</u>
The efficency of the single steps for separation of the reactant was controlled in the chemiluminescence assay: To 500 µl of the membrane fraction (0.3 mol/l $CaCl_2$, 0.13 mol/l glutamine) 500 µl of the eluted samples were added.

At pH 7 the substance could be eluted as an anion with 0.016 mol/l NaCl from the DEAE Sephacel. By gel filtration on Sephadex G25 we excluded substances with a molecular weight > 1000 D from the reactant solution. The enriched reactant was characterized by the UV-spectrum (Fig. 2) and the retention time in HPLC (Fig. 3). The yield of the total isolation procedure with respect to photon emission intensity was 50 %.

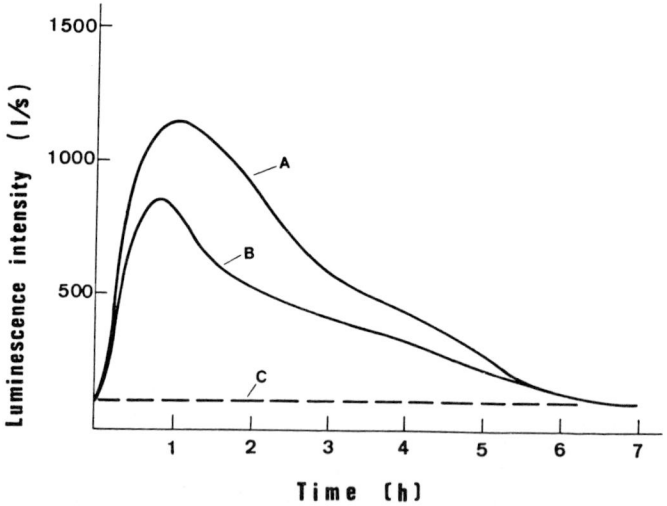

Fig. 1.: Photon emission of brain cell membranes. Comparison of the reactants from 100 000 g supernatant with ascorbate. 500 µl membrane fraction (1 mg protein) were suspended in double concentrated reaction buffer, ultrasonicated and mixed with 500 µl 0.11 mmol/l ascorbate in water (A) or with 500 µl 100 000 g supernatant from brain tissue homogenate (B). The dashed line (C) represents the pure membrane fraction with reaction buffer which is identical with the dark current of the photomultiplier.

Comparison of the isolated reactant with ascorbate

<u>Luminescence:</u> The search for a low molecular weight anion which could replace the 100 000 g supernatant in the chemiluminescence assay led us to ascorbate as a possible candidate. As shown in Fig. 1 ascorbate and the brain tissue supernatant showed both a similar course of photon emission with the membrane fraction in reaction buffer.

<u>Absorbance spectrum:</u> The UV-spectrum of the enriched reactant and of ascorbate are compared in Fig. 2 at pH 2 and pH 7. At pH 2 both compounds have a maximum at 240 nm and at pH 7 a maximum at 265 nm. The absorbance at pH 7 from the reactant solution correlated linearly with the efficiency in the luminescence assay. This ratio between the absorbance and the corresponding photon emission intensity was the same for the enriched reactant solution and for the ascorbate solution. The instability of the reactant fraction in the presence of Ca^{2+} decreased the absorbance at 265 nm (pH 7) which was correlated with a decreased photon emission intensity.

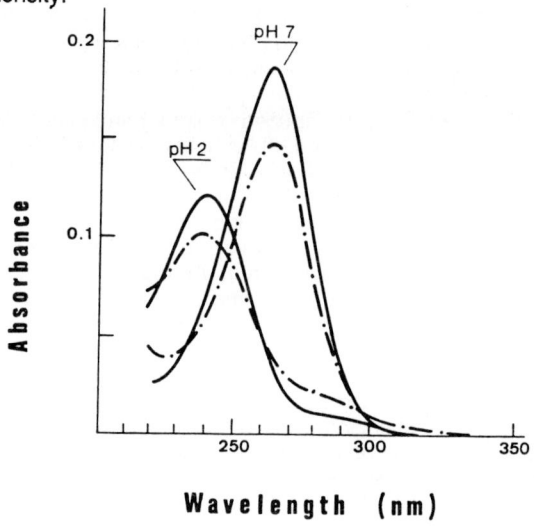

Wavelength (nm)

Fig. 2.: UV-absorbance spectra of ascorbic acid and the purified reactant from 100000 g supernatant of pig brain tissue homogenate (dashed lines). Ascorbic acid ($1 \cdot 10^{-5}$ mol/l) was measured in water; the purified eluant after a gel filtration in triethanolamine and DTT was diluted 1:10 with water. The pH was adjusted with HCL or NaOH. The spectra were measured in a double beam photometer with the corresponding blank solutions as a reference.

<u>HPLC-separation:</u> As shown in Fig. 3 both compounds, the reactant from brain tissue supernatant and ascorbate, had the same retention time (1.9 min) in the HPLC-system used.

<u>Stability:</u> Besides the biological activity, spectral and chemical similarities also physical properties like heat lability are common to both substances: Ascorbate and the isolated reactant from tissue homogenate lost their activity in the chemiluminescence assay if heated 30 minutes at 90°C. In addition both compounds are at low concentration labile in the presence of Ca^{2+} and could be stabilized by DTT.

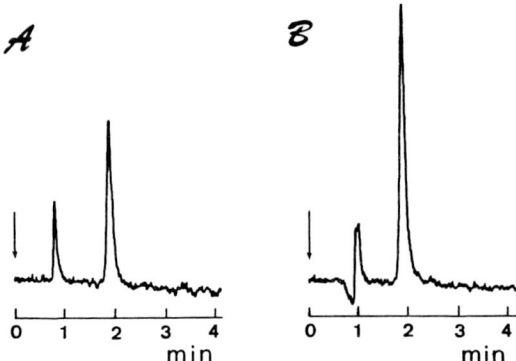

Fig. 3.: HPLC elution profiles of ascorbic acid (A) and purified reactant from 100000g supernatant of brain homogenate (B). Absorbance measured at 255nm.

Lipid peroxidation: Both compounds induced a lipid peroxidation which was detected by its malondialdehyde formation (2). In both cases the extent of the lipid peroxidation correlated with photon emission intensity.
Inhibition/suppression of luminescence: With increasing concentration of the ascorbate in the luminescence assay an increasing photon emission intensity could be observed up to a maximum at about 55 μmol/l ascorbic acid for a membrane fraction with 1 mg protein in the assay (1ml). At higher concentrations the luminescence intensity was decreased down to zero. This concentration- dependent course of luminescence intensity has been observed for the supernatant-derived reactant as well.

The similarities between ascorbic acid and the reactant from supernatant of brain tissue homogenate are very convincing. Due to these results we suggest that the non-enzymatic photon emission observed from disintegrated brain cells is a consequence of the endogenous ascorbic acid content of the cells. Together with an endogenous EDTA-extractable metal ion (2) the ascorbic acid could function as a prooxydant in the disintegrated non-vital cells. But the membrane related influences on the luminescence of brain cells are not restricted to the effect of endogenous metal ions (2).
Different influences of endogenous ascorbate on glial cells in vitro (prooxidant) and in vivo (antioxidant) are described and discussed in (2) as well.

REFERENCES
1. H. Reiber, Discrimination between types of low-level luminescence in mammalian cells: The biophysical radiation. J. Biolumin. Chemilumin.,4, 245-248 (1989).
2. H. Reiber and M. Uhr, Photon emission from brain cells. Subcellular localization, non-enzymatic reaction and kinetics. This volume.
3. Gebicke-Härter et al., Bulk separation an long-term culture of oligodendrocytes from adult pig brain. I. Morphological studies. J. Neurochem.42,357-368 (1984).
4. E. Nagy and I. Degrell, Determination of ascorbic acid and dehydroascorbic acid in plasma and cerebrospinal fluid by liquid chromatography with electrochemical detection. J. Chromatography, 497, 276-281 (1989).

THE EFFECT OF BETA-AGONISTS ON POLYMORPHONUCLEAR CELL (PMNL) FUNCTION IN VITRO

L. Ramage, I.A. Cree, A.L. Blair, D.P. Dhillon[1]

Department of Pathology, Ninewells Hospital and Medical School, and [1]Department of Respiratory Medicine, King's Cross Hospital, Dundee, DD1 9SY, Scotland

INTRODUCTION

The presence of B-adrenoceptors on human PMNL is well established and studies show that these receptors are exclusively B2-adrenoceptors. B2-Adrenergic agonists are commonly used bronchodilators in the treatment of asthma. Since these drugs will have effects on the inflammatory cells present in the bronchial mucosa, they may also have an effect on the disease itself (4).

In asthma, the ability of phagocytes to produce reactive oxygen species (ROS) is increased (2). Isoprenaline, Fenoterol, Adrenaline, and Noradrenaline have been found to inhibit human PMNL in vitro in a variety of assays (3,4). Since salbutamol and terbutaline are more commonly used therapeutically, we have investigated the effects of these drugs on human peripheral blood PMNL in vitro using lucigenin-enhanced chemiluminescence (CL).

MATERIALS AND METHODS

Cell preparation: PMNL were separated by density gradient centrifugation from freshly obtained heparinised venous blood from healthy male volunteers using Mono-Poly Resolving Medium (MPRM, Flow Laboratories, Irvine, Scotland). Following separation, the PMNL were washed twice in HBSS buffer prior to suspension at 1×10^6/ml in HBSS + 0.1% BSA.

Chemiluminescence Assay: CL assays were performed as described by Blair et al. (1). Briefly, 50 μl of the PMNL suspension was added to each well of white microtitre strips (Dynatech, Billingshurst, Sussex, England) with 50 μl of 5×10^{-4} mol/l lucigenin (Sigma) and 50 ul buffer with or without drug. The wells were allowed to equilibrate at 37°C for 15 min. Background CL readings were taken at 5 min intervals in an Amerlite microtitre plate luminometer (Amersham International, Amersham, Bucks., England) for a further 15 mins before the addition of 100 μl of 1×10^7 pre-opsonised zymosan particles/ml. Opsonisation of

zymosan was achieved by incubation of zymosan particles with 10%
serum for 30 min at 37°C prior to their addition to the wells. CL
readings continued to be taken at 5 min intervals for up to 120
min. All experiments were performed in triplicate wells.

Drugs: Salbutamol (Sigma), Terbutaline (Sigma), and Adrenaline
(Sigma) were dissolved in HBSS + 0.1% BSA to form 10 mmol/l stock
solutions. Aliquots of 1.5 ml were stored frozen at -20°C until
use. A 20 mmol/l stock solution of Propranolol was similarly
prepared for blocking experiments. All solutions were maintained
at pH 7.1.

Experimental Protocol: Experiments were performed to examine:
(1) the effect of varying concentrations of B-agonists, (2) the
effect of prolonged exposure of PMNL to B-agonists on CL, (3) the
effect of drug removal before addition of opsonised zymosan, (4)
whether the effects of B-agonists on PMNL could be reversed by a
B-adrenergic antagonist, and (5) whether B-agonists quenched
luminesence in a cell-free xanthine oxidase-hypoxanthine
reaction.

Data Analysis: The CL readings were transferred to an on-line
microcomputer for later analysis using Supercalc 5 (Computer
Associates, England).

RESULTS AND DISCUSSION

1. Varying Drug Concentration: Increasing inhibition of the CL
response was seen with increasing concentration of each drug
tested (table 1). Both salbutamol and terbutaline showed a
similar degree of inhibition of CL. There was considerable
variation between individuals in the degree of responsiveness of
PMNL to B-agonist mediated inhibition of CL (table 1). However,
there was always a clear dose response in each subject for any of
the two drugs tested. The variation in results may reflect
differing expression of B-adrenergic receptors on human PMNL,
rather than experimental variation.

Table 1. Dose relationship of total CL with varying
concentration of all three drugs tested. Results are
expressed as mean percentage of the total CL obtained
without drugs, with the range shown in brackets beneath.

Drug Conc mmol/l	Salbutamol	Terbutaline
10	34 (29-40)	22 (16-48)
1	50 (35-64)	58 (42-80)
0.1	75 (62-86)	75 (67-84)
0.01	83 (69-99)	82 (74-89)

2.   Length of Exposure of PMNL to Drug: PMNL were incubated with all three drugs at a concentration of 0.1mM for periods of 15 to 75 min before stimulation with opsonised zymosan. No major differences were seen in experiments using cells from three different individuals, although slightly less inhibition was seen at 75 min compared with 15 min in two subjects. This may be explained by modulation of B-adrenoceptors. Although this effect was minimal in vitro over the relatively short time period used, B-adrenoceptor tachyphylaxis in PMNL may occur in vivo.

3.   Effect of Drug Removal by Washing: Washing PMNL twice after exposure to any of the three drugs for 30 min to remove drug in solution showed complete loss of inhibition compared with control cells treated in exactly the same way. This result was unexpected and may reflect rapid receptor recycling or metabolism of the drug.

4.   Effect of Beta-antagonists: Cells were treated with 0.001 mmol/l Propranolol prior to the addition of Salbutamol and terbutaline at 0.1mmol/l. There was partial reversal of the CL suppression observed with B-agonist alone (fig. 1b), confirming the receptor-mediated nature of the suppression of phagocyte CL mediated by B-agonists.

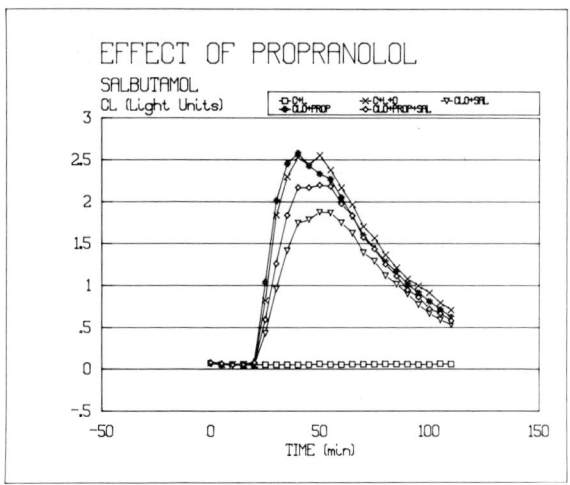

Fig. 1.   The effect of pre-treatment of PMNL with propranolol on salbutamol-mediated CL inhibition.

5.   Xanthine Oxidase: Some quenching was noted at the higher concentrations of B-agonist, but not at concentrations less than 0.1 mmol/l (fig. 2).

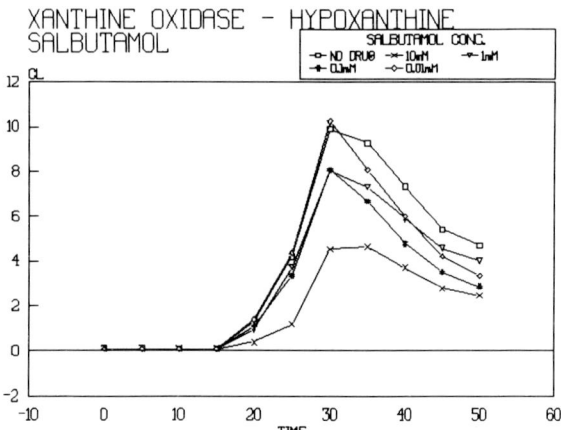

Fig. 2. The effect of varying concentrations on terbutaline on xanthine oxidase-hypoxanthine CL.

The inhibitory effect of B-agonists on PMNL is potentially useful in asthma patients. PMNL are part of the inflammatory infiltrate in the bronchial mucosa and release potentially damaging mediators and ROS. Therefore administration of B-agonists may modify the disease process. However, down-regulation of B-adrenoceptors may abrogate this effect in patients receiving B-agonists alone.

ACKNOWLEDGEMENTS

This work was supported by grant K/MRS/50/C983 from the Scottish Home and Health Department.

REFERENCES

1.   A.L. Blair, I.A. Cree and J.S. Beck. Measurement of phagocyte chemiluminescence using a microtitre plate luminometer. J. Biolum. Chemilum. 3, 67-70 (1989).
2.   M. Damon, P. Cluzel, P. Chanez and P.H. Godard. Phagocytosis induction of chemiluminescence and chemoattractant increased superoxide anion release from activated human alveolar macrophages in asthma. J. Biolum. Chemilum. 4, 279-286 (1989).
3.   J.A. Mack, C.P. Nielson, D.L. Stevens, R.E. Vestal. B-Adrenoceptor-mediated modulation of calcium ionophore activated polymorphonuclear leucocytes. Br. J. Pharmacol. 88, 417-423 (1986).
4.   R.E. Schopf, E.M. Lemmel. Control of the production of oxygen intermediates of human polymorphonuclear leukocytes and monocytes by B-adrenergic receptors. J. Immunopharmacol. 5, 203-216 (1983).

NEUTROPHIL CHEMILUMINESCENCE (CL) IN PSORIASIS USING A MICROTITRE PLATE LUMINOMETER

J. Mulvaney, D. Bilsland[1], I.A. Cree and J. Ferguson[1]

Department of Pathology and [1]Department of Dermatology, University of Dundee, Ninewells Hospital and Medical School, Dundee, DD1 9SY, Scotland

INTRODUCTION

Psoriasis is a disorder with a hereditary predisposition, characterised by an epidermal inflammatory exudate. Lesions in active forms are widely infiltrated by polymorphonuclear leukocytes (PMNL), whereas in chronic stable plaque psoriasis they are absent.

Peripheral blood neutrophils show increased locomotion, adherence and chemotaxis, phagocytosis, lysosomal enzyme release, cytotoxicity, Fc receptor activity and CL. Increased levels of membrane NADPH oxidase, myeloperoxidase (MPO), and glycolytic pyruvate kinase have also been reported. These indicate increased metabolic and phagocytic function of PMNL in active psoriasis.

However, several authors have failed to show significant differences in phagocyte activity of psoriatic PMNL compared to those of normal controls. Methodological differences may be partly to blame, but differences in disease activity may also play a role. It has been reported that treatment reduces PMNL CL stimulated by serum-opsonised zymosan in psoriatic patients after one month of treatment with PUVA.

Since previous assays of phagocyte CL in psoriasis have suffered from poor precision due partly to methodological problems, we decided to test a microtitre plate phagocyte CL assay in psoriasis and evaluate its use for serial follow up studies in patients receiving treatment.

MATERIALS AND METHODS

Patients: 9 patients, aged 25-68, 7 males and 2 females, suffering from plaque psoriasis, guttate psoriasis or both, were drawn from the population attending photobiology unit at

Ninewells Hospital, Dundee, beginning a course of Psoralen + UVA
(PUVA), PUVA with Tigason (a retinoid), or 312nm UVB photo-
therapy. Subjects participated after giving written informed
consent. Blood samples were drawn at two-weekly intervals during
treatment to assess its effect on PMNL CL responses in comparison
with disease activity.
Controls: 20 healthy controls with no history of psoriasis,
asthma, or minor intercurrent infection.
CL Assay: The assay (Blair et al., 1989) was performed using
PMNL obtained by density centrifugation from 5ml of venous blood
using the method shown in Fig. 1. SOZ was prepared by incubating
pooled normal serum with zymosan particles, washing and re-
incubating with serum. Aliquots of washed SOZ were stored -20 C
until use, when it was diluted to 1×10^7 particles/ml in HBSS with
0.1% BSA. PMA was prepared as a stock solution (1mg/ml) and
diluted to 2µg/ml in 5ml HBSS with 0.1% BSA. Assays were
performed in an Amerlite microtitre plate luminometer (Amersham
International plc, Amersham, UK). Each experiment was performed
in triplicate and all results were used in subsequent data
analysis using Supercalc 5 (Computer Associates, Slough,
Berkshire, U.K.).

Fig. 1. Method for microtitre plate CL assay.

5 ml heparinised venous blood
↓
Density centrifugation over Polyprep (Nycomed)
↓
Wash x2 in HBSS without phenol red and
resuspend to 1×10^6 PMNL/ml
↓
Add 100 µl of 2.5×10^{-4} mol/l lucigenin (Sigma)
to white plastic wells
↓
Equilibrate at 37 C for 15 min.
↓
Add 50 µl of cell suspension to wells
↓
Equilibrate at 37 C for 15 min.
↓
Take background readings for 15 min.
↓
Add stimulant (Buffer/PMA/Zymosan/Opsonised Zymosan)
Take readings every 5 min. for 105 min.

RESULTS AND DISCUSSION.

PMNL CL Response to PMA: PMNL of untreated patients with
psoriasis displayed significantly higher peak CL responses
compared to healthy control individuals when stimulated with PMA
(Fig. 2)($p < 0.02$, Mann-Whitney U-test). After 5 weeks of

treatment with PUVA and UVB, the total CL response decreased
significantly to control values. This coincided with clinical
improvement in all patients.

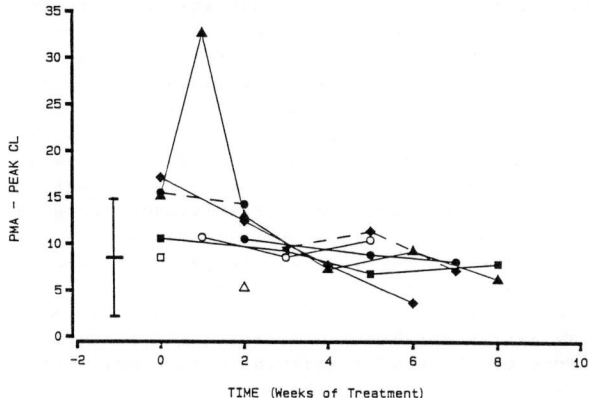

Fig. 2: Peak PMNL CL responses to PMA during treatment
with PUVA and UVB compared with the control mean +/- 2
standard deviations. Each line indicates a different
patient.

PMNL CL Response to SOZ: The peak CL response of PMNL from
psoriasis patients to SOZ was raised slightly above control
values before treatment (Fig. 3), but this was not statistically
significant. When compared to Fig. 1, it shows that the PMNL CL
responses of patients, though increased, lie largely within the
normal range.   The CL response to SOZ did not decrease
significantly after 5 weeks of treatment.

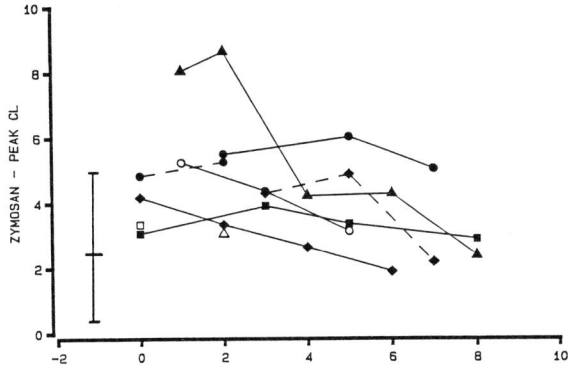

Fig. 3: Peak PMNL CL responses to Opsonised Zymosan during
treatment of psorasis with PUVA and UVB compared with the
control mean +/- 2 standard deviations. Each line
indicates a different patient.

Several reports suggest increased neutrophil responses in psoriasis which change with the activity of the disease and have been implicated in its pathogenesis (1,2). The current study also shows that PMNL of untreated patients with mild psoriasis display higher CL responses to PMA than normal controls. After treatment with PUVA and UVB, the CL responses from patients were similar to those obtained from controls. This coincided with clinical improvement and is in agreement with previous studies of CL in psoriasis.

The results show a marked difference between the CL responses of PMNL to stimulation by PMA and SOZ. In all patients, the CL response to SOZ was much more erratic compared with the close grouping of results from PMA stimulation. This may be explained by the ability of PMNL to regulate cell surface receptors, a phenomenon which occurs within a few minutes and can be triggered during cell separation. Since PMA stimulates the respiratory burst by activating Protein Kinase C (PKC) directly, it is not influenced by changes in receptor number but reflects changes in the enzymatic capacity of the pathway from PKC to superoxide production. Our observations suggest that both enhanced CL from PMNL in active psoriasis is due to receptor-independent mechanisms as well as increased numbers of cell surface receptors. Use of PMA stimulated CL may provide a more reliable method of assessing inflammation-induced changes in PMNL function in vitro.

The microtitre plate phagocyte CL assay appears to be a sensitive method of assessing PMNL function in psoriasis and produces results comparable to other CL assays.

ACKNOWLEDGEMENTS

This research was supported by grant K/MRS/50/C983 from the Scottish Home and Health Department.

REFERENCES

1.   R.E. Schopf, P. Altmeyer and M. Lemmel. Increased respiratory burst activity of monocytes and polymorphonuclear leukocytes in psoriasis. Brit. J. Dermatol. 107, 505-510 (1982).
2.   P. De Sole P. Polymorphonuclear chemiluminescence: some clinical applications. J. Biolum. Chemilum. 4: 251-262 (1989).
3.   A.L. Blair, I.A. Cree and J.S. Beck. Measurement of phagocyte chemiluminescence using a microtitre plate luminometer. J. Biolum. Chemilum. 3: 67-70 (1989).

THE EFFECT OF AMINOGLYCOSIDES ON HUMAN POLYMORPHO-NUCLEAR

LEUKOCYTE (PMNL) CHEMILUMINESCENCE

L.A. Pierce, A.L. Blair, I.A. Cree, W.O. Tarnow-Mordi[1]

Departments of Pathology and [1]Child Health, University of Dundee, Ninewells Hospital and Medical School, Dundee, DD1 9SY, Scotland

INTRODUCTION.

Chemiluminescence (CL) provides a reproducible method for the detection of the release of reactive oxygen species (ROS) from phagocytes during the respiratory burst which follows phagocytosis. Polymorphonuclear leukocytes (PMNL) show reduced CL responses to opsonised zymosan after treatment with Gentamicin in vitro (3). In this study, the effects of exposure of PMNL to four commonly used aminoglycosides on the CL response to opsonised zymosan were examined.

MATERIALS & METHODS.

PMNL Preparation: Fresh heparinised whole blood from healthy adult volunteers was separated by density gradient centrifugation using Mono-Poly Resolving Medium following the manufacturer's guidelines (Flow Laboratories, Irvine, U.K.). The resulting PMNL layer was removed and washed twice in HBSS buffer prior to resuspension in 1-2 ml HBSS. Cell purity and concentration were assessed using a Coulter counter with channelyser and the cells were subsequently diluted to 1×10^6 PMNL/ml in HBSS + 0.1% bovine serum albumin (BSA).

Zymosan: Zymosan was prepared according to the method of Lachmann and Hobart (1978) and diluted to 1×10^7 particles/ml in HBSS before opsonisation with homologous serum (37°C for 30 minutes).

CL Assays: CL assays were performed in triplicate using previously published methods (1,2). Briefly, 50µl freshly prepared 5×10^{-4} mol/l lucigenin in HBSS + 0.1% BSA was added to the wells of white plastic microtitre plates (Dynatech Laboratories, Billingshurst, Sussex, U.K.) and left to equilibrate for 30 minutes. PMNL at 1×10^6/ml were then added (50µl) and the plate transferred to a microtitre plate incubator (Amersham International) at 37°C. Drugs were added to test wells after 15 min equilibration at 37°C and control wells were made up to 250µl total volume with HBSS + 0.1% BSA. After a further 15 min equilibration period, during which background readings

were recorded, 100μl opsonised zymosan was added and readings
continued at 5 minute intervals for a further 90 minutes.
Dose Response: Drugs were made up in HBSS in a range of
dilutions, encompassing those found in patients. The pH was
constant at 7.1. Netilmicin, Gentamicin and Kanamycin were all
made up to a maximum of 200 μg/ml (final concentration) and
Streptomycin up to 400 μg/ml (final concentration).
Drug Exposure Time: Each drug was incubated with PMNL for
varying periods of time (15-60 min) before opsonised zymosan was
added.
Effect of Washing Cells: The effect of washing out the drug from
the cells was examined by washing cells twice in HBSS
following their incubation for 15 min. with drug or buffer, and
then stimulating the CL response with opsonised zymosan.
Xanthine Oxidase Assay: To determine whether observed effects on
PMNL CL responses were due to quenching, a non-cellular source of
ROS was used to generate CL in the presence and absence of drug.
50 μl of each antibiotic at varying dilutions was added to
microtitre plates containing 0.1 U/ml xanthine oxidase (50μl) and
5×10^{-4} mol/l lucigenin (100μl). 50 μl of hypoxanthine (10mg/ml)
was added to start the reaction, which was followed by CL
measurements every 5 minutes.
Data Analysis: Each experiment was performed three times using
triplicate wells. The results were transferred to an on-line
microcomputer for analysis.

RESULTS AND DISCUSSION.

All four aminoglycosides used suppressed the CL response from
PMNL in a dose-dependent manner (Fig. 1), although little
suppression occurred with Streptomycin at the doses shown. Since
all four drugs had similar therapeutic ranges (2-40μg/ml) and
suppression was not observed within this range for any of the
drugs used, this pharmacological effect is unlikely to produce
significant suppression of PMNL function in vivo.

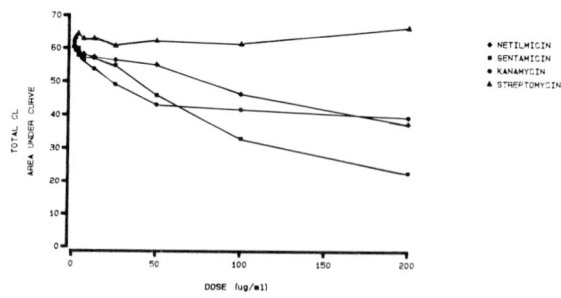

Fig. 1. Effect of variation of concentration of (a)
Gentamicin, (b) Netilmicin, (c) Streptomycin and (d)
Kanamycin on CL.

A concentration of 200µg/ml Gentamicin (final dilution) was used
in experiments to observe the effects of increasing the preincubation time of drug with the PMNL. Suppression of CL by
Gentamicin decreased as the incubation period of drug with PMNL
increased (Fig. 2). This suggests that these cells may be able
to metabolise the drug.

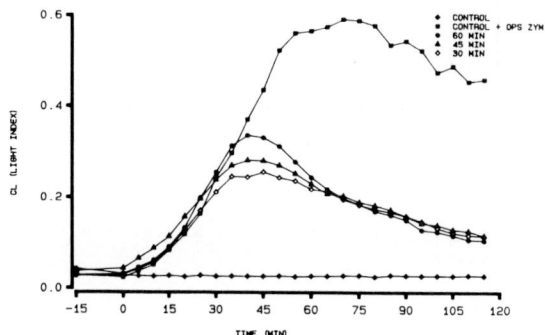

Fig. 2. Effect of varying pre-incubation time of
Gentamicin (200µg/ml) on phagocyte CL.

Washing out the aminoglycosides following exposure of the cells
to the drug for 15 min. changed the shape of the CL curve
produced, but did not reverse the suppressive effect of drug
exposure on the CL response (Fig. 3). This may indicate that the
drug has long-lasting effects on the cells which should be
investigated in patients receiving aminoglycosides.

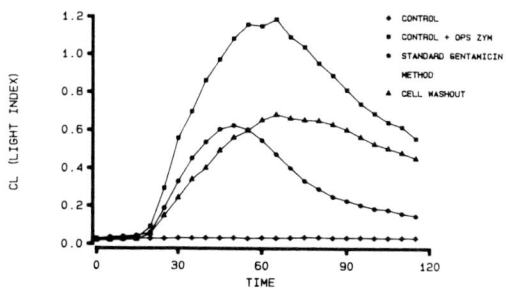

Fig. 3. Effect of washing PMNL following 15 min. exposure
to Gentamicin.

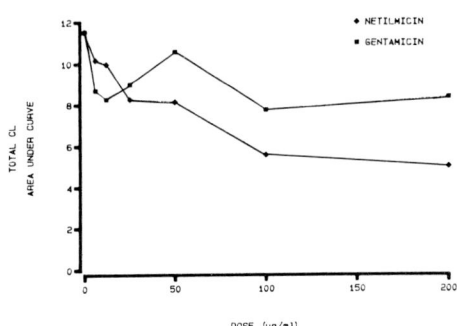

Fig. 4.   The effect of Gentamicin and Netilmicin on xanthine oxidase CL.

The xanthine oxidase experiments showed dose-related suppression of CL (quenching) at levels up to 200 µg/ml for both Gentamicin and Netilmicin (Fig. 4).   It is possible that they could exert a similar effect on PMNL oxidative responses, either by being bound to the cell membrane or internalised.

In conclusion, it seems from these results that aminoglycosides, with the possible exception of Streptomycin, exert _in vitro_ suppressive effect on PMNL CL responses which may be related to their ability to quench ROS production.   It is likely that these antibiotics bind to the cells, which may be able to metabolise them.

ACKNOWLEDGEMENTS

We wish to thank Dr PG Davey, Mr R Fawkes and Mr S MacPherson for their contributions to this research which was supported by grants K/MRS/50/C983 and K/MRS/50/C1211 from the Scottish Home and Health Department.

REFERENCES

1.   A.L. Blair, I.A. Cree, J.S. Beck and M.J.G. Hastings. Measurement of phagocyte chemiluminescence in a microtitre plate format. _J. Immunol. Methods_ 112, 163-168 (1988).
2.   A.L. Blair, I.A. Cree and J.S. Beck.   Measurement of phagocyte chemiluminescence using a microtitre plate luminometer. _J. Biolum. Chemilum._ 3, 67-70 (1989).
3.   P. Van der Auwera. The immunomodulating effects of antibiotics. _Curr. Opin. Infect. Dis_. 1, 363-374 (1988).

CHEMILUMINESCENCE PATTERN OF RAT ALVEOLAR MACROPHAGES
DURING MATURATION IN VITRO

J. Schmidt[*] and F. Tilkes

Institute of Hygiene, University of Gießen, FRG, [*]present address:
ASTA Pharma AG, Department of Pharmacology, Frankfurt, FRG

INTRODUCTION

Alveolar macrophages (AM) are derived from undifferentiated stem cells in the bone marrow. During maturation, AM-precursor cells leave the bone marrow and migrate as peripheral blood monocytes to their target organ. There is evidence of increasing differentiation and commitment during this process.
Jacobshagen and Andreesen (1) have shown that monocytes growing in teflon bags differentiate into mature macrophages over a time course of about 10 days. During this differentiation the luminol-dependent chemiluminescence (CL) decreases. In contrast, the lucigenin-dependent CL increases steadily with progression of the cells to more mature differentiation stages. In the present study, we investigated the relationship of the different CL-activities with respect to the differentiation process in AM during their cultivation in vitro.

MATERIALS AND METHODS

Cell preparation: Unstimulated AM of 12 weeks old Long-Evans rats were obtained by bronchoalveolar lavage with phosphate buffered saline (PBS) supplemented with EDTA (60 mmol/l), Heparin (40 U/ml), and Penicillin/Streptomycin (80 U/ml). The resulting cell suspension was washed (3 x) in Heparin-free PBS and adjusted in Dulbecco's minimum essential medium (MEM) for CL(Boehringer Mannheim, FRG) to 1 x 10^6 cells/ml. Cell suspensions (0.3 ml) were transferred into glass vials (Abbott Laboratories, USA) and allowed to attach for 90 min. Subsequently, the medium containing the non-adherent cells was removed and changed for MEM supplemented with Neomycin/Bacitracin and Amphotericin B (Serva, FRG). Depending on the experiment 5% fetal calf serum (FCS, Boehringer Mannheim, FRG) was added. The cell samples were maintained in an incubator at 38°C/5% CO_2/100% AH until measurements.
Chemiluminescence-measurements: CL-measurements were performed in a six channel biolumat (Berthold, FRG). Lucigenin (10,10'-dimethyl-9,9'-biacridinium-dinitrate, Sigma, FRG) was dissolved (10 mmol/l) in MEM and kept in the dark until use. The luminol-derivative DMNH (7-dimethylamino-naphtalene-1,2-dicarbonic-acid-hydrazide, Boehringer Mannheim, FRG) was prepared as a stock solution (10 mmol/l) in dimethyl sulfoxide and diluted in PBS (100 μmol/l).
After adding prewarmed lucigenin (0.1 ml) or DMNH (0.2 ml) to the samples, the cells were stimulated by application of 30 μl Zymosan A (Serva, FRG; 12.5 mg/ml). CL-generation was determined in triplicate

and the light emission was continously recorded over 30 minutes.
Values are means of triplicate samples from 2 - 3 independent
experiments.

RESULTS AND DISCUSSION

After zymosan-stimulation rat AM exhibited both lucigenin- and DMNH-
dependent CL-activities (Fig. 1-3).
CL-activities of AM cultivated up to 50 h in MEM ±5% FCS are depicted in **Fig. 1**. DMNH-CL was always lower than lucigenin-CL. There
was a steady increase of lucigenin-CL, which was particulary marked
in presence of FCS. After a short initial increase up to 14 h in
presence of FCS DMNH-CL progressively decreased until the end of the
cultivation period.
In **Fig. 2** CL-activities of AM cultivated up to 10 days are
illustrated. Every 24 h the medium of the monolayers was changed for
fresh medium and 24 h after the last medium change CL-activities
were measured. In presence of FCS, during the first 5 days lucigenin-CL dramatically increased. In contrast, there was a marked
decrease in DMNH-CL. In absence of FCS, only a decrease in DMNH-CL
was observed, whereas lucigenin values were relatively constant.

CL-measurements amplified by lucigenin and DMNH detect the generation of different reactive oxygen species (ROS) in phagocytes (2).
The acridinium salt lucigenin reacts with superoxide anion ($^{\cdot}O_2^-$),
the first primary reactive oxygen species generated by the phagocyte
specific membrane-bound NADPH-oxidase. The lucigenin-CL has been
shown to be very specific for $^{\cdot}O_2^-$ (3). DMNH is a derivative of the
commonly used CL-amplifier luminol. Comperative studies have shown
DMNH is less cytotoxic and more effective in measuring CL of
granulocytes and mononuclear phagocytes than luminol (4). Comparable
with luminol, DMNH as a cyclic hydrazide mainly detects the
metabolites of $^{\cdot}O_2^-$ generated by the phagocytic peroxidase/ myeloper-
oxidase-activities (2).
The marked changes of the CL-activities during the cultivation
period might be explained by a further "maturation process" of the
isolated macrophage population. In vivo, there is a continous
replenishment of the resident AM population by invading mononuclear
cells of monocytic origin. Monocytes possess a cyclic hydrazide
(luminol or DMNH) sensitive peroxidase system, which is lost during
in vitro maturation to macrophages (1). However, the $^{\cdot}O_2^-$-sensitive
lucigenin-CL is enhanced, if the $^{\cdot}O_2^-$-metabolizing peroxidase system
is diminished. In vitro, the replenishment by monocytes is fore-
stalled, allowing the maturation process to macrophages to predominate, which results in a loss of DMNH-CL and an increase of lucigenin-CL. The dramatic rise in lucigenin-CL in presence of FCS cannot
however be totally explained by the loss of the peroxidase system.
An explanation for this phenomenon might be, that macrophages are
primed by incubation in presence of FCS (5). Priming of phagocytes
results in an enhanced $^{\cdot}O_2^-$-generation after stimulation, realized in
our experiments as the increased lucigenin-CL.

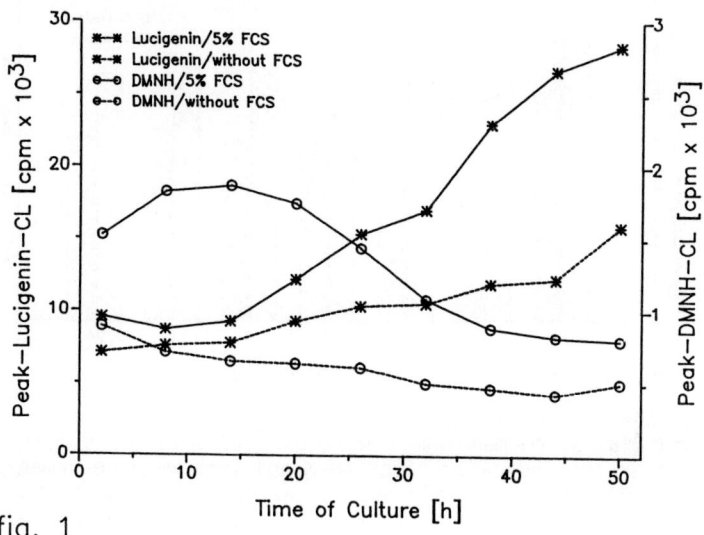

fig. 1

Fig. 1: CL-activities of AM after zymosan-stimulation cultivated up to 50 h

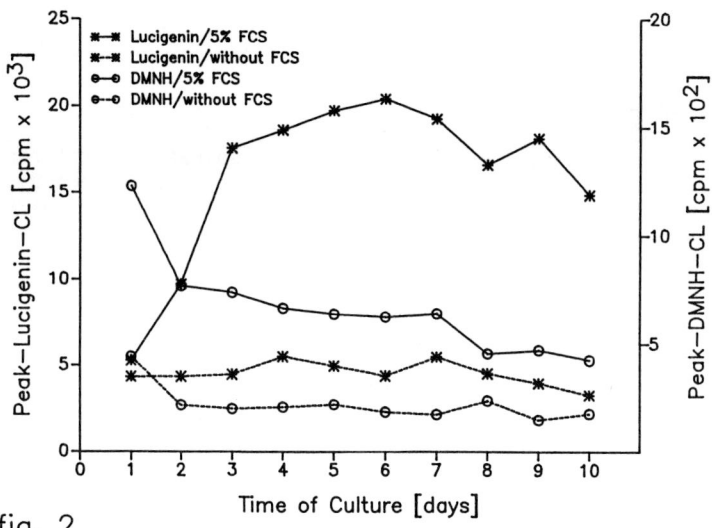

fig. 2

Fig. 2: CL-activities of AM after zymosan-stimulation cultivated up to 10 days

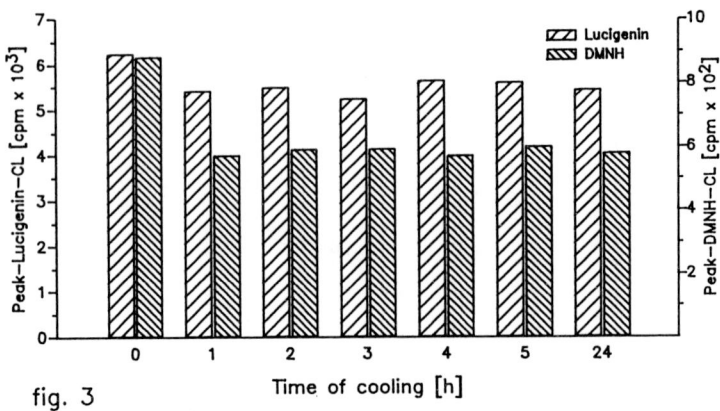

fig. 3

The CL-changes are inhibitable by cooling the cells to 4°C as depicted in **Fig. 3**. AM-monolayers were exposed after 20 h cultivation at 4°C. After the indicated times AM-monolayers were rewarmed at 38°C/-5% CO_2/100% AF for 90 min and CL was measured. There was in fact a slight decrease of CL-activities due to the cooling procedure, the CL values of the cooled samples remaining constant with up to 24 h cooling. This makes the AM-monolayers convenient for comparable CL-measurements over a whole working day despite a limited capacity of the luminometer.

ACKNOWLEDGEMENTS
We thank Sara L. Whitfield for help in preparation of the manuscript.

REFERENCES
1. U. Jacobshagen and R. Andreesen, Respiratory burst formation by human macrophages at different stages of maturation: dissociation of the generation of particular oxygen radicals. In: <u>Bioluminescence and Chemiluminescence</u>, J. Scholmerich et al. (Eds), Wiley, Chichester, p. 1-9 (1987)
2. R. C. Allen, Biochemiexcitation: chemiluminescence and the study of biological oxygenation reactions. In: <u>Chemical and biological generation of excited states</u>, W. Adam and P. Cilento (Eds), Academic Press, New York p.309-344 (1982).
3. H. Gyllenhammar, Lucigenin chemiluminescence in the assessment of neutrophil superoxide production. <u>J. Immunol. Meth. 97</u>, 209-213 (1987).
4. C. Eschenbach and U. Adrian, DMNH - a new sensitive indicator for measuring the chemiluminescence of granulocytes and monocytes. <u>Klin. Wochenschr. 63</u>, 1218-1225 (1985).
5 H. Hayakawa, K. Umehara and Q. N. Myrvik, Oxidative responses of rabbit alveolar macrophages: comparative priming activities of MIF/MAF, sera, and serum components. <u>J. Leuk. Biol. 45</u>, 231-238 (1989).

AN ELECTRONIC DATABASE OF 11,000 ENTRIES DEDICATED TO BIOLUMINESCENCE AND CHEMILUMINESCENCE REFERENCES

P. Stanley

Cambridge Research & Technology Transfer Limited,
48 Glisson Road, Cambridge CB1 2HF, U.K.

The scientific literature concerned with bioluminescence (BL) and chemiluminescence (CL) is widely dispersed through the disciplines and journals because of the wide range of topics encompassed within these headings. The following examples will suffice for illustration:- synthesis of luciferins; *lux* reporter genes and their applications; luminescence imaging; BL organisms in the sea; ATP and rapid microbiology; CL of phagocytosis; dioxetanes and blotting; CL immunoassay, etc..

This dispersal has retarded transfer of information and technologies between workers. Abstracting services and electronic databases can play some part in finding relevant papers but inevitably searching them effectively is quite a skilled process and can be expensive. Since the mid-1960's this author has been accumulating references on all aspects of BL and CL and previously had them in a database on a large computer installation. About 5 years ago they were transferred to a IBM-compatible PC and processed by a scientific references database (Paperbase DeLuxe, Wight Scientific, London, U.K.). Unlike most other products which require the use of a fixed amount of disc space for each reference (whether you use it or not) this programme uses only the amount of space required for each reference. This has meant that the 11,000 references are stored on less than 3 Mbytes of disc space. The disadvantage of the method is that for each search all references must be individually scanned and with a 386-25MHz machine the 11,000 references take approximately 70 seconds to process whereas with the fixed space per reference and the associated required indices the searching may take only a few seconds but in this latter format the penalty is that it may take some 10 Mbytes or more to store the 11,000 references and indices.

Each reference is stored with all authors (including their initials), the full title, journal and its volume, first and last pages of the paper and keywords. Up to 80 authors and 80 keywords may be entered per reference. Additionally you may add comments and or an abstract although long abstracts can be stored on a separate file to improve searching speed.

There are several modes for searching and these include by author name, keywords (including words in the title), year and journal. References which do *not* have a particular author, keyword, year or journal may be selected. Parts of words as well as wildcards may be used. Recently the programme has been enhanced so that a composite index of all authors, keywords, years and journals can be prepared and this makes searching far quicker although only whole indexed words can be sought. In addition searching in this mode does need a fully up to date index.

Of particular importance in developing and using this database has been the choice of keywords. This has proved to be crucial in terms of retrieving just the required references. Some keywords are collective in the sense that they permit retrieval of a broad range of references, e.g. *review* and others are generic, e.g. *luciferase* whilst others more specific, e.g.

Renilla BL studies. I have chosen only to use singular nouns where possible and also synthetic words, e.g. *nathist* for natural history aspects of BL. In addition I now add new keywords only after a great deal of consideration as problems of synonymy can prevent one from retrieving wanted references.

In Table 1 a list of numbers of references is presented.

KEYWORD(S)	Number
Phagocytosis	1150
Bacterial & Luciferase	680
Firefly & Luciferase	620
Chemiluminescence & Immunoassay	520
ATP & Rapid Microbiology	390
Lux or *Luc* genes	300
Aequorea or aequorin	210
Luminol (not phagocytosis studies)	140
Luminometers	100
Renilla	70
Lucigenin	60

TABLE 1. Number of references retrieved from database for keywords and combination of keywords. Note **&** means both keywords must be present; **or** means either keyword must be present; **not** means keyword must be absent.

The references have been gleaned from many sources including personally communicated information, reviews, *Medline* and *Chemical Abstracts*. What I would like to stress is that the title of a paper does not necessarily give one the clue that it is a relevant candidate for the BL and CL database. Reading the paper is ideal but the abstract if often sufficient however the former is often necessary when the paper contains a new method or special application involving BL or CL, e.g. use of luciferase reporter genes which is not mentioned in the title or abstract. As far as other workers are concerned such a reference may be very relevant.

Use of this database is made when preparing the Literature Search which appears in most issues of the *Journal of Bioluminescence and Chemiluminescence*.

Searches of the database are available on a commercial basis.

REFERENCE
Stanley. P.E. (1989) The scientific literature concerning bioluminescence and chemiluminescence. The use of a specialised database. In: *ATP Luminescence: Rapid Methods in Microbiology*, Stanley, P.E., McCarthy, B.J. and Smither, R. (Eds.), Blackwell, Oxford, Society for Applied Bacteriology, Technical Series Vol. 26, pp. 31-36.

ANALYSES OF DRUG BINDING TO SERUM MACROMOLECULES USING A BIOTIN-STREPTAVIDIN ENHANCED CHEMILUMINESCENCE WESTERN BLOT TECHNIQUE

H.V. SMITH, T.M. MAYAMBO[1], E.M. DUNLOP[1] AND P.H. HOLMES[1]

*Scottish Parasite Diagnostic Laboratory, Stobhill General Hospital, Glasgow, G21 3UW, UK and [1]
Department of Veterinary Physiology, University of Glasgow Veterinary School, Glasgow, G61, UK*

INTRODUCTION

African trypanosomiasis, caused by protozoan hæmoparasites of the genus *Trypanosoma*, is commonly a chronic disease of man and animals, and is transmitted by tsetse flies. Animal trypanosomiasis is a major economic handicap to the African continent, denying its ever-growing population much needed food and livestock products. It is estimated that the development of livestock and agriculture in tsetse infected areas could generate a further US $50 billion, annually (1). Chemotherapy and chemoprophylaxis are of particular importance as practical means of combatting trypanosomiasis of domestic animals, and over 25 million treatments are carried out annually in Africa, the majority being for the treatment and control of bovine trypanosomiasis. Isometamidium chloride (Samorin[R], RMB Animal Health Ltd.) administered intramuscularly at a dose of 2 mg/kg body weight for prophylaxis is currently the only available drug with both curative and prophylactic activity against salivarian trypanosomes in domestic animals (2). The precise duration of prophylaxis with isometamidium chloride is multifactorial and varies widely, and the major constraint in studying chemoprophylaxis, and hence the possible significance of drug resistance, has been the lack of suitable sensitive tests for the detection of low levels of this drug in serum and tissues. There are three major problems in the detection of isometamidium chloride in biological tissues and fluids, namely:- a) low drug concentrations in the plasma of treated animals, b) its apparently extensive binding to plasma and tissue macromolecules, and c) its instability in extreme pH conditions and high temperatures (2). Recently a sensitive competition ELISA, capable of detecting levels of 5-10 pg/ml has been developed (3). The aim of this communication is to address the phenomenon of extensive drug binding to serum macromolecules in treated animals by separating serum proteins on SDS-PAGE, and analysing the separated proteins for their ability to bind isometamidium by probing Western blots with a biotinylated polyclonal antibody to isometamidium, and detecting the binding of that antibody with streptavidin-peroxidase and an enhanced chemiluminescence detection system. In addition, we have developed a dot blot assay for the rapid semi-quantitative assessment of isometamidium in the sera of treated cattle which, with further refinement, could be employed in field studies.

MATERIALS AND METHODS

Materials Dimethyl sulfoxide (DMSO); luminol; NHS-biotin; streptavidin conjugated to horseradish peroxidase (STR-HRP) were all obtained from the Sigma Chemical Company (UK) Ltd. The 4-iodophenol was obtained from the Aldrich Chemical Company (UK) Ltd.

Biotinylated, hyperimmune sheep anti-isometamidium IgG Hyperimmune sheep anti-isometamidium serum was raised by multiple sub-cutaneous injections of isometamidium conjugated to porcine thyroglobulin emulsified sequentially in Freund's complete and incomplete adjuvant (3). The IgG fraction, following salt fractionation and column chromatography, was adjusted to 1 mg/ml and biotinylated according to standard procedures.

Samples Isometamidium conjugated to porcine thyroglobulin (PTG, 4 mg/ml), or egg albumin (EA, 4 mg/ml), EA and PTG alone, isometamidium spiked bovine serum albumin (BSA, 4 mg/ml), isometamidium spiked normal bovine serum (NBS, 50 µg of isometamidium per ml of serum) and sera from cattle treated with isometamidium (0.5 mg/kg body weight) were analysed for the presence of drug by an enhanced chemiluminescence Western blot system.

In addition, the sera of cattle, treated with 0.5 mg kg^{-1} body weight, administered intramuscularly on day 0, were analysed for the presence of drug by an enhanced chemiluminescence dot blot assay.

SDS-PAGE Samples were boiled in sample buffer and analysed on SDS-PAGE comprising of a 5% stacking and a 10% resolving gel using the discontinuous buffer system of Laemmli (4).

Western blot analyses Samples fractionated on SDS-PAGE, were electrophoretically transferred to pre-wetted Immobilon PVDF Transfer membrane (5), in a Bio-Rad Trans blot-cell at 30 V for 16 h, with cooling coils following which the blots were blocked in PBS containing 0.05% Tween-20 and 5% dried skimmed milk (PBS-T-DSM) for 1 h at room temperature.

Blots were probed with biotinylated anti-isometamidium IgG, diluted optimally in PBS-T-DSM, and isometamidium binding macromolecules were detected following the addition of streptavidin-horseradish peroxidase (STR-HRP), by placing the blots in 1 µg/ml STR-HRP in PBS-T-DSM for 30 min at room temperature, then washing the blots five times in PBS prior to the addition of the chemiluminescence detection system. All procedures were performed with constant shaking. The relative mobility of each band detected was calculated.

Dot blot analyses Five µl serum samples from isometamidium treated and untreated cattle, serially diluted, were spotted onto a pre-wetted PVDF membrane and air dried at room temperature for 30 min. The membrane was blocked in PBS-T-DSM, probed with biotinylated anti-isometamidium IgG, STR-HRP, and the enhanced chemiluminescence detection system.

Enhanced chemiluminescence detection system for horseradish peroxidase conjugates The enhanced chemiluminescence detection system described by Schneppenheim and Rautenberg (6) was used, with modifications. This was prepared by adding 40 mg luminol to 100 ml of 150 mM NaCl in 50 mM Tris-HCl, pH 8.0, to which was added 10 mg 4-iodophenol dissolved in 1 ml DMSO. Immediately before use 0.01% H_2O_2 (32 µl of 30%) was added.

Blots probed with STR-HRP were immersed in the chemiluminescence detection system for 1 min, excess solution drained off, and the wet blot placed either in a self sealing translucent polythene bag, or wrapped in a translucent polythene sheet and exposed to Dupont Cronex X-ray film for varying lengths of time from 10 s to 10 min depending on the intensity of the signal.

RESULTS AND DISCUSSION
Western blot analyses of isometamidium-protein conjugate
The specificity of the biotinylated purified antibody for isometamidium, and not the carrier protein (PTG) was demonstrated by the presence of bands in the lanes in which isometamidium-conjugated PTG and EA were electrophoresed, and the absence of reactivity in the lanes in which PTG or EA alone were electrophoresed (Fig. 1). Isometamidium remained conjugated to its carrier protein as witnessed by the relative mobilities of the developed bands, EA having a lower molecular weight than PTG, therefore migrating further from the gel origin. No drug-conjugated breakdown product of either conjugate was detectable. In addition, the separation and detection procedures used appear not to have affected the integrity of isometamidium specific epitope(s).

Detection of isometamidium in treated cattle
Macromolecule-bound isometamidium was detected in the sera of treated cattle on both occasions tested (2 and 16 days post treatment), with a maximum of one major (c. 140 kD) and two minor bands being detected. Previous work on the binding of isometamidium to macromolecules in rat serum implicated serum albumin, DNA, RNA, heparin and hyaluronic acid (7) and our inital observations suggested that the major band in cattle serum might also be albumin. In order to determine whether albumin is the major serum binding macromolecule, both BSA and NBS were spiked with isometamidium, electrophoresed, blotted and developed for the presence of drug. Only one isometamidium positive band (c. 140 kD) was observed in either BSA or NBS spiked samples (Fig. 2). No band was detectable when unspiked BSA or NBS were tested. These preliminary results are suggestive of the fact that the major drug-binding serum macromolecule in treated cattle has an apparent molecular weight of c. 140 kD and could be a dimer of albumin.

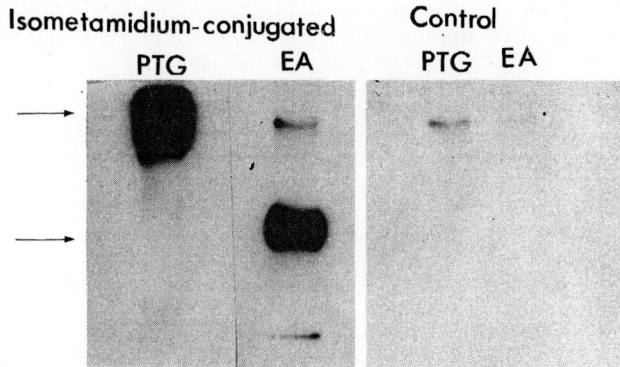

Figure 1. Enhanced chemiluminescence Western blot for the detection of isometamidium.
Lane 1. Isometamidium conjugated to porcine thyroglobulin. Lane 2. Isometamidium conjugated
to egg albumin. Lane 3. Porcine thyroglobulin. Lane 4. Egg albumin. Arrows indicate
isometamidium positive bands.

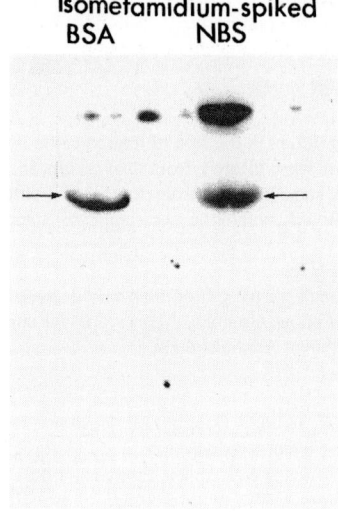

Figure 2. Enhanced chemiluminescence Western blot for the detection of the major isometamidium-
binding serum macromolecule. Lane 1. Isometamidium-spiked bovine serum albumin.
Lane 2. Isometamidium-spiked normal bovine serum. Arrows indicate isometamidium positive
band, with c. 140 kD apparent molecular weight, in each spiked sample .

Dot blot analyses of sera from treated cattle
Isometamidium was detectable in the sera of treated cattle by dot blot up to a dilution of 1:128
(rows A, B, C, D, E and H) but not in the sera of untreated cattle (rows F, G and I; Fig. 3). A
standard curve (row J) indicated that the end-point for detection of isometamidium by dot blot was
equivalent to 50 ng/ml of isometamidium (1:16 dilution of sample). Should such a level be of
clinical significance over time, the ease of use and portability of a dot blot assay should enhance its
field applicability. Alternatively, it may be possible to transport lysed blood samples from treated
cattle on PVDF membranes without affecting the integrity of the isometamidium epitope(s), for
their subsequent analyses, at a central facility.

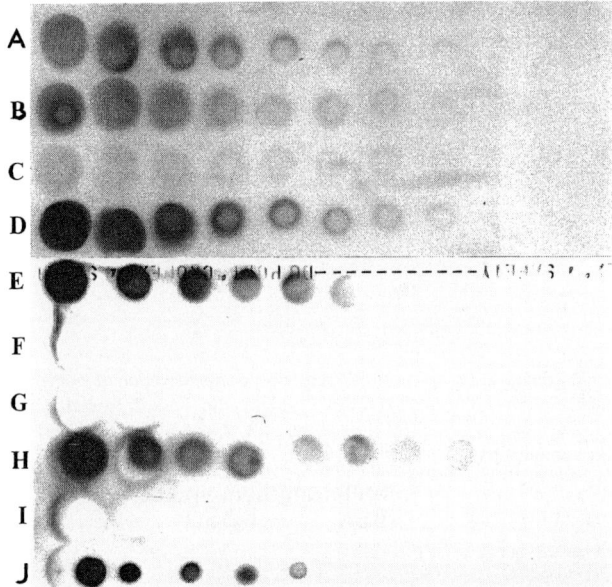

Figure 3. Detection of isometamidium in the sera of treated cattle by enhanced chemiluminescence dot blot technique. All samples were diluted, from neat serum, in a series of doubling dilutions. Rows A, B, C, D, E and H are samples from isometamidium treated cattle. Rows F, G and I are samples from untreated cattle. Row J contains an isometamidium standard.

ACKNOWLEDGEMENTS
We are grateful to Mrs. E.A. Gault for the gift of the purified sheep anti-isometamidium IgG. T.M. Mayambo is a British Council Fellowship holder. This work was supported, in part, by an award to P.H.H. by the Overseas Development Administration.

REFERENCES
1. J.C.M. Trail, K. Sones, J.M.C. Jibbo, J. Durkin, D.E. Light and M.Murray, Productivity of Boran cattle maintained by prophylaxis under trypanosomiasis. ILCA Research Report No. 9, International Livestock Research Centre for Africa, Addis Ababa, Ethiopia, (1985).
2. L.D.B. Kinabo and J.A. Bogan, The pharmacology of isometamidium. J. Vet. Pharmacol. Therapeutics 11, 233-245 (1988).
3. D.D. Whitelaw, E.A. Gault, P.H. Holmes, I.A. Sutherland, F.J. Rowell, A. Phillips and G.M. Urquhart, Development of an enzyme-linked immunosorbent assay for the detection and measurement of the trypanocidal drug isometamidium chloride (Samorin) in cattle. Res. Vet. Sci. In press (1990).
4. U.K. Laemmli, Cleavage of structural proteins during assembly of the head of bacteriophage T4. Nature 227, 680-685 (1970).
5. M.G. Pluskal, M.B. Przekop, M.R. Kavonian, C. Vecoli, and D.A. Hicks, Immobilon^{TM} PVDF transfer membrane: a new membrane substrate for western blotting of proteins. Biotechniques 4, 272-283 (1986).
6. R. Schneppenheim and P. Rautenberg, A luminescence western blot with enhanced sensitivity for antibodies to human immunodeficiency virus. Eur.J.Clin.Microbiol. 6, 49-51 (1988).
7. F.S. Phillips, S.S. Sternberg, A.P. Cronin, S.A. Sodergren and P.M. Vidal, Physiologic disposition and intracellular localisation of isometamidium. Cancer Res. 27, 339-349 (1967).

DETECTION OF SURFACE ASSOCIATED MOLECULES ON PARASITES USING A BIOTIN-STREPTAVIDIN ENHANCED CHEMILUMINESCENCE WESTERN BLOT TECHNIQUE: ANALYSES OF TROPHOZOITES OF *GIARDIA INTESTINALIS*

H.V. SMITH AND A.T. CAMPBELL

Scottish Parasite Diagnostic Laboratory, Stobhill General Hospital, Glasgow, G21 3UW, UK

INTRODUCTION

The protozoan parasite *Giardia intestinalis* is numerically the most important agent of parasitic diarrhoeal disease worldwide. It exists in two morphologically distinct forms, namely: a motile flagellated trophozoite which colonises the proximal small intestine and causes the disease state; and the infective form, a cyst, which is excreted in faeces into the environment (1). The trophozoite attaches onto enterocytes by means of a ventral disc and/or lectins expressed on its surface membrane. In addition, immunological interactions between the host and the parasite at this interface are thought to be important in controlling disease. For these reasons, a biochemical and immunochemical insight into surface expressed molecules of *G. intestinalis* trophozoites should lead to a clearer understanding of host-parasite interactions manifest at the host-parasite interface. Previous work in determining such parameters has relied on the extrinsic and/or intrinsic radiolabelling of parasite surfaces or detergent extracted surfaces followed by molecular sieving on SDS-PAGE gels and autoradiography to assess membrane composition. Whereas ^{125}I has been used extensively and successfully in parasitological research in the developed world, its use in underdeveloped countries has met with less success, partly because of its relatively short half-life, and partly because of import constraints. That a readily available, robust, sensitive non-radioiostopic assay is developed for such countries seems fundamental when considering that it is in such countries, where parasitic disease often is endemic, that research should be undertaken.

Numerous chemiluminescent and bioluminescent systems are available which enable the rapid quantitation of the widely used enzyme horseradish peroxidase (HRP) (E.C.1.11.1.7) (2), and which are far more sensitive than their colorimetric counterparts.

Herein, we describe a simple and sensitive method for the non-radioisotopic labelling and detection of surface-exposed parasite components which involves the covalent linking of biotin molecules to primary amines expressed on the surface of living *G.intestinalis* trophozoites, and detection of the products following SDS-PAGE and Western blotting using an enhanced chemiluminescence detection system.

MATERIALS AND METHODS

Materials Dimethyl sulfoxide (DMSO); D-biotin; luminol; streptavidin conjugated to horseradish peroxidase (STR-HRP), and streptavidin-FITC were all obtained from the Sigma Chemical Company (UK) Ltd. Sulfo-NHS-biotin was obtained from Pierce (UK) Ltd. The 4-iodophenol was obtained from the Aldrich Chemical Company (UK) Ltd.

Parasites *G.intestinalis* trophozoites (VNB1 isolate), were grown in axenic culture in TYI-S-33 medium, according to the method of Keister (3), and trophozoites in late log phase were harvested by chilling the culture tube in ice for 15 min, centrifuged at 800g for 5 min and washed three times in 10 mM phosphate-buffered saline, pH 7.4 (PBS) before being used. The viability of all trophozoite preparations used for surface labelling was at least 95% as determined both by the exclusion of 0.2% trypan blue, and motility.

Biotinylation of the parasite outer surface Surface expressed primary amines were labelled by incubating 10^8/ml washed trophozoites with 0.2 mg/ml sulfo-NHS-biotin in PBS for 30 min at 4^oC on a rotary shaker. The trophozoites were then transferred to a 15 ml conical tube and washed five times with cold PBS to remove any unreacted sulfo-NHS-biotin.

FITC labelling of biotinylated moieties expressed on the surface of living trophozoites A proportion of the biotinylated trophozoites obtained from the above procedures was incubated at 4^oC, for 20 min with streptavidin-FITC, at a concentration of 20 μg/ml in PBS, washed three times in PBS before being examined under a Zeiss epifluorescence microscope, in order to verify that the outer surfaces of living trophozoites had been biotinylated.

Detergent solubilisation of biotin labelled parasites Biotinylated trophozoites, at 10^8/ml, were resuspended in PBS containing 0.5% Triton-X-100, and incubated on crushed ice for 30 min before being centrifuged at 13,000 rpm for 10 min. The supernatant was used immediately or stored at -20^oC until used.

SDS-PAGE Triton X100 solubilized extracts were boiled in sample buffer and analysed on SDS-PAGE comprising of a 5% stacking and a 10% resolving gel using the discontinuous buffer system of Laemmli (4).

Western blot analyses of biotinylated surfaces Detergent-solubilised biotinylated trophozoite extracts fractionated on SDS-PAGE, were electrophoretically transferred to pre-wetted Immobilon PVDF Transfer membrane (5), in a Bio-Rad Trans blot-cell at 30 V for 16 h, with cooling coils following which the blots were blocked in PBS containing 0.05% Tween-20 and 5% dried skimmed milk (PBS-T-DSM) for 1 h at room temperature.

Biotin labelled components were detected following the addition of streptavidin-horseradish peroxidase (STR-HRP), by placing the blots in 1 μg/ml STR-HRP in PBS-T-DSM for 30 min at room temperature, then washing the blots five times in PBS prior to the addition of the chemiluminescence detection system. All procedures were performed with constant shaking.

Chemiluminescence detection system for horseradish peroxidase conjugates The enhanced chemiluminescence detection system described by Schneppenheim and Rautenberg (6) was used, with modifications. This was prepared by adding 40 mg luminol to 100 ml of 150 mM NaCl in 50 mM Tris-HCl, pH 8.0, to which was added 10 mg 4-iodophenol dissolved in 1 ml DMSO. Immediately before use 0.01% H_2O_2 (32 μl of 30%) was added.

Blots probed with STR-HRP were immersed in the chemiluminescence detection system for 1 min, excess solution drained off, and the wet blot placed either in a self sealing translucent polythene bag, or wrapped in a translucent polythene sheet and exposed to Dupont Cronex X-ray film for varying lengths of time from 10 s to 10 min depending on the intensity of the signal.

RESULTS AND DISCUSSION
Viability of trophozoites and localisation of biotin on living parasites

The viability of trophozoites following biotinylation was similar to that at the commencement of the experiments (>95%), and this procedure did not interfere with their growth curves when returned to *in vitro* culture. Surface fluorescence on living trophozoites was demonstrable only with sulfo-NHS-biotin treated trophozoites following the addition of streptavidin-FITC, indicating the successful biotinylation of living parasites. The biotin labelling method used in this study yielded stable compounds on viable parasites, which were demonstrable as surface associated fluorescence following the addition of streptavidin-FITC, indicating that the reactive biotin was surface accessible.

Western blot analyses of detergent solubilised trophozoite surfaces labelled with biotin.

Typically, a total of 15 discrete bands could be detected on a luminogram using the system described (Fig. 1), and this compares favourably with the previously published work of Clark and Holberton (7) who used the Bolton and Hunter reagent for ^{125}I labelling *Giardia* surfaces. Both these techniques utilise the reactive N-hydroxysuccinimide ester to link the label to target molecules. In addition, these workers reported the presence of minor components between 100 and 200 kD which were evident only after long exposures of autoradiographs, whereas in this study, these higher molecular weight bands were amongst the major bands detected using the short exposure times stated. Blotted unbiotinylated detergent extracted trophozoite surfaces did not react with STR-HRP, indicating that the banding pattern observed on X-ray film with the BEC system was the consequence of the selective binding of STR-HRP to biotin groups introduced into the trophozoite surface. In addition, the chemiluminescent reaction could be abolished by the addition of 0.1 mM D-biotin to the streptavidin reaction step (Fig. 1). Reproducible results of surface profiles can be obtained with 5×10^5 trophozoites loaded onto a gel, blotted and visualised in under 30 sec, which in our experience is less than the numbers necessary to provide similar data using a comparable radio-isotopic technique.

A further communication will address the direct comparison between the BEC system and ^{125}I labelling of *Giardia* trophozoite surfaces using the Bolton and Hunter reagent, as well as the use of the hydrazide derivative of biotin for extrinsically labelling surface exposed carbohydrates (Campbell and Smith, submitted).

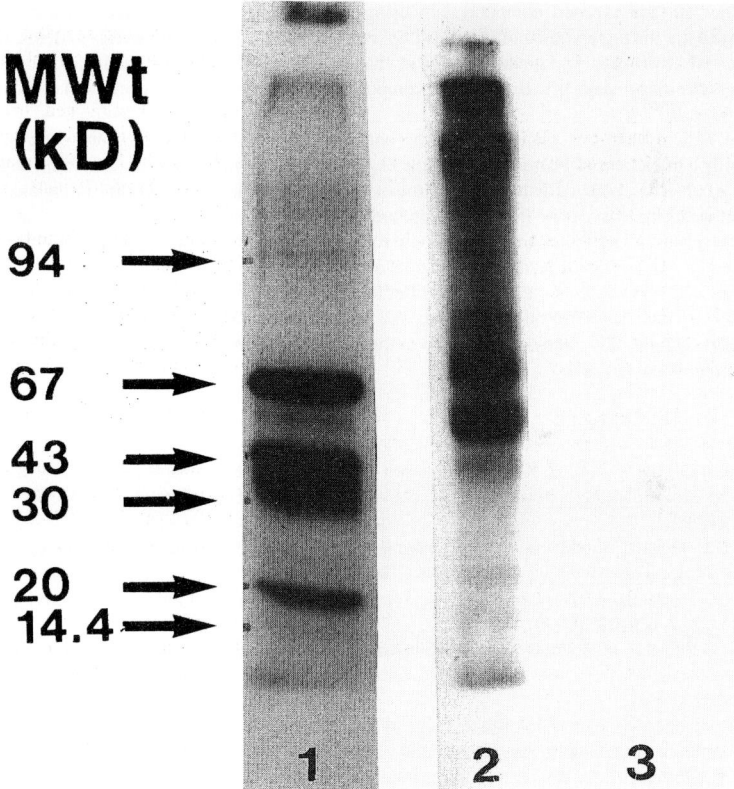

Figure 1. Enhanced chemiluminescence Western blot analyses of sulfo-NHS-biotin labelled *G. intestinalis* trophozoites.

Lane 1. Biotinylated protein standards (BIO-RAD).

Lane 2. BEC electroblot demonstrating sulfo-NHS-biotin labelled trophozoite surface components (1 min exposure).

Lane 3. BEC electroblot of sulfo-NHS-biotin labelled surfaces incubated with 0.1 mM D-biotin during the streptavidin-HRP reaction step. Relative mobilities are given in kilodaltons (kD).

We have described the use a non-radioisotopic system for the analyses of surface exposed molecules of *G. intestinalis*, which involves the biotinylation and detergent extraction of the surface associated molecules on living parasites, their separation and transfer to a teflon based membrane following SDS-PAGE electrophoresis and Western blotting. The detection system consists of streptavidin-horseradish peroxidase which, when diluted in a blocking reagent (PBS-T-DSM), binds specifically to the introduced biotin groups present on the separated transferred molecules bound on the transfer membrane. Following the addition of the enhanced chemiluminescence reagents, the transfer membrane is sealed in a thin gauge polythene bag and apposed and exposed to X-ray film in a lightproof cassette for varying time periods from 1 s to 10 min, depending upon the signal strength. We call this system the biotin-enhanced chemiluminescence (BEC) system.

In the present study we have expanded upon previous work for introducing specific labels into various surface exposed components by using an enhanced chemiluminescent detection system coupled to the high binding affinity of streptavidin to biotin. Such luminescent systems have the potential to provide even greater sensitivity than ^{125}I based assays and have the additional benefit of a much longer shelf life. Biotinylated compounds have few handling hazards and do not decay if stored at -20°C.

The BEC system has many applications and advantages. Exposure times are extremely short, usually a maximum of 10 min and multiple exposures can be made from the same blot without the use of unstable, potentially harmful radio-isotopes. In addition, *G. intestinalis* biotinylated surface expressed antigens can be immunoprecipitated and analysed by the BEC system with at least the same sensitivity attributed to radio-isotopic surface labelling methods (unpublished observations). Recently, we have expanded our observations using the BEC system, and have found it applicable for the analyses of the external surfaces of other parasites such as *Cryptosporidium parvum* oocysts and *Toxocara canis* second stage larvae, and their excretions/secretions, indicating that the BEC system should find many applications in the fields of human and veterinary infectious disease research and diagnostics.

REFERENCES

1. M.S. Wolfe, Current concepts in parasitology. Giardiasis. N.Engl.J. Med. 298, 319-321 (1987).
2. G.H.G Thorpe and L.J. Kricka, Enhanced chemiluminescent reactions catalysed by horseradish peroxidase. In: Methods in enzymology, (M.A Deluca and W.D. McElroy Eds), 133, part B, p. 331-353 (1986).
3. D.B. Keister, Axenic culture of *Giardia lamblia* in TYI-S-33 medium supplemented with bile. Trans.R.Soc.Trop.Med.Hyg. 77, 487-488 (1983).
4. U.K. Laemmli, Cleavage of structural proteins during assembly of the head of bacteriophage T4. Nature 227, 680-685 (1970).
5. M.G. Pluskal, M.B. Przekop, M.R. Kavonian, C. Vecoli, and D.A. Hicks, Immobilon^{TM} PVDF transfer membrane: a new membrane substrate for western blotting of proteins. Biotechniques 4, 272-283 (1986).
6. R. Schneppenheim and P. Rautenberg, A luminescence western blot with enhanced sensitivity for antibodies to human immunodeficiency virus. Eur.J.Clin.Microbiol. 6, 49-51 (1988).
7. J.T. Clark and D.V. Holberton, Plasma membrane isolated from *Giardia lamblia* : identification of membrane proteins. Eur.J.Cell Biol. 42, 200-206 (1986).

BIOLUMINESCENCE AND IMAGING

QUANTITATIVE PHOTON IMAGING IN THE LIFE SCIENCES USING INTENSIFIED CCD CAMERAS

Claire E. Hooper[1] and Richard E. Ansorge[2]

[1]*Robens Institute, University of Surrey, Guildford, Surrey GU2 5XH, UK*
[2]*Cavendish Laboratory, University of Cambridge, Madingley Road, Cambridge CB3 0HE, UK*
and Image Research Limited, St John's Innovation Centre, Cowley Road, Cambridge CB4 4WS, UK

INTRODUCTION

Low light level imaging technology first became available in the 1960's. Recent developments have been driven by the needs of astrophysics, high energy physics and night vision applications. Reynolds and co-workers[1] were the first to report the use of low light level imaging in a biomedical application, with the use of an image intensifier coupled to a vidicon camera. The more recent development of CCD-based imaging systems offers considerable advantages over analogue sensors such as the vidicon camera. Modern systems are based on cryogenic CCD cameras or intensified CCD cameras. These cameras offer much improved dynamic range and image quality including geometrical stability and sensitivity. Improvements in image processing hardware and software have increased the application of these systems in the biosciences.

Photon imaging is both simple and versatile. Chemiluminescence- and bioluminescence-based applications may be assessed qualitatively and quantitatively[2], both at the micro level (e.g. single mammalian cell) and at the macro level (e.g. microtitre tray, petri dish). As such it provides a powerful and sensitive tool for investigating processes at the cellular level (reporter genes, intracellular signalling) as well as the simultaneous measurement of multiple samples (e.g. immunoassays and phagocytosis).

EXPERIMENTAL CONSIDERATIONS

A number of important experimental considerations are involved in applying low light-level imaging as a technique for quantitative analysis of luminescent processes in biological systems, these include:

Sample Presentation - Luminescent samples are presented in a light tight box and usually viewed by the camera from above via a lens. Different geometries can be devised if required to view liquid samples from below e.g. using mirrors. Sample size is also an important consideration and the demagnification of the lens can be adjusted to allow the sample to fill the field of view of the camera (see Table 1). It should be noted that imaging systems have finite resolution and typically divide the image into 1/4 million elements i.e. 512 x 512 pixels, irrespective of the demagnification employed. Thus there is a loss of spatial resolution associated with large samples, which is illustrated in Table 1. A further consideration is sample temperature control e.g. for phagocytosis.

Spectral Emission- There is a range of luminescent chemistries, each having different characteristic emission wavelengths, e.g. firefly luciferase at 565 nm, ECL at 450 nm and dioxetane 477 nm. It is important to know the sensitivity of the camera being considered at the wavelength being used.

Table 1

Maximum Sample Area* (mm^2)	Demagnification	Sample Area corresponding to 1 pixel on image (μm^2)
126 x 94.5	9	246 x 185
70.0 x 52.5	5	137 x 103
14.0 x 10.5	1	27.3 x 20.5
1.4 x 1.0	1/10**	2.7 x 2.0

*   Maximum area viewed by detector for a given magnification ratio.
**   This value which is < 1 corresponds to magnification e.g. using a microscope.

Time dependence - The time over which light emission occurs can vary from under 1 second up to days. Most imaging systems are well adapted to quantify samples emitting light steadily over long periods of time. However real-time imaging systems are more suited to following rapidly changing photon fluxes.

Spatial Resolution - The resolving power of a low light-level imaging system is limited by the nature of the detector system, unlike a conventional microscope which is limited by the diffraction of light. In particular the resolving power of a CCD camera cannot be finer than the pixel structure of the CCD sensor. (The relationship between presentation and pixel size is clearly illustrated in Table 1). A point that may be overlooked is that the true resolving power of a pixel-based image sensor corresponds to a separation between features of two pixels, not one pixel; this is an example of the Nyquist sampling theorem[3].

A microtitre tray requires only a very coarse resolving power. A gel, on the other hand, typically requires a resolution of less than 0.5 mm. High resolution applications may benefit from a finer resolution imaging sensor (e.g. 1024 x 1024 elements); such systems are becoming available but are relatively expensive. For imaging over a large field of view, film (still) offers the highest resolution. The most sensitive imaging systems may be able to detect light emission from single point source in a large field of view (e.g. a petri dish), for example the detection of a single light emitting cells in a cellular monolayer may be useful to locate and quantify their light emission. However this is not strictly equivalent to resolving these cells from their neighbours. If detail within a cell needs to be observed, greater magnification with a microscope is certainly required.

Sensitivity - The intensity of light emission from different luminescent chemistries can vary by many orders of magnitude. However if it is desired to detect light from a single cell (or small colony or group of cells) or from very small amounts of analyte, a highly sensitive system is required. In situations where light emission is sustained for a period of time sensitivity can be improved by integrating light for a period of time. The gain in sensitivity is typically proportional to the square root of the integration time for intensified and peltier cooled CCD cameras (because the noise increases like the square root of time) and proportional to the integration time for liquid nitrogen cooled CCD cameras (because the noise is effectively independent of integration time), see Ansorge *et al.* this volume.

Dynamic Range - The dynamic range of an imaging system can be defined in several ways and care is required in comparing different instruments. The interesting number for practical applications is the ratio of the brightest and faintest parts of a field of view that can be detected and quantified within a given exposure. Consider a situation where a field of view contains both bright and very faint features. The faint features will require a longer exposure time than the bright features to obtain statistically valid data. Quantitative information for the bright features may be lost if the instrument suffers from saturation effects. In practice some instruments combine both long and short exposures to overcome this limitation, giving an effective dynamic range in excess of 10^4. A further consideration is that faint features close to bright ones will be harder to detect due to the spreading of light: thus resolution and dynamic range are also inter-related. Spreading of light will also occur due to diffraction and other effects of the optical system used.

IMAGE ACQUISITION AND ANALYSIS

A low light-level camera integrates the signal received from the sample for a period of time and stores the resultant image in digital form in a computer frame store. The frame store will have between 8 and 16 bits of memory defining each pixel of the image. The numbers in the frame store represent the integrated signals for the pixels. The precise interpretation of the numerical data requires a thorough understanding of the operation of the camera.

The digital image held in a frame store can be processed using a wide variety of digital techniques[2] which are generally available in image processing packages. the operations include:-

a) Numerical operations on an image, such as:

- Image arithmetic, e.g. addition, subtraction and overlay (useful for dual wavelength analysis and image comparison);
- Digital filtering, e.g. smoothing, edge finding and feature extraction (for counting and sizing cells and colonies).

b) Display manipulation operations which change the appearance of a digital image rather than the numerical values in the frame store, such as:

- Grey scale manipulation;
- False colour (contrast enhancement);
- Zoom and pan;
- Linear or logarithmic algorithm for mapping of 16 bit data on to standard 256 level (8 bit) display.

QUANTITATIVE IMAGE ANALYSIS

A digital image can be used for quantitative as well as qualitative work. A number of corrections to the values in each pixel are typically required. These corrections include:

- Dark field subtraction i.e. background subtraction of an image obtained with a dark field;
- Bright field correction using image obtained with uniform illumination over field of view;
- Possible geometrical corrections; these arise from optical effects, camera electro-optics and non-linearities in read-out electronics.

The response correction often referred to as 'flat fielding' is normally applied as a precursor to quantitative analysis to account for variations in detector response across the field of view. This correction is intended to allow for the different background level and gains in each element of an imaging detector and may be conveniently done by applying the following operation to each pixel of an image:

$$P_{corr} = S(P_{raw} - D) / (B - D)$$

where P_{corr} is the corrected image pixel value, P_{raw} is the uncorrected pixel value, D is the value of the pixel for a dark field, B is the value of the pixel for a bright field and S is a suitable scale factor typically the average bright field value. The corrected pixel values (P_{corr})

are a direct measure of the number of photons detected by the system per unit area per unit time (i.e. the detected photon flux).   To calculate absolute photon fluxes requires an understanding of the detection efficiency of the system used and loses due to optics and other geometrical effects.

A number of analysis methods may be applied to a corrected image for quantitative work. Essentially all methods involve summing or histogramming the signals in groups of pixels associated with features of interest. The way in which particular groups of pixels are defined depends on the nature of the problem. For example:

Microtitre Tray Analysis - the geometry of the wells is fixed, therefore a circular group of pixels can be assigned to each well and their contents summed to obtain a value for the light emission from each well.

Colonies on Petri Dishes - a threshold technique is used to locate bright spots associated with colonies. The pattern found will in general be random and 'feature extraction' software (either automatic or manually assisted) will be required to find (and hence count) and measure their sizes and levels of light emission. The growth of colonies at different stages can be compared using image superposition; images of a dish at different times may be overlaid with translation and rotation as necessary (Fig. 1).

Blots and Gels - e.g. Western blots. The lanes can be arranged to run horizontally or vertically across the image and a histogram of a horizontal or vertical slice of pixels through each lane will reveal the bands as peaks in the histogram. The areas under such peaks are a valid measure of the light emission from the bands (Fig. 3).

Time Dependent Measurements - images of the same object obtained at different times can be numerically compared to measure reaction kinetics.

Multi-wavelength - image ratio techniques may be used to compare images of the same sample obtained at different wavelengths, eg. dual wavelength fluorescence. Image overlay may be used to compare and contrast images e.g. luminescent and absorbence/transmission images.

APPLICATIONS OF QUANTITATIVE PHOTON IMAGING

Microtitre trays: Enzyme assays[2,4], immunoassays[5], gene probe assays and phagocytosis[6].

Petri dishes: Rapid microbiology - colonies of bacteria e.g.*lux* genes[7,8,9], ATP; Virology and molecular biology - cellular monolayers e.g. *luc*[10,11,12] & *lux*[13] reporter genes.

Gels and blots: 1-D & 2-D gels; Protein and nucleic acid analysis[14] - Western blots and DNA hybridisation techniques e.g. ECL[15,16] and dioxetane[17,18] chemiluminescence.

Microscopy: Tissue antigen analysis - antibody mediated detection of cell-bound antigens[19]. Analysis of cell structure and function [20], using *luc* & *phot* genes [21].

Intracellular signalling - calcium ion fluxes e.g. using aequorin and pholasin [6,21].

Cellular luminescence -   cellular metabolites(glucose and ATP), neutrophils and macrophages[22].

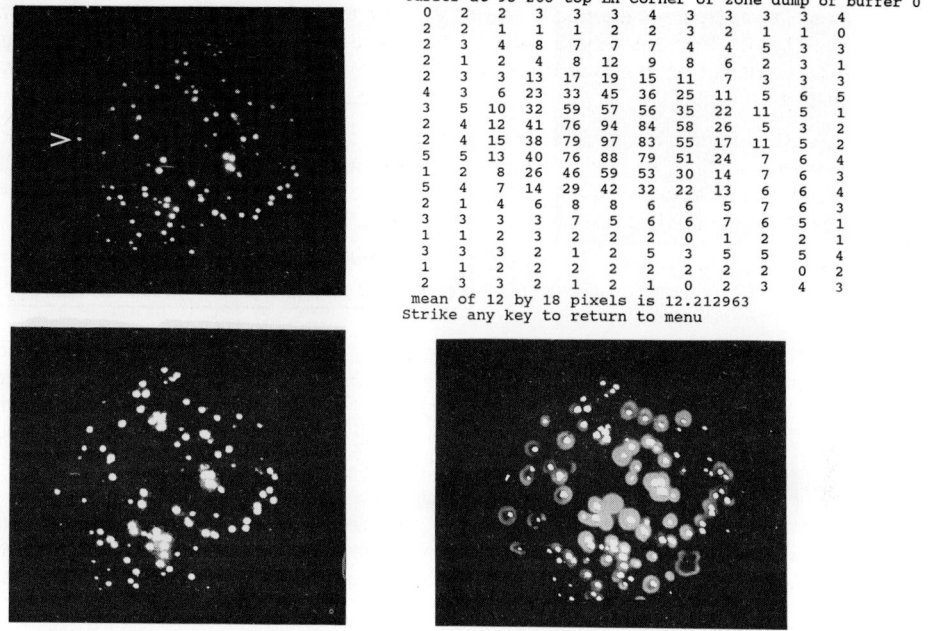

```
Cursor at 95 268 top LH corner of zone dump of buffer 0
 0   2    2    3    3    3    4    3    3    3    3   4
 2   2    1    1    1    2    2    3    2    1    1   0
 2   3    4    8    7    7    7    4    4    5    3   3
 2   1    2    4    8   12    9    8    6    2    3   1
 2   3    3   13   17   19   15   11    7    3    3   3
 4   3    6   23   33   45   36   25   11    5    6   5
 3   5   10   32   59   57   56   35   22   11    5   1
 2   4   12   41   76   94   84   58   26    5    3   2
 2   4   15   38   79   97   83   55   17   11    5   2
 5   5   13   40   76   88   79   51   24    7    6   4
 1   2    8   26   46   59   53   30   14    7    6   3
 5   4    7   14   29   42   32   22   13    6    6   4
 2   1    4    6    8    8    6    6    5    7    6   3
 3   3    3    3    7    5    6    6    7    6    5   1
 1   1    2    3    2    2    2    0    1    2    2   1
 3   3    3    2    1    2    5    3    5    5    5   4
 1   1    2    2    2    2    2    2    2    2    0   2
 2   3    3    2    1    2    1    0    2    3    4   3
mean of 12 by 18 pixels is 12.212963
Strike any key to return to menu
```

Figure. 1   Image of light emission from genetically engineered *E.coli* inoculated onto agar in a 90 mm petri dish. (a) light emission after 8 hours incubation. (b) Matrix of digital pixel values in a small region indicated by marker in (a) which contains one colony. The colony is seen as a cluster of larger numbers in the matrix. (c) light emission after 10 hours incubation. (d) light emission after overnight incubation (large and sometimes confluent grey colonies) with computer superposition of the colonies in image (c) as bright spots.

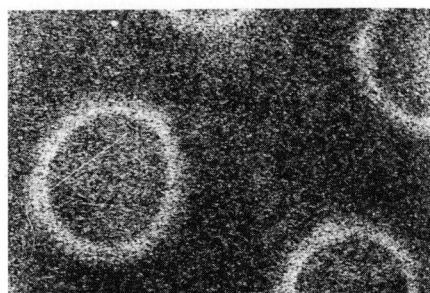

Figure. 2

Image of light emission from genetically engineered *E.coli* suspended in 10 µl drops on the surface of a petri dish. Each drop contains approximately 10 light emitting cells.The sample is viewed directly by placing the dish in contact with the input window of the intensified CCD camera (BIQ, Image Research Ltd, Cambridge). Light emission is concentrated at the edges of the drops due to total internal reflection.

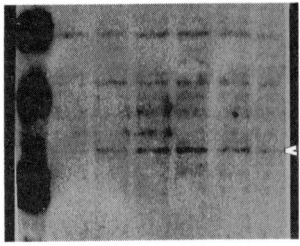
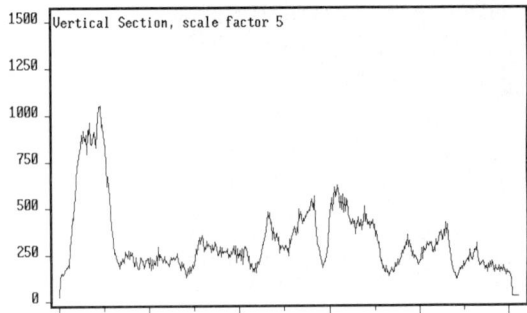

Figure. 3   Visualisation and quantification of specific oncogene product (ras protein) in mouse
               keratinocyte cell lines using ECL western blot analysis. Image of light emission
               integrated for 3 minutes using an intensified CCD imaging system (BIQ). (b) One-
               dimensional histogram of pixel values in a horizontal section 6 pixels wide through
               band corresponding to ras protein. The differing levels of light emission and gaps
               between lanes are clearly seen. (see Jones *et al.* this volume)

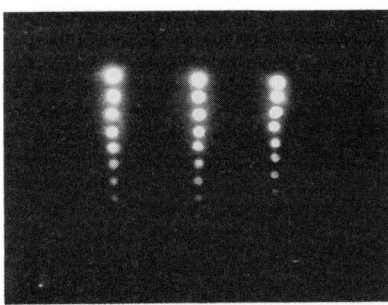
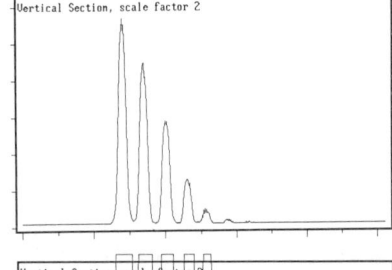
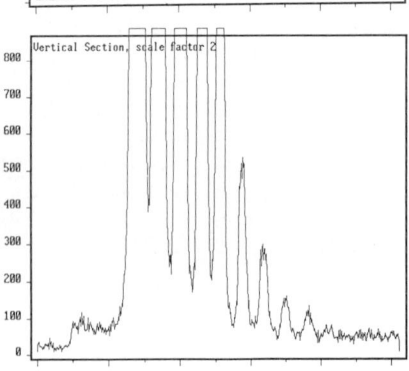

Figure. 4

Image of nylon membrane spotted with 3 series of
dilutions of biotinylated pBR322 DNA (ranging
from 200 pg to 0.4 pg per spot) labelled with
Avidin-Alkaline Phosphatase (image integration time
2 minutes). (a) Image of membrane after activation
with AMPPD substrate (Tropix Inc). (b) One-
dimensional histogram of pixel values in a vertical
section through the central row. (c) as (b) displayed
on an exaggerated scale to show fainter spots more
clearly.

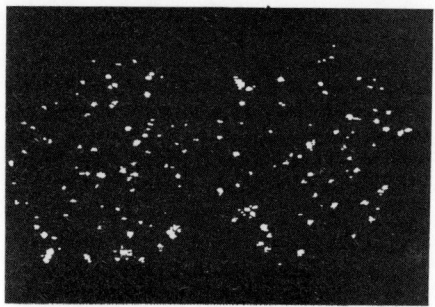 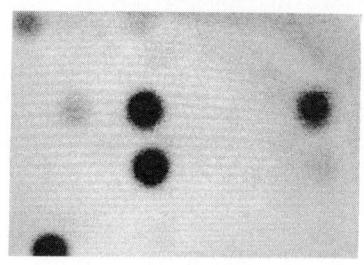

Figure. 5   Analysis of luciferase (*luc*) gene expression in vaccinia recombinant virus infected
                 cell monolayers in 5 cm petri dishes (12 hours post-infection). (a) Image of 2 dishes
                 viewed together using a lens coupled to an intensified CCD camera. Infected cells
                 are seen as white spots. (b) Image of portion of one dish viewed in direct contact
                 with the input window of the intensified CCD camera (shown as reverse video
                 display). Single infected cells are seen as dark regions. The increased sensitivity of
                 'contact' imaging allows cells to be detected at a very early stage of infection, when
                 DNA replication has only just begun (seen as pale grey areas).

DISCUSSION

Photon imaging is an emerging technique for the quantitative analysis of bioluminescence and chemiluminescence (1). Sensitive charge-coupled device (CCD) cameras, capable of detecting individual photons in two-dimensions, are now commercially available. Such systems include cryogenically cooled CCDs and more recently room temperature intensified CCD cameras. Continuing improvements in imaging systems and the clear downward trend in the cost will make the technology more accessible. A prospect for the near future is the availability of higher resolution (1024 x 1024 pixels) systems at affordable prices. It is likely that film will continue to be used for some applications as it offers the advantage of low unit cost and highest resolution. The disadvantages are inconvenience of use, lack of sensitivity and limited dynamic range. Real time intensified CCD imaging systems offer advantages in ease of use and have sensitivities at least as good as cryogenic CCD cameras for exposure times up to 15 minutes.

CCD imaging is a rapid and non-invasive method for the detection of low light-level processes. CCD based detectors have high resolution, large dynamic range and excellent image quality, which make them particularly suited to the investigation of processes and events at the sub-cellular level. Luciferase reporter genes are increasingly being used as specific markers for gene expression both in cell cultures and in experimental animals [10]. Imaging may be applied to the quantitative analysis of luciferase gene expression in single cells[11]. Another important application is the measurement of cell function e.g. intra-cellular signalling and covalent modification of proteins in single cells using bioluminescent markers (*luc* & *phot* genes, A.K.Campbell[6,20]). This new technique, combined with sensitive luminescent procedures, offers improvements and potentials in the fields of cellular and molecular biology and medicine.

ACKNOWLEDGEMENTS
Microbiology: J.Kuhn and S.Ulitzur, Biolume Ltd and Technion-Israel Institute of Technology Haifa,Israel.
Western blots: M.Stanley, P.Jones, I.Greenfield, Dept. of Pathology, University of Cambridge.
DNA blots: I.Bronstein and J.Voyta, Tropix Inc, USA.
Luc gene: H.Browne and A.Minson, Dept. of Pathology, University of Cambridge.
A.K.Campbell, University of Cardiff, U.K. and K.Wood, Promega Inc, U.S.A for useful discussions.

REFERENCES

1.   G.T.Reynolds, Application of photosensitive devices to bioluminescence studies. Photochem. Photobiol., 27, 405-421 (1978).

2.   C.E.Hooper and R.E.Ansorge, Quantitative luminescence imaging in the biosciences using the CCD camera: analysis of macro and micro samples.Trends in Analytical Chemistry, 9, 269-277 (1990).

3.   J.M.Blackledge, Fourier Transforms. In Quantitative Coherent Imaging:Theory, Methods and Some Applications, Academic Press, London. 42-62 (1989).

4.   D.H.Leaback, C.E.Hooper and R.Pirzad, The use of imaging luminometers for the simultaneous assay of multiple ATP samples by means of the firefly luciferase system. In ATP luminescence: Rapid Methods in Microbiology, Technical Series, Society for Applied Bacteriology Vol. 26, P.E.Stanley, B.J.McCarthy and R.Smither (Eds), 277-287 (1989).

5.   F.E.Maly, M.Vittoz, A.Urwyler, K.Koshikawa, L. Schleinkofer and A.L.DeWeck, A dual microtiter plate (192 sample) luminometer employing computer aided single-photon imaging applicable to cellular luminescence and luminescence. J.Immunol.Methods, 122, 91-96 (1989).

6.   A.K.Campbell, Chemiluminescence: Principles and applications in biology and medicine, Ellis Horwood, Chichester,(1989).

7.   S.Almashanu, B.Musafia. R.Hadar, M.Suissa and J.Kuhn, Fusion of luxA and luxB and its expression in E.coli, S.cerevisiae and D.melanogaster. J.Biolumin.Chemilumin., 5, 89-98 (1990).

8.   S.A.Jassim, A.Ellison, S.P.Denyer, G.S.Stewart, In vivo bioluminescence: a cellular reporter for research and industry. J.Biolumin.Chemilumin., 5, 115-122 (1990).

9.   C.M.Miyamoto, E.A.Meighen and A.F.Graham, Transcriptional regulation of lux genes transferred into Vibrio harveyi. J. Bacteriol., 172, 2046-2054 (1990).

10.   D.Rodriguez, J-R. Rodriguez, J.F.Rodriguez, D.Trauber and M.Esteban, Highly attenuated vaccinia virus mutants for the generation of safe recombinant viruses. Proc. Natl. Acad. Sci. USA, 86, 1287-1291 (1989).

11.   C.E.Hooper, R.E.Ansorge, H.M.Browne and P.Tomkins, CCD Imaging of luciferase gene expression in single mammalian cells. J. Biolumin. Chemilumin., 5, 123-130 (1990).

12.   O.Schwartz, J.Virelizier, L.Montagnier and U.Hazan. A microtransfection method using the luciferase-encoded reporter gene for the assay of human immunodeficiency virus LTR promoter activity. Gene, 88, 197-205 (1990).

13.   D.S Peabody, C.L.Andrews, K.W. Escudero, J.H.Devine, T.O.Baldwin and D.G.Bear, A plasmid vector and quantitative techniques for the study of transcriptional termination in E.coli using bacterial luciferase. Gene, 75, 289-296, (1989).

14.   R.Hauber, W.Miska, L.Schleinkofer and R.Geiger, New, sensitive, radioactive-free bioluminescence enhanced detection system in protein blotting and nucleic acid hybridization. J.Biolumin.Chemilumin., 4, 367-372 (1989).

15.   D.Pollard-Knight, C.A.Read, M.J.Downes, L.A.Howard, M.R.Leadbetter, S.A.Pheby, E.McNaughton, A.Syms and M.A.Brady, Nonradioactive nucleic acid detection by enhanced chemiluminescence using probes directly labelled with horseradish peroxidase. Analytical Biochemistry, 185, 84-89 (1990).

16.   I.Durrant, Light-based detection of biomolecules. Nature, 346, 297-298 (1990).

17.   R.Tizard, R.L.Cate, K.L.Ramachandran, M.Wysk, J.C.Voyta, O.J.Murphy and I Bronstein, Imaging of DNA sequences with chemiluminescence. Proc. Natl. Acad. Sci. USA, 87, 4514-4518 (1990).

18.   I.Bronstein, J.C.Voyta, K.G.Lazzari, O.Murphy, B.Edwards and L.J.Kricka, Rapid and sensitive detection of DNA in Southern blots with chemiluminescence. BioTechniques, 8, 310-314 (1990).

19.   E.Hawkins and R.Cumming, Enhanced chemiluminescence for tissue antigen and cellular viral DNA detection. J.Histochem. Cytochem., 38, 415-419 (1990).

20.   Y.Hiraoka, J.W.Sedat and D.A.Agard, The Use of a Charge-Coupled Device for Quantitative Optical Microscopy of Biological Structures. Science, 238, 36-41 (1987).

21.   A.K.Campbell, G.Sala-Newby, P.J.Aston, N.Kalshker, Y.Kishi and O.Shimomura, From luc and phot genes to the hospital bed. J. Biolumin. Chemilumin., 5, 131-139 (1990).

22.   P.Colepicolo, V.C.Camarero, M.T.Nicolas, J.M.Bassot, M.L.Karnovsky and J.W Hastings, A sensitive and specific assay for superoxide anion released by neutrophils or macrophages based on bioluminescence of polynoidin. Anal.Biochem., 184, 369-374 (1990).

VIDEO-IMAGING OF BLOOD MONOCYTE CHEMILUMINESCENCE : APPLICATION TO ASTHMA.

M. Damon, I. Vachier, Ch. Le Doucen, Ph. Godard[1] and J.C. Nicolas

INSERM Unite 58, 60, Rue de Navacelles and 1 Clinique des Maladies Respiratoires, Montpellier,34090, France

INTRODUCTION

Human alveolar macrophages have been shown to be involved in local inflammation partly through their ability to generate toxic reactive derivatives of oxygen in bronchial asthma (1). Large numbers of macrophages are daily extruded from the lung. The means by which these cells are replaced and the total population maintained, have been of interest for many years (2). Later studies have demonstrated that monocytes can enter the lung and become alveolar macrophages particularly during an inflammatory process (3,4). Monocytes are a prominent feature of the bronchial mucosal inflammatory cell infiltrate in chronic asthma (5). Under allergen or exercise stimulation, circulating monocytes could be considered as active mediator secreting cells and contributing to disordered airway physiology in asthma (6).

It has been shown that in vitro maturation of monocytes, which could be compared to in vivo transformation of these cells into macrophages in local inflammation, modifies certain functions of these phagocytes and particularly their ability to produce superoxide anion (7). Thus, the purpose of our research was to specify the role of circulating monocytes in asthma by studying superoxide anion (O_2^-) formation using conventional and video-imaging analyzers.

MATERIALS AND METHODS

Materials and reagents: Material for monocytes purification came from Becton Dickinson, Grenoble France. RPMI 1640 and heat inactivated fetal calf serum (FCS) were purchase from Gibco Laboratories Grand Islance NY. Percoll was obtained from Pharmacia (St Quentin, Yvelines, France). Lucigenin, phorbol myristate acetate (PMA) and dyes for staining came from Sigma (St Louis, USA). Anti-IgE came from Institut Pasteur (Paris, France) and IgE was obtained from patients having high IgE serum level

Selection of subjects.: The study included 5 healthy volunteers (HS), 5 patients with allergic rhinitis (AR) and 5 allergic asthmatic patients (AP) ranging in age from 18 to 45 years. None of the subjects were smokers. None were on medication. Asthma was diagnosed according to the American Thoracic Society (ATS) statement. Allergy was assessed by the presence of at least 3 positive skin tests and a high immunoglobulin E (IgE) blood level.

Blood monocyte preparation and stimulation : Mononuclear cells were isolated from heparinized (20U/ml) venous blood by centrifugation of samples over isotonic Percoll solution with a density of 1. 086 g/ml. Cells were washed in nutritive medium. Contaminating erythrocytes were lysed by incubation in a solution of 130 mmol/l NH_4Cl / Tris 10mmol/l / K_2CO_3 16 mmol/l pH 7.4. Total cells were recovered by centrifugation, and monocyte concentration was determined by neutral red and May-Grunwald-Giemsa stainings. Blood monocytes (BM) were purified by adherence on Petri dishes (35 x 10mm) : 10^6 cells were incubated with 2 ml RPMI containing 20% FCS in a humid atmosphere of 90% air and 5% CO_2 for 4 h at 37°C. Non-adherent cells were removed. Viability, evaluated by trypan blue exclusion test , was > 98%. Purity of monocyte monolayers, carried out by esterase staining on duplicates, was > 95%. For chemiluminescence analysis monocytes were removed by gentle pipetting and resuspended in RPMI, for video-imaging monocytes were maintained as monolayers (10^6 cells/1ml RPMI). O_2^- generation was induced by adding in the reaction mixture either PMA at final concentration of 10^{-7}mol/l or anti-IgE serum (10 µl) added after cell-preincubation in the presence of IgE ($240U/10^6$ BM) for 30 min at 37° C.

Lucigenin-enhanced chemiluminescence (CL) assays: CL was monitored in a model 1251 luminometer LKB Wallac (Wallac Co, Turku, Finland) connected to an Apple computer. Aqueous solution of lucigenin which was considered as a specific enhancer for superoxide anion determination, was used at a final concentratrion of 10^{-4}mol/l. 900 µl of BM suspension in RPMI were placed in luminometer vials, and first 100 µl lucigenin was added. Then background values were determined and 10 µl stimulus solution were added. Measurements were done every 1 min until maximal value (CL peak) was obtained. Controls were performed with cells incubated without stimuli. Results are expressed as mvolt/10^6 BM (mean ± SM).

Video-imaging assays : Photon numbers were evaluated by an ultrasensitive photon counting imaging camera equipped with a computer-assited image processor (Argus 100 from Hamamatsu Photonics, Japan). BM monolayers were maintained in adherence. Lucigenin was used as enhancer and the experiments were identical to those described for luminometer CL measurements. Light detection was performed every 5 min until maximal value was obtained (CL peak), by using a direct imaging measurement of Petri dishe area (Window = 170mm). Results are expressed as photon number/10^6 BM.

RESULTS AND DISCUSSION

Our data show that when BM are stimulated by PMA, the pattern of the cell response obtained either by luminometer analysis or by video-imaging, is quite identical (Fig.1a and b).

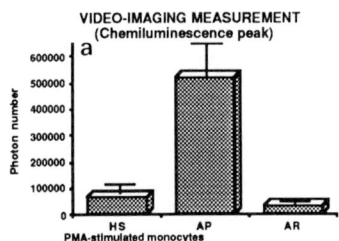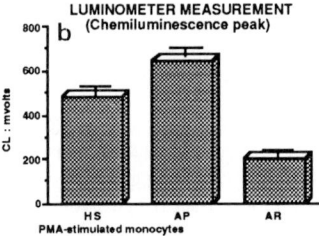

Fig. 1

BM from AP released higher quantities of O_2^- than those originated from the two other populations. The lowest response was observed for cells from AR.

On the other hand, when cells are stimulated by IgE/anti-IgE, only video-imaging system allows the chemiluminescence measurement.

The results are summarized on the following graphs. The background values obtained after 4 h adherence and 30 min IgE incubation indicate that BM from AP are activated and present a greater response than cells from HS or AR as compared to the control values (Fig.2a and b).

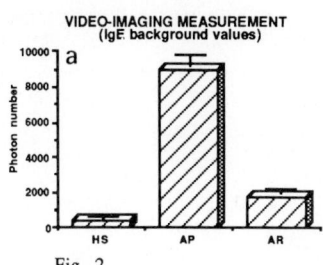

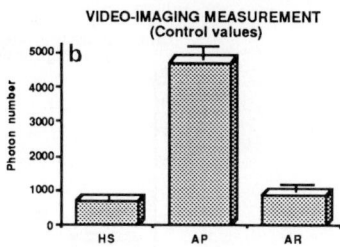

Fig. 2

These data are in agreement with those of Melewicz et al. (8) who showed that there is an increase in the number of monocytes with IgE receptors in allergic patients.

When anti-IgE was added to the medium, we observed a significant increase in the photon number, particularly in BM from AP (Fig.3).This observation was in agreement with those of Thorel et al. who showed that blood monocytes play an important role in allergy through their IgE-mediated generation of superoxide anion (9).

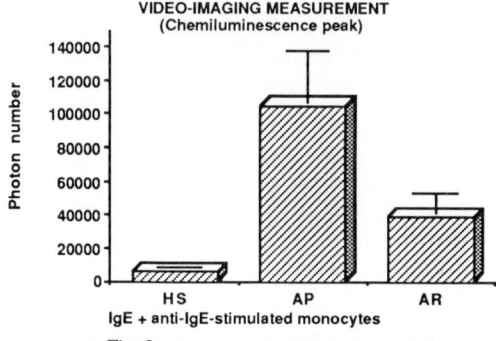

Fig. 3

These data confirm our previous results which demonstrated that human alveolar macrophages (AM) from AP in inter-crisis are more stimulable than macrophages from HS. However the intensity of BM responses to stimuli is greater than those from AM. This difference could be due to the effect of BM maturation during which certain BM functions are modified; this situation is indicated by the weak respiratory burst of resident macrophages and by the sharp decrease in the respiratory burst when monocytes differenciate into macrophages in vitro (10).

The comparison between the BM responses obtained for PMA stimulation and IgE/anti-IgE stimulation, has shown that the cells from AP were more stimulable than those from HS and AR. Moreover, when the BM ability to be stimulated by the IgE/anti-IgE reaction is compared to the responses induced by PMA, we observed that BM from AR were more sensitive to specific stimulation than the others. The percentages of the intensity of superoxide anion release induced by IgE/anti-IgE as compared to PMA, were 7, 22 and 105% for HS, AP and AR respectively (Fig 4). BM from AR could be considered as more sensitive to IgE-dependent stimulation than to PMA.

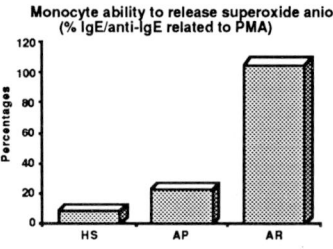

Fig.4

In conclusion, our results indicate that a video-imaging system allows a better evaluation of superoxide anion release by blood monocytes than luminometer analysis, particularly for IgE/anti-IgE stimulation. This could be due to the low affinity of IgE receptors which require a study on cells maintained in adherence. Thus we can consider that video-imaging system is essential to a better understanding of the activation capability of blood monocytes which are known to become adherent cells in the local inflammatory site.

REFERENCES

1.   M. Damon, M. Cluzel, P. Chanez and Ph. Godard, Phagocytosis induction of chemiluminescence and chemoattractant increased superoxide anion release from activated human alveolar macrophages in asthma. J. Biolumin. Chemilumin 4, 279-286 (1989).

2.   F.D. Bertanalffy, Respiratory tissue : structure, histopathology, cytodynamics. Part I. Review and basic cytomorphology. Int.Rev. Cytol. 16, 233-281 (1964).

3.   A. Blusse Van Oud Alblas, B. Van Der Linden-Schever and R. Van Furth. Origin and kinetics of pumonary macrophages during an inflammatory reaction induced by intra-alveolar administration of aerolized heat-killed BCG. Am. Rev. Respir. Dis. 128, 276-281 (1983).

4.   R. Furth. Phagocytic cells : development and distribution of mononuclear phagocytes in normal steady state and inflammation. In Inflammation : Basic Principles and Clinical Correlates.J.I.Gallin, I.M. Goldstein and R. Snyderman (Eds), Raven Press, New York, p 281-295 (1988).

5.   M.S. Dunnill. The pathology of asthma with special reference to changes in the bronchial mucosa. J. Clin. Pathol. 13, 27-35 (1960).

6.   S.T. Holgate. Macrophages and monocytes. In State of the Art Conference: Asthma-What are the important experiments? A.J. Woolcock (Ed), Am. Rev. Respir. Dis. 132, 730-744 (1988).

7.   R.A. Musson, L.C. McPhail, H. Shafran and R.B.Jr Johnston. Differences in the ability of human peripheral blood monocytes and in vitro monocyte-derived macrophages to produce superoxide anion : studies with cells from normals and patients with chronic granulomatous disease. J. Reticuloendothel. Soc. 31,261-266 (1982).

8.   F.M. Melewicz and H.L. Spiegelberg. Fc receptors for IgE on a subpopulation of human peripheral blood monocytes. J. Immunol. 125, 1026-1036 (1980).

9.   T. Thorel, M. Joseph, A. Tsicopoulos, A.B. Tonnel and A. Capron. Inhibition by nedocromil of IgE-mediated activation of human mononuclear phagocytes and platelets in allergy. Int. Archs Allergy appl. Immunol. 85, 232-237 (1988).

10.   S.J. Klebanoff. Phagocytic cells : products of oxygen metabolism. In Inflammation : Basic principles and Clinical Correlates. J.I. Gallin, I.M. Goldstein and R. Snyderman (Eds), Raven Press, New York, p 391-444 (1988).

RECENT DEVELOPMENTS IN LOW LIGHT-LEVEL IMAGING SYSTEMS

R.E.Ansorge[1,2], C.E.Hooper[3], W.W.Neale[1,2] and J.G.Rushbrooke[1,2]

[1]*Cavendish Laboratory, University of Cambridge, Madingley Road, Cambridge CB3 0HE, UK*
[2]*Image Research Limited, St. John's Innovation Centre, Cowley Road, Cambridge CB4 4WS, UK*
[3]*Robens Institute, University of Surrey, Guildford, Surrey GU2 5XH, UK*

INTRODUCTION

In this review paper we cover the following subjects:

1. The problems pertaining to digital low light-level imaging are briefly summarised.

2. We compare the performance and strategies employed by several major different instrumental approaches, viz.:

* analogue imaging photon detectors (e.g. IPD of ITL, PIAS of Hamamatsu);
* cooled CCD detectors (e.g. Astromed, Photometrics, Wright);
* intensified CCD cameras (e.g. Image Research).

3. We introduce the concept of contact imaging in biomedical instrumentation, as an alternative to the conventional lens-based imaging technique.

4. We introduce the instrument BIQ, an intensified CCD camera-based system manufactured by Image Research.

PROBLEMS OF DIGITAL LOW LIGHT-LEVEL IMAGING

There are two major requirements: any instrument has to be sensitive to single photons, and be truly photon counting; the actual position of any single detected photon must be determined, i.e. it is a true imaging instrument.

There are various practical concerns that must be faced:

* the means of sample presentation (e.g. a lens viewing the sample, both in a light-tight housing);
* the sensitivity of the instrument: in practise this pertains to the detective quantum efficiency at input, and to the next major concern, namely
* the intrinsic noise of the device;
* the overall ease of use (e.g. does the instrument display the acquired image in real time during acquisition; does it require time to cool down);
* the exposure time needed for various samples (e.g. are they short-lived light sources, or varying, or very faint).

The different instruments currently available manifest in an obvious way the different approaches to these concerns.

COMPARISON OF INSTRUMENTS

Analogue imaging photon detectors have been described in earlier articles (1,2,3), to which we refer the reader for details. Our main interest here is to describe the reasons for the two main distinct approaches to imaging with CCD's, mentioned above.

Cooled CCD imaging was developed mainly for research in astronomy, where long exposure times (e.g. the night-time hours) were needed to count photons from very faint, mainly steady, distant sources. Each photon has a good probability (called the quantum efficiency), which depends on the wavelength of the photon, of generating a photoelectron. The photoelectrons at any point of an image are accumulated in the pixel structure of the CCD. At the end of the exposure the charge in each pixel is read out and measured, effectively counting the number of liberated photoelectrons. Unfortunately, the CCD semiconductor material itself spontaneously creates noise electrons, fluctuations in which can mask faint signals. This is called dark current noise. Fortunately, there is a factor 2 reduction in the number of these noise electrons for every 8°C drop in temperature; by cooling to -140°C this gives an overall reduction factor 10^9. Such cryogenically cooled CCD cameras hence effectively have no such noise problem. There remains, however, a type of noise called read-out noise, which is introduced during the read-out process: this means that the read-out has to be carried out rather slowly, taking typically several seconds. A typical cooled CCD with carefully designed electronics can have read-out noise as low as 3 electrons per pixel which means that a signal as low as 6 electrons can be detected for a signal-to-noise ratio (SNR) of 2. Since the pixel can contain as many as 3×10^5 photoelectrons at saturation, the ratio between the maximum and minimum detectable signals (called the Dynamic Range) would be large, in this case 5×10^4.

The intensified CCD camera approach is derived from research in high energy physics, where, in effect, rapidly changing sequences of clusters of photons had to be recorded in real time. This requirement of a fast read-out rate (e.g. 25 Hz) meant that a cooled CCD could not be used. An image intensifier was therefore optically coupled in front of the CCD: the resultant amplified image signal is much stronger than the dark current noise even with a warm CCD, especially for short integration times, and very much stronger than the read-out noise.

The amplification used in the intensified CCD camera means that the CCD will saturate much sooner than a cryogenic CCD. One might therefore expect the dynamic range of such instruments to be very limited. However it must be remembered that the intensified CCD camera is read out at 25 Hz whereas a cryogenic CCD is read out only once at the end of an exposure, which takes several seconds. Successive frames from the intensified CCD camera can be either averaged, to produce a noise reduced image suitable for quantitating bright features or summed so as effectively to integrate photons detected during the summation period (which can be considerable). This produces an image suitable for quantifying faint features. In the BIQ instrument both strategies are combined so that the dynamic range of the instrument is about 10^4 for exposure times of 1 minute or more. The two approaches are represented schematically in Fig. 1. A comparison of the performances and properties of a cooled IPD, a cooled CCD, and the intensified CCD camera BIQ, is made in Table 1.

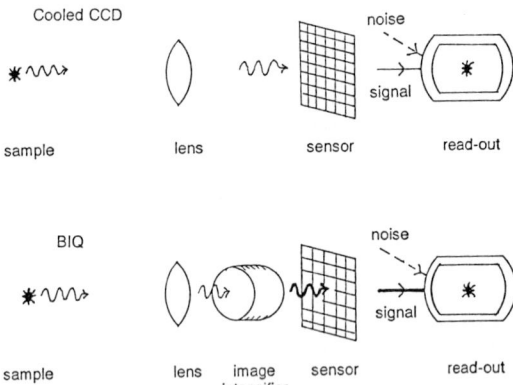

Fig. 1   Schematic representation of two approaches to quantitative CCD imaging: top with cooled CCD; bottom with intensified CCD (the BIQ).

The concept of Minimum Detectable Signal (MDS) is an important one for low light-level imaging, so we shall give two examples. Firstly, for the cooled CCD with 1 minute signal acquisition time, the entry in Table 1 is 0.5 photons/second/detector element. Since the quantum efficiency is 20% at 550 nm, the actual signal obtained is $0.2 \times 0.5 \times 60 = 6$ photoelectrons per detector element, and

Table 1   Comparison of Detectors [‡]

		Cooled IPD	Cooled CCD (4)	BIQ
Quantum Efficiency at:	450 nm	8%	5%	8%
	550 nm	14%	20%	10%
	750 nm	4%	50%	3%
Minimum Detectable Signal [*] at 550 nm (SNR = 2) for signal acquisition time of:	1 sec	7	30	1
	1 min	0.12	0.50	0.13
	1 hr	0.006	0.008	0.017
Amplification Factor (electrons per detected photon)		10^7	1	5×10^4
Typical Dynamic Range in a single exposure		few 10^3	5×10^4	10^4 [**]
Detector Element Size (μm x μm)		60 x 45	22 x 22	27 x 21
Sample Presentation		Lens	Lens	Lens or Contact
Operating Temperature		- 30°C	- 140°C	Room Temp.

[*] *in photons/second/detector element*   [‡] *data compiled from manufacturers' literature*
[**] *for 10 minute exposure*   *- ITL, Astromed and Image Research*

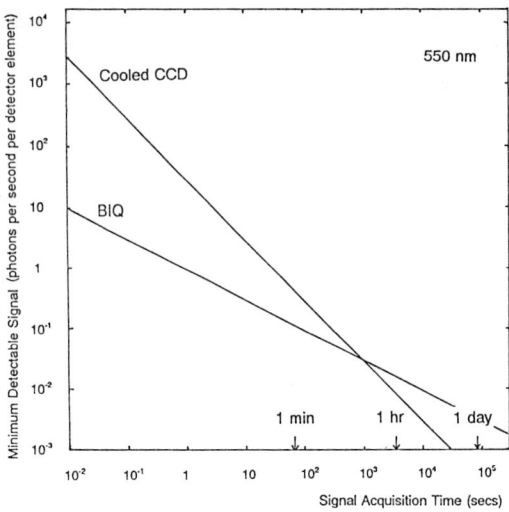

Fig. 2   Minimum Detectable Signal (MDS) measured in incident photons per second per
detector element, as a function of signal acquisition time in seconds, for a cooled
CCD system, and for an intensified CCD system (BIQ). A light source of
wavelength 550 nm is assumed, characteristic (say) of luciferin-luciferase-ATP
luminescence. To be detected above noise in the detector itself, it is assumed that
a signal-to-noise ratio (SNR) equal to 2 is required.

as a read-out noise of 3 electrons is expected, this gives SNR of 2, as required. Secondly for BIQ, with 1 hour acquisition time the equivalent entry in Table 1 is 0.017 photons/second/detector element: given that the quantum efficiency is 10% the signal is thus $0.1 \times 0.017 \times 3600 = 6$ photoelectrons/detector element. The measured dark current charge is 9 electrons/detector element/hour; this has a statistical (shot) type of behaviour and so fluctuates as the square root of this number; hence the dark current noise is 3 electrons, giving SNR = 2 as required.

Values of MDS are plotted against signal acquisition time in Fig. 2. Basically, BIQ is sensitive to smaller signals than a cooled CCD up to about 15 minutes imaging time, whereas the reverse is true for longer times. The Dynamic Range is conventionally defined as the ratio of the maximum signal to the MDS, as obtained in a single measurement. Sample presentation is usually by means of a lens, though the BIQ instrument offers the option of contact imaging, which is described in the next section.

CONTACT IMAGING

With a cooled CCD the sample has to be stood off from the vicinity of the CCD surface itself for reasons of heat insulation. A lens is therefore used to present the image of the sample to the surface of the CCD array. One can readily show that, for an isotropic point source in air, the fraction of the emitted light received at the image of the source is $1/(16\ F^2\ (m + 1)^2)$, where F is the ratio of the focal length of the lens to its effective diameter (e.g. for an f/1.2 lens we have F = 1.2), and m is the demagnification, i.e. the source size divided by the image size. Thus for m = 9 and F = 1.2, less than 0.05% of the light is captured). The intensified CCD camera BIQ offers the opportunity of placing the sample, say on a petri dish or microscope slide, in physical contact with the fibre optic input of the image intensifier, both at the same temperature. For a typical glass fibre optic faceplate, made of fibres of numerical aperture unity, one finds that this 'contact imaging' has a light gathering efficiency of the order of 20%. For comparison, life-size imaging with a lens (m = 1), offers at best an efficiency of about 2%.

SUMMARY

We have compared the performance of typical cryogenically cooled CCD imaging instruments with that of BIQ, a new intensified CCD imaging system developed for biomedical applications. BIQ has a better Minimum Detectable Signal response at imaging duration up to about 15 minutes, beyond which the cooled CCD approach is better. The conventional wisdom is that intensified CCD cameras have limited dynamic range. In BIQ, this limitation is overcome with sophisticated hardware and software for real-time frame averaging and photon integration which enables it to achieve a dynamic range of the order of 10^4, comparable to cooled CCD's. BIQ is a practical instrument with demonstrated performance in biomedical applications (5). It has performance the equal of any other system and also considerable advantages.

REFERENCES

1.   D.H.Leaback and C.E.Hooper, The use of an imaging photon detector in the simultaneous, rapid, determination of multiple, chemiluminescent and bioluminescent reactions in microlitre volumes. In Bioluminescence and Chemiluminescence: New Perspectives, J.Scholmerich, R.Andreesen, A.Kapp, M.Ernst and W.G.Woods (Eds), Wiley, Chichester, p.439-442 (1987).

2.   I.McWhirter, D.Rees and A.H.Greenaway, Miniature imaging photon detectors III. An assessment of the performance of the resistive anode IPD. J.Phys.E: Sci.Instrum. 15, 145-150 (1982).

3.   D.H.Leaback, C.E.Hooper, and R.Pirzad. In ATP Luminescence: Rapid Methods in Microbiology, P.E.Stanley, B.J.McCarthy and R.Smithers (Eds), (Technical Series, Society for Applied Bacteriology, Vol 26), Blackwell, Oxford, p.277-287 (1989)

4.   R.Haggart and D.H.Leaback, The photmetric properties of cooled slow-scan charge coupled devices as detectors in systems for luminogenic analysis assays. Delivered to this symposium.

5.   C.E.Hooper and R.E.Ansorge, Quantitative luminescence imaging in the biosciences using the CCD camera: analysis of macro and micro samples, Trends in Analytical Chemistry 9,269-277 (1990).

A BIOLUMINESCENCE METHOD FOR HIGH RESOLUTION IMAGING OF METABOLITE CONCENTRATIONS IN TUMOR CELL AGGREGATES

G. Schuler, S. Walenta, U. Karbach and W. Mueller-Klieser

Institute of Physiology & Pathophysiology, Pathophysiology Division,
University of Mainz, D-6500 Mainz, FRG

INTRODUCTION

Multicellular tumor spheroids are avascular "minitumors" that are cultured in vitro, and that show numerous analogies to tumor micro-regions in vivo. These analogies include tumor-typic proliferation gradients, differentiation phenomena, or the development of central necrosis at a certain spheroid size (for reviews see: 1,2). To investigate whether the cells in the spheroid center stop prolifera-ting and eventually differentiate owing to a hostile metabolic microenvironment, the concentration distribution of the energy-rich phosphate ATP has been assessed in spheroids using a novel technique of single photon imaging and quantitative bioluminescence. Also, the distribution of lactate has been determined in spheroids with this method.

MATERIALS AND METHODS

Cells Measurements were carried out on spheroids that were derived from a rat rhabdomyosarcoma BA-HAN-1 (3) and that were grown in Eagle's basal medium with 4.5 g/l glucose and 10 % (v/v) fetal calf serum. Spheroids were obtained from three BA-HAN-1 clones with different degrees of differentiation increasing from clone A to C (4). The present study has been restricted to clone C spheroids.

Spheroids Spheroid growth was initiated by incubating single cell suspensions in non-adhesive microbiological Petri dishes for 3-4 days. Emerging cell aggregates were then transferred into rotated suspension cultures where they were kept in a standard incubator atmosphere. The histological structure of spheroids at various stages of growth was analyzed in standard paraffin or cryostat sections. Further details on spheroid culturing and quantifying spheroid growth are published in previous reports (1,2,5).

Bioluminescence The distribution of ATP and lactate concentrations within spheroids of various sizes was assessed by a novel technique using single photon imaging and quantitative bioluminescence (6). The technique is based upon an imaging method with bioluminescence published previously (7-9). The new procedure allows for the detec-tion of metabolite concentrations in cryostat sections of tissues or spheroids with a spatial resolution at the cellular level using the ATP-dependent luciferase of fireflies or substrate-specific enzymatic reactions linked to the luciferase of marine bacteria. A simplified scheme of the respective biochemical reactions is illustrated in Fig. 1. The technique for imaging is shown schematically in Fig. 2. A cover glass with a frozen spheroid section adhered to its upper

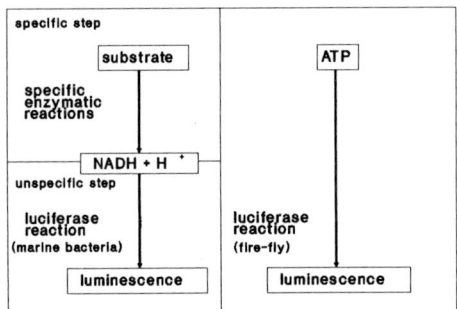

Fig. 1: Simplified scheme of biochemical reactions for measurements of metabolites with bioluminescence

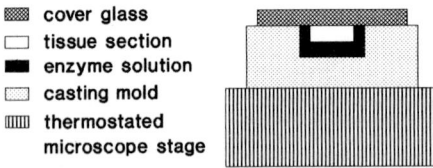

Fig. 2: Technical device for high resolution imaging of metabolites with bioluminescence

side is layed upside down upon a glass slide with a rectangular casting-mold. The mold is filled with a frozen cocktail of enzymes, and gelatine. The luciferase reaction with light emission is then started by raising the temperature of the whole array above the melting point of the section and the cocktail up to 10 $^{\circ}$C. For precise temperature control, the glass slides are located on a thermostated microscope stage. The spatial distribution of the bioluminescence intensity within the spheroid section is measured directly using an appropriate microscope (Axiophot, Zeiss, Oberkochen, FRG) and an imaging photon counting system (ARGUS 100, Hamamatsu, Herrsching, FRG). The intensity of the bioluminescence can be calibrated in mM with regard to tissue volume using heat-inactivated tissue homogenates with different metabolite concentrations that were determined with high performance liquid chromatography or standard enzymatic assays. Further details concerning the biochemistry, the experimental set-up, and the precision of measurement are given elsewhere (6-9).

RESULTS AND DISCUSSION

Clone C spheroids consisted of mononuclear cells and of multinuclear myotube-like giant cells. Whereas the former cells were mainly located in peripheral layers, the giant cells accumulated in the spheroid center. Unlike spheroids with a more restricted differentiation (1,2), clone C spheroids did not exhibit central necrosis up to large sizes (> 1 mm). Representative distributions of ATP and lactate concentrations on radial tracks through a clone C spheroid with a diameter of 520 µm is shown in Fig. 3. It is obvious that the concentration of both metabolites increases from the periphery to the center of the spheroid. ATP concentrations in central regions were as high as values that are commonly found in viable cells in culture (6). Accordingly, lactate concentrations are distinctly less than concentrations that have been shown to influence the growth behavior of cultured tumor cells (10). This was true for the entire range of spheroid sizes investigated which is demonstrated in Fig. 4 showing the central concentration of ATP and lactate for various sizes of clone C spheroids. The data suggest that mechanisms other

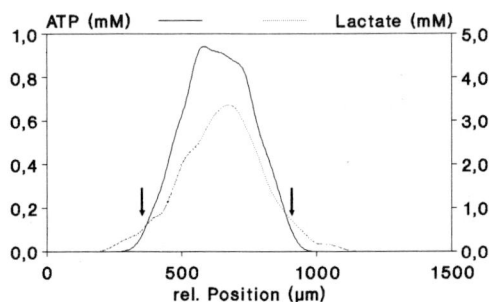

Fig. 3: Representative radial concentration profiles of ATP (solid line) and lactate (dotted line) in a clone C spheroid with a diameter of 520 µm (arrows: size of the spheroid)

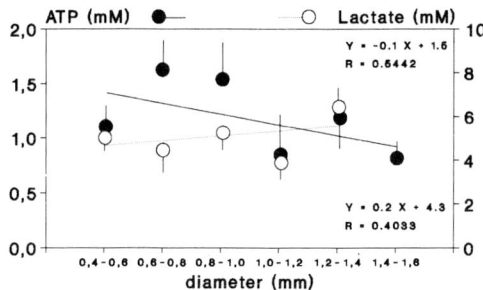

Fig. 4: Central ATP (dots) and lactate (circles) concentrations in clone C spheroids as a function of spheroid size (mean ± SD)

than a hostile metabolic microenvironment are involved in the induction of differentiation in the central portions of clone C spheroids. Among these mechanisms, cell-cell or cell-matrix interaction including growth factors may be most promising to be investigated in the near future.

ACKNOWLEDGEMENTS

This work was supported by the Deutsche Forschungsgemeinschaft (Mu576/2-3, Mu576/2-4) and by the Bundesministerium für Forschung und Technologie (01 ZO 8801).

REFERENCES

1. W. Mueller-Klieser, Multicellular spheroids. A review on cellular aggregates in cancer research. J. Cancer Res. Clin. Oncol. 113, 101-122 (1987).

2. R.M. Sutherland, Cell and environment interactions in tumor microregions: the multicell spheroid model. Science 240, 177-184 (1988).

3. C.D. Gerharz, H.E. Gabbert, R. Moll, W. Mellin, R. Engers and G. Gabbiani, The intraclonal and interclonal phenotypic heterogeneity in a rhabdosarcoma cell line with abortive imitation of embryonic myogenesis. Virchows Arch. B (Cell Pathol.) 55, 193-206 (1988).

4. U. Karbach, C.D. Gerharz, K. Groebe, H.E. Gabbert and W. Mueller-Klieser, Rhabdomyosarcoma spheroids: a model system for investigations on proliferation and differentiation. Cancer Res. (submitted).

5. R.M. Sutherland and R. E. Durand, Radiation effects on mammalian cells grown as an in vitro tumor model. Curr. Top. Radiat. Res. 11, 87-139 (1976).

6. S. Walenta, J. Doetsch and W. Mueller-Klieser, ATP concentrations in multicellular tumor spheroids assessed by single photon imaging and quantitative bioluminescence. Europ. J. Cell Biol. 52, 389-393 (1990).

7. W. Mueller-Klieser, S. Walenta, W. Paschen, F. Kallinowski and P. Vaupel, Metabolic imaging in microregions of tumors and normal tissues with bioluminescence and photon counting. J. Natl. Cancer Inst. 80, 842-848 (1988).

8. W. Mueller-Klieser, C. Schaefer, S. Walenta, E. K. Rofstad, B. M. Fenton, and R. M. Sutherland, Assessment of tumor energy and oxygenation status by bioluminescence, NMR spectroscopy, and cryospectrophotometry. Cancer Res. 50, 1681-1685 (1990).

9. W. Paschen, Regional quantitative determination of lactate in brain sections. A bioluminescent approach. J. Cereb. Blood Flow Metab. 5, 609-612 (1985).

10. W. Marx, W. Mueller-Klieser and P. Vaupel, Lactate-induced inhibition of tumor cell proliferation. Int. J. Radiat. Oncol. Biol. Phys. 14, 947-955 (1988).

APPLICATIONS OF THE DIRECT IMAGING OF FIREFLY LUCIFERASE EXPRESSION IN SINGLE INTACT MAMMALIAN CELLS USING CHARGE-COUPLED DEVICE CAMERAS

M.R.H. White, F.F. Craig, D. Watmore[1], F. McCapra[1] and A.C. Simmonds

Amersham International plc, Pollards Wood Laboratories, Little Chalfont, Bucks, HP7 9LL, UK and
[1]School of Chemistry and Molecular Science, The University of Sussex, Falmer, Brighton, BN1 9QJ, UK

INTRODUCTION

Since the gene for the firefly luciferase enzyme was first isolated it has become a widely used reporter for the quantitation of gene expression in mammalian cells [1], other eukaryotic cells [2,3], and prokaryotic cells [4]. Generally, luciferase analysis has involved lysis of the cells followed by luminometer based quantitation of the luminescence on addition of the substrates, ATP and luciferin. The application of low light level imaging technology to the analysis of luciferase expression in live, intact mammalian cells allows the visualisation of single cells expressing the luciferase reporter enzyme without the need for destruction of the biological sample [5,6]. We previously described the use of three different types of CCD camera for the imaging of single cells expressing firefly luciferase [5]. In this paper, we discuss the parameters affecting this assay and describe the use of novel luciferin derivatives to enhance luciferin entry into the cells. We also discuss the application of this assay for the single cell monitoring of trans-activation of the luciferase reporter gene by expression of the Human Immunodeficiency Virus (HIV-1) *tat* trans-activator protein.

MATERIALS AND METHODS

Expression of luciferase in cos cells: The luciferase expression vector pMW41 (in which the firefly luciferase gene is under the control of the cytomegalovirus immediate early enhancer/promoter) was transiently transfected into *cos7* cells by lipofection as previously described [5]. 48 hours after DNA addition, the cells were trypsinised, diluted 1 in 5 and replated onto 60mm petri dishes.
Analysis of luciferase expression: Luciferase expression was assayed 15-24 hours later, either from cell lysates in a Lumac M2010 luminometer, or in intact cells by CCD camera analysis as described [5]. Luciferin, dissolved in 3ml of assay buffer (25mmol/l glycylglycine pH7.75, 150mmol/l NaCl), was added to the petri dishes and the luminescent signal was integrated over consecutive 5 minute periods (or as described in Results) by either a Wright Instruments liquid nitrogen cooled CCD camera (camera 1 in ref. 5) or by a Peltier cooled CCD camera optically attached to an Olympus CK2 inverted microscope (camera 3 in ref. 5).
Analysis of luciferin esters: The luciferin esters were analysed essentially as for luciferin. The synthesis of the esters and further information on their analysis is described elsewhere [7].
Trans-activation by HIV-1 tat protein: Plasmid pHIV-L encodes 400bp of the HIV-1 LTR upstream of the luciferase gene [MW, manuscript in preparation]. HIV-1 *tat* expression was directed by plasmid pJKC63.4.1 which has been shown to express the *tat* protein strongly in mammalian cells [8]. pHIV-L (4µg) was transfected into *cos* cells with pRSV-CAT (0.6µg) and either pJKC63.4.1 (0.4µg) or a control plasmid (0.4µg). Luciferase assays were as above and the cell lysates were analysed for CAT activity to correct for variations in transfection efficiency.

RESULTS AND DISCUSSION

Detection of single intact cells expressing firefly luciferase: The liquid nitrogen cooled CCD camera was used to analyse the luminescence emitted from live *cos* cells transiently expressing firefly luciferase. This CCD camera visualises an active area of approx. 7 x 4.5 cm. On addition of luciferin to the cells the bioluminescent signal appears as a series of point light signals distributed within the area of the petri dish. The number of point light signals visualised was shown to increase with both cell density and transfection efficiency whereas the image intensity increased with the total level of luciferase expression directed by the transfected plasmid [5].

The luminescence from single magnified cells expressing luciferase can be clearly visualised using the Peltier-cooled CCD camera coupled to a microscope [5]. Cells transfected with the luciferase expression vector pMW41 were assayed with this camera after addition of 1mmol/l luciferin. A luminescence image (25 minute integration) and a backlit image (0.1s) of the cells were taken and the background pixel counts of each image file were set to zero. The image data for each file was adjusted so that the luminescent image was of higher intensity. The two images were then added and Figure 1 shows that the position of the superimposed luminescent signal (white) corresponded directly to the position of three magnified single cells (grey).

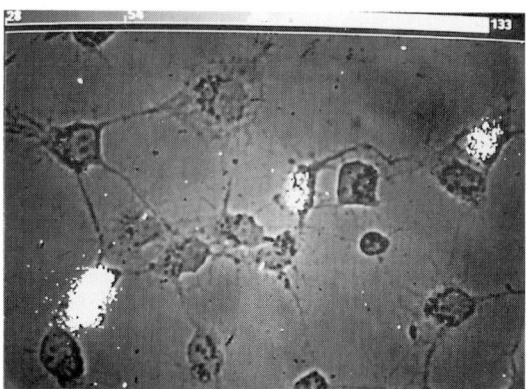

Figure 1: Image of the luminescent signal from luciferase expressing *cos* cells superimposed over a backlit image of the magnified cells.

Kinetics of the bioluminescent reaction: Repeated images of the luminescence from a single petri dish of luciferase expressing *cos* cells suggested that the punctate luminescent pattern could be maintained steadily over a period of several hours [5]. Analysis of successive images by measuring the average pixel intensity (over background) within a 65 x 65 pixel box (50% of image area) permits approximate quantitation of the total luminescent signal in each image. In a study over a

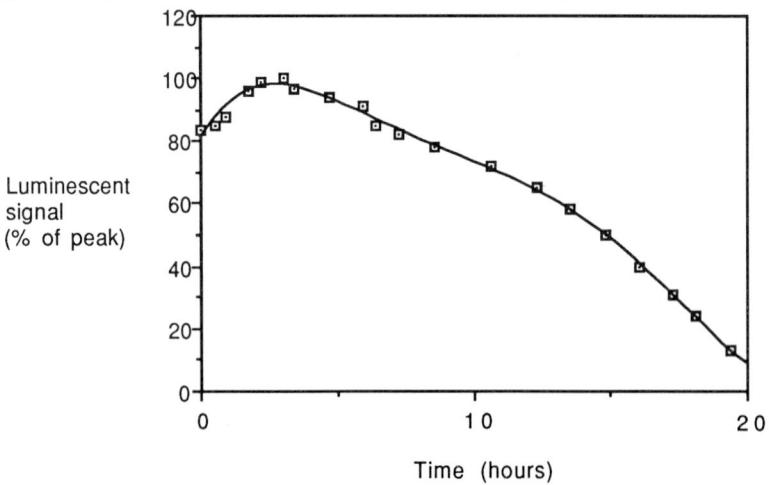

Figure 2: Lifetime of luminescence from *cos* cells expressing luciferase.

longer time-period, successive 5 minute integration readings were taken from a single petri dish of *cos* cells, transfected with pMW41, after addition of 1mmol/l luciferin (Figure 2). The timecourse showed a very steady luminescent signal for more than 10 hours. The lifetime of this luminescence was likely to be limited principally by the viability of the *cos* cells which were left in assay buffer at room temperature with sub-optimal carbon dioxide concentration. The luminescent signal appeared to die off more rapidly at lower luciferin concentrations, or when the level of luciferase expression in the cells was reduced (data not shown).

Luciferin entry into mammalian cells: Luciferin has a carboxylic acid group which may cause poor membrane permeability since it will be charged at physiological pH. We previously showed that luciferin concentrations of less than 2 mmol/l could decrease the luminescent signal from *cos* cells expressing luciferase [5]. We therefore investigated the use of luciferin esters which should show enhanced membrane permeability without affecting cell integrity and could be hydrolysed to luciferin by intracellular esterases. This approach was previously used successfully for the fluorescent calcium indicators [9]. Luminometer analysis indicated that the esters were neither luciferase substrates nor significantly contaminated with luciferin [7]. In this study the luminescent signal from luciferase expressing *cos* cells was quantitated, after addition of each ester to the cells, by successive 5 minute integrations with the liquid nitrogen cooled CCD camera and estimation of the mean signal within a 65 x 65 pixel area (Table 1).

Luciferin Derivative	Final Concentration mmol/l	Average Luminescent Signal	
		Peak	Total
Luciferin	1.0	88	462
	0.1	67	257
	0.01	13	106
Ethoxyvinyl	0.1	93	179
ester	0.01	28	89
2-hydroxyethyl	0.1	98	404
ester	0.01	74	311
3-hydroxy-n-propyl	0.1	56	241
ester	0.01	41	226
Ethyl	0.1	90	463
ester	0.01	65	349

Table 1: Quantitative assays of luciferin esters with *cos* cells expressing luciferase (n=2).

Table 1 shows both the peak luminescent reading over a 5 minute integration period and the total luminescence recorded over a half hour assay period following substrate addition to the cells. Although some esters gave increased luminescence when compared with equivalent concentrations of luciferin, the luminescence directed by the esters was not higher than that obtained with 1mmol/l luciferin. Analysis of the luminescence lifetime showed that the timecourse of the signal varied significantly between different esters [7]. All of the esters studied inhibited purified luciferase [7] and this may have contributed to the complex kinetics of the luminescent signal observed.

Trans-activation of luciferase by HIV-1 *tat* expression: The HIV-1 *tat* gene expresses a trans-activator protein which activates genes downstream from the HIV-1 long terminal repeat (LTR). The *tat* protein interacts with RNA molecules containing the *tar* region which lies downstream from the site of start of transcription in the HIV-1 LTR [8]. Co-transfection of a reporter gene under the control of the HIV-1 LTR together with a *tat* expression vector can give up to several hundred fold activation of the reporter gene activity. Trans-activation of luciferase expression by the HIV-1 *tat* protein was previously shown to occur either when *tat* was expressed from a co-transfected plasmid, or when *tat* was expressed following HIV-1 infection of the cells [10]. In this study, plasmids pHIV-L and pJKC63.4.1 (see Materials and Methods) were co-transfected into *cos* cells. Cell lysate analysis showed that luciferase expression was increased more than 80 fold in the presence of the *tat* expression vector pJKC63.4.1. The transfected *cos* cells were assayed by CCD camera analysis on addition of 1mmol/l luciferin to the cells (Figure 3).

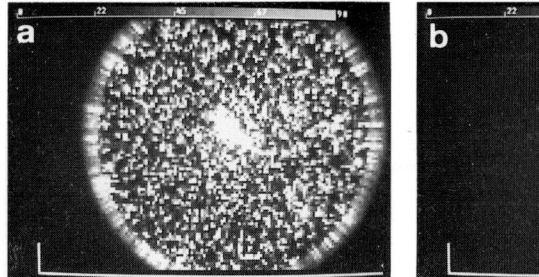
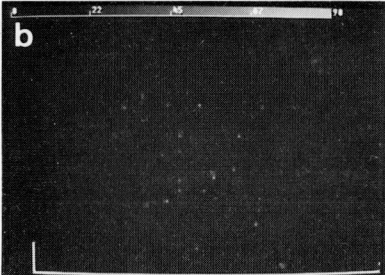

Figure 3: Direct imaging of luciferase expression in cos cells transfected with pHIV-L. A: + *tat* (pJKC63.4.1 *tat* expression vector). B: -*tat*.

These results show that trans-activation of luciferase by HIV-1 *tat* can be readily detected in intact *cos* cells. Taken together with the previous study [10], this suggests that the infection of single, live, mammalian cells with HIV-1 virus could be monitored by CCD camera imaging. This approach to the study of HIV-1 infection of luciferase marker cell lines may allow improved methods for detection of infectious viral particles and screening of potential therapeutic agents which may block HIV-1 infection.

ACKNOWLEDGEMENTS
We thank M. Brady, C. Mundy, D. Pollard-Knight, and M. Evans for helpful advice and also Z. Boniszewski for help with CCD cameras and data analysis. The HIV-1 *tat* expression vector pJKC63.4.1 was generously provided, together with helpful suggestions, by Dr J. Karn of the Medical Research Council, Laboratory of Molecular Biology, Cambridge. Much of this work would have been impossible without the support, help and advice of Dr D. Chiswell.

REFERENCES
1. J.R. De Wet, *et al*. Firefly luciferase gene: structure and expression in mammalian cells. Mol. Cell Biol. 7, 725-737. (1987).
2. H. Tatsumi, *et al*. Synthesis of enzymatically active firefly luciferase in yeast. Agric. Biol. Chem. 52, 1123-1127. (1988).
3. D.W. Ow, *et al*. Transient and stable expression of the firefly luciferase gene in plant cells and transgenic plants. Science 234, 856-859. (1986).
4. J.R. De Wet, *et al*. Cloning of firefly luciferase cDNA and the expression of active luciferase in *E. coli*. Proc. Natl. Acad. Sci. USA 82, 7870-7873. (1985).
5. M.R.H. White, *et al*. Imaging of firefly luciferase expression in single mammalian cells using highly sensitive charge-coupled device cameras. Technique, in press.
6. C.E. Hooper, *et al*. CCD imaging of luciferase gene expression in single mammalian cells. J. Biolumin. Chemilumin. 5, 123-130. (1990).
7. F.F. Craig, *et al*. Membrane-permeable luciferin esters for assay of firefly luciferase in live intact cells. Manuscript submitted.
8. C. Dingwall, *et al*. Human immunodeficiency virus 1 *tat* protein binds trans-activation-response region (TAR) RNA *in vitro*. Proc. Natl. Acad. Sci. USA 86, 6925-6929. (1989).
9. R.Y. Tsien, A non-disruptive technique for loading calcium buffers and indicators into cells. Nature 290, 527-528. (1981).
10. O. Schwartz, *et al*. A microtransfection method using the luciferase-encoding reporter gene for the assay of human immunodeficiency virus LTR promoter activity. Gene 88, 197-205. (1990).

THE POINT SPREAD FUNCTION IN SINGLE PHOTON EMISSION IMAGING (SPEI) OF NADPH ASSOCIATED CHEMOLUMINESCENCE

G. Bernroider and K. Überriegler

Institute of Zoology, University of Salzburg,
A-5020 Salzburg, Austria

INTRODUCTION

The formation of images from single photon emissions (SPEI) is a non-deterministic event which may best be understood by the notion of probability. The path of a photon which is emitted from a low level light source and eventually strikes a target cannot be predicted, as it has an "amplitude" to go 'anywhere', but also a preference to go 'somewhere'. With recent progress in the technology of Channel Electron Multipliers (CEMS) very low light emissions have become accessable for optical imaging. The challenge now is to understand the formation of source specific patterns from basically disordered environments. As quantum entities, photons have an uncertainty in position, time, energy and momentum, their emission and progression is 'lumpy' and their arrival at the target may not be sensed directly but must be extracted from random noise. These imaging proplems have renaissanced a body of methods called "Quantum Optics" which uses principles and concepts of Quantum Electrodynamics. Both concepts replace deterministic descriptions for electromagnetic wave propagation with probability laws (3). As compared with SPEI, classical images may be considered as the realisation of an infinity of photonic events and the process of image formation may be described by a continuous version of an imaging equation which relates the image -i(x)- to an object -o(x)- by a transfer-function -s(x)-. Single photon emission images result from discrete phenomena and can be built up by a stepwise superposition of a finite number of photons. Thus, there is a correspondence of discrete images due to single photonic events to continuous images due to an infinity of events. SPEI images may be regarded as extremely 'spikey' versions of classical pictures.

An additional correspondence exists between the concept of probability on one hand and physical quantities such as intensity of source and images on the other. This can best be seen from the following equation

$$P(y_j) \propto P(x_i) \tag{1}$$

which states that the probability of observing a photon y_j in the image is proportional to the probability of emission of a photon at object location x_i. Thus probabilities correspond to 'brightness' within sources and images. The fundamental law of image formation which will be used throughout this paper can now be derived easily. If we let $P(y_j \mid x_i)$ denote the conditional probability of observing a photon say within $y_j + dy$ in the image, provided it has been emitted within $x_i + dx$ in the object, the law of probability partition $P(A) = \sum P(A \mid B)$. $P(B)$ lets us set

$$P(y_j) = \sum_i P(x_i) \cdot P(y_j \mid x_i) \tag{2}$$

With the above mentioned correspondence between optical 'brightness' and probabilities in mind, we finally arrive at

$$i(y_j) = \sum_i o(x_i) \cdot s(y_j ; x_i) \tag{3}$$

By the above equation we relate the image (i.e. 'brightness' at y_j) with the object o(x) by a conditional probability with its physical equivalent denoted as s(y;x). The function s(y; x) represents the optical transfer which characterizes the (unpredictable) path of photons through imaging devices such as lenses.

Optical transfer usually imposes a translocation of object points within the image leading to a 'spread' of point locations. It is therefore also called 'point-spread-function'. In the present paper we demonstrate methods how to pursue object and transfer characteristics for very low light emissions by simulating all or parts of equation (3). We start with various modelistic assumptions about probability laws underlaying object and spread distributions. Realistic experiments employing NADPH associated chemoluminescence and 2-dim photon counting have been used to constrain the possible geometries that can arise from single photon sources and transfer functions. Some of the theoretical background required for the simulation part of this work comes from statistical optics (2,3). SPEI for enzymatic chemoluminescence and applications for the study of steroid hormone metabolism in brain tissue have been published previously (1). As our results indicate a strong resemblance between simulated and realistic images from chemoluminescent sources as low as 10 cps/mm^2 , we expect that computer simulations of this type may give insight into the physico-chemical nature of very low light emissions.

MATERIALS AND METHODS
NADPH-associated chemoluminescence
Coupled enzymatic reactions with NAD(P)H specific flavin mono-nucleotide that produce FMNH$_2$ by the activity of a bacterial oxidoreductase as a first step and a luciferase driven oxidation with FMNH$_2$ as a substrate as a second step have been used to produce light emissions in the range of 10 cps from 1 mm^2 sources and 550 nm wavelength. MR (NADPH monitoring reagent, LKB Wallac No. 1243-104, containing Luciferase from Beneckea harveyi, NADPH:FMN oxidoreductase, FMN, 5 x 10^8 mol), bovine serum albumin and decanal (1.0 mg in 0.5 ml isopropanol), LP (0.1 M potassium phosphate buffer, pH 7.0) and ST (0.025 umol NADPH standard) were mixed together in aliquots· of MR(100):LP(300):ST(40). Single 'drops' of 4 ul reagent were pipetted onto non reflecting plastic wells and placed under a completely darkened 2-D photon counting tube. Emissions were detected by a digital image processing system with different integration time and conditions (ranging from 1 to 30 minutes). The photon counting tube used for the present study was a 2-stage multi-channel plate enhanced bi-alkali photocathode, coupled to a low-lag vidicon (system ARGUS from Hamamatsu).
As one of the principal problems that we have been concerned with, is the generation of space-time resolved photonic emission and the detection of minimium signal strength that is necessary to gradually form images, we have used various 'shapes' and concentrations of NADPH-standard drops. See figure 1a for an example.
Computer simulations
The following version of the imaging equation -3-, suggested by FRIEDEN (3), can be seen as a simple addition of random variables x$_i$ for points within an image, x$_o$ for points within an object and x$_s$ for contributions of the spread function. This presents an easy way to (Monte-Carlo) simulate SPEImages:

$$x_i = x_o + x_s \qquad (4)$$

This very convenient relation is in fact obtained from equation -3-, because images can be regarded as "convolutions" of object functions with spread functions. Provided that all events are mutually independent, convolution turns into a simple addition of random variables (2,3). Now we have a literal translation of a single photon imaging process into a random superposition of single point locations. With the help of a uniformly random number generator almost any shape of density function donating points for object and transfer function may be generated by rejecting points outside a given domain or model (e.g. halfspheres and cylinders). NADPH drops have been simulated by the half-sphere model for probability density functions. To simulate the point-spread the best model for the objects (drops)as described above, has been found to be the 'uniform cylindric tube' type of function. The number of points or events arising from the chosen proabability law was adjusted to the empirically observed number of photon counting events from NADPH assoc. chemoluminescence. The diameter of both, halfspheres for objects and cylinders for point-spread, has been chosen to be equal. In other words, the 2-dim domain within random numbers are generated are of identical size, but different shapes. Once, a unifrom random number has been generated from $[0,1]$, the number is transformed into x,y,z coordinates and, for half-spheres, the result constrained by $x^2 + y^2 + z^2 \leq r^2$. Similar criteria may be used to generate other shapes.
RESULTS AND DISCUSSION
SPEImages from NADPH standard solutions (4 ul of a 5 x 10^{-6} mol) have led to total counts in the

range of 10.000 to 20.000 photons within 3 to 6 min of photon integration time. These experiments indicate a signal strength of 10 to 100 photons per sec and mm². Emission intensity with NADPH assoc. luminescence can be adjusted by different concentrations of NADPH standard within constant volumes. This is apparent from figure 1. The evolution of images formed by single photon integration requires some time as photons gradually build up the final image by 'random superposition' (versus light interference in classical images). Figure 3a shows an example of the time resolved formation of a one drop image of NADPH. Detection of photons within the image is demonstrated for two different sizes of counting frames. That is, two different sizes of targets are allowed for photons to arrive. A larger test-frame with a relative size of 0.33 (number of photons/number of test-points) obviously collects more events than does a smaller frame of relative size 0.61. Both frames collect photons arising from the same source or object. Also, the initial rate of the emission process, within the first 3-4 minutes of integration is almost identical. Thus, an additional and source independent contribution of photons contributing to background noise can be expected.

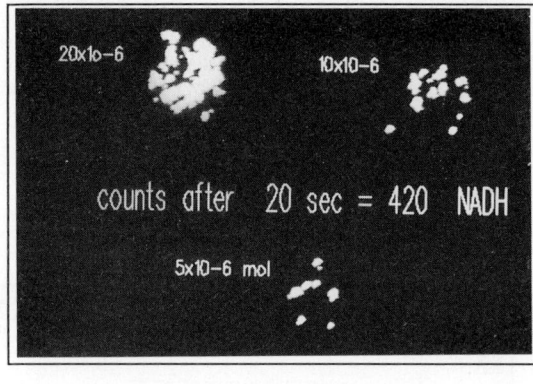

Fig.1 Single photon emissions from 3 drops of NADH standard solutions with concentrations 20,10 and 5 x 10⁻⁶ mol (4ul volume each).

A comparison of computer simulations with NADPH-SPEImages has therefore caused us to add an extra feature to our simulation. This is the apparent additive noise contribution. For the generation of images such as in figure 2 we have superimposed uniform, uncorrelated random noise onto the image. A closer look at the simulated images shows that the observed output signal $i(x)$ departs from the input signal $o(x)$ in more than one way: As can be expected from the generation of image points by simple random additions (equation 4), the spread function $s(x)$ diffuses the signal from a half-sphere shaped model towards a Gaussian distribution, increasing the variance of the resulting image (the sum of two independent random processes is even more random). As the resulting image becomes more blurred than either spread or object function, we would expect a decreased number of photons within the image plane that correspond to object locations. Thus relative detection, i.e. number of image events/source events will never approach 100%, but will depend on the point spread function of the underlaying imaging process. Our results from simulations under the above mentioned conditions indicate that relative photon detection approaches 75% for signals as low as 10 cps/mm². Detection efficiency constitutes a limit which is gradually approached as the

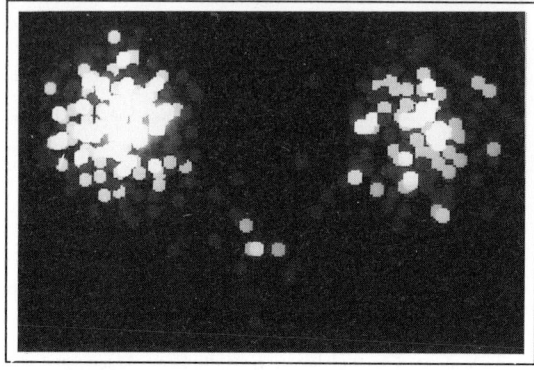

Fig.2 Monte-Carlo simulated SPEImage of NADPH-associated chemoluminescence. 3 different concentrations of NADPH yielding 20, 108 and 193 counts within the image plane.

number of emission increases (such as a relative frequency converges to its corresponding probability). Thus, detection efficiency does not depend on signal strength but is charcterized by the point-spread function.

Our simulations also suggest that the observed random noise in SPEI must be independent from signal and transmission charcteristics, but a considerable number of events originate from signal independent sources. The most obvious candidate will probably be ion feedback within single multi channel plates. This phenomenon produces positive ions which feed-back to channel input (refer to (4) for more details). The two curves in Figure 3b demonstrate the difference between background-noise added signals and 'noise-stripped' signals (hatched curve). With low emissions (up to 300 in fig.3b) or during the initial rate of the process, one cannot separate noise from signal specific events. This difference gradually evolves as the emissions increase. We can therefore anticipate that unlike detection effiency, noise separation does reflect signal intensity properties.

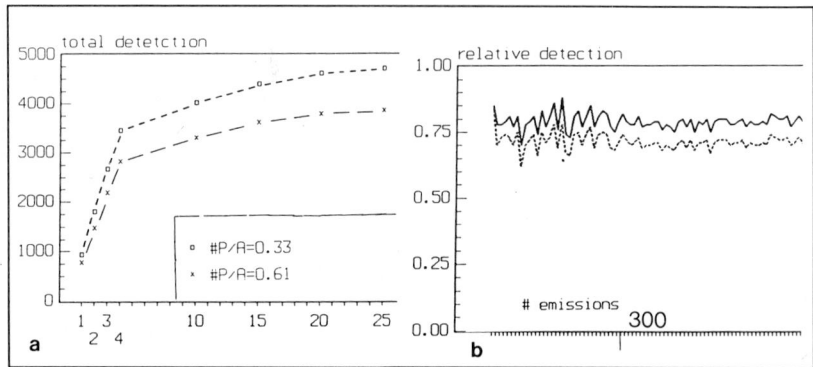

Fig.3a Total detection versus time of integration
   3b Relative detection (image / object) versus total emissions

Above considerations are of more than technical interest, because they indicate a way to 'understand' the formation of patterns as ordered entities from principally unpredictable environments. If we can understand some of the typical transfer characteristics emerging in SPEImaging, we may finally pursue in-situ chemical issues by visualization of single photonic emissions.

We gratefully acknowledge help and cooperation from Martin Lechner and Marianne Steiner, Ferdinand Österreicher and Lynne Susan Bernroider.

REFERENCES

1. G. Bernroider, M. Lechner, WD. Krautgartner, M. Steiner-Flieser, G.N. Nyman, L. Schleinkofer and   H. Adam,   Single Photon Emission Imaging of Enzymatic Chemoluminescence. Proc. Biomedical Micro, London, (1987).
2. G. Bernroider and K. Überriegler, The Simulation of Photon Images from Objects with Very Low   Light Emissions. Proc. 14th ÖAGM Meeting, Salzburg, ÖCG Publ.Vienna (1990).
3. B.R. Frieden, Probability, Statistical Optics and Data Testing, Springer (1983).
4. E.A. Kurz, Channel Electron Multipliers, American Lab, March (1979).

The Photometric Properties of Cooled Slow-Scan Charge Coupled Devices as Detectors in Systems for Luminogenic Assays

R. Haggart and D.H. Leaback[1]

*Astromed Ltd, and ¹Biolink Technology Ltd, The Innovation Centre,
Cambridge Science Park, Cambridge CB4 4GS, UK*

Introduction

Cooled slow-scan CCD detectors have been used to quantitate and detect luminogenic assays using different solid phase formats (Leaback and Haggart, 1989) and for the detection of cellular antigens (Haggart and Leaback, 1989).

The photometric properties of cooled slow-scan CCD based imaging systems were investigated; linearity of response, "flatfielding" and light detection efficiency. These are compared with the photometric properties of a conventional tube luminometer, the Biocounter M2010, based on a photomultiplier detection system. The Biocounter has been used in the past as a reference system to compare assay results with those obtained using the CCD based systems. The light detection efficiency of the Biocounter was measured using the luminol light standard of Lee et al. (1966).

Calibration of the Biocounter allowed the calibration of a blue "ß-light". This is a sealed glass tube, filled with a tritiated gas. The inner surface of the glass is coated with a phosphor. ß-emission from the tritated gas causes the phosphor to scintillate at the emission wavelength of the phosphor. The "ß-light" has a stable light emission and was used to measure the light detection efficiency of different CCD based systems. The "ß-light" and a sucession of ND filters where used to evaluate the linearity of response of both detection systems.

Materials and methods

The Lumac Biocounter M2010 was calibrated using the secondary light standard of Lee et al. (1966), based upon the light emission from the peroxidase catalysed chemiluminescent reaction between luminol and hydrogen peroxide. Luminol was purified by recrystalisation from NaOH by the method of Stott and Kricka (1986). Calibration of the Biocounter was by the method of O'Kane et al. ,(1986). Light emission was plotted against time and the total light detected determined. The light gathering efficiency was calculated; as 1ml of luminol (OD=1; 347nm) emits, on average, 9×10^{14} photons (O'Kane et al. ,1986).

The spectra of luminol and a blue "ß-light" was ascertained using a Baird Atomic Nova 1 spectrofluorimeter. The spectra were corrected using the emission spectrum of quinine sulphate, and data from Miller (1970). The spectral characteristics of a typical bialkali photocathode photomultiplier tube, which is used in the Biocounter, was obtained from Thorn EMI Ltd. The spectral characteristics of a typical CCD were obtained from EEV Ltd.

A blue "ß-light", the light emitted attenuated using a 4ND filter, was presented to the Biocounter and the light emission recorded. The light emitted from the "ß-light" could be calculated; knowing the efficiency of the Biocounter, the attenuation of light due to the 4ND filter, and a correction factor to account for the difference in efficiency of the PMT at the wavelengths of light emitted from the luminol reaction and the "ß-light". The linearity of response of the Biocounter and CCD systems were evaluated using a blue "ß-light" and a succession of ND filters. The linearity of the CCD systems was also determined by incrementing the exposure time of the "ß-light" to the detector.

The calibrated blue "ß-light" was used to determine the light detection efficiency of three slow-scan CCD based low light level detection systems (Fig.1). Light emitted from the "enhanced chemiluminescence" reaction (Thorpe and Kricka, 1986) measured in Biocounter and the CCD-TSS was used to check the calibration.

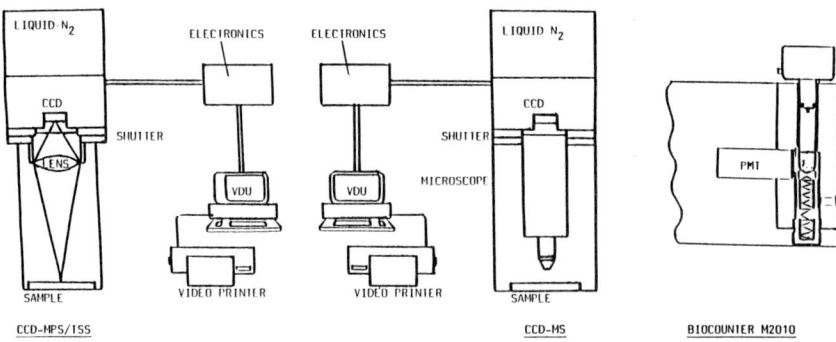

Fig 1 Schematic diagram of the Biocounter M2010 and the different CCD based detection systems

"Flatfielding" images from the different CCD based detection systems were obtained using a homogeneous light emitting solution of luminol presented in a specially constructed tray or in a black microtitre plate (Dynatech Ltd.). The light emitting solution was based on the enhanced chemiluminescence reaction. A thin film (0.02mm) of a homogeneous solution of sodium fluorescein (10uM) was used to flatfield the CCD fluorescence microscope system.

Results

The light detection efficiency of the Biocounter was calculated from the total light emitted from the luminol standard recorded by the Biocounter (average, 6×10^{12} RLUs) divided by the average light emitted from the luminol reaction (9×10^{14} photons); an efficiency of 0.7%.

The efficiency of a typical bialkali photocathode at detecting light from the luminol reaction and the blue "ß-light" was 18% and 15% respectively.

The efficiency of a typical CCD at detecting light emitted from the luminol reaction and the blue "ß-light was 3.3% and 5.5% respectively.

The light emitted from the blue "ß-light" attenuated by a 4 ND filter, and recorded by the Biocounter was 4323 RLUs/s. The 4ND filter attenuated the light emitted by a factor of 32000. The correction factor for the difference in the spectral characteristics of the Biocounter PMT at detecting light from the blue "ß-light and the luminol reaction was calculated to be 1.2. The total light emitted from the blue "ß-light" was calculated to be 2.3×10^{10} photons/s.

Linearity measurments indicated the Biocounter could detect light linearly over 5 decades (Fig.2); the CCD 4 decades (Fig.3).

The calibrated blue "ß-light" was used to determine the light detection efficiencies of the three CCD systems.

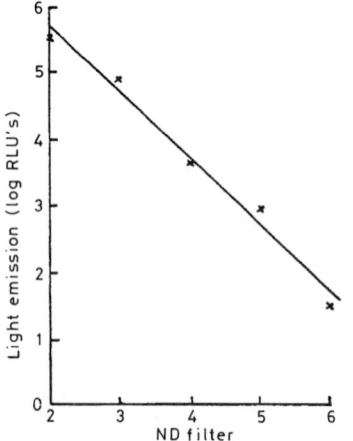

Fig 2 Evaluation of the linearity of the Biocounter M2010 using blue "ß-light".

The efficiencies of the CCD-Microtitre Plate System (CCD-MPS), and the CCD-Tissue Section System (CCD-TSS) were 0.03%, 0.2% respectively. The efficiency of the CCD-Microscope System (CCD-MS) depended upon which objective was used. The efficiency of the CCD-MS for a x10, x20, x40 and x100 objectives was 3.45%, 10.28%, 18.1% and 17.28% respectively.

Light emitted from the "enhanced chemiluminescence" reaction carried out in the Biocounter

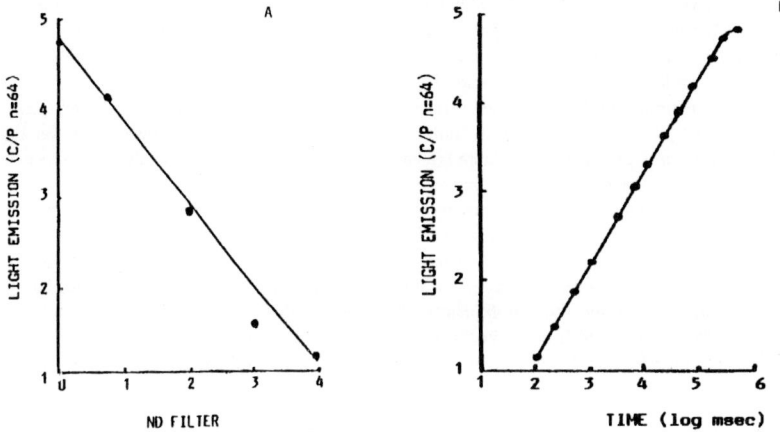

Fig 3 Evaluation of the linearity of the CCD
   a) Blue "ß-light" and different ND filters.
   b) Varying the exposure time to the blue "ß-light".

M2010 and the CCD-TSS suggested that the CCD-TSS had a light detection efficiency of only 0.13%. The difference between the efficiency of a typical CCD at detecting light from the enhanced chemiluminescence reaction and the blue "ß-light" is 1.5 . If this is used as a correction factor; the efficiency of the CCD-TSS with respect to the "ß-light" is 0.2%.

"Flatfielding" experiments revealed that the light intensity from the light emitted from a homogeneous light emitting solution varied from 9%, over the field of view, for the CCD-TSS and up to 20% for the CCD-MPS. The fluorescence emitted from the thin film of sodium fluorescein suggested variations of light intensities, over the field of view, of up to 13.6% with the CCD-MS.

Discussion

The luminol standard allowed the light detection efficiency of the Biocounter M2010 to be determined. The measurement was easy to perform and there was no need for costly instrumentation normally associated with the calibration of detectors (Selinger, 1978).

The calibration of the Biocounter allowed the calibration of a blue "ß-light". The blue "ß-light" was used to determine the light detection efficiencies of the different CCD based imaging systems. "ß-lights" give adequately stable light emission due to the long half-life of the tritiated gas (12.5 years), and were easily presented to the different CCD based systems.

"Flatfielding" experiments indicated unevenness of the light detected over the CCD image. This is most likely due to the optics of the CCD systems. Flatfielding correcting factors can be used to manually correct results from subseqent images, or relevant computer software (Astromed Ltd.) can automatically correct each image with reference to a flatfield image.

The Biocounter could detect light emitted linearly over 5 decades and the CCD over 4 decades. The results suggest the Biocounter has a greater dynamic range than the CCD based systems. The CCD is capable of accumulating up to 500,000 photoelectrons, but the analogue to digital converter can only cope with 16 bits of information (16 bits = 65335) hence the dynamic range of 4 decades. It is possible to alter the gain of the amplifier in the electronic system so that one registered count is the equivalent of 2,4,8 or 16 photoelectrons; increasing the dynamic range to 5 decades.

The light detection efficiency of the different CCD systems was depended on the optics. Systems used to detect samples in microtitre plates or to image large areas of a microscope slide required the sample to be distanced from the detector to enable a demagnified image of the whole area to be seen. Demagnifications of around 11.5 are needed to image a microtitre plate and 3.4 to image half the area of a microscope slide. The light gathering efficiencies of such systems are lower than the Biocounter. The lower efficiency is probably due to the the distance of the sample from the detector and the loss of light passing through the compound lens. In the Biocounter the sample is presented in close proximity to the detector with no complex focusing lens.

1:1 imaging increases the light detection efficiency of the CCD-TSS system to 2.5% (unpublished results). The use of such a lens system limits the imaging area to the dimensions of the light sensitive area of the CCD; in this case approximately 9x12.5mm. Larger CCDs, although expensive are available, thus increasing the imaged area.

The CCD detector mounted on a microscope has good light gathering efficiency due to the high numerical aperture of the objective lenses. The highest efficiency of 18.1% (as measured using the blue "ß-light") was obtained using a x40 objective (NA 0.95).

The quantum efficiency of the CCD is lower in the blue end of the spectrum (approx. 3%, luminol spectrum) compared with the PMT in the Biocounter (approx. 18%, luminol spectrum). Coating the CCD with a fluorophore which emits in the higher wavelengths where the CCD is more efficient would increase the efficiency in the blue area of the spectrum to around 20%. Thinning the CCD and back illuminating can increase the efficiency to around 80% in the blue area of the spectrum (Astromed, 1990).

The photometric properties of the cooled slow scan CCD based systems have been described and compared with a conventional luminometer based upon a PMT detector. Both detection systems have shown good linear response over similar dynamic range. The efficiency of both detection systems has been determined. The efficiency of a typical CCD detector is lower than a PMT in the blue area spectrum. However, the detection efficiency of the CCD based systems, depending on optics, can be better than the Biocounter M2010. If increased sensitivity in the blue area of the spectrum is required a coated CCD or a thinned, back illuminated CCD are available. If luminescent measurements are required in the green area of the spectrum (firefly bioluminescence) or the red area of the spectrum (dioxetane chemiluminescence plus enhancer), the efficiency of a standard CCD compared to the Biocounter PMT at 537nm is 28% (CCD), 8% (PMT); at 622nm 40% (CCD) and 1% (PMT).

The Biocounter is limited to detecting assays presented in specific cuvettes. The CCD based systems have the ability of producing quantitative 2-D images. They are versitile and can be used to detect luminogenic assays presented in different solid phase formats: microtitre plates, tissue sections, and cells.

References

Astromed Ltd., Technical Data Sheet (1990)

R. Haggart and D.H. Leaback, Anal. Chim. Acta. 227 p257-265 (1989).

D.H. Leaback and R. Haggart, J. Biolumin. Chemilumin. 4 p512-522 (1989).

J. Lee, A.S. Wesley, J.F. Ferguson III, and H.H. Selinger, In: Bioluminescence in Progress, F.H. Johnson and Y. Haneda (Eds.), Princeton Univ. Press, Princeton, p35-43 (1966).

J.N. Miller, Standards in Fluorescence Spectrometry, J.N. Miller (Ed.), Chapman and Hall, p79-113 (1981).

J.D. O'Kane, M. Ahmad, I.B.C. Matheson, and J. Lee, Methods in Enzymology, M.A. Deluca and W.D. McElroy (Eds.) 133 p109-128 (1986).

H.H. Selinger, Methods in Enzymology, M.A. Deluca (Ed.), 57 p560-600 (1978).

R.A.W. Stott and L.J. Kricka, In: Bioluminescence and Chemiluminescence; New Perspectives, J. Scholmerich, R. Andereesen, A. Kapp and W.G. Woods (Eds.), Wiley, Chichester, p237-240 (1987).

G.H.G. Thorpe and L.J. Kricka, Methods in Enzymology, M.A. Deluca and W.D. McElroy (Eds.), 133 p331-353 (1986).

Session V
ANALYTICAL APPLICATIONS

ANALYTICAL APPLICATIONS OF LUMINESCENCE DETECTION IN LIQUID CHROMATOGRAPHY AND FLOW INJECTION ANALYSIS

P.J. Worsfold

Department of Environmental Sciences, Polytechnic South West, Drake Circus, Plymouth PL4 8AA, U.K.

INTRODUCTION

The major attractions of chemiluminescence (CL) and bioluminescence (BL) reactions for analytical applications are the excellent detection limits and wide dynamic range that can be achieved with relatively simple batch instrumentation. The most well known analytical uses of CL and BL are as a means of detection in immuno-assay and for the determination of ATP. The salient feature of both applications is the selectivity engendered by the immunological interaction and the bioluminescent enzyme respectively.

In order that CL and BL reactions can be more widely exploited within analytical chemistry, better sample handling and on-line separation techniques are required and to this end both flow injection analysis (FIA) and high performance liquid chromatography (HPLC) offer attractive possibilities. They are both unsegmented, liquid phase, continuous flow techniques that are used in conjunction with flow-through detectors and are therefore compatible with CL and BL reactions.

This paper describes the basic principles of FIA and HPLC in combination with CL and BL detection and presents an overview of current and potential applications of the combined techniques.

FLOW INJECTION ANALYSIS

FIA is a simple yet elegant analytical technique in which a discrete liquid sample is injected into a liquid carrier stream which trans-ports it to a flow-through detector (1,2). A block diagram of the various components is shown in Fig. 1. An important feature of FIA is that during transport the sample can undergo on-line physical and/or chemical treatment. It is now widely used for automated wet chemical analysis in the laboratory and has considerable potential for process analytical applications.

In its simplest form FIA is used solely to transport a sample to a detector. In this situation the instrumentation consists of a reservoir of an appropriate liquid carrier stream, a means of propulsion (typically` a peristaltic pump operating at 1-4 ml/min), a sample injection valve (typically a six port rotary PTFEvalve with 0.02-0.5 ml sample loop) and a flow-through detector (typically optical or electrochemical). The components are connected by PTFE

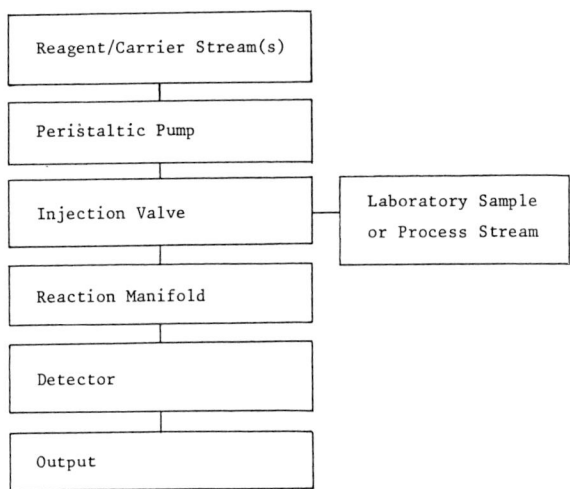

Fig. 1 Block diagram of a flow injection analyser

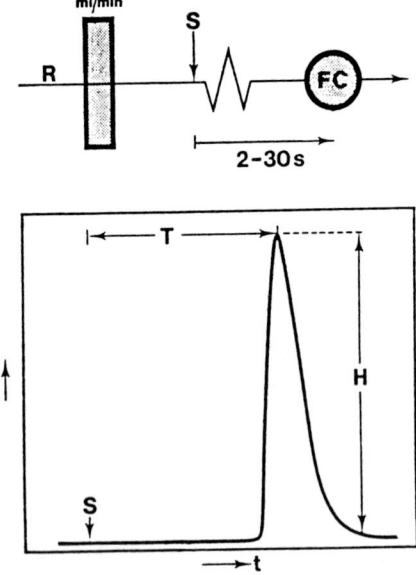

Fig. 2 Single-line FIA manifold and typical peak output

tubing (0.5-0.8 mm i.d.) and the detector monitors an inherent
property of the sample. Sample residence time within the manifold
is typically 2-30 s (Fig. 2).

For on-line physical treatment, procedures such as dialysis, gas
diffusion and solvent extraction can be readily incorporated. For
on-line chemical derivatization additional reagent lines and mixing
pieces can be added to the basic manifold and solid-phase reactors
can be incorporated. The latter feature is particularly important
for applications requiring immobilized enzymes, e.g. BL reactions.

The major attraction of using FIA specifically as a tool for
monitoring CL and BL reactions is the rapid and reproducible mixing
between analyte and reagent that can be achieved in close proximity
to the detector. This is due to the controlled physical dispersion
of the sample within the confines of the manifold tubing, the length
of which can be easily modified to suit the kinetics of a particular
reaction.

CL AND BL APPLICATIONS OF FIA

The first CL reaction that was adapted to a flow-injection procedure
was the copper(II) catalyzed oxidation of luminol for the
determination of hydrogen peroxide in 1979. Since that time several
variations on the original manifold have been reported for the
determination of a range of metal ions and other catalysts (3). A
typical manifold for the determination of cobalt is shown in Fig. 3
and a recorder trace of the output, demonstrating the rapid sample
throughput that is possible, is shown in Fig. 4. The usual practice
is to maximize the signal to noise ratio by simplex optimization of
a number of key variables such as reagent concentrations, flow
rates, pH, sample volume and mixing coil lengths.

The basic two-line manifold shown in Fig. 3 can be readily extended
to accomodate more complex reaction chemistries and difficult sample
matrices. For example, Fig. 5 shows a four-line manifold
incorporating an on-line dialyser and immobilized enzyme reactor for
the determination of glucose in blood plasma (4). The production of
hydrogen peroxide by the action of glucose oxidase on the analyte
and its subsequent detection via the luminol reaction is a well
known procedure, but the attractions of the FIA method are the ease
with which it can be automated at relatively low cost and the high
sample throughput and reproducibility that can be achieved. Fig. 6
shows another manifold design incorporating an ion-exchange column
for the determination of transition metals (see below).

The incorporation of immobilized enzyme reactors within the FIA
manifold can be extended to include luciferases, providing that they
are immobilized in close proximity to the detector. This is
relatively easy to achieve and procedures have been reported for
ATP (5), NAD(P)H (6) and a range of secondary enzymes and
metabolites. A FIA manifold for the determination of NADH is shown
in Fig. 7; important features are the merging zones design whereby
NADH and decanal are simultaneously injected into separate carrier

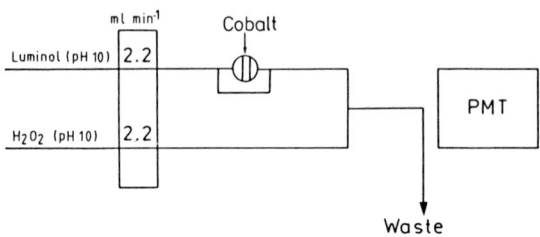

Fig. 3 FIA manifold for the determination of cobalt

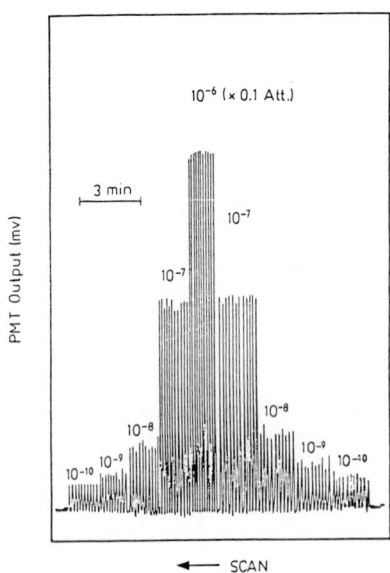

Fig. 4 Typical FIA output for the determination of cobalt

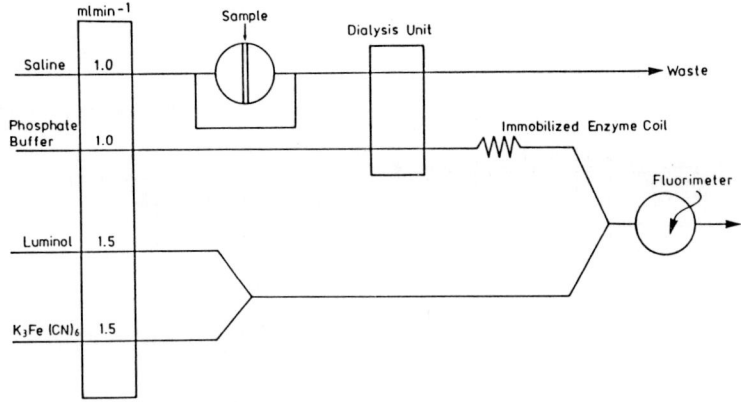

Fig. 5 FIA manifold for the determination of glucose in
       plasma

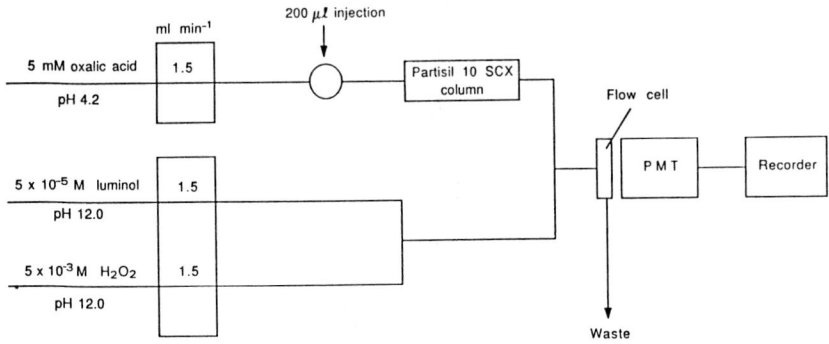

Fig. 6 HPLC/FIA manifold for the simultaneous determination
       of transition metal ions

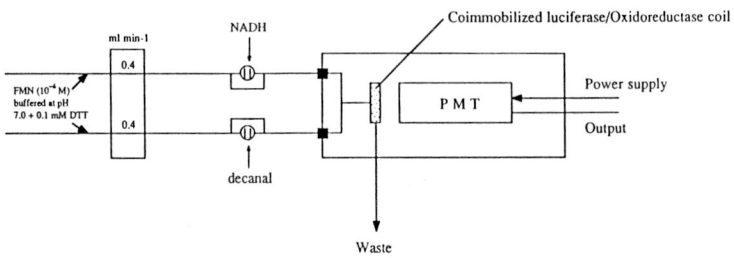

Fig. 7 FIA manifold for the determination of NADH

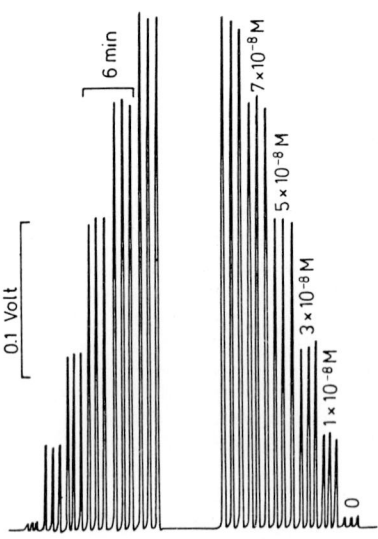

Fig. 8 Typical FIA output for the determination of NADH

streams and merged downstream to minimize reagent and sample
consumption and the co-immobilization of enzymes within the flow-
through detector. A typical recorder output for the determination
of NADH is shown in Fig. 8. The fact that the immobilized enzymes
are continuously in contact with an unsegmented liquid carrier
stream, which can be optimized to maximize enzyme stability, is an
advantage over air segmented systems and immobilized enzyme probes.

Another aspect of FIA that is useful for expanding the range of
analytical applications of CL is the ease with which potential CL
reactions can be investigated and optimized. One example is the
direct determination of morphine in forensic samples by oxidation
with acidic potassium permanganate. Similar procedures have been
reported for a range of pharmaceuticals such as tetracycline and
benzodiazapines. Another example, in the environmental area, is the
direct determination of tertiary amines in sea water by oxidation
with sodium hypochlorite in the presence of rhodamine B as a
sensitizer (7).

Clearly FIA is a useful tool for both the fundamental investigation
of CL and BL reactions and the quantitative analysis of real
samples. For complex matrices however, selectivity can still be a
problem and to this end optimized FIA procedures are best used for
post-column derivatization in conjunction with HPLC.

HPLC
The combination of HPLC separation and CL detection is potentially a
powerful analytical system but there is often a conflict between the
optimum mobile phase conditions required for efficient separation
and the optimum reaction conditions for maximizing CL emission. For
example the basic pH required for the luminol reaction would rapidly
degrade silica based columns, organic modifiers in the mobile phase
quench CL emission from the luminol reaction and nucleophilic
solvents such as water attack aryl oxalates.

Due to the rapid and transient nature of most CL emissions
derivatization is always performed post-column, although for
peroxyoxalate reactions pre-column derivatization of the analyte may
be necessary to produce a suitable fluorophore for post-column
energy transfer. In order that adequate chromatographic resolution
is maintained in the post-column region the detector must minimize
band broadening and the derivatization reaction should be relatively
fast (less than 30 s). In practice the eluant from the HPLC column
can be introduced directly into a FIA manifold (replacing the
injection valve) without significantly impairing resolution.

CL AND BL APPLICATIONS OF HPLC
The CL reagents most commonly used for post-column derivatization
are luminol and the peroxyoxalates TCPO and DNPO (8). The luminol
reaction has been used in conjunction with low capacity cation
exchange resins for the determination of transition metal ions that
catalyze the reaction (9). One possible manifold for such a
determination is shown in Fig. 6. In this particular case 5 mM

oxalic acid at pH 4.2 is used as the mobile phase and the pH is adjusted post-column. The oxalic acid provides good separation of copper, cobalt and iron(II) without seriously suppressing the CL emission. The luminol reaction can also be used to monitor species that suppress the CL emission, such as organic polyacids, and species that displace one of the reactants from a solid support.

The peroxyoxalate reaction is the most extensively used for post-column derivatization. It is applicable over a wide pH range; TCPO being the more stable and DNPO the more sensitive of the two commonest reagents. Polyaromatic hydrocarbons (PAHs) can be determined directly by energy transfer but many more species can be determined indirectly by derivatization to a suitable fluorophore.

The lucigenin reaction, the ruthenium tris-bipyridine system and electrogenerated CL have also been used in conjunction with HPLC. In addition, novel CL reactions, such as that between morphine and acidic permanganate (10), have been developed to solve specific analytical problems where sensitivity is the primary requirement.

The continuing evolution of robust and high efficiency HPLC columns should make CL detection an increasingly viable and attractive proposition and it will not be long before dedicated CL detectors for HPLC are more widely available.

REFERENCES

1.   J. Ruzicka and E.H. Hansen, Flow Injection Analysis, Wiley, Chichester (1988).
2.   M. Valcarcel and M.D. Luque de Castro, Flow Injection Analysis: Principles and Applications, Horwood, Chichester (1987).
3.   A. Townshend, Solution Chemiluminescence - Some Recent Analytical Developments, Analyst 115, 485-500 (1990).
4.   P.J. Worsfold, J. Farrelly and M.S. Mathasu, A Comparison of Spectrophotometric and Chemiluminescence Methods for the Determination of Blood Glucose by FIA, Anal. Chim. Acta 164, 103-109 (1984).
5.   P.J. Worsfold and A. Nabi, Bioluminescent Assays with Immobilized Firefly Luciferase based on FIA, Anal. Chim. Acta 179, 307-313 (1986).
6.   A. Nabi and P.J. Worsfold, Bioluminescent Assays with Immobilized Bacterial Luciferase using FIA, Analyst 111, 1321-1324 (1986).
7.   J.S. Lancaster, P.J. Worsfold and A. Lynes, FIA Procedure for the Determination of Tertiary Amines in Water and Sea Water with CL Detection, Analyst 114, 1659-1661 (1989).
8.   J.W. Birks, Chemiluminescence and Photochemical Reaction Detection in Chromatography, VCH, Weinheim (1989).
9.   B. Yan and P.J. Worsfold, Determination of Cu(II), Co(II) and Fe(II) by Ion Chromatography with CL Detection, Anal. Chim. Acta, in press (1990).
10.   R.W. Abbott, A. Townshend and R. Gill, Determination of Morphine in Body Fluids by HPLC, Analyst 112,397-400(1987).

THE FORENSIC APPLICATION OF LUMINOL AS A PRESUMPTIVE BLOOD TEST

T.E. YESHION

Florida Department of Law Enforcement
P. O. Box 151776, Tampa, Florida 33684

The presumptive determination of a substance as blood on evidence recovered from the scene of a crime may be extremely critical in a criminal investigation. This analysis not only serves to reconstruct the circumstances of the crime, it also provides an element of legal proof, even if only to demonstrate chemical indications for the presence of blood. The presumptive tests also serve as screening tools to distinguish between non-blood substances (rust, paint, cosmetics, food and other reddish-brown materials) which may be visually confused for bloodstains. In most cases, a second opportunity to conduct tests at the scene before the evidence is disturbed will not present itself. Therefore, it is essential that the investigator know which tests are available, how to apply each one and how to accurately interpret the results. Emphasis must also be placed on understanding the mechanism of reaction for each test and on one's ability to effectively communicate technical details to a lay jury.

Since the turn of the century, many catalytic presumptive blood tests based upon the peroxidase-like activity of hemoglobin, have been employed. Among these, the Phenolphthalin, Tetramethylbenzidine and Luminol Tests are the most preferred. Unlike the chromogen based presumptive tests, luminol (5-Amino-2,3-dihydro-phthalazine-1,4-dione) is unique in that its reaction with blood results in the production of light. Luminol is best employed to detect trace quantities of blood which are not visible to the naked eye; e.g., areas intentionally wiped clean of blood, washed clothing, dark surfaces, cracks and crevices, plumbing segments and large areas to be screened. The patterns of deposited and removed bloodstains revealed by luminol are of greatest forensic significance and through these findings, even sequential reconstruction of criminal activity is possible.

Luminol was first synthesized in 1902.[1] However, it was not until 1928 that the chemiluminescent properties of luminol after oxidation in an alkaline solution were described.[2] In 1934, the name "luminol" was given to this compound.[3] In 1936, confirmation of previously reported findings was made and it was observed that reactions with hematin were the most intense.[4]

The first proposal for the use of luminol in medico-legal investigations as a presumptive test for blood was reported by Walter Specht in 1937.[5] Specht's experiments were studied further and confirmed in 1939.[6] Most notable was the finding that dried, decomposed and generally older bloodstains elicited a much more brilliant and longer lasting reaction with luminol than did fresh blood. It was also observed that the luminescence could be reactivated by applying fresh luminol spray after allowing the previous applications to dry and that hematin could be detected in a dilution of $1:10^8$. While luminol is extremely sensitive, it will react non-specifically to any blood regardless of its species origin.

The luminol test is a chemiluminescent reaction based upon its oxidation in an alkaline solution in the presence of an oxidizing agent ($NaBO_3$) and a peroxidase system as found within the hemoglobin molecule of blood. The result of this oxidation reaction is observed as blue luminescence. Blood, or more specifically animal peroxidase, is extremely stable to heat and time. Stains many years old will give excellent reactions. (This author has successfully tested forensic exhibits more than 24 years old.) In fact, luminol will react better to old stains than it will to fresh liquid blood. The drying of a stain exposes the heme molecule to allow it to catalyze the reduction of peroxy compounds (sodium perborate) through its peroxidase-like activity. This in turn, catalyzes the oxidation of luminol. The suggested mechanism of reaction for luminol is as follows[7]:

The luminol test employs the use of sodium carbonate (Na_2CO_3) to create an optimal alkaline solution of pH 10.4-10.8[8], sodium perborate ($NaBO_3$), as an oxidizing agent, luminol as the substance to be oxidized with subsequent light emission, and distilled water. Blood is the peroxidase system which catalyzes the reaction. Although hydrogen peroxide as an oxidizing agent yields a more intense reaction, it is shorter lived than when using perborate. To obtain maximum efficiency, it is necessary to add the chemicals in a particular order to the distilled water. $NaBO_3$ is more soluble in water than in Na_2CO_3 while luminol is more soluble in Na_2CO_3. Therefore, the following classical formula is recommended for stability and efficiency:

I   3.5g $NaBO_3$
II 0.5g Luminol + 25.0g Na_2CO_3
III 500ml Distilled water

Add I + III and shake well. Add II to the solution and shake. Allow sediment to settle and decant into spray bottle. To retain full potency of the solution, it should be made up just prior to use. While the liquid reagent should not be used for more than 8 hours after mixing, the dry chemicals can be maintained indefinitely without loss of activity when stored as described above.

Although not a panacea in criminal investigations, luminol is an extremely useful scientific tool to obtain clues as to what may have happened at the scene of a violent crime. In the author's experience, luminol has been responsible for many on-the-spot confessions and verified theories of criminal activity.

ADVANTAGES OF THE LUMINOL TEST

- Luminol is highly sensitive to trace quantities of blood.
- Results are immediately observed.
- Applied as a spray, luminol can screen large surfaces for the presence of blood, quickly and easily.
- Luminol is non-corrosive and non-staining to target surfaces.
- Luminol is relatively non-destructive to blood. It does not hinder serological determinations; e.g., species orgin determinations and ABO blood groupings. However, it can have deleterious effects on certain electrophoretic systems for polymorphic enzyme typing.

- Luminol reveals trace bloodstains otherwise unnoticeable to the naked eye. It also reveals patterns of placement and removal of bloodstains and chemical oxidants used for cleaning.
- Luminol will not react with body fluids other than blood.
- Observed reactions can be permanently recorded for study and court presentation.
- Luminol is not mutagenic.
- Luminol is an established test with an established track record in forensic investigations. It is accepted by both the scientific and legal communities as a valid presumptive test.

DISADVANTAGES OF THE LUMINOL TEST

- Application of the luminol spray must be conducted in the dark making maneuvers in unfamiliar and often crowded areas awkward.
- Luminol is not specific for blood. The test will react with certain vegetable peroxidases, chemical oxidants and copper-based metals.
- Recording the observed reactions via photography can be tricky and at times difficult to perform in confined and dark areas.
- Luminol does not do well on unsheltered outdoor scenes.
- To properly apply the test and accurately interpret the results, training from a qualified forensic scientist or investigator with case experience is required.

The reason luminol and other screening tests are categorized as presumptive blood tests is because of their lack of absolute specificity to blood. The false positive reactions which occur are generally due to:

- PLANT PEROXIDASES: Plants and vegetables which contain a high concentration of peroxidase (apple, cabbage, cauliflower, horseradish, onion, turnip) cause reactions to the presumptive blood tests. The plant peroxidase appears to be more closely associated with the particulate matter of the plant (which can be observed under a microscope) rather than the juice extract.[9] These reactions are usually very slow in developing and very weak. They can be eliminated by a dry heat treatment (100° C).

- CHEMICAL OXIDANTS: Oxidizing agents such as household cleaners and antiseptics cause reactions to the presumptive blood tests in the absence of blood. When using a multi-step method of testing as with phenolphthalin and tetramethylbenzidine, this category of false positive reactions can be detected by observing a color change prior to the addition of H_2O_2. This reaction cannot be differentiated with luminol alone, but using phenolphthalin as a backup test to luminol, the false positive interference is detected. This information can be used to the advantage of the criminal investigator by showing that a cleaning agent (oxidant) was used to intentionally clean or mask the presence of blood in a particular area. Luminol will reveal the location as well as any notable patterns, provided that blood is still present in detectable levels or that an oxidizing agent was used or a combination of both.

- CHEMICAL CATALYSTS: This group, consisting primarily of copper and its alloys, and nickel salts in a strong solution, give positive reactions to benzidine-like reagents. Usually a weak coloration occurs before the addition of H_2O_2, but is then quickly removed once added. Then, a slow color develops. On filter paper rubbings of the suspected stain, this can be observed as an outer ring of color which gradually extends toward the center. A normal reaction would start from the center and burst outward. With luminol, this reaction can be differentiated from one with blood by the appearance of short, choppy, twinkling luminescence as opposed to a smooth, longer lasting glow observed with a reaction to blood. Luminol will react directly to copper and its alloys, nickel and iron, whereas a solution of these metals and catalysts is required to obtain a reaction to phenolphthalin or tetramethylbenzidine.

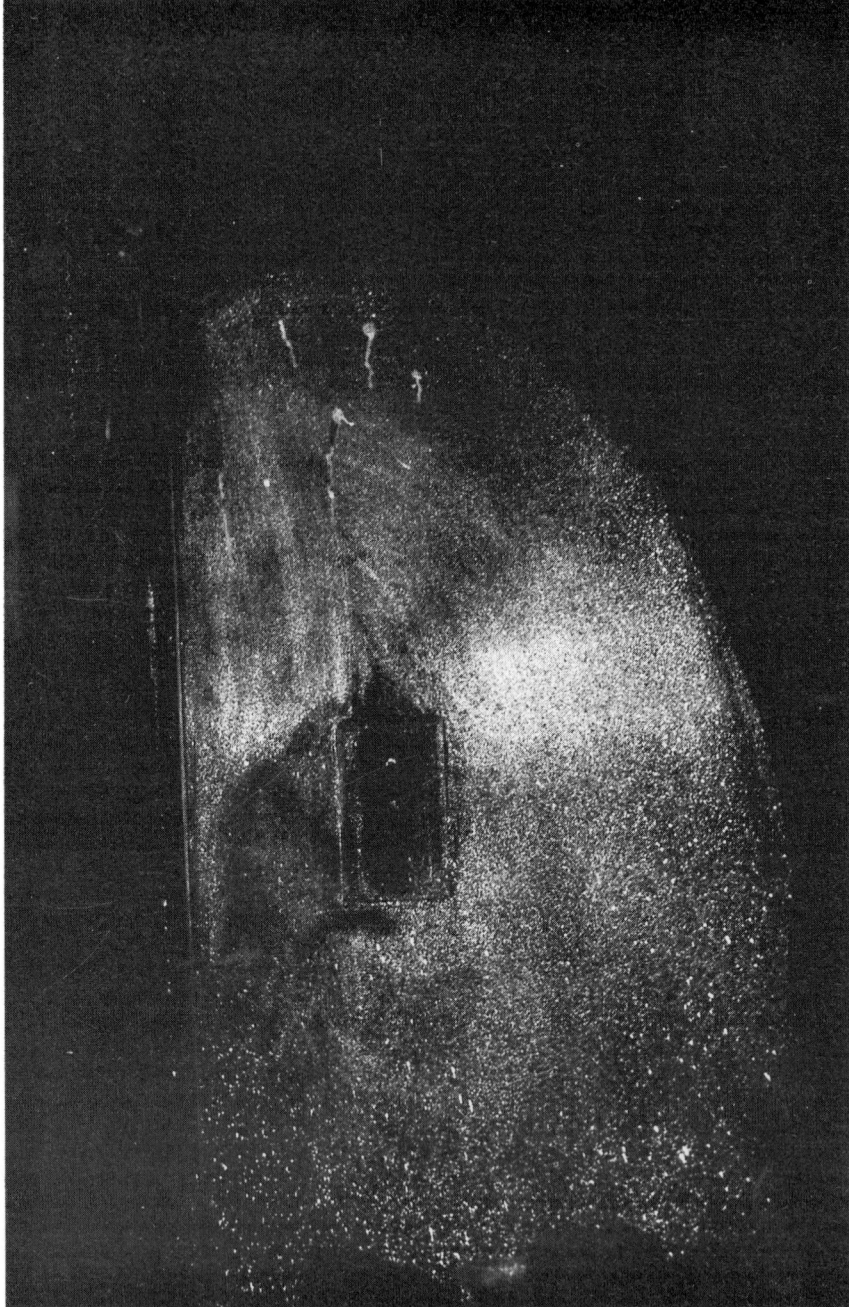

Figure 1.   Luminol applied at the scene of a homicide indicates blood remaining on wall
                 near lightswitch after clean-up.

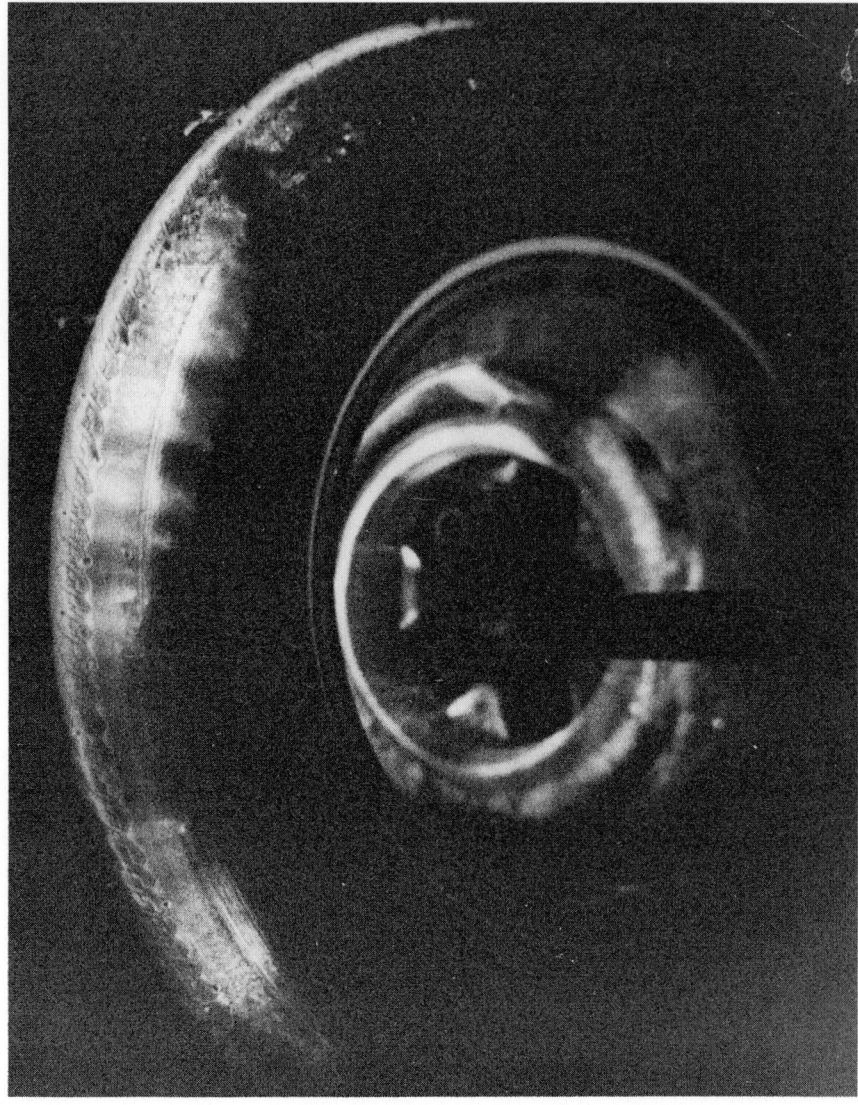

Figure 2.   Luminol applied to a tire from a vehicle involved in a homicide reveals chemical
                   indications for blood.   Sufficient stain was recovered from the tread to
                   demonstrate human blood of the same ABO Type as the victim.

- BLOOD CONTAMINANTS: Due to the high sensitivity of these tests, trace quantities of blood transferred from one area to another will be detected as a strong reaction. Extreme care must be taken to avoid cross contamination.

Luminol is applied in the dark as a spray directly on the evidence to be screened. (Figures 1 and 2). This is an exception to all other presumptive blood tests which are applied to filter paper rubbings or swabbings of the suspected stain so as to avoid any interference with subsequent serological analysis. Prior to spraying, the area must first be examined for other physical evidence including the processing for latent fingerprints. Any visible bloodstains must first be removed and packaged for subsequent serological analysis. At times, however, luminol may reveal certain bloodstains which would have gone completely unnoticed by an examiner. This may be blood soaked within the fibers of a dark colored carpet, dried blood crust in a crevice of furniture, etc.. In these instances, the blood can be collected after luminol has already been applied and analyzed by a forensic serologist to determine:

- that blood is present
- its species origin
- its international blood group (A, B, O, AB)
- the polymorphic enzyme system types and
- the DNA types.

While luminol sprayed onto bloodstains will not compromise the preliminary tests, species origin or ABO determinations, it may compromise the polymorphic enzyme and DNA systems. Therefore, it is critical to remove any visible bloodstains prior to applying luminol.

A negative result with any of the presumptive, catalytic blood tests is conclusive evidence that detectable levels of heme are absent while a positive result only indicates its presence. The ability of luminol, however, to screen large areas quickly and easily with ultra-sensitivity to its target as well as to reveal latent patterns, makes the luminol test an extremely valuable investigative tool.

References:
1. A. Schmitz, Uber das hydrazid der trimensinsaure und der hemimellitsaure in Ber. Dtsch. Chem. Ges., T. Curtius and A. Semper(Eds.), Heidelberg, 46:1162(1913).
2. H.O. Albrecht, Uber die chemiluminescenz des aminophthalsaurehydrazid. Z Physiol. Chem., 136:32(1928).
3. E. Huntress, L. Stanley and A. Parker, The preparation of 3 aminophthalhydrazide for use in the demonstration of chemiluminescence, J. Am. Chem. Soc., 56:241-242(1934).
4. K. Gleu and K. Pfannstiel, Uber 3-aminophthalsaurehydrazid. J. Prak. Chem., 146:137(1936).
5. W. Specht, Die chemiluminescenz des hamins ein hilfsmittel zur auffindug und erkennug forensisch wichtiger blutspuren. Angew. Chemie., 50:155-157(1937).
6. F. Proescher and A. M. Moody, Detection of blood by means of chemiluminescence. J. Lab. Clin. Med., 24:1183-1189(1939).
7. J. I. Thornton and R. S. Maloney, The Chemistry of the luminol reaction - Where to from here?, CAC Newsletter, September, pp. 9-17.
8. W. R. Seitz, Chemiluminescence detection of enzymically generated peroxide. Methods Enzymol., 57:445-462(1976).
9. B. J. Culliford, The Examination and Typing of Bloodstains in the Crime Laboratory, U. S. Gov't. Printing Office, Wash., D. C., p.49(1971).

ASSAYS OF FREE FATTY ACIDS AND GLUCOSE BY HORSERADISH PEROXIDASE CATALYSED LUMINOL REACTION

B. Näslund, P. Arner, K. Bernström, J. Bolinder, L. Hallander[1] and A. Lundin

Clinical Research Centre and Department of Medicine, Karolinska Institute, Huddinge University Hospital, S-141 86 Huddinge, Sweden and
[1] *Department of Pharmacology, Karolinska Institute, S-104 01 Stockholm, Sweden*

INTRODUCTION

The fat cell takes up fatty acids and glucose from the blood. Glucose is the precursor of α-glycerophosphate. Triacylglycerols are formed from α-glycerophosphate and fatty acids. When energy is needed the triacylglycerols are hydrolysed by lipase to glycerol and free fatty acids (FFA). The amount of metabolites released from fat cells can be studied by incubation of isolated fat cells (1) or by *in vivo* microdialysis (2-4).

In microdialysis a probe with a semipermeable membrane is inserted in the tissue of the patient. A solution is pumped through the probe. Metabolites in the interstitial fluid will diffuse through the membrane and can be measured in the dialysate. Only low molecular weight molecules pass through the membrane and the recovery depends on the dimensions of the membrane surface and on the perfusion speed. Concentrations in the dialysate may be as low as 5 % of the concentration in the interstitial fluid and typical sample volumes are in the microliter range. Thus the measurement of metabolites in the dialysate as well as in incubations of isolated fat cells require sensitive analytical methods. The metabolites of most interest are glycerol, glucose and FFA. Previously a kinetic assay of glycerol based on ATP monitoring by the firefly luciferase reaction was developed for lipolysis measurements in fat cell incubations (5, 6) and in dialysates from microdialysis of adipose tissue and blood (3). In our laboratory this method is routinely used for up to 400 glycerol assays per day.

The present paper summarizes the recent development of assays for glucose and FFA in small biological samples. These assays (cf. principles indicated below) are based on endpoint formation of hydrogen peroxide by oxidase reactions from glucose by glucose oxidase (GOD) and from FFA by acyl-CoA oxidase (ACO). Subsequently the resulting hydrogen peroxide is measured by a horseradish peroxidase (HRP) catalysed luminol reaction in the presence of diethylenetriamine-pentaacetic acid (DTPA). The two reactions are automatically performed in the luminometer with 25 samples in a single run. In the glucose assay the rate of the equilibration of α - and β - glucose is increased by aldose mutarotase. To bind the fatty acids released during the incubation of isolated fat cells bovine serum albumin (BSA) is used. Therefore, the FFA assay involves a pretreatment of the samples with sodium dodecyl sulphate (SDS) to liberate FFA from BSA before the acyl-CoA synthethase (ACS) step.

Principles of glucose assay:

All steps performed in the luminometer:

Step 1 a $\quad$ β-D-glucose + O_2 + H_2O $\xrightarrow{\text{GOD}}$ D-gluconic acid + H_2O_2

Step 1 b $\quad$ α-D -glucose $\xleftrightarrow{\text{mutarotase}}$ β-D-glucose

Step 2 $\quad$ 2 H_2O_2 + luminol $\xrightarrow[\text{DTPA}]{\text{HRP}}$ 3-aminophthalic acid + N_2 + 2 H_2O +light

<u>Principles of FFA assay:</u>
Pretreatment steps 1-3 performed outside the luminometer:

Step 1	$BSA\text{-}(FFA)_n + n\ SDS \longrightarrow$	$BSA\text{-}(SDS)_n + n\ FFA$

Step 2 a $FFA + ATP + CoASH \xrightarrow{\quad ACS \quad} acyl\text{-}CoA + AMP + PP_i$

Step 2 b $PP_i + H_2O \xrightarrow{\quad pyrophosphatase \quad} 2\ P_i$

Step 3 $CoASH + NEM \longrightarrow$ covalent binding of CoASH

Step 4 $acyl\text{-}CoA + O_2 \xrightarrow{\quad ACO \quad} 2,3\text{-}trans\text{-}enoyl\text{-}CoA + H_2O_2$

Step 5 $2\ H_2O_2 + luminol \xrightarrow[DTPA]{\quad HRP \quad} 3\text{-}aminophthalic\ acid + N_2 + 2\ H_2O + light$

MATERIALS AND METHODS

<u>Instruments</u>: A 1251 luminometer (Bio-Orbit Oy, Turku, Finland) with a sample carousel for 25 samples, temperature control and three dispensers was used. Cuvettes were obtained from Boehringer Mannheim Scandinavia AB (Bromma, Sweden). The luminometer was connected to a Z-88 computer for data collection. Data were transferred to and calculated on a MacIntosh computer. The microdialysis probe (10 x 0.5 mm; cut off point MW 20,000) and the CMA/100 Microinjection pump were obtained from Carnegie Medicin AB (Stockholm, Sweden).

<u>Reagents:</u> The glucose standard and reagents for the spectrophotometric glucose assay were obtained from Apoteksbolaget AB (Stockholm, Sweden). The NEFA-C test kit for the spectrophotometric FFA assay was obtained from Wako Pure Chemical Industries Ltd. (Osaka, Japan). Glucose oxidase (GOD; type V), acyl-CoA oxidase (ACO; from *Candida lipolytica*), acyl-CoA synthethase (ACS; from *Pseudomonas* spec.), horseradish peroxidase (HRP; type VI, dissolved in double distilled deionized water with 50% (v/v) propylen glycol), ascorbate oxidase and adenosine deaminase were obtained from Sigma Chemical Company (St. Louis, MO). All these enzymes, if not otherwise stated, were dissolved in buffer with 50% (v/v) propylen glycol and stored at -20°C. Aldose mutarotase (from hog kidney; the ammonium precipitate was collected by centrifugation and dissolved in double distilled deionized water) and pyrophosphatase (from yeast; the precipitate was dissolved in 10 mmol/l Tris acetate) were obtained from Boehringer Mannheim GmbH (Mannheim, Germany). Luminol was obtained from Biothema AB (Dalarö, Sweden) and stored as a 100 mmol/l solution in dimethyl sulfoxide at +4°C. The Tris buffer used was 50 mmol/l Tris phosphate buffer, pH 7.75, when not otherwise stated.

<u>Microdialysis</u>: Detailed descriptions of the microdialysis device (2) and the experimental protocol (4) have been published previously. In brief, a microdialysis probe was inserted in the abdominal subcutaneous adipose tissue of non-obese healthy subjects and perfused using Ringer´s solution. The perfusion speed was 5 µl/min. At this speed the *in vitro* recovery of glucose was 20 %, which remained constant up to at least 20 mmol/l of glucose (4). The subjects were investigated in the morning after an overnight fast. After a 30-45 min equilibration period 15 min dialysate fractions were collected. After two baseline samples a 75 g oral glucose load was given and the glucose level in the dialysate was followed for 165 min.

<u>Fat cell incubations:</u> Fat cells were isolated (1), and incubated in a shaking waterbath (37 °C) in Krebs-Ringer phosphate buffer, pH 7.4, in a total volume of 500 µl of a solution containing: 1% (w/v) BSA, 1 mg/ml glucose, 0.1 mg/ml ascorbic acid, 1 U/ml adenosine deaminase with and without 0.1 µmol/l isoprenalin. When the shaker is turned off the fat cells float to the surface and cell free samples can be collected from the bottom of the tubes.

<u>Glucose assay:</u> Endpoint incubation was performed in cuvettes containing 170 µl of Tris buffer with mutarotase. Twenty µl of dialysate sample (diluted 1:20 in Ringer's solution) or glucose or H_2O_2

standard (0.5-50 µmol/l diluted in Ringer's solution) were added. After transferring cuvettes to the luminometer 10 µl of GOD was automatically added resulting in the following concentrations (in the 200 µl incubation mixture): 0.05-5 µmol/l glucose, 40 U/ml GOD, 60 U/ml mutarotase in Tris buffer. An incubation for 10 min at 25°C was used. The H_2O_2 formed was determined by the HRP catalysed luminol reaction by measuring the total light emission. After 10 min 790 µl of the luminol reagent (200 µmol/l luminol, 50 mmol/l DTPA in Tris buffer),and then, to start the reaction, 10 µl of the HRP reagent (0.9 mg/ml HRP in Tris buffer) were automatically added resulting in the following final concentrations in 1 ml: 160 µmol/l luminol, 40 mmol/l DTPA, 9 µg/ml HRP and 50 mmol/l Tris buffer and a concentration of glucose from 0.01 to 1 µmol/l.

The luminometer program was as follows: After 5 min of preincubation at 25 °C cuvette no 1 entered the injection position and received oxidase from the first dispenser. After 2 min cuvette no 2 received oxidase and so on. After the addition of oxidase to cuvette no 6, 10 min had passed since oxidase was added to the first cuvette and luminol reagent was added from the second dispenser. The light reaction was immediately started by the addition of HRP from the third dispenser. Then oxidase was added to cuvette no 7 and the luminol and HRP reagents to cuvette no 2. Repeating the procedure in this manner result in 10 min between the addition of oxidase and the luminol reaction. The light emission was integrated for 106 s and the values (in mVs) were first corrected for individual instrument blanks (signal before addition of HRP reagent to each individual cuvette), and then for the reagent blank measured in a separate cuvette with Ringer's solution as sample blank.

FFA assay: In this assay (7) a cell free sample was taken from a fat cell incubation and diluted 1:4 in 50 mmol/l Tris acetate buffer, pH 8.0, with 0.25% (w/v) Triton X-100. A 25 µl aliquot of the diluted sample was treated for 20 min at 37°C with 25 µl 0.2 mmol/l SDS. Endpoint incubation was performed by addition of 25 µl ACS reagent mixture (10 min at 37°C). The final concentrations (in the 75 µl incubation mixture) were: 0.67-67 FFA µmol/l, 0.5 mmol/l CoASH, 2 mmol/l ATP, 5 mmol/l $MgCl_2$, 20 mmol/l KCl, 0.1 U/ml ACS, 1 U/ml pyrophosphatase, 1.5 U/ml ascorbate oxidase, 0.067 mmol/l SDS, 0.25% (w/v) Triton X-100, and 50 mmol/l Tris acetate buffer, pH 8.0. Unreacted CoASH was covalently bound by addition of 25 µl of 5 mmol/l N-ethylmaleimide (NEM). Aliquots (20 µl) were added to cuvettes and automatically analysed with the same luminometer program as for glucose. In the luminometer acyl-CoA was oxidised by ACO (the final concentration in 200 µl Tris buffer was 0.08 U/ml) and the hydrogen peroxide formed was measured by the HRP catalysed luminol reaction as described above.

Other assays: The glycerol concentration was measured in a glycerokinase reaction using the firefly-luciferase reaction as previously described (5, 6). Spectrophotometric assay of FFA were performed using the NEFA-C test kit (8). The glucose reference method was performed at the Department of Clinical Chemistry by a routine method involving GOD and peroxidase.

RESULTS AND DISCUSSION

In the glucose assay a 100% recovery of added glucose was obtained as compared to a H_2O_2 standard. Standard curves for glucose and hydrogen peroxide were identical. Microdialysate obtained from adipose tissue did not interfere with the glucose analysis. With a sample volume of 1 µl dialysate the amount of glucose to be analysed was covered by the linear range of the assay, i. e. 0.01-1 nmol in the cuvette.The results from an experiment using the microdialysis technique were compared to glucose values obtained by a routine spectrophotometric assay (involving GOD and peroxidase) using a 60 fold higher sample volume. The correlation coefficient between the two methods was 0.984. Compared to earlier published luminometric methods for glucose determination the present assay is more convenient. This is due to the fact that the GOD and the luminol reaction are automatically performed without changing medium between the different steps in the procedure. The cost of reagents and disposables is 15 cents per assay.

When fat cells are incubated, BSA has to be added to bind the fatty acids, which are released from the cells. Dilution of the samples to a concentration of 0.25% BSA and a pretreatment with the detergent SDS was necessary. In the FFA assay pyrophophatase was included to hydrolyse PP_i thereby avoiding a reverse reaction. Unreacted CoASH was bound to NEM which otherwise would interfere with the HRP-luminol reaction. For the same reason ascorbate oxidase was included in the ACS incubation. The recoveries of different free fatty acids (C_{10}-C_{18} with the exception of $C_{18:3}$) were above 90% as compared to a hydrogen peroxide standard (7). The results obtained with the luminometric assay of FFA were compared to results obtained with the spectrophotometric NEFA C test. With a sample volume of 25 µl, which is recommended for serum samples, the detection limit was 80 µmol/l for the NEFA method. Therefore we had to modify this method by increasing the sample volume. With 200 µl of fat cell sample it was possible to detect 10 µmol/l FFA in the sample corresponding to 2 nmol of FFA. It should however be pointed out that the absorbancy change at this level was below 0.01 unit. Using a 32 fold smaller sample volume (6.25 µl out of which only one fifth was used in the luminometer) it was possible to measure 5 µmol/l FFA in the original sample by the luminometric method. The coefficient of correlation between the two methods was 0.997. Compared to spectrophotometric assays of FFA with a detection limit normally at 1-10 nmol in the original sample, the luminometric method is much more sensitive with the analytical range between 0.05-5 nmol in the original sample. The cost per assay is approx. 50 cents. A bioluminometric method reported earlier (9) has similar sensitivity, but involves 7 enzymatic steps and has a rather limited linear range. Among other things we have used our luminometric methods to measure the simultaneous release of FFA and glycerol from isolated fat cells. The concentrations of the two metabolites increased in parallel in the medium as would be expected. Furthermore the rates of the release were stimulated by isoprenalin, which is a powerful lipolytic agent.

In summary we have developed simple and sensitive luminometric assays for glycerol, glucose and free fatty acid. These techniques are now used in conjunction with microdialysis and isolated fat cell techniques to study lipid metabolism in human adipose tissue. Such studies include the mechanisms of hormone regulation and diseases such as diabetes and obesity.

ACKNOWLEDGEMENTS

This investigation was supported by grants from the Swedish Medical Research Council (No. MRF 19X-01034), the Swedish Diabetes Association, the Nordic Insulin Foundation, the Karolinska Institute, the Osterman and Stohne Foundations, the Swedish Athletics Confederation.

REFERENCES

1.   M. Rodbell, Metabolism of isolated fat cells. I. Effects of hormones on glucose metabolism and lipolysis. J. Biol. Chem. 239, 375-380 (1964).

2.   U. Tossman and U. Ungerstedt, Microdialysis in the study of extracellular levels of amino acids in the rat brain. Acta Physiol. Scand. 128, 9-14 (1986).

3.   P. Arner, J. Bolinder, A. Eliasson, A. Lundin and U. Ungerstedt, Microdialysis of adipose tissue and blood for in vivo lipolysis studies. Am. J. Physiol. 255 (Endocrinol. Metab. 18): E737-E742 (1988).

4.   J. Bolinder, E. Hagström, U. Ungerstedt and P. Arner, Microdialysis of subcutaneous adipose tissue in vivo for continuous glucose monitoring in man. Scand. J. Clin. Lab. Invest. 49, 465-474 (1989).

5.   A. Lundin, P. Arner and J. Hellmér, A new linear plot for standard   curves in kinetic substrate assays extended above the Michaelis-Menten constant: Application to a luminometric assay of glycerol. Anal. Biochem. 177, 125-131 (1989).

6.   J. Hellmér, P. Arner and A. Lundin, Automatic luminometric kinetic assay of glycerol for lipolysis studies. Anal. Biochem. 177, 132-137 (1989).

7.   B. M. A. Näslund, K. Bernström, A. Lundin and P. Arner, Free fatty acid determination by peroxidase catalysed luminol chemiluminescence. J. Biolum. Chemilum. 3, 115-124 (1989).

8.   W. G. Duncombe, The colorimetric micro-determination of nonesterified fatty acids in plasma. Clin. Chim. Acta 9, 122-125 (1964).

9.   H. Kather and E. Wieland, Free fatty acids: Luminometric method. In Methods of Enzymatic Analysis, H. U. Bergmeyer, J. Bergmeyer and M. Grassl (Eds), Verlag Chemie, Weinheim, p. 25-34 (1985).

LUMINESCENT ASSAY FOR TOTAL PEROXYL RADICAL-TRAPPING CAPABILITY OF PLASMA.

Timo Metsä-Ketelä

Department of Biomedical Sciences, University of Tampere,P.O.B. 607 33101 Tampere, Finland.

INTRODUCTION

The involvement of free oxy-radicals and lipid peroxidation in the development of several disease states including cancer, rheumatoid athritis, atherosclerosis and consequent post-ischemic reoxygenation injury of heart and other organs has been proposed (1). In addition to the classical antioxidant vitamins A, C and E the human plasma contains several other compounds which can be referred as antioxidants (2). Wayner et al. (3) have developed a method and introduced a term Total (peroxyl) Radical-trapping Antioxidant Parameter (TRAP). They exposed plasma sample to peroxyl radicals generated by thermal decomposition of an azo compound, ABAP. They measured the induction period of oxygen comsumption caused by antioxidants with an oxygen electrode. In the present study, a new accurate luminescent method for TRAP was developed.

MATERIALS AND METHODS

Reagents ABAP (2,2-azo-bis(2-amidinopropane hydrochloride)) was purched from Polysciences (Warrington, PA, USA), luminol (5-amino-2,3-dihydro-1,4-phthalazinedione) from Sigma (St Louis, MO, USA) and linoleic acid from Nu Chek Prep (Elysian, MN, USA). Linoleic acid was purified free of lipid hydroperoxides with a silicic acid column and stored in benzene at - 20°C. The sodium salt of linoleic acid was prepared by addind 2 M NaOH to the suspension of linoleic acid and saline until the suspension became clear. The pH of this solution was adjusted between 8 and 9. Trolox was a generous gift from F. Hoffmann-La Roche Ltd, (Basel, Switzerland).

Assay protocol The reaction was initiated by mixing 450 ul of oxygen-saturated sodium phosphate buffered (100 mM, pH 7.4) saline with 50 ul of 400 mM ABAP (prepared in 100 mM phosphate buffered saline) ,25 ul of 160 mM sodium linoleate and 50 ul of 10 mM luminol. The cuvette was placed in the temperature controlled sample carousel of the luminometer (37°C). The computer automatically measures the chemuluminescence (CL) of each cuvette in intervals from 50 ms (one cuvette, maximal frquency) to about 40 sec (six couvettes). Simultaneously with the chemiluminescence the computer reads the time for the measurement. The sample was injected at 15 min, when the chemiluminescence was already stabilized. The sample or standard in a volume of 20 ul was

injected directly into the cuvette and the computer continued
measuring the chemiluminescence at the earlier intervals.
TRAP of unknown sample can be calculated from the duration of
extinction of chemiluminescence and from the fact that trolox,
like other phenolic antioxidants, are known to trap two radicals
per molecule (3).

RESULTS AND DISCUSSION

Figure 1 shows the time response curve and the effect of linoleic
acid on ABAP induced chemiluminescence. The chemiluminescence is
greatly potentiated by linoleic acid. This could at least partly
be explaned by an inhibiton of the termination reaction of two
ABAP derived peroxyl radicals.

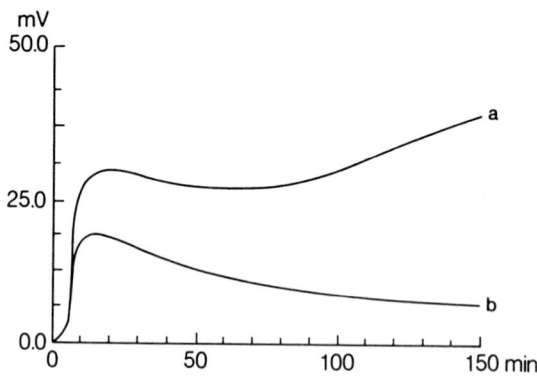

Figure 1. Effect of sodium linoleate on ABAP induced
chemiluminescence. a) 4 umol/tube linoleic acid, b) control.

Figure 2 shows the response of a water soluble tocopherol ,
derivative, trolox on the chemiliminescence. The extinction and
the recovery of CL are complete and very rapid.The the duration of
extiction caused by an antioxidant correlates with the amount of
the antioxidant The linear regression line for trolox is y =
131.7x + 43.2, r = 1.00. Consequently, the production rate of
peroxyl radicals was 0.911 nmol/min. The additon of linoleic acid
do not affect the duration of extinction caused by an antioxidant.
The effect of human plasma on CL has been shown in figure 3. The
peroxyl radical production rate was controlled by an addition of 5
nmol oftrolox. The TRAP-value for sample a is 854 uM and for
sample b 1245 uM.

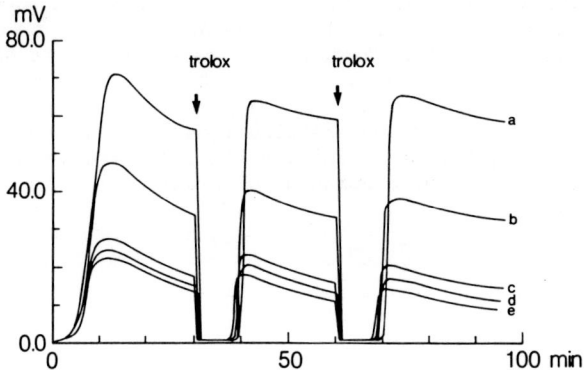

Figure 2. Effect of linoleic acid on the duration of extinction of CL induced by trolox. a) 8 umol/tube, b) 4 umol/tube, c) 1 umol/tube, d) 0.5 umol/tube and e) control. Trolox (5 nmol) was added twice.

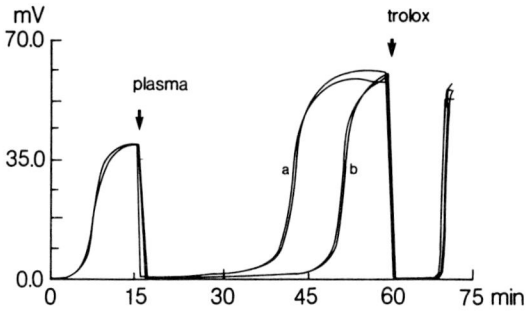

Figure 3. Determination of human plasma TRAP.

ACKNOWLEDGEMENTS
The study was supported by a grant from Finnish Academy of Sciences.

REFERENCES

1) C.E. Cross, B. Halliwell, E.T. Borish, W.A. Pryor, B.N. Ames, R.L. Saul, J.M. McCord, D. Harman, Oxygen radicals and human disease. Ann. Intern. Med. 107, 526-545.

2) B. Halliwell and J.M.C. Gutteridge, The antioxidants of human extracellular fluids. Arch. Biochem. Biophys.280, 1-8 (1990).

3) D.D.M. Wayner, G.W. Burton, K.U. Ingold, L.R.C. Barclay and S.J. Locke, The realative contributions of vitamin E, urate, ascorbate and proteins to the total peroxyl radical-trapping antioxidant activity of human blood plasma.

CHEMILUMINESCENT REACTIONS OF LUMINOL OXIDATION INDUCED BY DIFFERENT PEROXIDASES. STOP-FLOW MEASUREMENTS.

B.B. Kim, V.V. Pisarev, A.M. Egorov

Chemistry Department, M.V. Lomonosov Moscow University, Moscow, 199899, USSR

INTRODUCTION

In recent years, enzyme immunoassay with chemiluminescent detection has gained popularity. In particular, the reaction of enhanced chemiluminescence - cooxidation of luminol and enhancer (e.g. certain phenols, aromatic amines) by hydrogen peroxide, catalyzed by Horse radish peroxidase (HRP) is widely applied (1). With this in view the investigation of peroxidases from different sources is of great interest. In this work we explored the mechanism of luminol oxidation by H_2O_2 in presence of peroxidases from the *Medicago Sativa* L-1 (cell culture, LP-1), Horse radish and fungus *Arthromyces Ramosus* (ARP). Some catalytic properties of these enzymes have been previously reported (2,3,4). It has been shown that enzyme preparations differ in their catalytic activities towards luminol. In this work we studied the mechanism of luminol oxidation induced by several peroxidases by stop-flow method.

MATERIALS AND METHODS

Reagents. HRP_b, HRP, basic isozymes, RZ 3.0; HRP_{ac}, HRP, acid isozymes, RZ 2.8, (Agrotekhnika, USSR); HRP_d, HRP, dimeric form, basic isozymes, RZ 3.1 (NPO Biolar, USSR); ARP, RZ 1.8, acid isozymes (Suntory Ltd., Japan); LP, basic isozymes, isolated and purified as reported elsewhere (2); luminol (Amersham, Great Britain); other reagents (Reakhim, USSR).

Assay protocol. Isoelectrofocusing was performed on a Mini-IEF instrument (Bio-Rad, USA). The dependence of chemiluminescence intensity on pH and substrate concentrations was studied on a 1251-OO2 luminometer (Wallac, Finland). Spectrophotometric measurements were made on a DU 8B instrument (Beckman, USA). FPLC (Pharmacia, Sweden) was used for chromatography. Molecular weights of the peroxidases were determined by gel filtration on a precalibrated TSK G3000SW column (LKB, Sweden) and by SDS PAGE electrophoresis.

The presteady-state kinetics was studied by the stop-flow technique on a RA-401 spectrophotometer (Union Giken, Japan) as described elsewhere (3). The spectral properties of the oxidized peroxidase forms were studied in the range 360-480 nm.

RESULTS AND DISCUSSIONS

Biochemical properties of the peroxidases are summarized in Table 1. All the peroxidases under investigation, are monomeric (except HRP_d) and have similar molecular weights. Peroxidases have a typical spectrum of hemin-containing proteins with the Soret band at 350-425 nm and the absorption maximum near 403 nm. The minor variations of the molar absorption coefficient values may be assigned to the differences in protein surrounding of the hemin. According to isoelectrofocusing data enzyme preparations were not homogeneous and consisted of several isozymes, which are usually divided into acid (pI>7) and basic (pI<7) forms. Peroxidases were shown to differ in their thermostability. The rate constant of thermoinactivation was notably higher for ARP compared with HRP.

Table 1. Biochemical properties of peroxidases.

PROPERTY		PEROXIDASE			
	HRP_b	HRP_{ac}	HRP_d	ARP	LP
Molecular weight, Da	44,000	42,000	88,000	41,000	48,000
λ_{max}, nm	403	403	403	403.5	403
$\xi * 10^{-5}$, l/(mol*cm)	1.02	1.15	1.05	1.02	1.02
Isoelectric points	7.2, 8.7, 9.05	3.5, 4.5	-	3.6, 3.8	8.9, 9.2 9.3
Rz	3.0	2.8	3.1	1.8	2.5
k_{inact}, min^{-1}	$1.2*10^{-3}$	-	-	$7*10^{-3}$	$3.6*10^{-3}$

Influence of pH on the light intensity from the reaction of luminol oxidation induced by peroxidases under investigation is shown in Fig.1 and Table 2. The curves are complicated; it may reflect a complex nature of acid-basic equilibria in solution and differences in pK of the enzyme groups important for catalysis.

Study of the presteady-state kinetics was based on the assumption that the Chance (5) scheme was valid for all the peroxidases under investigation:

$$E + H_2O_2 =======> E_1 \qquad\qquad k_1$$
$$E_1 + LH^- =======> E_2 + LH^{\cdot} \qquad k_2$$
$$E_2 + LH^- =======> E + LH^{\cdot} \qquad k_3,$$

where E is peroxidase; E_1, E_2 are compounds I and II of the enzyme; LH^- is the ionized form of luminol; $LH^{\cdot}$ is the luminol radical; k_1, k_2, k_3 are the second-order rate constants.

The values, obtained by the stop-flow method are given in Table 2. The rate constants depend on the source and isozyme composition of the peroxidases. The latter fact was discussed earlier, basing on the stationary kinetics data (6). The most significant differences (up to 3 orders) are observed for the k_3 values. This constant is very important for peroxidase action because it usually limits the rate of the enzyme turnover. Dimeric form of the peroxidase, HRP_d, is more active than monomeric. This phenomenon

and its presumable physiology role have been recently discussed
(7).

Table 2. Catalytic properties of the peroxidases in the reaction of luminol oxidation.

PROPERTY	ENZYME					
	HRP_d (3) luminol	p-I-phenol	HRP_{ac} luminol	HRP_b luminol	LP luminol	ARP luminol
pH-optimum	8.8	8.5^3	8.2	8.7	8.0	8.8
k_1, 1/(mol*s)	$1.9*10^7$	$1.9*10^7$	$1.6*10^6$	$9.4*10^6$	$9.6*10^6$	$5.3*10^6$
k_2, 1/(mol*s)	$8.0*10^5$	$2.8*10^7$	$2.0*10^6$	$3.7*10^5$	$4.2*10^6$	$2.4*10^7$
k_3, 1/(mol*s)	$1.2*10^4$	$3.4*10^6$	$5.3*10^3$	$3.9*10^3$	$1.2*10^4$	$2.2*10^6$
$V_{g,LH}$, mol/(1*s) [1]	3	871^3	1.3	1	3	308
I_{cl}, steady state [1,2]	1.6	1400^3	1.3	1	32	512

[1] The values are normalized to those, obtained for HRP_b induced luminol chemiluminescence

[2] The values obtained at optimal conditions (pH, substrate concentrations) for each peroxidase.

[3] For the reaction of enhanced chemiluminescence.

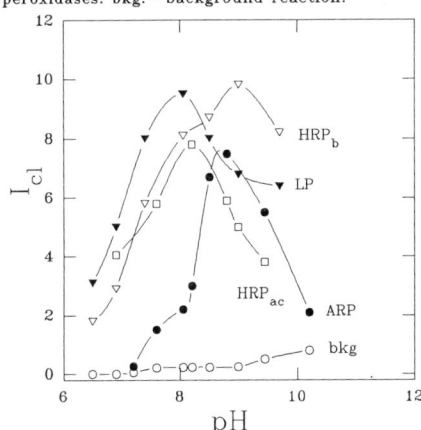

Figure 1. Dependence of light emission from the reaction of luminol oxidation on pH for different peroxidases. bkg.- background reaction.

The stop-flow method allows to explain the high activity of ARP in luminol oxidation. Compare the rate constants of the reactions of the oxidized forms of ARP and HRP with substrates (Table 2). One can see that k_2 and k_3 values for ARP - induced luminol oxidation are notably higher, than those for HRP and close to the values, characterizing p-iodophenol oxidation in presence of HRP. Thus luminol is a "good" substrate for ARP and its oxidation needs no presence of the second substrate, enhancer.

From the steady-state approximation one can calculate $V_{g,LH}$, the stationary rate of generation of luminol radicals by the formula:

$$V_{g,LH} = \frac{2*k_1*k_2*k_3*[LH^-]_o*[H_2O_2]_o*[E]_o}{k_2*k_3*[LH^-]_o + k_1*(k_2 + k_3)*[H_2O_2]_o} \qquad (I)$$

where $[LH^-]_o$, $[H_2O_2]_o$, $[E]_o$ are the initial concentrations of luminol, hydrogen peroxide and peroxidase.

Assuming that the enzymatic steps are rate-limiting for luminol oxidation, we can conclude that light intensity, I_{cl}, must be proportional to $V_{g,LH}$.. The results of calculations and experiments, are given in Table 1 (the values are normalized to those, obtained for HRP-induced chemiluminescence). They show that, except LP, for all peroxidases there is a correlation between the two parameters.

ACKNOWLEDGMENT.
We thank Dr. Teruo Amachi (Suntory Ltd., Japan) for providing *Arthromyces ramosus* peroxidase, Dr. I. Gazarian (Moscow University) for providing peroxidase from *Medicago Sativa* cell culture and Dr. S. Vlasenko (Moscow University) for the help with stop-flow measurements and for productive discussions.

REFERENCES
1. Kricka, L. J. and Thorpe G.H.G. (1990). Bioluminescent and Chemiluminescent detection of Horseradish peroxidase labels in ligand binder assays. In: Luminescence Immunoassay and Molecular Applications, K. Van Dyke and R. Van Dyke (Eds.), CRC Press, Boca Raton, p.78-98 (1990).
2. Gazarian I.G., Veryovkin A.N., Pisarev V.V., Kim B.B. and Yegorov A.M. Catalytic properties of *Medicago Sativa* L-1 peroxidase. Abstracts of the II International Symposium on Molecular and Physiological Aspects of Plant Peroxidases, Lublin (Poland) 27-29 August 1990, p.125 (1990).
3. Vlasenko, S.B., Arefyev, A.A., Klimov, A.D., Kim, B.B., Gorovits, E.L., Osipov, A.P., Gavrilova, E.M., Yegorov, A.M. An Investigation on the Catalytic Mechanism of Enhanced Chemiluminescence: Immunochemical Application of this Reaction. J. Biolum. Chemilum. **4**, 164-176.(1989).
4. Shinmen, Y., Asami, S., Amashi, T., Shimizu, S., Yamada, H. Crystallization and Characterization of an Extracellular Fungal Peroxidase. Agric. Biol. Chem. **241**, 247-249. (1986).
5. Dunford, H.B., Peroxidases., Adv. Inorg. Biochem. **4**, 41-51, (1990).
6. Jansen E. H.J.M., Berg R. H. van den B., Zomer G. Horseradish peroxidase as label in chemiluminescent immunoassay. In: Luminescence Immunoassay and Molecular Applications, K. Van Dyke and R. Van Dyke (Eds.), CRC Press, Boca Raton, p.58-75 (1990).
7. Nakajima R., Hoshino N., Yamazaky I. Oxidative decomposition of oxyperoxidase during peroxidase reactions - effect of localization of the enzyme. Abstracts of the II International Symposium on Molecular and Physiological Aspects of Plant Peroxidases, Lublin (Poland) 27-29 August 1990, p.23-25 (1990).

A NOVEL ASSAY OF O_2^- GENERATION IN IN SITU STOMACH OF EXPERIMENTAL RAT: ISCHEMIA-REPERFUSION

Minoru Nakano and Atsushi Takahashi

College of Medical Care and Technology, Gunma University
Maebashi, GUNMA, JAPAN

Introduction

We have first demonstrated the O_2^- generation in in situ lungs of rats, who treated with drugs to induce experimental acute distress syndrom under the contineous infusion of MCLA (a cypridina luciferin analog), using a sensitive photon counter to detect MCLA-dependent luminescence from lung surface (1). Recent studies have implicated reactive oxygen metabolites (O_2^-, H_2O_2 and $\cdot OH$) in ischemia-induced gastric mucosal injury (2, 3). However, there is no direct evidence for the generation of O_2^- in ischemia or reperfusion after ischemia. If O_2^- generates in gastric mucosa of rats in ischemia-reperfusion, it would be detected as MCLA or another cypridina luciferin analog (CLA)-dependent light emisssion from the surface of gastric mucosa.

The objectives of the present work are 1) to determine whether ischemia and reperfusion of stomach produce O_2^-, 2) to determine whether SOD or allopurinol (an inhibitor of xanthine oxidase) modifies the ischemia and reperfusion induced O_2^- generation, 3) to know the relationship between the O_2^--induced luminescence and gastric mucosal injury (or leukocyte infiltration).

Materials and Methods

Surgical procedure and photon counting system-With male wister rat (about 300g) anesthetized with pentobarbital, a microcannula was inserted into right juglar vein and placed in superior vena cava to be used for CLA infusion. Left femoral artery and vein were also cannulated for monitoring arterial blood pressure and SOD injection, respectively. A midline abdominal incision was made, and gastroesophagal junction and duodenum at a site of 0.5 cm distal to the pylorus were ligated. A vascular front area of stomach was incised, and gastric mucosa was outwardly placed on the abdominal wall. A. and V. gastric sinistra were clamped with esophagus by a forceps for preparing ischemia. Area interupting the photon counting was covered with a black cloth. Single photon counting system was essentially the same as described previously (1). Preparation of experimental rats-Leukocyte depleted rats were prepared according to the method described previously (1). To prepare allopurinol treated rat, allopurinol (50 mg/kg) was orally administered for 3 days perior surgery. Just before the experiment, additional allopurinol (50 mg/kg) was intravenously injected to the rat. Histologic technique and morphologic evaluation-Samples of stomach isolatd from the experimental rats were quickly placed in 10% normal formalin. The tissue was sectioned at a small omentum site, embeted in paraffin, sectioned and stained with hematoxylin-eosin. These sections were examined by light microscopy (x 1000), with respect to leukocyte infiltration (a number of leukocytes between lamina propria and mucosal base), edema formation (thickness between lamina propria and mucosal base) and erosion formation. Histological evaluation was done according to the procedure described by Mangino et al (4).

Results and Discussion
1) O_2^- generaion and leukocyte infiltration
        Typical chemiluminescence patterns in three stages (preischemia during ischemia and after reperfusion) are shown in Fig. 1A. When both left gastric artery and vein were clamped surgically during 30 min at the stage at which a constant (CLA-dependent) luminescence (a non-specific luminescence; NSCL) was observed, the luminescence fell down to about 1/2 of NSCL. The release of the clamp caused a rapid raiseof the luminescence, which decreased after peaking and then again went up to 1.3 to 1.5 times that of NSCL. Then the increased luminescence continued for about 80 min and fell to NSCL level at about 130~150 min after the reperfusion. Such a long lasted chemiluminescence over NSCL observed after the reperfusion

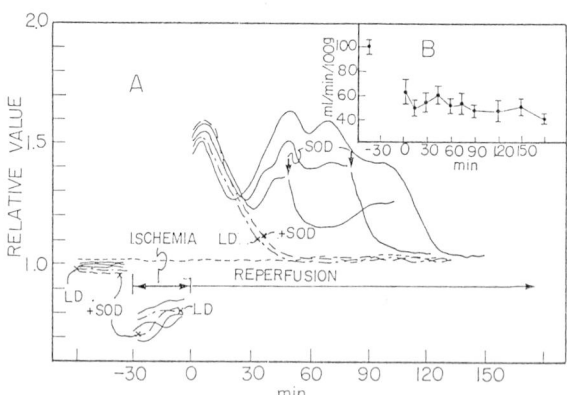

Fig. 1A. CLA-dependent chcmiluminescence from the surface of gastric mucosa of rats without and with leukopenia(LD) and the effect of SOD on the luminescence. The 0.1m mol/ℓ CLA at a Flow rate of 5 ml/kg/h was continuously injected during the time observed. The 10 mg of SOD/h(SOD) were continuously infused with the CLA solution or injected in bolus at the time indicated by arrow. Fig.1B. Mucosal blood flow of pre-ischemia and 30 min of ischemia followed by reperfusion.

completely abolished by a contineous infusion of SOD and CLA and rapidly fell down by a single injection of SOD, indicating the participation of O_2^- in the long lasted chemiluminescence(O_2^--induced luminescence). Under the same experimental conditions, the O_2^--induced luminescence was not detected from the surface of gastric mucosa of leukocyte depleted rat. These results clearly indicate that O_2^-, which could be detected by the chemiluminescence method, mainly generates by activated leukocytes at the place where SOD is easily reactable. As shown in Fig. 2B, the fundic mucosal blood flow after the reperfusion was approx. 30% of that observed before ischemia and continued to slowly fall through the 180 min observed period. On the histological examination of submucosal area the leukocyte infiltration occured even at 15 min of reperfusion, increased at 75 min and decreased to the level of 15 min of reperfusion. The comparison of the chemiluminescence intensity (Fig. 1A) with the leukocyte infiltration

may give us a cease of their oxidative metabolism at 150 min after
reperfusion, probably by a lack of activator or by undergoing their death.
(deta not shown)

2) Effects of Postischemic O_2^- generation and histological change on the
   duration of ischemia
      Typical chemiluminescence patterns obtained by reperfusion after
ischemia of 5 min and 30 min are shown in fig.3.   For the convenience,
relative value of O_2^-
generaion was calculated
by integration of
chemiluminescence
intensities (over NSCL)
from 25 min to 150 min
after reperfusion (Fig.
3A). With such a
expression, O_2^-
generated by
reperfusion after
ischemia of 30 min were
found to be about 2
times those for
ischemia of 15 min,
while little or no O_2^-
generated by
reperfusion after
ischemia of 5 min
(Fig.3B).   To know the
relationship between
O_2^- generation and
histological change in
gastric mucosa, three
indexes, such as
leukocyte infiltration,
edema formation and
erosion formation, were
chosen for the latter
and were compared with

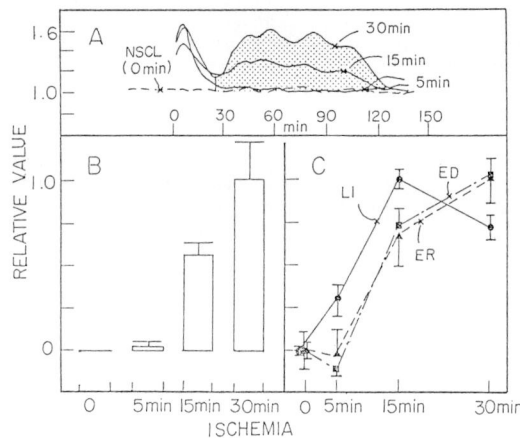

Fig.2A,B.   Effect of postischemic O_2^--induced luminescence
on the duration of ischemia.   Experimental procedure were
essentially the same as in the legend for Fig.1. Relative
O_2^- generation was expressed as shaded area (A).
Fig.2C.   Histological changes at 2.5 h of reperfusion after
the certain period of ischemia. LI, leukocyte infiltration;
ED, edema formation; ER, erosion formation.

relative O_2^- generation after reperfusion or duration of ischemia.   The
histological examination was carried out with stomach isolated from rats
at 2.5 hr after reperfusion.   As shown in Fig.3C, leukocyte infiltration
was obvious in 15 min-ischemia rats with an opposite relation with O_2^-
generation while edema and erosion fomation were parallel each other and
increased with increasing the duration of ischemia and O_2^- generation
(Fig.3B).   Slightly histological change, identical to non-ischemia rats,
were observed in 5 min-ischemia rats.

3) Effects of allopurinol treatment on O_2^- generation and tissue damage
      A typical CLA-dependent chemiluminescence pattern in three stages
(pre-ischemia, ischemia of 30 min and reperfusion) from surface of gastric
mucosa of allopurinol treated rat are shown, comparing with that for
control, non-treated ischemia-reperfusion rat, in Fig.4A.   This indicated
that regardless to the drug treatment, the luminescence patterns were
essentially the same each other except for the luminescence from 25 min

to 150 min after reperfusion.

　　As shown in Fig. 4B, average of relative O_2^- generation, monitored by the integrated light intensity (over NSCL, intensity) from 25 min to 150 min after reperfusion, in alloprinol treated rats was significantly low, i.e., about 40 % of that in the controls.　On the histological examination of mucosal membrane isolated from rats at 2.5 h after reperfusion, number of leukocytes infiltrated in submucosal area was not obviously reduced (about 68 % of that in the controls), but edema and erosion formations were greatly improved by allopurinol treatment (Fig. 4C).

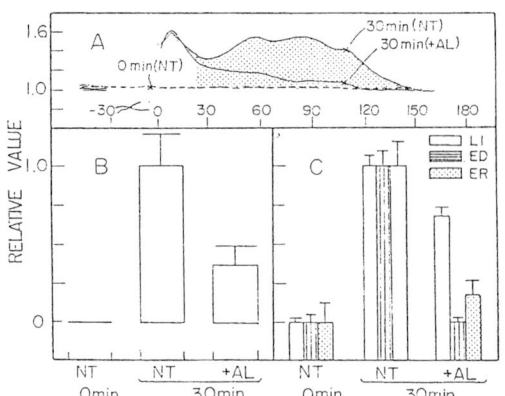

　　From the results obtained here, mucosal injury by infiltrated leukocytes would be mediated exclusively by neutrophil drived potentoxidants, such as O_2^-, H_2O_2, $\cdot OH$ and HOCl.

Fig. 3A, B.　Effect of allopurinol(+AL) on postischemic O_2^- -induced luminescence. The experimental procedure was essentially the same as in the legend for Fig.1. Duration of ischemia was 30 min. Relative O_2^- generation was expressed as shaded area (A).
Fig. 3C.　Histologiucal change in gastric mucosa at 2.5 h after 0 or 30 min of ischemia in rat treated or untreated with allopurinol (+AL). Abbreviations were the same as those in Fig. 2C.

References
1) A. Takahashi, M. Nakano, S. Mashiko, and H. Inaba.　The first observation O_2^- generation in in situ lungs of rats treated with drugs to induce experimental acute respenatory distress syndrome. FEBS Lett. 261, 369–372 (1990)
2) M. Itoh and P. H. Guth. Role of oxygen-derived free radicals in bemorrhagic shock-induced gastric lesions in the rat. Gastroentorology, 88, 1162–1167(1985)
3) W. A. Perry, S. Wadhwa, D. A. Parks, W. Pickard and D. N. Granger. Role of oxygen radicals in ischemia induced lesions in the car stomack, Gastroenterolog. 90, 362–367(1986)
4) M. J. Mangino, C. B. Anderson, M. K. Murphy, E. Brunt and J. Turk. Mucosal arachidonate merabolismand intestinal ischemia-reporfusion injury. Am. J. Physiol. 257, G299–G307(1989)
5) D. A. Parks and D. N. Granger. Ischemia-induced vascular clanges; role of xauthine oxidase and hydroxyl radicals. Am. J. Physiol 245. G285–G289(1983)
6) D. A. Paterson, B. Kelly and J. M. Gerrard. Allopurinol can act as an electron transfer agent is this relevant during reserfusion injury? Biochim Biophys. Res. Commun. 137, 76–79(1986)
7) P. C. Moorhouse, M. Grrotverd, B. Hallowell, J. G. Quinlam and J. M. C. Gutteridge. Allopurinol and oxypurinol are hydroxyl radical scavengers. FEBS Lett. 213, 23–28(1987)

LUMINESCENCE DETECTION OF SUPEROXIDE RADICALS WITH THE PHOTOPROTEIN POLYNOIDIN

M.T. Nicolas, P. Colepicolo*, V. Camarero*, M. Damon**, J.C. Nicolas** and J.M. Bassot

Bioluminescence et Technologie Appliquée à la Microscopie,CNRS, 105 Blvd Raspail, 75006 Paris (France)
*Universitad de Sao Paulo, Instituto de Quimica, Sao Paulo, CP 20780 (Brazil)
**INSERM, U58, 60 Rue de Navacelles, 34090 Montpellier (France)

INTRODUCTION

Polynoidin is the key molecule of the bioluminescent system of scale-worms. In vivo, the system emits series of flashes of 80 ms. The microsources or photosomes (1) are made of 20 nm tubules of endoplasmic reticulum organized in a paracrystalline array. They are excitable (2,3) and bear the photoprotein.

Polynoidin has been isolated and purified by chromatography (4). Its luminescence is specifically triggered by activated oxygen species produced, for ex. by Fenton reagent or by the xanthine/xanthine-oxidase system. Assays with scavengers strongly suggest that the superoxide radical is involved rather than the hydroxyl. Moreover, superoxide dismutase (SOD) inhibits the light emission, and KO2 dissolved in DMSO, an aprotic solvent, triggers it (4).

Thus polynoidin appears as a potential probe to measure reactive oxygen species (ROS) and more specifically the superoxide radicals produced for ex. during inflammatory processes or phagocytosis. ROS are generally detected by luminol. Other luminescent probes, such as lucigenin (5), pholasin (6) (a protein tightly bound to Pholas luciferin and which functions like a photoprotein (7)) and Cypridina luciferin analogs (8) are more specific for superoxide anions.

This paper recalls data obtained on mice neutrophils and macrophages (9) and presents preliminary results obtained on human neutrophils. They confirm the interest of polynoidin, even in crude extracts, as a specific probe for the detection of superoxide radicals generated during physiological processes.

MATERIAL AND METHODS

Specimens and preparation: Common scale-worms (Harmothoe lunulata, annelids, polynoïnae) were collected at low

tide in Brittany. All the elytra from one individual were
carefully removed under anaesthesia and homogenized in
500µl of PBS buffer (pH 7.4). The luminescent activity of
the batch was estimated by Fenton reagent (4).
Neutrophils and macrophages from mice (9): Mice were
injected intraperitoneally with sodium caseinate (12% for
neutrophils, collected after 5 hours; 1.2% for
macrophages, collected after 5 days). Cells were kept in
Dulbecco buffered saline, pH 7.0, until used at
concentrations of 1.6 1Q6 to 15 1Q5 cells/ml. 10-7M
phorbol myristil acetate (PMA) was used for stimulation.
Human neutrophils: Peripheral blood samples were obtained
by venipuncture (10U heparin/ml blood). Neutrophils were
isolated by centrifugation over a percoll discontinous
isotonic gradient (density 1.097 g/ml and 1.086 g/ml).
Cells were washed twice, resuspended in medium "199" at
3   105 neutrophils/ml.
Luminescence assays: PMA was used as stimulus at 10-7M
final concentration. The detectors used were luminol (10-
4M), lucigenin (10-4M), pholasin (10-9M) and polynoidin
(estimated concentration 10-9M). Light emission was
measured on a LKB luminometer (Wallac model 1251), at
37°C,with continuous stirring, during 10 s, every min.
The background (=light emission of unstimulated cells)
was also recorded.

RESULTS AND DISCUSSION
Following the stimulation of polymorphonuclear (PMN)
cells by PMA, a light emission was recorded in presence
of polynoïdin as well as with luminol, lucigenin or
pholasin. The polynoidin signal was much weaker than the
luminol one. However, its background ratio was excellent.
    In mice, the time course of the polynoidin signal was
faster and its amplitude 3 times higher for neutrophils
than for macrophages (Fig.1). The intensity was directly

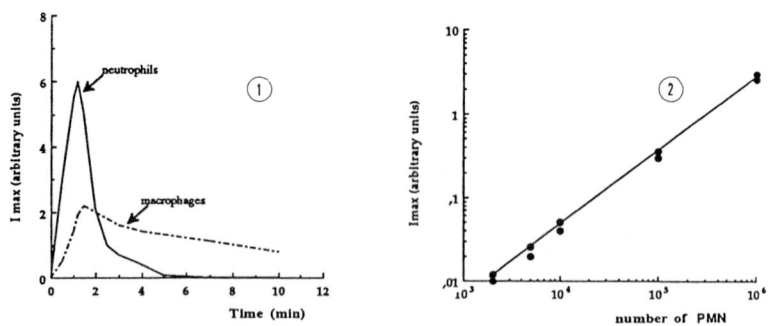

Figure 1. Light emission of polynoidin in the
presence of mice blood white cells stimulated by PMA.
Figure 2. The intensity of the light emitted by
polynoidin is proportional to the number of PMN.

proportionnal to the number of cells (Fig.2). The
specificity of polynoidin for superoxide radicals was
ascertained. In presence of SOD (10U) the luminous signal
of activated PMN was abolished. 50 mM mannitol (a
scavenger of OH.) increased the signal by 20%, and 200U
of catalase (which destroys H2O2) by 80%.
Diethylthiocarbamic acid (1μM), an inhibitor of SOD,
produced a 2 fold increase of the signal, which is
probably due to the endogenous SOD activity present in
the scales' crude extract (9).

  The preliminary results with PMA stimulated human
neutrophils confirmed those obtained in mice. The
polynoidin and pholasin responses had a similar shape,
with a sharp peak and rapid decrease (Fig.3 and 4)
whereas lucigenin and luminol gave a slow-rising signal
(5) (Fig.5 and 6). The fastest response was that of
polynoidin, which reached its maximum 1 min only after
stimulation. The decline of the signal was not due to the
exhaustion of the photoprotein since the solution emitted
an amount of light almost as important as the control
upon addition of Fenton reagent.

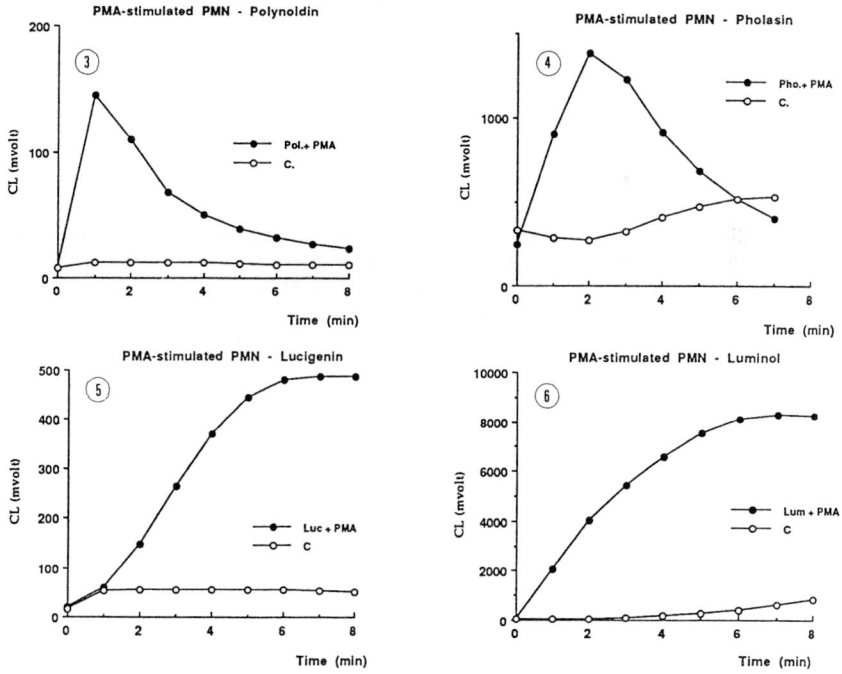

Figures 3 to 6. Amount of luminescence emitted by
polynoidin, pholasin, lucigenin and luminol in the
presence of 10 5 human PMN cells. Measures are made
every min for 10 s. White circles (c): unstimulated
cells. Blak circles: cells stimulated by PMA.

Thus, polynoidin appears to be a specific probe for the superoxide anion. It reveals the earliest event happening during neutrophils stimulation. Crude extracts of scale-worms allow simple and reproducible assays. However, future investigations based on this photoprotein require its complete purification and its cloning.

AKNOWLEDGMENTS:
We thank Dr. A.K. Campbell for his gift of pholasin, C. Le Doucen and the staff of Roscoff Biological Station. The work was supported by CNRS (US45) and INSERM (U58).

REFERENCES:
1. J.M. Bassot and A. Bilbaut, Bioluminescence des élytres d'Acholoe (IV). Biol. Cell. 28, 163-168 (1977).
2. J.M. Bassot, A transient intracellular coupling explains the facilitation of responses in the bioluminescent system of scale-worms. J. Cell Biol. 105, 2235-2243 (1987).
3. J.M. Bassot and G. Nicolas, An optional dyadic junctional comples revealed by fast-freeze fixation in the bioluminescent system of the scale worm. J. Cell Biol. 105, 2245-2256 (1987).
4. M.T. Nicolas, J.M. Bassot and O. Shimomura, Polynoïdin a membrane photoprotein isolated from the bioluminescent system of scale-worms. Photochem. Photobiol. 35, 201-207 (1982).
5. R.C. Allen, Lucigenin chemiluminescence. In: Bioluminescence and Chemiluminescence. M.A. de Luca and W. Mc Elroy (Eds), Acad. Press, N.Y., p.63-73 (1981).
6. T. Müller, E.V. Davies and A.K. Campbell, Pholasin chemiluminescence detects mostly superoxide anion released from activated human neutrophils. J. Biolum. Chemilum. 3, 105-113 (1989).
7. J.P. Henry and A.M. Michelson, Studies in bioluminescence (VIII). Biochimie 55, 75-81 (1973).
8. M. Nakano, Determination of superoxide radical and singlet oxygen based on chemiluminescence of luciferin analogs. Methods Enzymol. 186, 585-591 (1990).
9. P. Colepicolo, V. Camarero, M.T. Nicolas, J.M. Bassot, M.L. Karnovsky and J.W. Hastings, A sensitive and specific assay for superoxide anion released by neutrophils and macrophages based on bioluminescence of polynoidin. Anal. Biochem. 185, 369-374 (1990).

Session VI

FIREFLY LUCIFERASE-ATP:
APPLICATIONS AND ASSAYS

EVOLUTION OF BIOLUMINESCENT ATP ASSAYS

E. Schram

Vrije Universiteit Brussel, 1640-Sint-Genesius-Rode, Belgium

IMPORTANT MILESTONES

Photometry belongs to the most sensitive analytical techniques and has been used in many different ways. Chemiluminescence and bioluminescence are also based on the measurement of light but possess the additional advantage that they do not require sophisticated light sources, optical systems, spectral analyzers, etc. Bio- and chemiluminescence produce their own light and instrumentation reduces to a sensitive detector. However, contrary to other photometric measurements which rely on physical light-matter interactions, the amount of light produced in bio- and chemiluminescence depends on chemical reactions whose rate and efficiency are likely to be influenced by various factors.

Bioluminescence has been applied as early as 1928 for the analysis of biological processes by Harvey who used luminous bacteria to measure the oxygen production by plants kept in anaerobic conditions. The first analytical applications of firefly luminescence date back to the early fifties after the initial discovery of McElroy in 1947 that ATP is an essential element in this reaction. In a survey published in 1954 Strehler and Totter (1) describe the assay not only of ATP but also of related substances and enzymes. With some improvements these methods are still in use nowadays. The next important step was the elucidation of the structure of firefly luciferin and its synthesis by White, McCapra and Field in 1963 (2). In the meantime an impressive amount of data was accumulated by the group of McElroy, soon joined by DeLuca, on the kinetics and physical properties of firefly luciferase. These studies have recently culminated with the synthesis of recombinant luciferase.

At the end of the nineteen-sixties space research has oriented firefly luminescence into new paths, when it was realized that ATP could be used as a measure of biomass and for the detection of life in general (3). Part of the present methodology was developed at that time in the NASA laboratories and can be found in detailed accounts published by this organism, in U.S. Patent 3.745.900 (1973) and in the proceedings of two seminars held in 1975 and 1977 (see reference 4 for more details). The methodology developed at that time includes the since then classical use of a non-ionic detergent (Triton) and apyrase for the selective elimination of somatic ATP prior to the extraction of bacterial ATP. Several methods have been used for the latter purpose (DMSO, inorganic acids as TCA, PCA and HNO_3, or boiling buffer) but all suffer from the same drawback in that they all result in a significant dilution of the sample with a concomitant reduction of sensitivity. A major simplification was therefore achieved by the introduction of cationic extractants (U.S. Patent 4.303.752, 1981). However, their exact composition was not revealed by manufacturers and for many years very little was to be found in scientific literature about their action.

Firefly luminescence was the first bioluminescent system to become chemically elucidated, soon followed by bacterial luminescence, and its broad analytical potentialities were early recognized and applied. Probably therefore it has long remained one of the most popular and most widely applied bioluminescent reactions until, in the course of the last decade, quite a number of new systems were developed. The firefly system itself has further led to unexpected developments: the isolation of the luciferase gene and its expression in foreign cells has indeed supplied genetic engineers with a new and original reporter-gene while production of rec-luciferase will safeguard our needs for this enzyme in future years. Quite another apparently serendipitous discovery, i.e. the synergistic effect of firefly luciferin in the HRP catalyzed luminol reaction, has led to the development of the now well-known enhanced luminescence phenomenon which has found wide applications in luminescent immunoassays.

BASIC BIOCHEMISTRY

Whenever considering the application or development of ATP assays some basic concepts should be kept in mind (4) :

1°) the observed level of luminescence depends on the rate of the reaction; at high enzyme concentrations a higher signal is obtained but luminescence decreases more rapidly because of substrate (ATP) depletion;

2°) the end-product of the reaction dissociates but slowly and is responsible for end-product inhibition of luciferase; at higher ATP concentrations this results in a gradual decay of the luminescence, followed by steady-state luminescence at a lower level;

3°) pyrophosphate, at higher concentrations, inhibits the formation of luciferyl-AMP; at lower concentration (10^{-6} M) it favours the dissociation of the end-product and reduces the level of enzymatic inhibition;

4°) a parameter that in our opinion is still controversial is temperature. Values of 23°C and even 25°C are still quoted as optimal in many papers although, according to our experiments, luminescence efficiency starts decreasing above 20°C. Anyway, as for all enzymatic reactions, assays should be performed at controlled temperature. Working at 20°C will evidently often imply cooling of the measuring instrument;

5°) firefly luciferase is a lipophilic enzyme that is quite sensitive to ionic environments; internal standardization is therefore to be advocated whenever variations in sample composition are likely to affect the luminescence efficiency.

REAGENTS

a) Luciferase-luciferin reagent

Purified reagents have not been commercially available until the end of the sixties (DuPont 1968) but even thereafter crude luciferase preparations have still been used for many years, chiefly for economical reasons. Such preparations contained contaminating enzymes as nucleotide diphosphate kinase, adenylate kinase, etc. which were responsible for a rapid conversion of ATP with a concomitant decay of the luminescence. This difficulty could be circumvented by measuring the height of the light peak observed immediately after addition of the luciferase reagent. Such procedure is however not practical for kinetic measurements involving the monitoring of the ATP over prolonged periods of time and precludes the subsequent addition of an internal standard. In 1976 Lundin and Thore (5) showed that luciferase preparations could be obtained that were sufficiently free of contaminating enzymes to produce a constant light response. The practical concentration range was furthermore improved by the addition of pyrophosphate at low concentration, as well as of L-luciferin, a competitive inhibitor of D-luciferin (6). The luciferase concentration in commercial reagents was adapted in such a way that the luminescence decay does not exceed one percent per minute.

Recent efforts have concentrated on the eliminiation of the residual ATP that was always found even in purified luciferase preparations and which contributed to the blank value. Keeping luciferase preparations (especially concentrated solutions) at room temperature for some hours has often proved efficient in lowering this blank value. Reagents are now being supplied whose blank value is lower than the background signal of available luminometers. The sensitivity of the assays can therefore be increased by turning back to the peak method and using increased enzyme concentrations. (Reagents with high luciferase content are presently available from Sigma and Amersham). It should be noted at this stage that residual ATP in the luciferase reagent is not the only source of ATP likely to increase the blank value. Traces of ATP may contaminate buffers, glassware and pipets and special care is essential when trying to achieve mamximum sensitivity. Buffer solutions in particular should be treated with apyrase and sterilized (the latter treatment will inactivate apyrase).

b) Extraction and destruction of somatic ATP

Triton has been used since the beginning for the selective lysis of somatic cells and is also present in commercial reagents. It does not affect luciferase but its selectivity should not be considered absolute and one should keep aware of the fact that some somatic cells are more resistent to lysis (in particular lymphocytes) and that some bacteria may loose part of their ATP. Apyrase remains the cheapest way to destroy somatic ATP before the extraction of bacterial ATP but is difficult to inactivate. It has therefore been proposed to use membrane ATPase from brain or kidney which can be inhibited by orthovanadate without affecting luciferase significantly (4). Filtration is another easy way to eliminate unwanted apyrase prior to the extraction of bacterial ATP (see further)

c) Extraction of bacterial ATP

The methods proposed initially, and still used as a reference were based on the use of DMSO, inorganic acids (TCA, PCA, HN03) or boiling buffer. All these techniques involve significant dilution of the sample and hence a reduced sensitivity.

Towards the end of the seventies a new type of reagent was introduced, based on the use of ethoxylated amines and quaternary ammonium salts (NRB, Lumac). These bactericidal substances render membranes permeable to small molecules as nucleotides while leaving larger molecules as enzymes inside the cell. Their effect is much concentration dependent. Satisfactory results were also obtained with chlorhexidine, a biguanidine. All these substances can be used without significant dilution of the sample but produce a more or less rapid inhibition of firefly luciferase, the strongest inhibitory action being due to quaternary ammonium salts. An important step was therefore set when it was discovered that lipids could protect luciferase against the inhibitory action of anionic detergents as desoxycholate (7). During early work on firefly luciferase lipids had actually been found to be present in tail extracts but were at that time eliminated by aceton extraction. Lecithin has now been found to be an efficient protecting agent not only against anionic detergents as desoxycholate but also against cationic detergents as those mentioned above. Commercial lecithin consists of a mixture of substances and identification of the active component(s) is still in progress. Protection by lipids is also encountered with other membrane enzymes as membrane ATPase and the effect is therefore not unexpected. Other data were obtained in the meantime showing the efficient protection of firefly luciferase against benzethonium chloride by Tween (polyethylene sorbitan), a non-ionic detergent (Eur. patent 0309184 A2, 1988). The use of Tween-SD had been advocated before for the inactivation of quaternary ammonium compounds in the study of their lethal effects.

The above findings (together with the production of recombinant luciferase) are among the most important improvements registered in recent years as far as reagents is concerned and they have opened the door to improved methodologies regarding biomass assays. Not only can a constant light signal be obtained but internal standardization has now also become possible. However, as for the extraction of somatic ATP, the efficiency of bacterial extractants should be considered critically.

Several studies (8) have indeed indicated that extraction is not equally efficient for all bacterial strains. Further studies are also needed on the efficiency of extraction as a function of storage, temperature, growth stage, stress, etc.

METHODOLOGY

Space precludes a discussion on particular methodologies. A few specific aspects should nevertheless be mentioned.

a) Peak vs. steady state luminescence

Thanks to the introduction of purified reagents, peak measurements were gradually replaced by constant light measurements. However, as mentioned earlier, the quality of reagents has still been improving as far as residual ATP is concerned, and peak measurements can now again be considered, where the highest possible sensitivity is to be attained. Present reagents allow the measurement of 10^{-13} M ATP in a 100 ul sample (in biomass assays this corresponds to ca. 100 bacterial cells/ml).

b) Separation and concentration procedures

Biomass determinations have often implied a concentration procedure when the level of suspended cells was too low for direct determination. Filtration is obviously the method of choice in this case and has been used long ago already for plankton measurements. Filtration serves another purpose when suspended material or inhibitory substances have to be eliminated as is the case with various foodstuffs and biological samples (see applications). Double filtration, (coarse filter followed by bacteriological filter) has been advocated for specific applications. One of the main problems remains the clogging of the filters and centrifugation seems therefore the only outway in some cases. Gradient centrifugation has also been proposed for the concentration of bacteria. Unfortunately both filtration and centrifugation are cumbersome methods when simple and rapid microbiological tests are aimed at; in the case of filtration this disadvantage can however be compensated by some level of automation.

c) Enzyme immobilization

Firefly-luciferase has been immobilized on various supports as Sepharose, Nylon, collagen, cellulose porous glass, either alone or in combination with other enzymes in the case of coupled reactions. A biosensor equipped with a glass fiber light-guide has also been described. Although immobilized enzymes offer advantages in automated systems they do not necessarily reduce the cost of the assays in the particular case of firefly luciferase, since the firefly system still requires the injection of luciferin which is until now the most expensive component; the stability of the immobilized enzyme is also limited in time. An attractive approach has been proposed by Lundin (9) in order to simplify the procedure in the screening of bacteria. In order to avoid multiple pipetting stages the firefly reagent and the ATP standard are delivered in dried form on paper discs which can be introduced directly in the luminescence test tube. The presence of such paper discs has only a minor influence on the light signal.

ANALYTICAL APPLICATIONS OF FIREFLY LUCIFERASE

When looking at the recent evolution in the field of bioluminescent ATP assays one comes to the conclusion that interest is chiefly centering on biomass assays. This trend has even led to the development of what is now called "rapid microbiology" (10). Some parameters deserve particular attention in this respect and are dealt with in the following paragraphs.

a) Sensitivity of assays

An ever recurring problem is the way to express this sensitivity. As mentioned earlier the sensitivity of ATP assays depends among other factors on the accepted decay rate for the luminescence. For a decay rate of ca. 1% and with currently available instrumentation the limit of sensitivity lies around 10^{-11} M for 100 ul samples, i.e. 5 pg or ca. 1000 bacteria in the sample tube. Inasfar as the blank value of the luciferase reagent permits, higher sensitivities (10^{-13} M) can be attained at higher luciferase concentrations by registering the peak of luminescence. Pursuing such higher sensitivities is however only justified if all sources of contaminating ATP are under control.

It should be made clear that for the assay of enzymes and substrates the high sensitivities mentioned above are usually not required.

b) Specificity

Biomass assays based on ATP luminescence do not provide any information as to the specificity of the organisms involved. In rapid microbiology such tests can therefore only be used for screening purposes, and should, as the case may be, be implemented by identification procedures. Beside classical bacteriological procedures new tools have now become available which are once more based on luminescence. Luminescent DNA probes specific for bacterial r-RNA are indeed being developed that will facilitate the rapid identification of specific strains.

c) Reliability

Bacterial biomass assays are based on the assumption that the average ATP content of bacteria is more or less constant. Sources of error may have to do with uncomplete extraction and also with variations in physiological state. Further studies are definitely desirable in this respect. Most important for the future development of rapid microbiology is the level of correlation between ATP content and the number of CFU's (colony forming units). The discrepancies could well be much greater than the fluctuations of ATP/cell, due to aggregation of cells or chain formation and ATP content could hence be a more reliable parameter in this case. It remains that data are rather searce on this aspect of the problem.

Biomass assays have been applied in the past to bacteriuria screening and to antibiotics susceptibility tests. The results have been reviewed several times. Another field meeting with renewed interest is that of bacteriological milk tests. Assays have until now been hampered by the high non-bacterial ATP content of milk. In milk containing 400.000 somatic and 100.000 bacterial cells per ml (which can be considered as representative figures) the somatic ATP amounts to 2000 times the bacterial ATP. It was further shown that an important part of non-bacterial ATP is bound to casein micelles. In order to get rid of this ATP it has therefore proved necessary to combine the classical Triton/apyrase treatment with filtration. Small-size filters were developed in our laboratory for this purpose, which, after filtration of the milk sample, can be introduced directly into the luminescence tubes where they are extracted and assayed for ATP. Combination of the filtration technique with the use of the newly developed reagents makes the method much more promising and attractive than it used to be.

The assay of substrates and enzymes with firefly luciferase has not undergone major changes in the last years. If not used on a large basis they have nevertheless proved their efficiency and usefulness in many cases; such methods are being used routinely in our laboratory for the measurement of e.g. AP4A formation after heat shock and for the assay of lypolysis based on glycerol production.

INSTRUMENTATION

From the foregoing it is clear that in order to cope with most applications of FF luciferase the ideal instrument should include a minimal number of features :

a) temperature control around 20°C, this means possibility of both cooling and heating depending on room temperature;

b) provision for both peak and steady state luminescence measurements;

c) provision for kinetic measurements of several samples in parallel.

Useful accessories include the automatic injection of reagents and standards.

New instruments have recently been developed for the routine assay of large number of samples, based on the use of microtiter plates. The growing demand for biomass assays in the food industry has also prompted the development of instruments with automatic sample handling.

REFERENCES

1.   B.L. Strehler and J.R. Totter, Determination of ATP and related compounds: firefly luminescence and other methods, Methods of Biochemical analysis, 1, 341-356 (1954)

2.   E.H. White , F. McCapra and G.F. Field, The structure and synthesis of firefly luciferin, J. Am. Chem. Soc., 85, 337-343 (1963)

3.   G.V. Levin, J.R. Clendenning, E.W. Chappelle, A.H. Heim and E. Rocek, A rapid method for the detection of microorganisms by ATP assay, Bioscience, 14, 37-38 (1964)

4.   E. Schram and A. Weyens-van Witzenburg, Improved ATP methodology for biomass assays, J. Biolum. and Chemilum., 4, 390-398 (1989)

5.   A. Lundin, A. Rickardson and A. Thore, Continuous monitoring of ATP converting reactions by purified firefly luciferase, Anal. Biochem., 75, 611-620 (1976)

6.   A. Lundin, Applications of firefly luciferase, in Luminescent Assays, M. Serio and M. Pazzagli (Eds.), Raven Press, New York, p. 29-45 (1982)

7.   N.N. Ugarova and A.F. Dukhovitch, functions of lipids in firefly luciferase, in Bioluminescence and Chemiluminescence: new perspectives, J. Schölmerich et al. (Eds.), Wiley, Chichester, p. 409-412 (1987)

8.   J.M. Steijns, E. Dirix and H. Vanstaen, ATP extraction and its relevance as applied to rapid microbial counting in food samples using bioluminescence techniques, in ATP luminescence, rapid methods in microbiology, P.E. Stanley, B.J. McCarthy and R. Smither (Eds.), Blackwell, Oxford, p. 183-188 (1989)

9.   A. Lundin, ATP assays in routine microbiology: from visions to realities in the 1980s, ibid., p. 11-30

10.   P.E. Stanley, B.J. McCarthy and R. Smither (Eds.), ATP Luminescence, rapid methods in microbiology, Blackwell, Oxford (1989)

LUMINOMETRIC ATP MONITORING OF OXIDATIVE PHOSPHORYLATION AND ENERGY METABOLITES IN NEEDLE BIOPSIES FROM SKELETAL MUSCLE

A. Lundin, R. Wibom[1], K. Söderlund[1] and E. Hultman[1]

Clinical Research Centre, Department of Medicine and [1]Department of Clinical Chemistry II, Karolinska Institute, Huddinge University Hospital, S-141 86 Huddinge, Sweden

INTRODUCTION

There are two types of muscle fibres converting ATP into mechanical work. In fast twitch fibres used for short term work the required ATP comes mainly from glycolysis in the cytoplasm. In slow twitch fibres used for long term work the ATP comes from oxidative phosphorylation in the mitochondria. The ATP generated in the mitochondria is partially used to form phosphocreatine (PCr) by the creatine kinase (CK) reaction. PCr is used as a short term energy buffer and as a means of energy transportation to the myofibril. When ATP is consumed in the myofibrils it is continuously regenerated by the CK reaction at the expense of PCr. In the mitochondrion the enzymes of β-oxidation of fatty acids, the Krebs cycle and the respiratory chain are responsible for the ATP production from lipids and carbohydrates. Depending on the activity of different enzymes the mitochondrial ATP production rate (MAPR) with various substrates may vary in different individuals and inborn metabolic errors have been described. Measurements of MAPR with different substrates is also of obvious interest in sports medicine. Until recently such measurements were based on monitoring of the oxygen consumption rate in isolated mitochondria in the presence of various substrates. This is not only a rather indirect measurement but also requires at least 1 g of tissue. This has seriously limited the use of the technique. The assay of ATP and PCr in individual fibres has been possible only with complicated methods based on enzymatic cycling or with HPLC working at the detection limit of this technique.

Under appropriate analytical conditions the light emission from the firefly luciferase reaction is proportional to the ATP concentration and affected neither by gradual inactivation of luciferase nor by consumption of ATP in the luciferase reaction (1). Thus ATP converting reactions can be luminometrically monitored in much the same way as NAD(P)H converting reactions are spectrophotometrically monitored. A firefly reagent for ATP monitoring has been previously used for assays of several enzymes, e. g. CK (2), and metabolites, e. g. ATP/ADP/AMP (3), PP_i (4) and glycerol (cf. references in ref. 5), and for monitoring of photophosphorylation (6).

The present paper summarises the development of methods based on luminometric ATP monitoring for measurement of oxidative phosphorylation in the presence of various substrates (7, 8) and of ATP and PCr (9). A micromethod was developed requiring only 40 mg needle biopsy samples for preparation of isolated mitochondria and allowing luminometric assays of MAPR. Most of the sample is actually used for spectrophotometric measurements of protein and glutamate dehydrogenase (GDH; a marker enzyme for mitochondria). Luminometric assays of ATP and PCr were performed in single fibres (or even parts thereof). The assay of PCr is based on endpoint formation of ATP from PCr and ADP by the CK reaction. Some physiological and clinical results will be given.

MATERIALS AND METHODS

Assay of mitochondrial ATP production rate (MAPR): Procedures for tissue sampling, preparation of mitochondrial suspension and for measurement of GDH and MAPR have recently been described in detail (7, 8). Tissue samples from human muscle weighing as little as 40 mg were obtained by a needle biopsy technique under local anaesthesia. After homogenisation mitochondrial suspensions were prepared by differential centrifugations. The micropreparative technique developed was rapid and produced a high yield of reasonably pure mitochondria. The measurement of MAPR was performed in a 1251 Luminometer (BioOrbit Oy, Turku, Finland) using ATP Monitoring Reagent and ATP Standard (BioOrbit Oy, Turku, Finland). The firefly reagent (AMR) was reconstituted in 47.5 ml of a solution containing 180 mmol/l sucrose, 35 mmol/l potassium phosphate buffer and 1 mmol/l EDTA. The limited buffering capacity made it necessary to adjust the solution to pH 7.5 with KOH before as well as after the reconstitution. The ATP Standard was reconstituted with 10 ml dist. water to give a concentration of 0.01 mmol/l. Since commercially available ADP is normally contaminated with 1% ATP specially purified ADP was used (7). Six substrate solutions were used (final concentrations in mmol/l as shown within brackets): 1) pyruvate (1) plus malate (1), 2) palmitoyl-L-carnitine (0.005) plus malate (1), 3) α-ketoglutarate (10), 4) succinate (20) plus rotenone (0.1), 5) succinate (20), and 6) pyruvate (1) plus palmitoyl-L-carnitine (0.005) plus α-ketoglutarate (10) plus malate (1). Six cuvettes containing reconstituted AMR (0.95 ml) were supplied with substrate (0.035 ml), ADP (0.005 ml; concentration in cuvette 0.03 mmol/l) and finally 0.01 ml mitochondrial suspension (diluted 500 times in reconstituted AMR before the addition). A seventh cuvette with all additions except substrate was used as blank. The cuvettes were preincubated (25 $^{\circ}$C, 5 min) in the luminometer after the addition of the mitochondrial suspension. An automatic assay was then started measuring the light emission from all seven cuvettes repeatedly for approx. 20 minutes. Finally the luminometer added 0.2 µmol/l ATP Standard. The rate of the ATP formation was calculated as the rate of the increase of the light emission divided by the sudden increase at the addition of the ATP Standard multiplied by the concentration of the standard. The rate of the MAPR by oxidative phosphorylation was obtained by subtracting the rate of ATP formation in the blank cuvette, which was due to contaminating adenylate kinase (AK) activity.

Assay of ATP and PCr in individual muscle fibres: Biopsy samples from human quadriceps femoris muscle were obtained by the percutaneous needle biopsy technique. After freeze-drying single muscle fibres were dissected under microscope and each fibre was divided in two parts that were weighed using a quartz fibre fish-pole balance. One part was used for determining the type of the fibre by a histochemical technique. The other part of the fibre weighing 1-4 µg was extracted in 0.1 ml 2.5% trichloroacetic acid followed by neutralisation with 0.01 ml 2.2 mol/l $KHCO_3$. Extracts were stored at -80 $^{\circ}$C until analyzed. The same luminometer and firefly reagent as for assay of MAPR was used but the pH of the AMR was adjusted to 6.7 (pH optimum of CK). The ADP reagent (BioOrbit, Turku, Finland) contained 0.05 mmol ADP, 0.1 mmol AMP and 20 nmol P^1, P^5-di (adenosine-5') pentaphosphate (DAPP) reconstituted in 1.5 ml dist. water. CK (Boehringer-Mannheim, cat. no 127566) was dissolved in dist. water (10 mg/ml). Cuvettes containing AMR (0.95 ml) and extract (0.025 ml) were loaded into the luminometer and an automatic assay started. The assay included: 1) preincubation (25 $^{\circ}$C, 10 min), 2) measurement of light corresponding to ATP level, 3) addition of ADP reagent (0.015 ml), 4) measurement of light emission (increased due to contaminating ATP in the ADP reagent), 5) addition of CK (0.010 ml), 6) measurement of light emission after attaining the endpoint of the CK reaction, 7) addition of ATP Standard (0.01 ml) and 8) measurement of light emission after addition of ATP Standard. Contaminating AK activity gives a continuous increase of the light emission after addition of ADP reagent. Thus the increase of the light emission resulting from the CK reaction was calculated by extrapolating the light emission data at the endpoint back to the time of CK addition. Each step was carried out in all cuvettes before the next step was started. With 25 samples the entire assay takes approx. 35 min. Contaminating ATP in the reagents is compensated for by running a blank containing neutralized extraction medium. The effect on the luciferase activity by the ADP and CK reagents is compensated for by running three blanks with ATP Standard addition before as well as after addition of these reagents and calculating the ratio between the two values. Detailed descriptions of programs for running the automatic luminometer and for calculating the results can be obtained from one of the authors (R. W.).

RESULTS AND DISCUSSION

The assay of MAPR was optimized using rat muscle mitochondria (7). The ADP concentration had to be optimized taking into consideration the AK blank in the firefly reagent as well as the mitochondria. A 0.03 mmol/l ADP concentration was optimal for the MAPR from oxidative phosphorylation. Increasing the ADP concentration even up to 0.15 mmol/l caused no further increase of MAPR from oxidative phosphorylation but increased the AK activity more than 4-fold. At 0.03 mmol/l ADP the AK activity is low enough to be measured and subtracted as a blank. It might be possible to reduce the AK activity by inhibitors as AMP and DAPP. However, MAPR depends on many enzymes and potential interference from the inhibitors would be difficult to control. In general optimal substrate concentrations could be chosen for the assay. However, for α-ketoglutarate and succinate no optimal concentration was found within the physiological range and 10 and 20 mmol/l, respectively, were chosen. In the presence of 0.1 mmol/l rotenone 20 mmol/l succinate was optimal. The substrate concentrations chosen for measurements of MAPR in mitochondria from rat muscle were also used for human muscle. In addition a mixture of pyruvate, palmitoyl-L-carnitine, α-ketoglutarate and malate believed to give close to maximal MAPR in humans was used. The coefficient of variation between duplicate determinations of MAPR expressed per kilogram muscle was found to be approx. 10%. In preparations of muscle mitochondria from rats but not from humans a 10% reduction per hour of the the MAPR activity was observed and compensated for (7, 8). MAPR was measured using the six substrate conditions in healthy human subjects (n=21) divided into sedentary, moderately active and highly active. With most of the substrates but not all the MAPR was correlated to the degree of activity in the subject groups and mainly reflected the level of mitochondrial protein. In more recent studies effects of intensive training periods on the relative rates with different substrates have been found. Differences in relative and absolute rates have also been found in patients with inborn metabolic errors.

The development of the assay of ATP and PCr in a single cuvette was mainly based on the optimized assay of CK isoenzymes (2). It was not necessary to choose optimum conditions with respect to the CK reaction, since adequate levels CK could be added to achieve the endpoint in the CK reaction. The assay buffer was chosen because it gave a very good stability of the reconstituted AMR, when used in the assay of MAPR. As in the assay of CK isoenzymes (2) contaminating AK in the reagents was inhibited by AMP and DAPP. Effects of remaining AK activity was compensated for by extrapolating light emission data obtained at the endpoint back to the time of CK addition. In contrast to the otherwise similar endpoint assay of ATP/ADP/AMP (3) the automatic luminometer performed each step in the analytical procedure in all cuvettes before beginning the next step. This way of running the luminometer is advantageous if the individual steps are time-consuming. The luminometric assay of ATP and PCr was compared to corresponding spectrophotometric and HPLC assays. These experiments had to be performed with whole muscle samples rather than individual fibres due to the limited sensitivity of the non-luminometric methods. Coefficients of correlation were 0.99 for ATP and 1.00 for PCr. The CV of the luminometric method was 4% for ATP and 3% for PCr using duplicate determinations of the same fibre extract. When using separately extracted pieces of the same fibre corresponding figures were 5% for ATP and 7% for PCr. The latter measurement includes the error involved in weighing the small pieces of fibres used for the assay. Measurements of ATP and PCr in single fast and slow twitch fibres during and after a 50 seconds electrical stimulation of human muscle were performed. During the stimulation the ATP level dropped slightly but started to go back to the original level as soon as the stimulation ceased. The PCr level dropped almost to zero during the stimulation and went up to a level slightly above the original level within 15 minutes after turning off the stimulation. The initial rate of rebuilding the PCr pool was lower in fast twitch fibres, since they have fewer mitochondria than slow twitch fibres (glycolysis is only active during stimulation).

Over the years we have developed a panel of luminometric assays for energy related metabolites like ATP/ADP/AMP (3), ATP/PCr (9), PPi (4), glycerol, fatty acids and glucose (cf. references in ref. 5) and for photophosphorylation (6) and oxidative phosphorylation (7, 8). These assays have been used in studies on energy metabolism in photosynthetic bacteria and in cancer, sperm, fat and muscle cells. The sensitivity of luminometric assays has been found to greatly simplify such studies due to the small amount of biological material required. This is particularly important when using human material as in the present work or when monitoring the interstitial space by the microdialysis technique (5).

ACKNOWLEDGEMENTS

This work was supported by grants from the Swedish Medical Research Council (02647) and from the Swedish Sports Confederation (75/85, 68/86 and 86/87).

REFERENCES
1.   A. Lundin, Analytical applications of bioluminescence: The Firefly system. In Clinical and Biochemical Luminescence, L. J. Kricka and T. J. N. Carter (Eds), Marcel Dekker, New York, p. 43-74 (1982).
2.   A. Lundin, B. Jäderlund and T. Lövgren, Optimized bioluminescence assay of creatine kinase and creatine kinase B-subunit activity. Clin. Chem. 28, 609-614 (1982).
3.   A. Lundin, M. Hasenson, J. Persson and Å. Pousette, Estimation of biomass in growing cell lines by adenosine triphosphate assay. Meth. Enzymol. 133, 27-42 (1986).
4.   P. Nyrén and A. Lundin, Enzymatic method for continuous monitoring of inorganic pyrophosphate synthesis. Anal. Biochem. 151, 504-509 (1985).
5.   B. Näslund, P. Arner, K. Bernström, J. Bolinder, L. Hallander and A. Lundin, Assays of free fatty acids and glucose by horseradish peroxidase catalyzed luminol reaction. In this volume.
6.   A. Lundin and M. Baltscheffsky, Measurement of photophosphorylation and ATP-ase using purified firefly luciferase. Meth. Enzymol. 57, 50-56 (1978).
7.   R. Wibom, A. Lundin and E. Hultman, A sensitive method for measuring ATP-formation in rat muscle mitochondria. Scand. J. Clin. Lab. Invest. 50, 143-152 (1990).
8.   R. Wibom and E. Hultman, ATP production rate in mitochondria isolated from microsamples of human muscle. Am. J. Physiol. 259 (Endocrinol. Metab. 22), E204-E209 (1990).
9.   R. Wibom, K. Söderlund, A. Lundin and E. Hultman, A luminometric method for the determination of ATP and phosphocreatine in single human skeletal muscle fibres. In press.

ATP TUMOR CHEMOSENSITIVITY ASSAY

P.E. Andreotti, J.T. Thornthwaite[1] and I.S. Morse

MCL Technologies, Fort Lauderdale, Florida, U.S.A.
[1]Baptist Hospital, Miami, Florida, U.S.A.

INTRODUCTION

An ATP Microbioluminometry Assay (MBA) system has been developed for human tumor chemosensitivity testing. The MBA system is·simillar to bioluminescent methods described previously (1-3), and is based on concepts for nonclonogenic chemosensitivity assays (4,5).

Chemosensitivity assays are used in the selection of treatment protocols, and are particularly relevant for cases involving protocol failures, tumors of unknown primary site, rare tumors, drug resistance, or drug stimulation of tumor growth (6). Clonogenic assays which measure drug effects on proliferating tumor stem cells have been widely investigated (7). Nonclonogenic systems including *in vivo* subrenal capsule assays, tetrazolium dye colorimetric assays, thymidine incorporation assays, and differential staining and fluorescence methods using visual analysis have been developed to overcome technical problems and limitations associated with clonogenic assays (4-6). Based on >2000 clinical correlations, the predictive accuracy of chemosensitivity assays is estimated at 66% "true positive" and 92% "true negative". These results are reasonably similar to estrogen receptor assays for breast cancer (4,8).

ATP bioluminescent assays are a practical alternative with potential advantages for tumor chemosensitivity testing. The sensitivity of the method allows fewer cells to be used and hence more drugs to be tested against a specimen; overall cell kill in dividing and nondividing cell populations can be evaluated; cytostatic and cytocidal drug effects can be measured, and objective quantitative results can be obtained in 7 days using a relatively simple procedure that does not require single cells or radioactive isotopes.

MATERIALS AND METHODS

Solid tumor specimens and lymph node metastasis used in MBA assays were washed twice in RPMI-1640 with 10% fetal calf serum (FCS), 25 µg/ml gentamicin, 100 µg/ml streptomycin and 100 U/ml of penicillin (RPMI-10%+). Excess fat was removed and the tumor was minced into 0.5 - 2.0 mm^3 pieces. Approximately 0.5 - 1.0 g of tumor was then dissociated into a cell suspension by digestion for 8-16 hrs at 37°C in RPMI-10%+ with collagenase, DNase I and dispase (Boehringer Mannheim). Ficoll-hypaque density gradient centrifugation was used to enrich the viability of dissociated tumor cells, and isolate normal erythrocytes (RBC), peripheral blood leukocytes (PBL), lymphoma cells, and metastatic pleural fluid cells used in MBA assays.

Cell preparations were washed twice and resuspended for assay in RPMI-1640 with 20% FCS, 25 µg/ml gentamicin, 100 µg/ml streptomycin and 100 U/ml of penicillin (RPMI-20%+). Assays were performed using $2.0 \times 10^4 - 5.0 \times 10^4$ viable cells per 0.2 ml culture in RPMI-20%+ containing 100%, 30% or 10% of the estimated peak plasma concentration (PPC) of each test drug. Drug concentrations were assayed in triplicate with maximum ATP inhibitor control cultures containing 0.8% NaN_3 and control cultures without inhibitor. Cells were cultured in 96 well flat-bottom microtiter plates (Corning) coated with a polymerized matrix (MCL Laboratories) to inhibit attachment of anchorage dependent cells. After incubation for 7 days at 37°C in a humidified 95% air/5% CO_2 atmsphere, the cultures were extracted for luminometry using 0.1 ml of extraction reagent (MCL Laboratories). Counting was performed for 20s in a Packard Picolite 6500 luminomter or MCL Microluminometer using 0.075 ml samples injected with 0.05 ml of 5.0 µg/ml luciferase (Amgen) in 0.025 M Tris buffer pH 7.8 containing 0.2 mM luciferin (Sigma). Results were determined as "strongly sensitive" for >70% inhibition, "partially sensitive for 50%-70% inhibition, "resistant" for -20%-50% inhibition, and "stimulatory" for >-20% inhibition by the formula:

$$\% \text{ Inhibition} = 1.0 - \frac{PPC - MI}{OD - MI} \times 100$$

where PPC = mean counts for each test drug concentration, MI = mean counts for maximum ATP inhibitor controls, and OD = mean counts for controls without inhibitor.

RESULTS AND DISCUSSION

Cell preparations used for chemosensitivity assays are comprised primarily of tumor cells, but may also include subpopulations of erythrocytes, leukocytes, and other anchorage dependent normal cells, e.g. fibroblasts and endothelial cells. Culture systems which prevent the attachment and growth of anchorage dependent cells are used for both clonogenic and nonclonogenic assays to eliminate measuring drug effects on normal cells (3-5,7). Fig 1 shows the kinetics of ATP production by different human cell types in the MBA system using 1×10^5 cells/culture in polymer-coated plates. This cell concentration is approximately twice the number of viable tumor specimen cells generally assayed in the MBA system. Normal RBC and PBL do not show significant ATP levels after 6 days in culture, and monocyte adherence is not readily observed. Similarly, significant ATP levels are not evident with normal WI-38 fibroblasts (ATCC CCL75) which do not attach and proliferate as in uncoated plates. In contrast, exponential growth comparable to that measured in uncoated plates is detectable with the colon adenocarcinoma cell line SW620 (ATCC CCL227). These and other results from additional experiments performed with normal and tumor cell lines indicate that subpopulations of normal blood cells and anchorage dependent normal cells are not the principal target cells measured in 7 day MBA cultures.

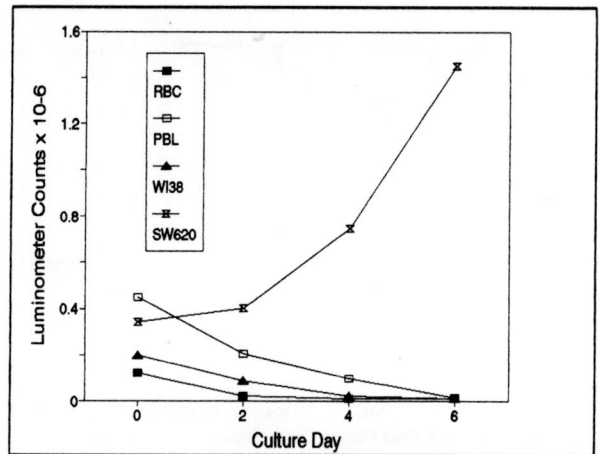

Figure 1. Kinetics of human cell ATP production in MBA system.

MBA assays have been performed with experimental variation on 19
solid tumor specimens (11 tumor cell types), 3 lymph node metastasis, 1 pleural fluid metastasis and 2 lymphomas. Cumulative results
indicate the coefficient of variation for triplicate cultures is
less than 20%, with 92% (23/25) of specimens evaluable. Two solid
tumor specimens were not evaluable due to low cell viability after
enzyme dissociation. In all cases evaluated, clearly differentiated
patterns of sensitivity and resistance have been determined, including studies with lymphomas which are poorly clonogenic and not
readily tested with assays dependent on cell proliferation (6,8).
Initial clinical correlations are being determined based on results
calculated for 10% and 30% PPC cultures.

Fig 2 shows results for a primary breast tumor specimen and a "stem"
cell line derived from the specimen after 5 weeks of culture in
RPMI-20%+. The results illustrate patterns of chemosensitivity detectable with the MBA system, and the ability to measure drug effects on the total tumor cell population that may not be evident
with assays dependent on cell proliferation. The primary tumor cells
were strongly sensitive to 4-hydroxycyclophosphamide (4-HC) and CAF
(4-HC, adriamycin, 5-FU), whereas adriamycin (ADR) and 5-FU alone
were stimulatory or resistant at higher concentration. The lack of
sensitivity with ADR and 5-FU demonstrates that 4-HC is the active
drug in CAF. The stem cell line was strongly sensitive to both 4-HC
and 5-FU, but resistant or only partially sensitive to ADR. Stimulatory effects similar to that observed with the primary tumor cells
have been documented, although the mechanism and clinical relevance
of this phenomenon is unclear (9). These results may reflect the
ability of the MBA system to quantitate drug effects on G_0 tumor
cells, e.g. recruitment, which have different chemosensitivity and
may be tumorigenic in vivo (4).

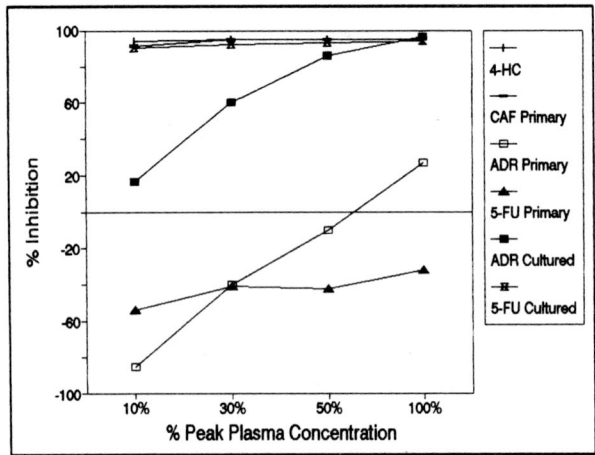

Fig 2.   MBA results with primary tumor cells and cultured cell line.

Previous reports have demonstrated the utility of ATP assay systems for chemosensitivity testing (1-3). Our MBA system has been developed for nonclonogenic assays to provide additional information about drug effects that cannot be determined with clonogenic assays. The MBA system can also be used to quantitate clonogenic assays.

REFERENCES

1.   L. Kangas, M. Gronroos and A.L. Nieminen, Bioluminescence of cellular ATP: A new method for evaluating cytotoxic agents in vitro. Medical Biology. 62, 338-343 (1984).

2.   H.S. Garewal, F.R. Ahmann, R.B. Schifman and A. Celniker, ATP assay: Ability to distinguish cytostatic from cytocidal anticancer drug effects. JNCI. 77, 1039-1045 (1986).

3.   B.U. Sevin, Z.L. Peng, J.P. Perras, P. Ganjei, M. Penalver and H.E. Averette, Application of an ATP bioluminescence assay in human tumor chemosensitivity testing. Gynecologic Oncology. 31, 191-204 (1988).

4.   L.M. Weisenthal and M.E. Lippman, Clonogenic and nonclonogenic in vitro chemosensitivity assays. Cancer Treatment Reports. 69, 615-632 (1984).

5.   J.B. Mitchell, Potential applicability of nonclonogenic measurements to clinical oncology. Radiation Res. 114, 401-414 (1988).

6.   W.L. McGuire, d.H. Kern, D.D. Von Hoff and L.M. Weisenthal, In vitro assays to predict drug sensitivity and drug resistance. Breast Cancer Research and Treatment. 12, 7-21 (1988).

7.   S.E. Salmon, Chemosensitivity testing: Another chapter. JNCI. 82, 82-83 (1990).

8.   D.D. Von Hoff, Commentary: He's not going to talk about in vitro predictive assays again, is he?. JNCI. 82, 96-101 (1990).

9.   P. Vichi and T.R. Tritton, Stimulation of growth in human and murine cells by adriamycin. Cancer Research, 49, 2679-2682 (1989).

APPLICATIONS OF LUMINESCENCE IN CLINICAL MICROBIOLOGY
J G M Hastings

Department of Clinical Microbiology,
Queen Elizabeth Hospital, Edgbaston,
Birmingham B15 2TH

INTRODUCTION

Despite several decades of research and development luminescence-based technologies have yet to make a major impact in microbiology, particularly in the clinical field. Early problems associated with reagents and an absence of sensitive detectors have been resolved with the availability of purified reagents in kit form and the development of sensitive, automated instrumentation. Despite this progress, ATP-based techniques have yet to make a significant impact in commercial terms.[1]

One criticism of the early attempts at promoting the application of luminescence in microbiology is that there appears to have been no clear critical analysis of the particular advantages and disadvantages of the technique. The commercial development which followed resulted technique instrumentation and reagent presentation which failed to produce a successful product. However, some recent developments in reagents and instrumentation may at last give luminescence the chance to become widely applied in microbiology practice.

ATP BIOLUMINESCENCE IN CLINICAL MICROBIOLOGY

Of the various suggested applications for ATP bioluminescence in the clinical microbiology field,only two have received detailed evaluation.

Screening for bacteruria. Although there are a number of approaches to ATP screening for bacteruria, most reports describe protocols based on assaying extracted bacterial ATP following the selective destruction of somatic ATP. In order to appreciate the problems of introducing a technique to screen urine samples for significant bacteruria it is important to understand why it is regarded as an advantage to screen such samples.

There are two main areas of diagnostic microbiology where urine screening is being performed, in the doctors' office and in the hospital laboratory. In the doctor's office the physician requires a screening device, which is rapid, (10-15 min), simple (preferably with no pipetting steps) and inexpensive. In the laboratory, the clinical microbiologist wants a urine screening technique which allows him to screen multiple urine samples rapidly for little cost.

In the laboratory context the approach to bioluminescence urine screening has been to use large, automated luminometers with automatic injection facilities for reagent addition. This reduces hands-on time to a minimum, but the major disadvantage is that each urine sample urine screened takes several minutes, and screening 30-40 urines in one batch can take over an hour. Such an approach fits poorly into the work patterns of many microbiology laboratories.

The recent availabilty of luminometers using a microtitration plate format may help to circumvent this problem. By using microtitration trays it is possible to set up to 96 urine samples in a single tray; reagent additions for destruction of somatic ATP and assay of bacterial ATP have to be done manually, but this is simple and rapid with modern multi-channel pipettes and other automated devices. The advantage in terms in laboratory work flow is that all 96 urines can then be analysed in parallel very rapidly, as microtitration plate luminometers can read a complete tray in only a few minutes. Furthermore, the microtitration tray approach requires only small volumes of reagents which allows a significant cost saving. The disadvantage is in terms of sensitivity; the criteria for significant bacteruria has traditionally been set as 10^5 organisms/ml of urine. The microtitration plate system for urine screening, can achieve this. However, some clinical microbiologists are proposing that a cut off level of 10^4 or even 10^3 organisms/ml may be desirable. Some automated tube luminometers can achieve such sensitivity.

Sophisticated luminometers, either tube-based or microtitration tray based, are inappropriate for urine screening in the physician's office. The requirement here is for a small, inexpensive, single chamber luminometer; these are already available. However, the problem with using ATP bioluminescence in this area has been in reagent presentation. Complicated pipetting steps are not easily performed in the doctor's office. Recently, Amersham International, (Little Chalfont, UK) have been developing an ATP bioluminescence assay device in which the reagents are presented on a small stick. For commercial reasons the details of this ATP stick technology are not yet available. However, in a small study using a prototype of this device at the Royal Hallamshire Hospital, Sheffield, UK, the ATP stick performed better than both nitrite or leucocyte esterase sticks, both of which are commonly used by general practioners. (Table) Whether such a urine screening device will ever succeed commercially remains to be seen; the cost per test may be the deciding factor.

Table

Screening for urinary tract infections
(159 urine samples)

	LE*stick (%)	Nitrite (%)	ATP stick (%)
Sensitivity	85	54	94
Specificity	70	97	95
PPV+	60	91	91
NPV+	90	80	97
Efficiency	75	82	95

* Leucocyte esterase
+ PPV, positive predictive value; NPV, negative
  predictive value.

Antibiotic susceptibility tests are performed on the majority of
bacterial isolates in microbiology laboratories. The traditional
overnight disk diffusion technique remains the most widely used
method.

Rapid, automated systems based on optical density measurements
have become popular in the USA and some European countries.
However, it is the automation rather than the rapid result which
is seen as the main advantage of this approach. Rapid antibiotic
susceptibility tests are in fact a misnomer as the result is still
not available till some 24 hours after the original sample was
taken, and the benefit to the patient of such rapid results is
still unclear. The issue remains controversial. Against this
background, it is not surprising that ATP bioluminescence has been
difficult to promote as a technique for rapid antibiotic
susceptibility testing.

The technique clearly works and results by bioluminescence are
equivalent to standard overnight data! However, methodoligies
remain cumbersome and even the more stream-lined protocols,
designed around microtitration plate luminometers, have not been
successfully introduced into routine laboratories.

The area of antibiotic susceptibility testing in which
bioluminescence might be more successful commercially is in the
testing of atypical micro-organisms, particular mycobacteria.
Traditional susceptibility tests on mycobacteria take from 2–6
weeks before results are available. Using ATP bioluminescence, it
is possible to perform these assays in 3–5 days.[2,3]

PORTABLE LIMINOMETERS

One of the problems faced by those promoting rapid microbiology tests is in devising simple protocols allied to a small portable instrument, which allows the test procedure to be carried on site by relatively unskilled staff and provides an instant result on which action can be taken. Rapid ATP bio-luminescence tests have come up against this problem. For example, as illustrated by several papers in these Proceedings, ATP bioluminescence can be successfully applied to mastitis-testing in cattle. Unfortunately, the technique has not been widely accepted because the user requires the test to be performed on site("cow-side") rather than in the laboratory.

Thus, the development of small portable luminometers, which can be operated by battery and which are packaged along with reagents and other equipment in a simple suitcase, is a major advance. These instruments will hopefully facilitate the application of bioluminescence techniques to a much wider area of microbiology, allowing simple rapid assays to be performed in the field.

Two instruments are currently available, one from Bio-Orbit (Turku, Finland), which uses a photomultiplier tube and a second from Biotrace Ltd (Bridgend, UK) which is a solid-state device. Both these companies are promoting their instruments and reagents for use in hygiene monitoring. This application is designed to allow the user to measure the amount of somatic and microbial ATP on surfaces. Using the portable luminometer it is possible to provide an instant result which should equate with the general cleanliness of the areas being investigated. It is hoped that this technique will become widely accepted as a means of assessing cleaning programmes in food preparation areas and other facilities where high standards of hygiene are important.

IN-VIVO BIOLUMINESCENCE

Since Ulitzur and Kuln[4] demonstrated the transfer of the luminescent phenotype into bacteria, there has been considerable interest in the possible applications of in-vivo bioluminescence. By transforming certain key bacteria it has been shown that luminescence may be used as a bio-sensor to detect and monitor biocides and antimicrobial agents. Using bacteriophages carrying the lux gene and specific for particular bacterial species, it is possible to detect very low numbers of specific bacterial species, such as Salmonella species or Listeria monocytogenes in foods and clinical specimens.

The possible applications of lux gene technology in microbiology
are enormous and are discussed in a recent review.[5] The linking
of ATP bioluminescence, a rapid but non-specific indicator of
microbial presence, with assays utilising lux bacteriophages,
rapid and specific indicators of bacteria, remains an exciting
prospect for the future. Such developments should be aided by the
convenient, portable luminometers now available.

REFERENCES

[1]J G M Hastings: Luminescence in clinical microbiology. In
Bioluminescence and chemiluminescence: New perspectives, J
Scholmerich et al (Eds), Wiley, Chichester, p.453-461 (1987).

[2]B Beckers, H R M Lang, D Schimke, A Lammers, Evaluation of a
bioluminescence assay for rapid antimicrobial susceptibility
testing of mycobacteria. Eur J Clin Microbiol 4, p.556-561
(1985).

[3]J G M Hastings, P Sparham, R C Spencer, P F Wheat, Susceptibility
testing of mycobacteria by ATP bioluminescence. Abstracts of the
28th Interscience Conference on Antimicrobial Agents and
Chemiotherapy, p.326 (1988).

[4]S Ulitzur and J Kuln, Introduction of lux genes into bacteria, a
new approach for specific determination of bacteria and their
antibiotic susceptibility. In Bioluminescence and
chemiluninescence: New perspectives, J Scholmerich et al (Eds),
Wiley, Chichester, p.463-472 (1987).

[5]S A A Jassim, A Ellison, S P Denyer, G S A B Stewart, In vivo
Bioluminescence: A cellular reporter for research and industry.
J Biolumin, Chemilumin, 5, 115-122 (1990).

DENATURATION OF FIREFLY LUCIFERASE

C.Y. Wang and J.D. Andrade

Department of Bioengineering, 2480 MEB University of Utah, Salt Lake City, UT 84112, USA

INTRODUCTION

In order to improve the stability, duration and intensity of the light output of the firefly luciferase-luciferin system, we are studing the relationship between enzyme activity and conformation. Because the activity of recombinant luciferase is five times higher than natural purified luciferase, it is important to explore both firefly luciferases. Ultraviolt difference spectroscopy and fluoroescence are used to study the denaturation of luciferase. We have found that both luciferases denature at low urea concentration (50% denaturation), at about 1M, and at very low guanidinium chloride concentration (50% denaturation), about 0.5M. The light output of the luciferase-luciferin system at different urea concentrations was also measured to check luciferase activity. Scanning calorimetry also shows that luciferase is thermally labile. Denaturation temperature is about 40°C. The pH is also critical to the conformational stability of luciferase.

MATERIALS AND METHODS

Luciferase from Photinus pyralis was obtained from Sigma Chemical Company. The Sigma luciferase was chromatographically prepared, crystallized and lyophilized with 90% purity. The luciferase was dissolved in 0.5M, pH 7.7 Trizma succinate buffer, purchased from Sigma. The pH was measured by Corning pH meter 245 and pH indicator sticks (EM Laboratories Inc.). Recombinant firefly luciferase was prepared from a procaryotic recombinant host harboring the luc structural gene by Amgen Biologicals. It was purified by sequential chromatography to 95% purity. This recombinant luciferase is dissolved in 0.5M, pH 7.7 Trizma succinate buffer to obtain a homogeneous solution.

Urea and Guanidinium Chloride were purchased from Research Plus Inc. (absolute grade). The urea and GdnCl 8M stock solutions were prepared by weight (1) and then adjusted to pH 7.7. The stock solutions were then diluted to the desired concentration. Perkin Elmer UV spectrophotometer (model C688-0002) was used to measure the UV absorbance from 200nm to 350 nm. The concentration of protein used in the UV absorbance studies was 0.75mg/ml (A_{1cm}=0.75 mg/ml at 280 nm) (2). An ISS Grey 200 multifrequency phase fluorometer was used to measure tryptophan flourescence of the protein solution, excited at 280nm. The concentration of protein used in fluorescence detection was 0.3mg/ml. The temperature was controlled at 22°C in both experiments. The time required to reach equilibrium at the different denaturant concentrations was also measured. Further details on the assay and methods used for measuring the light emission of the luciferase-luciferin system are available (2). The pH of the solutions were adjusted to 7.2 and the experiments were repeated.

Thermal stability was determined by differential scanning calorimetry. The concentration of protein was 0.5mg/ml. The buffer used was 0.5M, pH 7.7 Glycylglycine. The scanning range was from 30°C to 70°C. UV absorbance was also used to determine the thermal denaturation curve.

RESULTS AND DISCUSSION

Both luciferases are not denatured by urea concentrations lower than 0.1M, at pH 7.7. They begin to denature at urea concentrations larger than 0.1M. When the urea concentration is larger than 3 M, both luciferases appear to be maximally "denatured" (Fig.1.a,b). There are no significant differences between the urea denaturation curves of these two luciferases. The activity (light output) measurements also support the above results (fig.2). Both luciferases begin to denature when the concentration of guanidinium chloride is greater than 0.1M, at pH 7.7, and are maximally denatured at 3M GdnCl (Fig. 3.a,b). When the pH is decreased to 7.2, the urea concentration for 50% denaturation decreases from 1.8M to 1M and GdnCl concentration for 50% denaturation decreases from 0.5M to 0.4M. The light intensity decreased 30% (2).

Scanning calorimetry curves show that the Sigma luciferase denatures at about 40°C and aggregates at 54°C. Thermal denaturation curves determined by UV absorbance at 280nm also show that both luciferases denature at about 40°C.

Firefly luciferase (photinus pyralis) is composed of 550 amino acids. The amino acid sequence was determined by the analysis of cDNA and genomic clones by Wood et al (3). Sixty percent of its amino acids are nonpolar (4), so it aggregates easily in aqueous solution. Urea and GdnCl can cause denaturation of both the Sigma and the recombinant luciferase. The conformational stability is poor, possibly because luciferase contains only six to seven sulfhydryl groups (5) which are important in protein structure and activity. Anions such as sulfate and chloride will inactivate the activity of luciferase (6). Both luciferases are thermally labile. They denature at low temperature, 40°C, probably due to their weak non-covalent interactions. At higher temperature they aggregate very quickly and lose their activity completely. This work is ongoing (7).

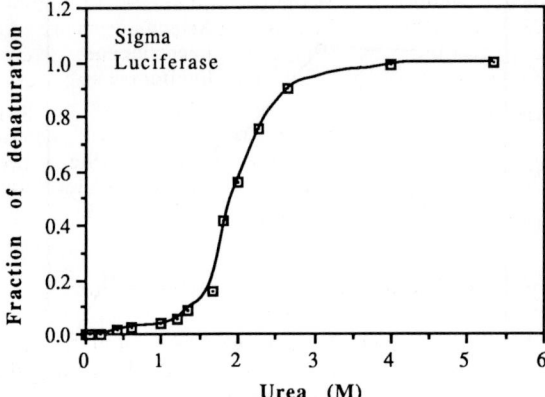

Fig.1.a Tryptophan fluorescence of Sigma luciferase (0.3mg/ml in 0.025M Glycylglycine) as a function of urea concentration (excitation wavelength 280nm, emission wavelength 340nm). Curve assumes a simple two-state denaturation model(2). The urea concentration for 50% "denaturation" is 1.8M.

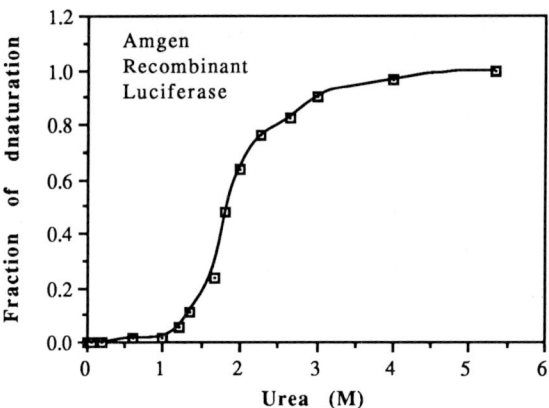

Fig.1.b Tryptophan fluorescence of recombinant luciferase (0.3mg/ml in 0.05M Trizma succinate) as a function of urea concentration (excitation wavelength 280nm, emission wavelength 340nm). Curve assumes a simple two-state denaturation model(2). The urea concentration for 50% "denaturation" is 1.8M.

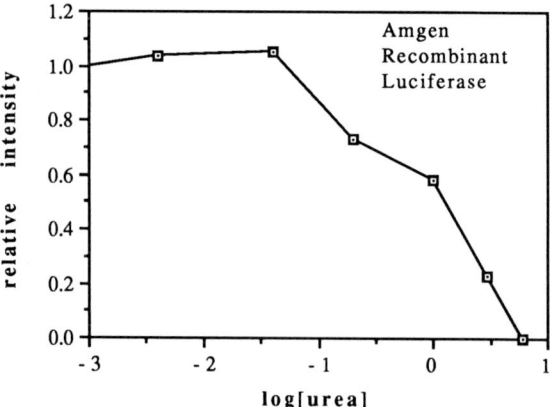

Fig.2 Relative intensity (based on solution without addition of urea) of the recombinant luciferase-luciferin system at different urea concentrations.   Light output appears consistent   with   the   fluorescence "denaturation" results, i.e. the enzyme is apparently   inactivated   at   high   urea concentration.   The data for Sigma luciferase is very similar (not shown).   The apparent small   increase   in   light   output   at   urea concentrations lower than 0.1M may be a real effect.   This is being further studied.

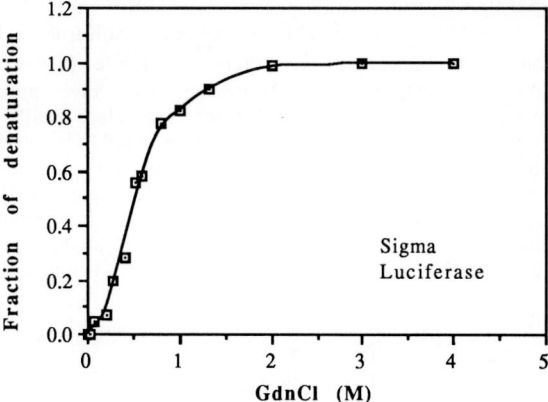

Fig.3.a Tryptophan fluorescence of Sigma luciferase (0.3mg/ml in 0.025M Glycylglycine) as a function of guanidinium chloride concentration (excitation wavelength 280nm, emission wavelength 340nm). Curve assumes a simple two-state denaturation model(2). The guanidinium chloride concentration for 50% "denaturation" is 0.5M.

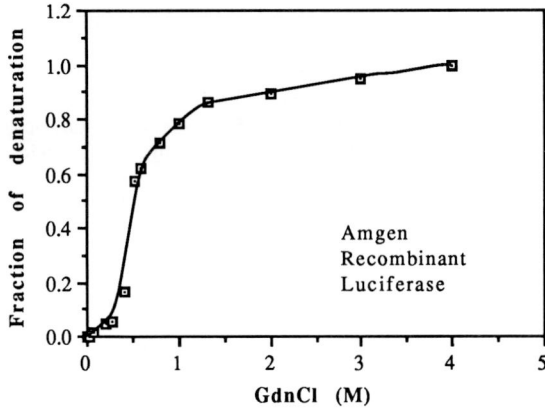

Fig.3.b Tryptophan fluorescence of recombinant luciferase (0.3mg/ml in 0.05M Trizma succinate) as a function of guanidinium chloride concentration (excitation wavelength 280nm, emission wavelength 340nm). Curve assumes a simple two-state denaturation model(2). The guanidinium chloride concentration for 50% "denaturation" is 0.5M

SUMMARY

Luciferase is hydrophobic and unstable in aqueous solution. Both the Sigma and recombinant luciferases are thermally labile ($T_d=40^{\circ}C$). Other hydrophobic proteins, such as lysozyme, can have much higher denaturation temperature.(8). The luciferases are also easily denatured by urea and guanidinium chloride. The anions easily inactivate luciferase possibly because they first denature the active sites of luciferase. It is important to find a stabilizer to protect luciferase from denaturing in aqueous solution. The stabilization of the enzyme by nonionic surfactants, such as PVP and PEG, has been studied (6). Other hydrophobic proteins may be helpful to enhance the stability of luciferase.

ACKNOWLEDGEMENTS

This work is supported by Protein Solutions Inc., Salt Lake City, Utah, USA.

REFERENCES

1. C. Nick Pace, B.A. Shirley and J.A. Thomson, Measuring the Conformational Stability of a Protein. In Protein Structure, T.E. Creighton(Eds), IRL Press, p.316(1989).
2. P.Y. Yeh, M.S, Thesis, Department of Materials Science and Engineering, University of Utah, Salt Lake City, Utah(1989).
3. J.R. DeWet, K.Y. Wood, M. Deluca, D.R. Helinski and S. Subramani, Firefly Luciferase Gene: Structure and Expression in Mammalian Cells, Mol. Cell. Biol. 7, p.725-737(1987).
4. C. R. Cantor and P. R. Schimmel, Biophysical Chemistry, 1, San Franciso, W.H. Freeman and Company, p.52(1980).
5. M. Deluca, G.W. Writz and W.D. McElory, Role of Sulfhydryl Groups in Firefly Luciferase. Biochem. 3, p.935-939(1960).
6. L.J. Kricka and M. De Luca, Effects of Solvents on Catalytic Activity of Firefly Luciferase. Archives of Biochemistry and Biophysics 217, p.674-681(1982)
7. C.Y. Wang, M.S. thesis, Department of Bioengineering, University of Utah, Salt Lake City, Utah, in preparation.
8. A.P. Wei, J.N. Herron and J.D. Andrade, The Role of Protein Structure in Surface Tension Kinetics. In From Clone to Clinic, D.J.A. Crommelin(Ed) in press(1990)

BIOLUMINESCENT REAGENTS AND "RAPID MICROBIOLOGY" METHODS

L.Yu.Brovko, N.N.Ugarova and A.F.Dukhovich

Chemistry Department, Lomonosov Moscow State University
Moscow 119899 USSR

INTRODUCTION

Measurement of microbial cells is of great importance in medical laboratories, food hygiene, fermentation industry and environmental studies. The conventional methods for microbial cell counting are time consuming and make no distinction between dead and viable cells. Out of the indirect cell measurements through the quantity of metabolic products, the bioluminescent ATP assay is the most rapid and sensitive method for determining the number of bacterial cells in a sample (1-3). The variants of bioluminescent biomass assay have been reported (4, 5). The authors did not always realized possible sources of draw-backs of this method, which worsen the detection limit up to 10^5 cells/ml. The purpose of this work was to elevate the sensitivity and accuracy of the bioluminescent assay of microorganisms. To this end, the methods of sample pretreatment and ATP release from cells as well as the composition of the bioluminescent ATP reagents, earlier developed in our laboratory (6, 7), have been optimized.

MATERIALS AND METHODS

Reagents. Bioluminescent ATP-Reagents, MICROLUM and IMMOLUM developed in our laboratory were used (Table 1). Titanium (IV) hydroxide was obtained as described before (8) and stored in the dark at 20°C. Microorganisms and somatic cells were cultivated on the standard culture media and counted with a particle counter.

Assay protocol. ATP-assay was performed using 0.2-1.0 ml ATP Reagent, 0.01-0.1 ml ATP sample and 10 μl of 1-10 μmol/l ATP Standard (3). For ATP extraction 0.5-0.9 ml DMSO, was used for each 0.1 ml cell suspension. When bacterial cells were preconcentrated by filtration or by immobilization, the filter or the precipitate was placed into 0.3-1 ml DMSO.

Cell concentrating. The cell suspension (1-20 ml) was filtered through a membrane filter (type A, Whatman) or through 0.3 μM Sinpore membrane. The filter should not go dry. For cell immobilization 0.1 ml Ti(OH)$_4$ in 1% NaCl was mixed with 1-20 ml cell suspension, agitated (10 min, 20°C) and centrifuged (8 000 rpM, 10 min, 20°C). The supernatant was discarded.

Bacterial ATP assay in human tissue samples. A sample of

Table 1. Properties of ATP reagents

	"MICROLUM" soluble	"IMMOLUM" immobilized
Firefly luciferase *L.mingrelica*		
Luminescence intensity at 0.01 μmol/l ATP (LKB-1250 luminometer)	> 300 mV	> 50 mV
Range of measurable ATP quantities, pmoles	0.01-1 00	0.1-10 000
Decline in luminescence intensity, in min	< 4%	< 2%
Stability during storage	1 year (-20°C)	1 year (4°C)
Stability after reconstitution with water at +4°C at +20°C	5-6 days 2-3 days	8-10 days 3-5 days
Number of assays per 1 vial depending on packing	10-200	10
Sensitivity to detergents and luciferase inhibitors	low	very low
Time of ATP measurement	< 1 min	< 1 min
Use of Internal Standard	possible	possible

0.5 g of human tissue was homogenized with 4.5 ml sterile physiological solution and incubated at 37°C for 5 h. To eliminate non-bacterial ATP, 0.01 ml 0.1 mol/l $NaIO_4$ in 5% Triton X-100 solution was added to 1 ml homogenate and incubated for 1 min. Then 0.2 ml of the incubated mixture was added to 0.8 ml DMSO and used to assay ATP.
Susceptibility Assay to Antibiotics. The pus sample (0.5 ml), was inoculated into 4.5 ml culture medium which contained either no antibiotic (control) or its single concentration and incubated with agitation (150 rpm) at 37°C for 5 h. 0.1 ml aliquots were taken, placed into 0.5 ml DMSO, mixed thoroughly and tested for ATP. Simultaneously, the samples were tested for the susceptibility to antibiotics by the standard agar diffusion method.
       RESULTS AND DISCUSSION
Cell concentrating. Filtration of samples lowered ATP content by 45+3% irrespective of the type of the used membrane and of the cell concentration if it varied within 0.005-2 mg/ml as well as irrespective of the sample volume, 1-20 ml. The cell filtration permitted to lower the biomass detection limit down to 1 μg/ml. Upon the cell immobilization on $Ti(OH)_4$ the percentage of the retained ATP content was as high as 85+8%. This lowered the biomass detection limit approx. 40 times (down to 0.5 μg/ml).
Removal of non-bacterial ATP. The treatment with Triton X-100 + apyrase (5) seemed to be ineffective for tissue samples because the important part of somatic cells remained undestroyed. We proposed a simpler and cheaper method to remove the non-bacterial ATP. When the tissue homogenate was treated with 1 mM $NaIO_4$ + 0.05% Triton X-100, for 1 min, up to 96-98% of somatic ATP destroyed while the microbial cells remained intact. The following

addition of DMSO (up to 80% v/v) stopped ATP oxidation by
NaIO$_4$ and the bacterial ATP was simultaneously extracted.
Assay of bacterial contaminations in human tissues. The
treatment with 1 mM NaIO$_4$ + 0.05% Triton X-100 decreased
the somatic ATP content about 50-100 times but the
residual somatic ATP concentration remained rather high.
This is why the direct assay of bacterial ATP in the
samples was impossible for the diagnostically important
range of bacterial contaminations (10^3-10^7 cells/g of
tissue). To increase the level of bacterial cells, we
preincubated the contaminated tissues homogenate at 37°C
for 5 h. After this, the non-bacterial ATP was destroyed
and the bacterial ATP was assayed. We found that for
different strains (*E.coli*, *St.aureus*, *Klebsiella sp.*, *Enterobac-
ter sp.*, *Proteus sp.*), the concentration of bacterial ATP, in
tissue homogenate after its preincubation at 37°C for 5h,
depended directly on the bacterial cells content in the
initial sample.
Susceptibility to antibiotics was defined for the
microflora taken from patients treated against pyo-infla-
mmatory infections of soft tissues. Fig.1 show three
types of dependencies ATP content-time: for monoculture
(Fig.1a); for monoculture, when the patients were treated
with antibiotics previously (Fig.1b), and for the mixed
culture (Fig.1c).

ATP, μmol/l

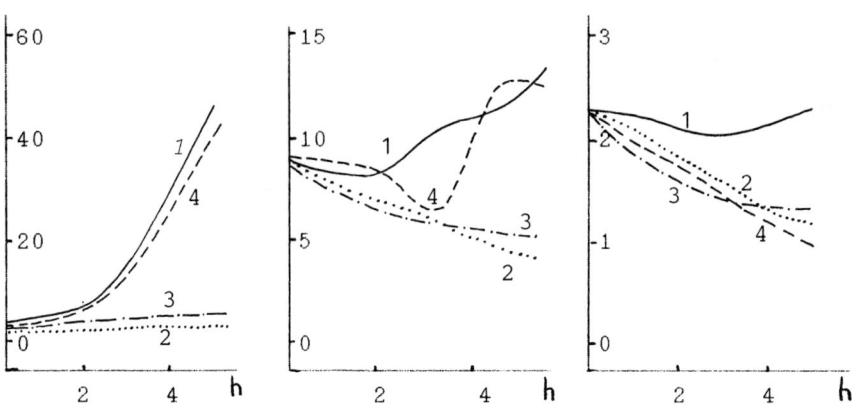

Fig.1. Bacterial ATP growth in clinical samples incubated
without antibiotic (1) or with 5 μg/ml, erythromycin (2);
8 μg/ml gentamycin (3); and 2 U/ml penicillin (4); (a) -
monoculture (*St.aureus* or *E.coli*); (b) - monoculture (*St.
aureus* or *St.saproph*) when patients previously were
treated with antibiotics; (c) - mixed culture (*E.coli* +
St.aureus + *Proteus mirabilis*). Incubation was at 37°C in
beef-peptone-extract broth with agitation.
Nevertheless, in all cases the difference between the ATP
content in the control and the samples in the presence of

antibiotics was observed in 3 h cultivation. As a
quantitative parameter defining the susceptibility to
antibiotics, we offered the ratio between the ATP content
in the presence of antibiotics ($[ATP]_{sample}$) and the ATP
in the control in 3 h incubation (U, %), which was
calculated by the formula:

$$U, \; (\%) = (1 - \frac{[ATP]_{sample}}{[ATP]_{control}}) \times 100\%$$

Comparison of U calculated by this formula with the
results obtained by the agar diffusion method shows that
for antibiotic resistant strains, U < 20% whereas for
antibiotic-sensitive strains, U > 20%. The clinical data
showed good correlation between the bioluminescent 3 h
susceptibility test and the standard agar diffusion
method.

REFERENCES
1. E.W.Chappelle and C.V.Lewin, Use of the firefly bioluminescent reaction for rapid detection and counting
   of bacteria. Biochem.Med., 2, 41-52 (1968).
2. Ch.Harris and D.Kell, Estimation of microbial biomass.
   Biosensor, 1, 17-84 (1985).
3. N.N.Ugarova, L.Yu.Brovko, I.Yu.Trdatyan and E.I.Raini-
   na, Bioluminescent assays in microbiology. Prikl.
   Biochim.Mikrobiol. (Russ), 23, 14-24 (1987).
4. Lundin and A.Thore, Comparison of methods for extraction of bacterial adenine nucleotides determined by
   firefly luciferase assay. Appl. Microbiol., 30, 713-
   721 (1975).
5. Ph.E.Stanley, A review of bioluminescent ATP techniques in rapid microbiology. J.Biolum.Chemilum. 4, 375-
   380 (1989).
6. N.N.Ugarova, L.Yu.Brovko and N.V.Kost, Immobilization
   of luciferase from the firefly Luciola mingrelica -
   catalytic properties and stability of the immobilized
   enzyme. Enzyme Microb.Technol. 4, 224-228 (1982).
7. N.Yu.Filippova, A.F.Dukhovich and N.N.Ugarova, New
   Approaches to the preparation and application of
   firefly luciferase. J.Biolum.Chemilum. 4, 419-422
   (1989).
8. J.Kennedy, Facile methods for the immobilization of
   microbial cells without disruption of their life
   processes. In Immobilized Microbial Cells, K.Venkatsu-
   bramanian (Ed), Am.Chem.Soc., Washington, D.C., p.119-
   120 (1979).

THE APPLICATION OF AUTOMATED AND SENSITIVE LUMINOMETRY FOR MICROBIAL DETECTION IN FOOD AND BEVERAGES AND OTHER APPLICATIONS USING SELECTIVE AND STABLE BIOCHEMICAL PROCEDURES

W. Gehle, R. Presswood[1] and D.A. Stafford

Dynatech Corporation, Medical and Diagnostic Products Group, 14340 Sulleyfield Circle, Chantilly, VA 22021, USA; [1]IMSCO, 3 Dundee Office Park, Andover, MA 01810, USA

INTRODUCTION

The production of biochemical reagents and automated sensitive luminometers has provided a good background within which to develop microbiological contamination detection opportunities. Such opportunities have been described(1) and developed(2) and the applications reviewed(3). The enumeration of yeasts using the ATP released and subsequent detection using firefly luciferase kits has been shown to be an effective method for measurement of food or beverage spoilage(3). It has also been shown that to increase the sensitivity of the luciferase enzyme an understanding of the lipid nature of the enzyme is required(4). Using differential centrifugation coupled with stabilizing additives a 10-100 improvement in sensitivity can be achieved. The kinetics of this protein lipid complex have been identified(5) and may be used in applications for kits to determine the lower levels of contamination of bacteria and yeasts in food and beverage samples.

The use of a highly sensitive luminometer with sophisticated software allows the processing of multiple samples and reagent injections. Dynatech and IMSCO have developed a 24 sample mini well plate luminometer for detecting low concentrations of ATP. The incorporation of up to three injectors enables a versatile and selective extraction procedure and bioluminescence emissions from samples contaminated with micro-organisms.

MATERIALS AND METHODS

The instrumentation constructed of an automated luminometer, the MICROSURE 1000, is available in the USA from IMSCO, Andover, Massachusetts and in Europe and Asia from Dynatech. It is controlled by an external microcomputer executing proprietary software; up to three external injectors can be used to inject into the sample well, somatic and microbial extracts as well as luciferase/luciferin cocktails. The reagents used were proprietary firefly luciferase and somatic and bacterial extractants(2)(3). The enzyme was reconstituted from lyophilized powders with distilled water filtered through a 0.2um pore size membrane. Once dissolved, the enzyme could be used immediately for routine tests, or stored at -20°C for future use.

RESULTS AND DISCUSSION

With an average reagent background of 7.75×10^{-17} moles of ATP at 2 standard deviations a sensitivity of about 15 Colony Forming Units (CFU's), for the detection of microbes such as <u>Lactobacillus</u> and <u>Aerobacter</u> can be achieved. These results were obtained with a laboratory temperature of about 30°C using an off-the-shelf bottle of bacterial extractant and freshly constituted enzyme. With cooler temperature controlled environments sensitivities of better than this can be obtained.

The results being obtained in practice and with improvements being made in the extraction of luciferases(4) and in their prolonged stability the sensitivity to a single bacterial CFU is well within the practical goals of food microbial detection.

Other applications for the pharmaceutical industry and for urine testing also enable more informed interpretation of results to be made. The increase in sensitivity originates in two main areas. The first is reagent signal to noise ratio. The second is in the instrumentation, the alignment of the photomultiplier with the sample and the control software. Having the sample and the phototube as close as possible to each other contributes to the sensitivity enhancement. The phototube output is digitized every 10 msec without intervening signal damping or time constant. The system noise is of very high frequency compared to the kinetics of the peak height assay; mathematical techniques can be used to eliminate the noise. The assay is based upon peak height determination, and instrument drift would therefore not be relevant; in any event we have a low drift instrument. Thus the instrument provides for sample proximity to photomultiplier tube, low noise and low drift.

Fig. 1

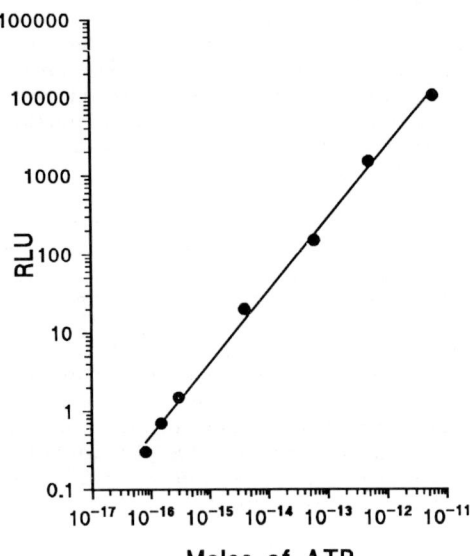

Moles of ATP

The software also allows for evaluation during calibration of from 4 to 22 reagent backgrounds. The software is written to report CFU's directly, either bacteria or yeasts. With software modification the enumeration of epithelial cells can be made in urines. One of the somatic extractants used by us with contained ATPase, and produced by IMSCO, has an advantage for the detection of microbial contamination in tomato paste and fruit juices in that it is active at acid pH. These applications are directed to both freshly manufactured as well as warehoused products. The current combined chemical and instrument advantages are being improved upon by incorporating advantages developed elsewhere to improve both the stability and sensitivity of the luciferase enzyme to reliably provide single CFU measurements.

With regard to other applications more recently, the new enzyme preparation has been used successfully to determine single epithelial cells in urine samples together with less than 50 bacterial cells. Using the selective somatic extractant we have provided a high degree of specifity for urine screening. An antibiotic sensitivity test has also been developed which can be completed within 4 hours of sample taking. Using E.coli as a contaminant in urines with routine highly purified enzymes and with a ambient laboratory temperature of about 30°C the following calibration curves were obtained. With the proprietary somatic extractant, to remove epithelial ATP the limit of sensitivity of the test method is reduced from 2×10^4 cells/sample to 2×10^2. Better sensitivity may be obtained to 5×10 with cooled enzyme and samples.

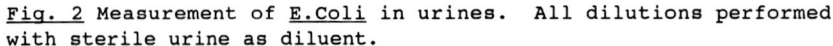

Fig. 2 Measurement of E.Coli in urines. All dilutions performed with sterile urine as diluent.

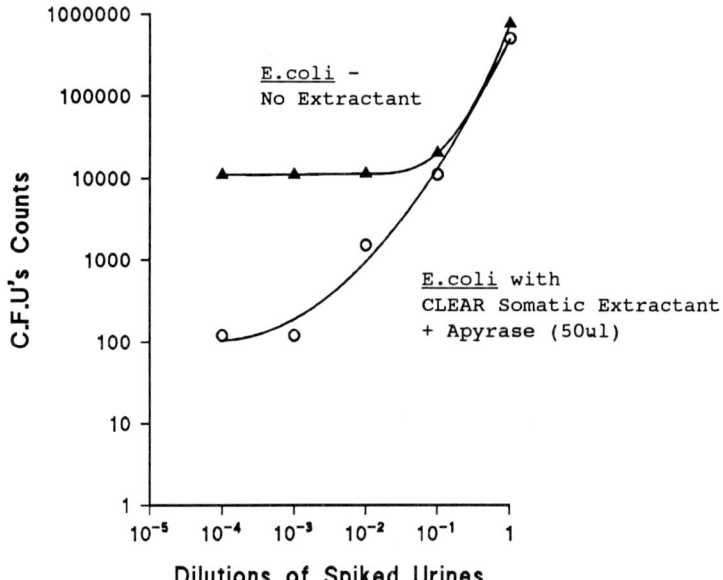

REFERENCES

1.   P.E. Stanley and B.J. McCarthy, Reagents and instruments for assays using ATP and luminescence: present needs and future possibilities in rapid microbiology. In ATP Luminescence-Rapid Methods in Microbiology, P.E. Stanley, B.J. McCarthy and R. Smither (Eds), Blackwell Scientific Publications, Oxford, p.73-80 (1989).

2.   G.B. Sala-Newby, C. Goodfield, I.R. Johnson, P.G. Massey, D.A. Stafford, A.N. Hampton and A.K. Campbell, Ultra-sensitive detection of microbes by ATP analysis. In ATP Luminescence-Rapid Methods in Microbiology, P.E. Stanley, B.J. McCarthy and R. Smither (Eds), Blackwell Scientific Publications, Oxford, p.262-260 (1989).

3.   R.P. Presswood, Microbial estimation in beverages. In ATP Luminescence-Rapid Methods in Microbiology, P.E. Stanley, B.J. McCarthy and R. Smither (Eds), Blackwell Scientific Publications, Oxford, p.161-166 (1989).

4.   N.Yu. Filippova, A.F. Dukhovich and N.N. Ugarova, New approaches to the preparation and application of firefly luciferase. J.Biolumin.Chemilumin. 4, 419-422 (1989).

5.   N.N. Ugarova, A.F. Dukhovich, S.V. Shvets, N.Yu. Philippova and I.V. Berezin, Kinetics of the inactivation of the protein-lipid complex, firefly luciferase, by sodium deoxycholate and its reactivation by phosphatidylcholine. Biochem. et Biophys. Acta. 921, 465-472 (1987).

DETERMINATION OF O_2^- GENERATED BY MICROSOMAL MEMBRANES USING CYPRIDINA LUCIFERIN ANALOG-DEPENDENT LUMINESCENCE

Satoshi, Koga and Minoru, Nakano[1]

Taiho Pharmaceutical Co.,Ltd., Kandanishiki-cho, Chiyoda-ku, Tokyo 101,Japan and [1]College of Medical Care and Technology, Gunma University, Maebashi, Gunma 371, Japan

INTRODUCTION

It has been found that the lipid peroxidation in a reconstructed system, in which NADPH, Fe^{3+}-ADP, phospholipid liposomes, cytochrome P-450 and its reductase were present, is significantly inhibited by a catalytic amount of SOD(1). This indicates the participation of O_2^- generated by enzymatic reduction of cytochrome P-450 in the presence of O_2 in the lipid peroxidation. However, a dramatic inhibitory effect of SOD on the lipid peroxidation has not been observed with the system in which NADPH, Fe^{3+}-ADP, and intact microsomes(2). Such a discrepancy may be due to a structural complexity of microsomes.

The present work was undertaken to prove the behavior of O_2^- in the lipid peroxidation using intact microsomes as sources of phospholipid and O_2^- generating system and cypridina luciferin analog (MCLA) as a chemiluminescence probe for O_2^- assay.

MATERIALS AND METHODS

Microsomes were prepared from rat liver according to the method described Levin et al.(3). Fe^{3+}-ADP complex was prepared by the method as previously described(4). The complete system contained microsomes(1mg of protein/ ml), 10μmol/l Fe^{3+}-ADP complex, 1mmol/l NADPH, and 0.1mol/l Tris-HCl buffer at pH 7.4 in total volume of 1.0ml. Xanthine oxidase system contained 0.3mmol/l hypoxanthine, 2.5μg/ml xanthine oxidase and 0.1mol/l Tris-HCl buffer at pH 7.4 in total volume of 1.0ml. Lipid peroxidation and 2-methyl- -6-(p-methoxy-phenyl)-3,7-dihydroimidazo[1,2-a]pyrazin-3-one(MCLA)-dependent chemiluminescence were determined by the method as previously described(1). All incubation experiments were carried out at 37 ℃.

RESULTS AND DISCUSSION

Inhibition of lipid peroxidation by MCLA and other agents

To investigate possible inhibitor on the lipid peroxidation in the complete system, each of the compounds was added to the system prior to the reaction and lipid peroxidation was monitored by the formation of TBARS. The results obtained are shown in Table 1, cypridina luciferin analogs, such as MCLA and CLA, desferrioxamine (a powerful iron chelating agent) and Cu-(salicylate)₂ (a O_2^- scavenger) at their low concentrations possessed powerful inhibitory effects on the lipid peroxidation, while SOD did not inhibit the lipid peroxidation. In addition to these agents, aminopyrine and aniline (substrates of reduced cytochrome P-450) at 5mmol/l possessed substantial effects on the protection of lipid peroxidation.

These results suggested that O_2^-, scavenged by MCLA(CLA) or Cu-(salicylate)$_2$, but not by SOD, is involved in iron-induced phospholipid peroxidation.

Table 1.
Effects of various inhibitors on the NADPH-dependent
lipid peroxidation

Incubation Systems	MDA(nmol/min/ml)
Control (-NADPH)	0.49
+ NADPH(1 mmol/l)	4.65
+ NADPH + SOD (0.5µmol/l)	4.33
+ NADPH + ceruloplasmin(0.05mg/ ml)	4.14
+ NADPH + MCLA (40µmol/l)	0.78
+ NADPH + CLA (40µmol/l)	0.76
+ NADPH + Cu-(salicylate)$_2$(40µmol/l)	0.99
+ NADPH + aminopyrine (5mmol/l)	1.63
+ NADPH + aniline (5mmol/l)	0.80
+ NADPH + desferrioxamine(30µmol/l)	0.48

Extramicrosomal O_2^- generation

To make sure of the reactivity of O_2^- with MCLA, MCLA at low concentration(10µmol/l) was exposed to a well known O_2^- generating system, hypoxanthine and xanthine oxidase, and the MCLA-dependent chemiluminescence was detected by a Luminescence Reader. As shown in Fig.1-A, the luminescence appeared just after the addition of xanthine oxidase and gradually decreased after peaking, but it was not significantly influenced by Fe^{3+}-ADP. With MCLA at low concentration, at which iron-induced phospholipid peroxidation would not be inhibited, a system excluding iron from the complete system (NIC system) also emitted light, just after the addition ofNADPH, and decreased exponentially after peaking, while the complete system had a short lasted emission with lower light yield compared with NIC system(Fig.1-B). In both cases, additional MCLA greatly recovered the reduced luminescence. These results suggest that MCLA undergoes a partial decomposition according to non-emissive process during iron-induced lipid peroxidation. To know the behavior of Fe^{3+} added to microsomes, MCLA at high concentration (40µmol/l),at which iron-induced phospholipid peroxidation could be completely inhibited, was added to NIC system and the complete system, and chemiluminescences in both systems were detected.As shown in Fig.1-C, MCLA-dependent luminescence in NIC system lasted for a long period after the addition of NADPH, while the luminescence in the complete system decreased rapidly after a initial burst and the reduced luminescence was promptly recovered to the level of the luminescence in NIC system by addition of desferrioxamine (a strong chelating agent of Fe^{3+}). This indicates that MCLA-dependent luminescence (a O_2^- induced luminescence) is quenched by Fe^{3+}, probably coordinated with phosphate moiety in microsomal phospholipids. Under the same experimental conditions, save that iron concentration varied, MCLA-dependent luminescence reduced with increasing iron concentration. The plot of integrated light intensity during 15min against iron concentration gave a hypobolic curve(Fig.1-D). On the assumption that iron coordinates with phosphate moiety in microsomes, a minimal chelating concentration of Fe^{3+} was graphically calculated to be about 6 µmol/l per 1.4 mmol/lphosphate.

Less involvement of extramicrosomal O_2^- for the lipid peroxidation

To investigate the ability of extramicrosomal O_2^- to promote lipid peroxidation, intact microsomes chelated with iron was exposed to O_2^- generated by a hypoxanthine/xanthine oxidase system and the lipid peroxidation was monitored by TBARS formation. However, TBARS formation (0.59nmol/min/ml) in the system containing 0.3 mmol/l hypoxanthine, 10µg/ml xanthine oxidase and

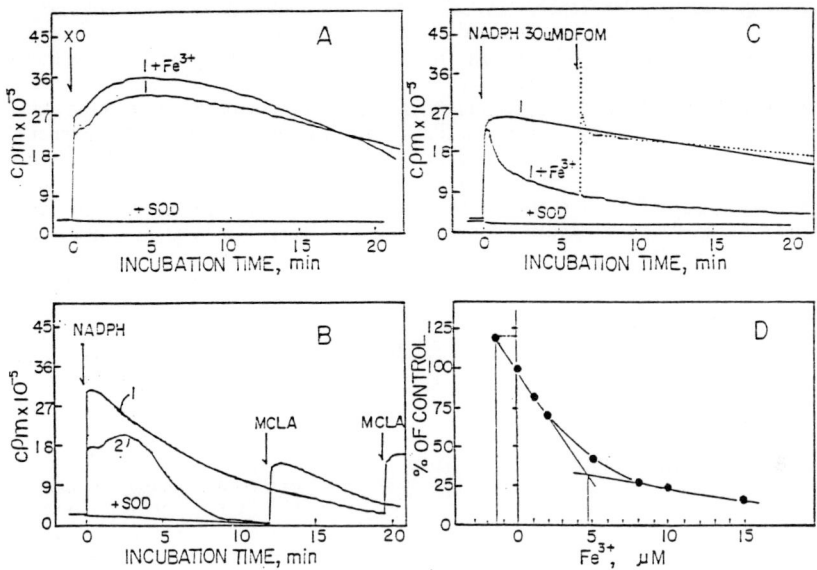

Fig. 1. MCLA-dependent chemiluminescence

(A) MCLA-dependent chemiluminescence in hypoxanthine/xanthine oxidase system with(1+Fe^{3+}) or iron-ADP(1).
(B) MCLA-dependent chemiluminescence in NIC system(1) and complete system(2, for lipid peroxidation) using low level of MCLA(10μmol/1).
(C) MCLA-dependent chemiluminescence in NIC system(1) and complete system(1+ Fe^{3+}, for lipid peroxidation) using high level of MCLA(40μmol/1).
(D) Effect of Fe^{3+}-ADP on the MCLA-dependent chemiluminescence(integrated light intensity) in complete system using high concentration of MCLA(40μmol/1).

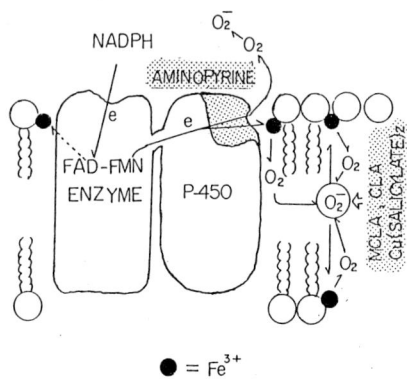

Fig. 2. Possible mechanism of NADPH-dependent lipid peroxidation

microsome(1mg of protein/ml)-Fe^{3+}(10μmol/1)-ADP(0.17mmol/1) was less than threefold that excluding iron (0.22nmol/min/ml), indicating less involvement of extramicrosomal O_2^- for the iron induced lipid peroxidation.

From the result obtained here, we tentatively proposed the mechanism of the NADPH-dependent lipid peroxidation as shown in Fig.2.

REFERENCES
1. H.Minakami, H.Arai, M.Nakano, K.Sugioka, S.Suzuki and A.Sotomatu ,A new and suitable reconstructed system for NADPH-dependent microsomal lipid peroxidation. Biochem.Biophys.Res.

Commun. 153, 973-978 (1988).
2. J. L. Poyer and P. B. McCay, Reduced triphosphopyridine nucleotide oxidase-catalyzed alterations of membrane phospholipids. J. Biol. Chem. 246, 263-269 (1971).
3. W. Levin, D. Ryan, S. West, and A. Y. H. Lu, Preparation of partial purified, lipid-depleted cytochrome P-450 and reduced nicotinamaid adenin dinucleotide phosphate-cytochrome c reductase from rat liver microsomes. J. Biol. Chem. 249, 1747-1754(1974).
4. K. Sugioka and M. Nakano, Mechanism of phospholipid peroxidation induced by ferric ion-ADP-adriamycin-co-ordination complex. Biochem. Biophys. Acta. 713, 333-343(1982).

CHEMILUMINESCENT DETERMINATION OF SUPEROXIDE ION BY USING IMMOBILIZED SUPEROXIDE DISMUTASE

K.Goda, S.Ikeda and R.Kishimoto

Department of Nutrition, Kobe-Gakuin University, Kobe, 673, Japan

INTRODUCTION

Reduction of O_2 during aerobic cellular metabolism may result in the production of toxic oxygen intermediates such as O_2^- and so on. Enzymes which scavenge them such as superoxide dismutase(SOD), catalase, and peroxidase are primary line of defence against their toxicity.

The determination of O_2^- is important in biological systems and in several applications. Some methods are already available for its determination, for examples, O_2^- mediated reduction of ferri-cytochrome c, ESR measurement and others. A rapid and simple method, however, valuable.

This deals with the preparation and evaluation of immobilized enzyme, SOD, membrane for the chemiluminescence assay method. The luminol enhanced chemiluminescence(CL) mediated from O_2^- was measured. SOD was used as the enzyme for immobilization. Two immobilization methods were used, which were based on the co-polymeization with other protein, BSA, using cross-linking reagent, glutaraldehyde, and the physical entrapment in a polyacrylamide gel membrane. The procedure is both rapid and simple. The inhibited CL with SOD was determined and calibration experiments for the response of immobilized enzyme to O_2^- were performed.

MATERIALS AND METHODS

Reagents Bovine erythrocyte Cu,Zn-SOD was used as the enzyme(SOD) for immobilization, and xanthine oxidase, XO(milk XO), as the enzyme for O_2^--generating system, xanthine as the substrate of XO. BSA and glutaraldehyde were used for co-polymrization procedure, and N,N,N,'N'-methylenebis[acrylamide](Bis), N,N,N,'N,'-tetramethylethylenediamine(TEMED) and ammonium persulfate for polyacrylamide gel entrapment method. [1,2]

Method The CL measurements were performed by the single photon counting method. Mixing of solutions was made by stopped-flow method. The SOD membrane was overlapped on to the surface of glass spacer inserted into 1-cm quarzt cuvettes, and retained with a rubber "O" ring.

Preparation On a glass plate, 4µl of BSA(2%,w/v) and 4µl of SOD(0.4 %, w/v) were mixed with 1µl of glutaraldehyde(1%, v/v) as the standard mixture. The solution was spread over and allowed to crosslink for several hours at room temperature, open to air.

The enzyme membrane formed was detached from the plate and attached

on the glass spacer. For physical entrapment procedure, 25µl of SOD
(0.4%, w/v), 70µl of acrylamide, 10µl of Bis(2.0%, w/v), 70µl of TE
MED(5.0%, w/v) and 10µl of ammonium persulfate(1.0%, w/v) were mixed, and spread on a glass plate. The solution was allowed to poly-
merize for 10-30 minutes at room temperature. The SOD entrapped
polyacrylamide membrane was attached on the glass spacer like the
former.
The luminol enhanced CL derived from O_2^- was measured with and without SOD membrane for a same sample. The inhibited CL by SOD was
calculated, which is the difference between CL with and without SOD.

RESULTS AND DISCUSSION
The rate constants(k and k_{SOD^-}) of the spontaneous dismutation and
SOD-catalyzed dismutation of O_2^- at neutral pH range is ascertained
to be k=0.39 $M^{-1}s^{-1}$ [3] and k_{SOD} 10^9 $M^{-1}s^{-1}$ [4] for each. An advantage of having the immobilized SOD is that the inhibited CL is base-
ed on SOD-accelerated dismutation of O_2^-, which means the inhibited
CL to be attributed entirely derived from O_2^-, and SOD membrane repeatedly to be usable.
The actual CL curves with and without SOD membrane are shown in Fig.
1, in which xanthine-XO system was used as the O_2^- -generating source
and its concentration was calculated from another method [5].
The effect of variations of the amounts of SOD, BSA and glutaraldehyde used for SOD membrane preparation was investigated. The amount
of one component was varied while other two remained essentially
constant. The enzyme membrane response for CL inhibition was plotted
against the varied component. The response(the inhibited CL inten-

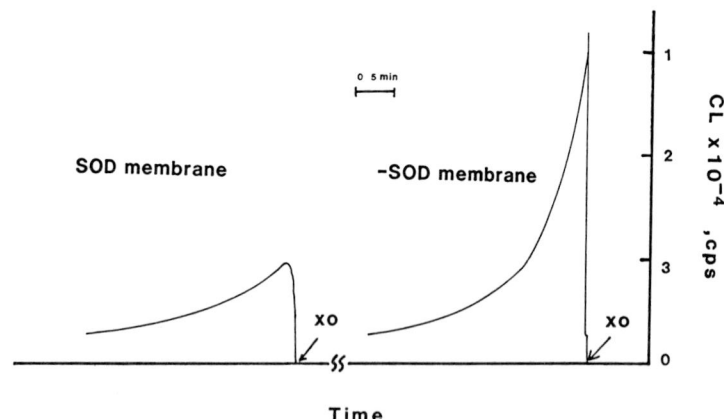

Time

Fig.1. Typical chemiluminescence pattern for O_2^-
from xanthine-xanthine oxidase system. The reaction mixture contained xanthine, 40µmol/1, XO, 2
units/ml and luminol, 40µmol/1 in 5 mmol/1 phosphate buffer, pH 7.4.

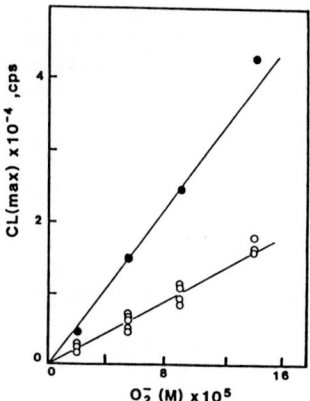

Fig.2. Relations between chemiluminescence intensity and superoxide concentrations with immobilized SOD.

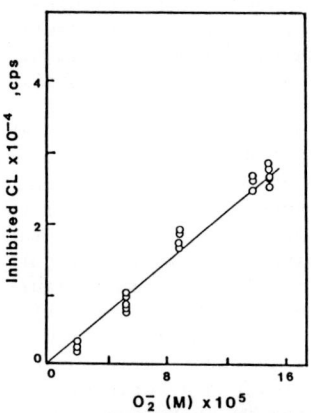

Fig.3. Relations between chemiluminescence intensity and superoxide concentrations with immobilized SOD.

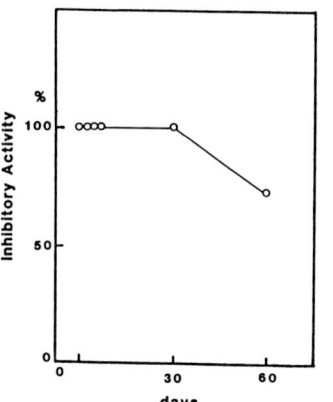

Fig.4. Long-term stability of immobilized SOD activity.

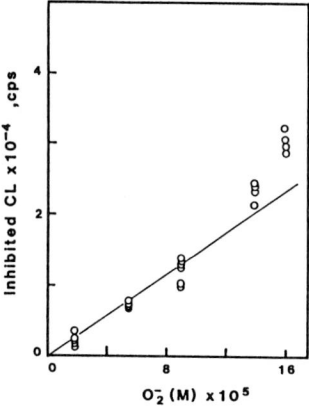

Fig.5. Relations between chemiluminescence intensity and superoxide concentrations with immobilized SOD(poly-acrylamide).

sity) was maximun around the concentration of 0.4% (w/v) for SOD (4 mg/ml), 2% (w/v) for BSA (20 mg/ml) and 0.1% (v/v) for glutaralde-hyde. However, the physical properties of the membrane prepared at this concentration of glutaraldehyde was a little bit soft and weak , which means to be less cross-linked. Then, the preparation was made at 1% (v/v) of glutarladehyde.

As shown in Fig.2, CL intensity at various concentrations of O_2^- with (○) and without (•) SOD membrane was plotted. From the Fig., the inhibited CL intensity was calculated and plotted for the concentra-tion of O_2^- (Fig.3). As seen in Fig.3, it indicates a linear respon-se range of 1×10^{-5} mol/l to 15×10^{-5} mol/l in the superoxide.

The pH which the SOD membrane exhibits maximun CL inhibition was de-termined in a series of buffer at the constant O_2^- concentration. The result is shown in Table 1. In this study, pH 7.4 was mainly selected as the pH operation.

The long-term stability of SOD membrane is illustrated in Fig.4. As shown in the Figure, the enzyme activity is kept mostly stable for 1 month, and after that period, it gradually decreases during the storage in buffer.

The same result was obtained for the physically entrapped SOD in polyacrylamide gel membrane (Fig.5).

The application of this SOD membrane method to biological O_2^- gener-ating systems will be expected.

Table 1. Effect of pH on the inhibitory activity of immobilized SOD for the chemiluminescence. O_2^-, 6×10^{-5} mol/l.

pH	Inhibition	
	%	ratio
7.4	61.8	100
8.0	56.7	91.8
9.0	40.9	66.2

REFERENCES

1. L.D.Bower, Applications of immobilized biocatalysts in chemical analysis. Anal.Chem.,58, 513A-530A (1986).
2. W.C.White and G.G.Guilbault, Lysine specific enzyme elec-trode for determination of lysine in grains and foodstuffs. Anal.Chem., 50, 1481-1486 (1978).
3. B.H.Bielski, Reevaluation of the spectral and kinetic prop-erties of HO₂ and O₂⁻ free radicals. Photochem.Photobiol., 28, 645-649 (1974).
4. I.Fridovich, Superoxide dismutase. Adv.Enzymol., 41, 35-97 (1974).
5. Y.Sawada and I.Yamazaki, One-electron transfer reactions in biochemical systems. VIII. kinetic study of superoxide dis-mutase. Biochem.Biophys.Acta, 327, 257-265 (1973).

SUPEROXIDE RADICAL PRODUCTION IN HUMAN BLOOD IS
STIMULATED BY PERTUSSIS AND INHIBITED BY CHOLERA
TOXIN : EVIDENCE FOR G-PROTEIN FUNCTIONAL ACTIVITY
RELATED TO SUPEROXIDE GENERATION

K. Hohlmaier, J. Gürenci and H. Holzmann

Zentrum für Dermatologie, Johann Wolfgang von Goethe-
Universität, 6000 Frankfurt am Main 70, Theodor-Stern-
Kai 7, West Germany

INTRODUCTION

The major sources of superoxide radical (SOR) production in peripheral blood are polymorphnuclear
leucocytes (PMNs) and monocytes (MCs) .
The primary biochemical events after stimulation of PMNs
by bacterial chemotactic factors like FMLP
have been thoroughly examined (1). The presence and
involvement of GTP-dependent regulatory proteins could be
shown in plasma membranes by immunoblotting (2),
radioactive labeling with 32-P-ADP and
determination of FMLP-dependent GTPase activity .
More indirectly, the effect of functional modification of G-proteins by ADP ribosylation can be
measured in whole cells after treatment by bacterial
exotoxins like pertussis or cholera toxin .
Such investigations confirmed that adenylate cyclase is
regulated by 2 clearly distinct entities of trimeric
proteins named Gi and Gs referring to their inhibitory
or stimulatory control of adenylate cyclase activity .
Here, pertussis toxin inactivates Gi and cholera toxin
activates Gs . In neutrophilic granulocytes, Gi appears
to be the maior G-protein; although coupled to phospholipase C, its regulatory role is not very well established in native human PMNs . Nevertheless, in spite of
the complex network of biochemical steps (phospholipase C activation with subsequent production of
inositol phosphates and phospholipase A2 activation with
diacylglycerol release and proteinkinase C activation)
in superoxide production with G-proteins being an early
step of the transduction and amplification cascade, it
has been demonstrated that (i) phospholipase C activity
(3) and (ii) even FMLP-induced superoxide production in
human PMNs (4) can be inhibited by pretreatment with
pertussis toxin . Other reports (5) showed that cholera

toxin, too, is able to modify a Gi-like protein in
neutrophilic HL-60 cells . In this study, we investigated the effect of pretreatment with cholera toxin and
pertussis toxin on net production of reactive
oxygen metabolites (in our detection system mainly
superoxide radicals, as can be shown by addition of
superoxide dismutase) . Due to technical reasons,
we did not use isolated PMNs or MCs, but freshly
isolated heparinized whole blood and determined the
effects of the G-protein modifying toxins on basal
and FMLP-, TPA- and Calcium ionophore A 23187-stimulated
SOR production .

MATERIALS AND METHODS

Reagents : Stimulators used were : N-formyl-L-
methionyl-L-leucyl-L-phenylalanine (FMLP), calcium ionophore A 23187, Tetradecanoylphorbolacetate (TPA) and bis-
methylacridiniumnitrate (lucigenine), all purchased from
Sigma . Cholera and pertussis toxin were from LIST
Biological Laboratories, Campbell, CA. (USA) . Heparin
used was Liquemin 25000 from Hoffmann-La Roche, Grenzach,
West Germany; Minimal Essential Medium Dulbecco was from
Biochrom KG, Berlin, West Germany .
[32-P]-NAD was from Amersham; all other reagents
(salts for buffers, nucleic acids, gel electrophoresis
chemicals were either from Sigma, Serva (Heidelberg,
West Germany) or Boehringer (Mannheim, West Germany)) .
SDS-Gels were autoradiographed on Whatman 3M paper .
Methods : Blood was taken from a peripheral vein of
healthy volunteers in a heparinized syringe (approx.
10 U/ml blood) . After 90 minutes of incubation
with cholera toxin, pertussis toxin or vehicle, 10 ul of
blood were added to an incubation mixture
of 300 ul Minimal Essential Medium Dulbecco (MEM),
100 ul stimulator in MEM (or MEM only), 100 ul
of luminescence enhancer (lucigenine) and incubated
at 37 °C in a 6-channel-chemiluminescence recorder unit
(biolumat 9505 by Berthold, Wildbad, West Germany) .
Chemiluminescence intensity (cpm) was recorded over
35 minutes; the time integrals were determined automatically and used for calculation of stimulation
and inhibition ratios (i.e. cpm (treatment)/cpm
(appropriate control)) .
HL-60-promyelocytic leukemia cells were a friendly
gift of Dr. Habenicht, Dept. of Medicine, Heidelberg .
The cells were cultured, submitted to neutrophilic

differentiation by dimethylsulphoxide and plasma
membranes prepared as described elsewhere (7) .
After activation of cholera and pertussis toxin
by dithiothreitol, plasma membranes were incubated
with the toxin and 32-P-NAD to label G-proteins according
to (7) . Membranes were submitted to SDS gel
electrophoresis (8), dried in a Bio Rad slab gel drier
and exposed to Kodak XR Omat film to visualize radioactive banding patterns .

RESULTS AND DISCUSSION

The results obtained by us are, clearly, limited by the
complex system used . That means effects on
SOR-production always reflect contributions of PMNs
and MCs together as well as the influence of soluble
factors possibly produced in blood during the incubation time . Nevertheless, measurements of toxin
effects in isolated PMNs or MCs are restrained by their
limited viability and could not be performed
reproducibly after another 90 minutes incubation
with either toxin .
Altogether, more than 80% of chemiluminescence activity in whole blood are produced by neutrophilic
granulocytes . Therefore, we -very carefully- interpret our results as effects of the neutrophil .
Due to the very slow kinetics (not shown), "non-
specific" effects (e.g. interaction with ConA-
receptors) seem very unlikely to us .

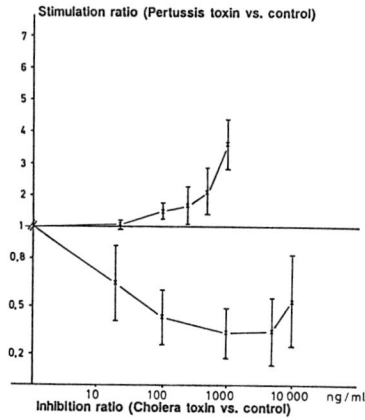

Fig.1: Effect of toxin
treatment on lucigenin-
enhanced
chemiluminescence

Different concentrations
of cholera or pertussis
were incubated at 37 ℃
with whole blood; the
ratio cpm treated vs.
untreated control was
determined .

To our knowledge, effects of G-protein modifying toxins

on SOR production of neutrophilic granulocytes
in their "native" environment (whole blood) have not
been performed .
Contrary to other studies (4) with neutrophils, we
observe a dose-dependent stimulation of SOR production by pertussis toxin in a dose range of 25-
1000 ng/ml up to 4fold (Fig. 1) .

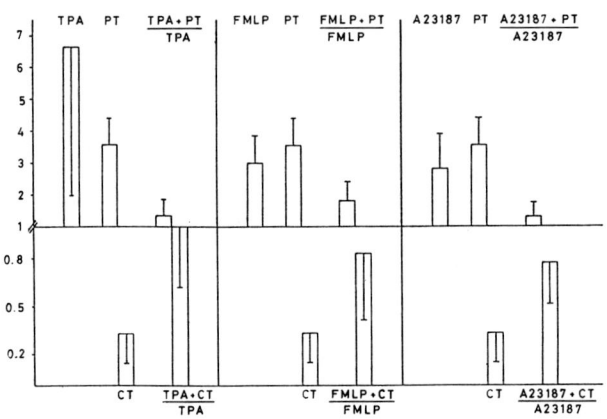

Fig.2: Effect of different stimulators on lucigenin-
       enhanced chemiluminescence after pretreatment with
       cholera or pertussis toxin for 90 min. at 37 °C .
       Concentrations were 1000 ng/ml for either toxin .

Cholera toxin inhibits SOR production already at low
(10 ng/ml) concentrations; at 1000 ng/ml inhibition
is 50% . Whereas FMLP stimulation is influenced in
a similar way, Ca-ionophore A 23187-induced super-
oxides are (slightly) inhibited by cholera toxin;
TPA-stimulated SOR production is not influenced by
the toxins (Fig. 2) . Therefore (Fi. 4), toxin effects
appear to be mediated by phospholipase C/inositol-
phosphates rather than by proteinkinase C .

Fig.3: ADP-ribosylation pattern
of HL-60 neutrophilic cells

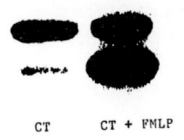

CT   CT + FMLP

-- 45 kDa

-- 40 kDa

Plasma membranes of HL-60-cells
(100 ug of protein) were incubated
with cholera toxin (9 ug, CT) or cho-
-lera toxin + FMLP (10 uM) . 32-P-NAD
(5 uCi) was added, ADP ribosylation
performed for 20 min. at 37 °C .
ADP-ribosylated proteins were separated on 10% polyacrylamide slab gels,
dried and autoradiograms prepared .

In HL-60 cells, cholera toxin labeling of a 40 kDa
protein is greatly enhanced after FMLP-stimulation (Fig.
3 and (6)) . This has been confirmed by others in HL-60
cells (5) and NG 108-15 neuroblastoma cells (7) .
From this we conclude that superoxide production
in human blood (by PMNs and MCs) which is differen-
tially influenced by cholera and pertussis toxin
is very probably regulated by (i) either two different G-
proteins or (ii) one G-protein with different
function/sensitivity to bacterial toxins .
Further investigation of their molecular identity
could provide insights in functional aberrations of PMNs
in diseases like psoriasis or palmoplantar pustulosis .

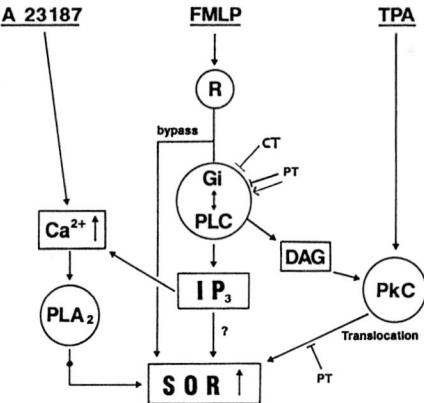

Fig.4: Schematic presentation of signal transduction
pathways contributing to superoxide generation
in human neutrophil granulocytes

Fig. legend : A 23187 = Calcium ionophore; FMLP = formyl-
methionyl-leucyl-phenylalanine; TPA = Tetradeca-
noylphorbolacetate; R = FMLP-receptor; Gi = "inhibiting"
signal-transducing protein; PlC = phospholipase C;
PlA2 = Phospholipase A2; DAG = diacylglycerol; PkC =
Proteinkinase C; IP3 = inositol-1,3,5-triphosphate;
SOR = superoxide radicals; PT = pertussis toxin;
CT = Cholera toxin; ⟶⊣ = inhibitory; ⟶➤ = stimulatory

REFERENCES

1. Rossi, F.: The O2-forming NADPH oxidase of the
phagocytes : nature, mechanisms of activation and
function . Bioch. Biophys.Acta 853, 65-69 (1986)
2. Gierschik, P., Sidiropoulos, D., Spiegel, A.,
Jakobs,K.H.: Purification and immunochemical
characterization of the maior pertussis-toxin-sensi-
tive guanine nucleotide binding protein of bovine
neutrophil membranes . Eur.Journ.Bioch. 165, 185-194
(1987)
3. Ohta, H., Okajiama, F., Ui, M.: Inhibition by
islet-activation protein of a chemotactic peptide
induced early breakdown of inositol phospholipids
and Ca mobilization in guinea pig neutrophils .
Journ.Biol.Chem. 260, 15771-15778 (1985)
4. Christiansen, N.O.: Pertussis toxin inhibits the
FMLP-induced membrane association of protein kinase C in
human neutrophils . Journ. Leukoc. Biol. 47,
60-63 (1990)
5. Iiri, K., Tohkin, M., Morishima, N., Ohoka, Y., Ui,
M.: Chemotactic peptide receptor-supported
ADP-ribosylation of a pertussis toxin substrate GTP-
binding protein by cholera toxin in neutrophil-
type HL 60 cells . Journ.Biol. Chem. 264,
21394-21400 (1989)
6. Hohlmaier, K.: Regulation of adenylate cyclase
in neutrophilic HL-60 leucemia cells . M.D. Thesis,
p. 33 (1988)
7. Klinz, F.-J., Costa, T.: Cholera toxin ADP-ribosy-
lates the receptor-coupled form of pertussis toxin-
sensitive G-proteins . Bioch.Biophys.Res.Comm. 165,
554-560 (1989) .
8. Lämmli, U.K.: Cleavage of structural proteins
during the assembly of bacteriophage T4 . Nature 277,
680-685 (1970)

Bioluminometric Determination of Aldehydic Lipid Peroxidation Products During the
Oxidation of Low Density Lipoproteins (LDL)

E. Wieland, A. David, H. Kather*, and V.W. Armstrong

*Department of Clinical Chemistry, University Hospital, Göttingen, D-3400 FRG, and Institute for
Myocardial Infarction Research, University Hospital, Heidelberg, D-6900, FRG*.

INTRODUCTION

Plasma low density lipoproteins (LDL) are a well established risk factor for cardiovascular
diseases. However, the exact mechanism whereby LDL exert their atherogenic influence are still
being elucidated. One of the early events in atherosclerosis is the accumulation of cholesterol
laden foam cells in the subendothelial space. These cells originate mainly from monocyte-
macrophages. Recently oxidative modification of LDL has been implicated as a factor in the
pathogenesis of atherosclerosis (1). Oxidation of low density lipoprotein lipids initiates a series
of events that finally lead to an enhanced uptake of these particles in macrophages. A variety of
aldehydic lipid peroxidation products are formed during the in vitro oxidation of LDL of which
hexanal is the most abundant aldehyde associated with LDL lipids (2). Aldehydes generated by
peroxidation of LDL lipids can form Schiff-base adducts with lysine residues of the apoprotein
B100 thus causing its recognition by the scavenger receptor of macrophages (2).
Convenient and reliable methods to estimate the degree of LDL lipid peroxidation were as yet
unavailable, although the thiobarbituric acid assay (TBA) is commonly used to estimate lipid
peroxidation in the form of malondialdehyde (MDA) eqivalents, most of the aldehydes
determined by this method are generated during the acid heating step of the assay itself. In
addition this method has several well known drawbacks concerning specificity and sensitivity (3).
We describe here a sensitive and specific bioluminescence assay for the direct determination of
saturated straight chain aldehydes formed from polyunsaturated fatty acids during peroxidation.
The method uses the bacterial NAD(P)H-linked luciferase system which requires a long chain
aliphatic aldehyde in order to produce light in the reaction with $FMNH_2$ and luciferase (4). We
have applied our method to follow LDL oxidation in vitro as well as to estimate aldehydes
present in human plasma.

MATERIALS AND METHODS

Bioluminescence Assay. Luciferase (1.0 mU/ml), NAD(P)H:FMN oxidoreductase (210
mU/ml) and NADH (2,3 mmol/l) were dissolved in a potassium phosphate buffer (0.03 mol/l)
containing 0.4 mmol/l DTT. Aldehydes were dissolved in H_2O containing Triton-X-100 (10%).
4-hydroxynonenal was a gift from Professor Dr. H. Esterbauer, University of Graz, Austria. The
reaction was started by adding a 100 ul sample to 200 ul of the assay cocktail. Light emission
was recorded for 30 sec. at 37°C using a Berthold LB 9505 (Wildbad, FRG) luminescence
analyzer.
LDL Oxidation. LDL (d = 1.019-1.063) were isolated by ultracentrifugation from plasma of
healthy blood donors. 200 ug/ml LDL protein were incubated at 37°C

for different times with 5 umol/l $CuSO_4$ in Ham's F-10 medium. LDL samples were directly measured without extraction.

Plasma samples. EDTA plasma was collected from 10 healthy blood donors aged 22-28 years. After separation of plasma and cells samples were kept on ice and 40 umol/l butylated hydroxy toluene (BHT) was added in order to prevent oxidation during extraction procedures. Extraction was performed using the Bligh and Dyer procedure (5). Samples were dried under N_2 and redissolved in H_2O/Triton-X-100 (10%). Aldehydes and cholesterol were measured in redissolved samples. Aldehydes were quantitated using hexanal as an external standard. The aldehyde concentration is expressed per mg cholesterol per redissolved sample.

Other Assays. The Thiobarbituric acid (TBA) assay was used according to Ohkawa (6) with malondialdehyde as an external standard. Cholesterol was determined using the CHOD-PAP method.

RESULTS AND DISCUSSION

Aldehydes belong to the few chemically well defined products of lipid peroxidation. They are long lived and can provide evidence that lipid peroxidation has occured. Methods to measure aldehydes using dinitrophenylhydrazone (DNPH) or cyclohexanedione (CHD) are time consuming and not applicable to routine analyses (7). The more convenient TBA method is unreliable and not specific (3). In contrast bioluminescence methods are reliable specific and sensitive. The bacterial NAD(P)H-linked luciferase system has been used for the analysis of various compounds and its applicability to biological samples has been verified (8). Its suitability for the determination of aldehydes has been long known (4). However, the system has never been applied to aldehydes derived from the peroxidation of polyunsaturated fatty acids in biological samples. Since we required a reliable and sensitve assay to estimate LDL lipid peroxidation we have developed the method presented here. The reaction scheme is given in Fig.1. Aldehydes generated during the peroxidation of polyunsaturated fatty acids are introduced into the NAD(P)H-linked luciferase reaction.

Fig. 1 Reaction Scheme

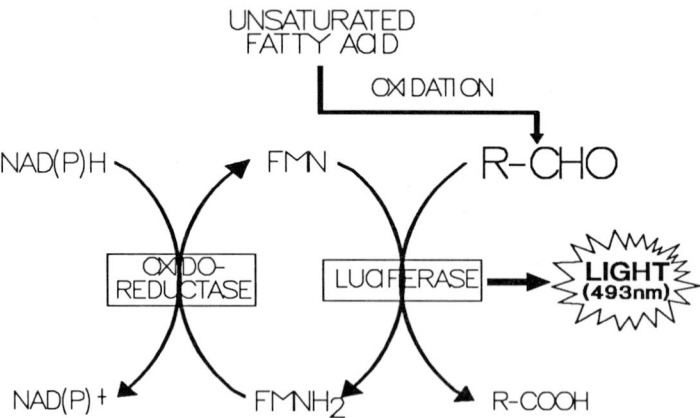

Fig. 2 Concentration curve for hexanal. Based on a signal to noise ratio of 1, the detection limit is 1 umol/l. Light production is linearly related to hexanal concentrations up to 1000 umol/l.

Fig. 2 Concentration Curve for Hexanal

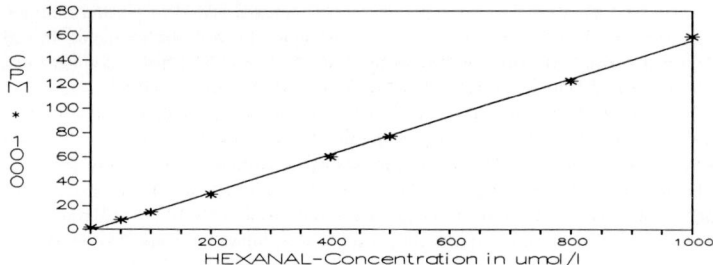

Fig. 3 shows the kinetics of aldehyde formation during $CuSO_4$ mediated oxidation of LDL in vitro. Simultaneous measurement of TBARS compared well with the bioluminometric aldehyde method.

Fig. 3 Time Course of LDL Oxidation by $CuSO_4$

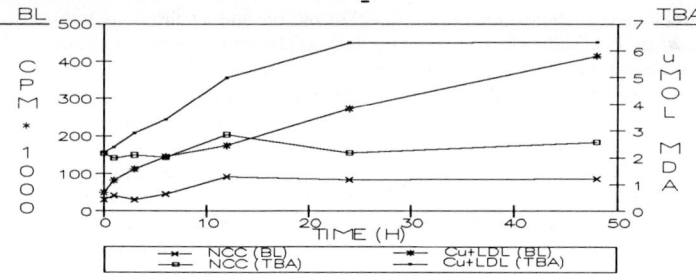

As previously published the light production was maximal with middle to long chain aliphatic aldehydes (4). Neither alkenals, branched-chain aldehydes or organic aldehydes reacted to an appreciable extent in the assay. Using 10 umol/l hexanal within-day imprecision of the method averaged 2.8% (n=20) and 5% from day to day (n=5). Application of the method to plasma samples is shown in the following Table (Table 1).

Table 1.Aldehydes Contained in Plasma Samples

Sample No.	Cholesterol (mg/dl plasma)	Aldehydes (umol/l)	umol Aldehydes/ mg Cholesterol
1	154	271	0.33
2	153	179	0.23
3	135	274	0.30
4	119	249	0.37
5	222	272	0.16
6	219	258	0.20
7	156	175	0.18
8	165	251	0.21
9	190	169	0.20
10	150	261	0.23

Plasma was obtained from healthy blood donors. Cholesterol was determined in plasma and extracted samples.

Aldehyde concentrations observed in plasma ranged between 0.16 and 0.37 umol aldehyde/mg cholesterol expressed as hexanal equivalents. 10 umol/l hexanal added to plasma samples were quantitatively recovered (98-102%, n=4). Plasma samples need to be extracted in order to avoid interferences from antioxidants. The assay is standardized with hexanal since this is the major aldehyde formed from C 18:2 during LDL oxidation (2). Since we did not determine the aldehyde composition of extracted plasma samples we can not exclude that aldehydes others than hexanal contributed to the light signal when these samples were measured.

In summary, we describe a highly specific and sensitive bioluminescent assay for aldehydes based on the NAD(P)H-linked luciferase system in which aliphatic aldehydes specifically react with the luciferase enzyme. The assay has been sucessfully applied to follow oxidation of LDL in vitro and to estimate the amount of aldehydes contained in human plasma. The method should be suited to investigating a possible relation between plasma lipid peroxidation and atherosclerosis. Whether the method is also applicable to LDL directly obtained from human plasma by precipitation is currently under investigation.

REFERENCES

1.   D. Steinberg, S. Parthasarathy, T. Carew, J.C. Khoo, and J.L Witztum. Beyond Cholesterol. Modifications of Low-Density Lipoprotein that increase its atherogenicity. New Engl. J. Med. 320, 915-924(1989).

2.   G. Jürgens, H.H Hoff, G.M. Chisholm, and H. Esterbauer, Modificaton of human serum low density liporotein by oxidation-Characterization and pathophysiologcal implications. Chem. Phys. Lip.45, 315-336 (1987).

3.   H. Kosugi, T. Kojima, and K. Kikugawa, Thiobarbituric acid-reactive substances from peroxidized lipids. Lipids 24, 873-881 (1989).

4.   P.E. Stanley, Determination of subpicomole levels of NADH and FMN using bacterial luciferase and the liquid scintillation spectrometer. Anal Biochem. 39, 441-453 (1971).

5.   E.G. Bligh and W.J. Dyer. A rapid method of total lipid extraction and purification. Can. J. Biochem. Physiol. 8, 911-917 (1959).

6.   H. Ohkawa, N. Ohishi, and K. Yagi, Assay for lipid peroxides in animal tissues by thiobarbituric acid reaction. Anal Biochem. 95, 351-358 (1979).

7.   H. Esterbauer and H. Zollner, Methods for determination of aldehydic peroxidation products. Free Rad. Biol. Med. 7, 197-203 (1989).

8.   E. Wieland, H. Kather, P.D. Niedmann, F. Diedrich,and D. Seidel, Luminescence in the study of lipid metabolism. J. Biolum. Chemilum.4, 436-445 (1989).

ACKNOWLEDGEMENTS

We thank Professor Dr. H. Esterbauer for the generous gift of 4-hydroxynonenal and the Department of Transfusion Medicine, Univ. Göttingen for providing us with plasma samples. We are grateful to Dr. P. Schuff-Werner for providing the laboratory equipment for luminescence measurements.

QUALITATIVE AND QUANTITATIVE MEASUREMENT OF URINARY PORPHYRINS BY APPLICATION OF PEROXYOXALATE-CHEMILUMINESCENSE

H. Brandl, S. Albrecht[1] and R. Köstler[2]

College Kaltenkirchen, [1]Institute of Clinical Chemistry of the Medical Academy Dresden and [2]Department of Dermatology of the Countryhospital Dresden-Friedrichstadt, Germany

INTRODUCTION

The number of diseases by chronic hepatic porphyrias is about to rise. Therefore, simple, promptly realizable but highly sensitive porphyria-screening tests become more and more significant.

Special porphyrins are known as excellent sensitizers in peroxyoxalate-chemiluminescence-systems. On the basis of substituted aryloxalates we developed a qualitative chemiluminescence-test for the determination of porphyrins in urine. [1,2]

The system oxalic acid / carbodiimide / peroxide now also permits a simple quantitative analysis of porphyrins in protic solvents (e.g. urine) down to the concentration of about 250 µg/L by using a suitable photodetector with a maximum in a narrow of 630 nm.

MATERIAL AND METHODS

Qualitative determination of urinary porphyrins [1]: 500 µL of the urine sample are mixed with 5 mL of the chemiluminescence reagent (500mg bis(2,4-dinitrophenyl)oxalate and 5 mL peracetic acid (40 p.c.) in 50 ml absolute ethanol).

Porphyrinurias may be reliable discerned by their orange or red chemoluminescence (λ_{max} 610-650 nm) beginning with a total concentration of porphyrins of 250 µg/L urine.

Quantitative determination of urinary porphyrins: All chemiluminometric measurements were made on a Clinilumat LB 9502 (Fa. Berthold, Germany). Dosed by injector, 300 µL of a solution of bis(cyclohexyl)-carbodiimid in abs. ethanol (50 g/L) were added to 100 µL of urine sample mixed with 100 µL of 0,2 mol/L $K_2(COO)_2$-solution and adjusted to pH=1 (HCl). Measurement was started immediately after injection and continued for 10 s.

Reaction principle:

$$(O_2N)_2(C_6H_3)\text{-}O\text{-}\overset{\overset{O}{\|}}{C}\text{-}\overset{\overset{O}{\|}}{C}\text{-}O\text{-}(C_6H_3)(NO_2)_2 + H_2O_2 + P \text{ --------->} \tag{1}$$
$$2(O_2N)_2(C_6H_3)OH + 2CO_2 + P^*$$

$$HO\text{-}\overset{\overset{O}{\|}}{C}\text{-}\overset{\overset{O}{\|}}{C}\text{-}OH + H_2O_2 + P \xrightarrow{\text{carbodiimide/H}^+} 2CO_2 + 2H_2O + P^* \tag{2}$$

$$P^* \text{ --------------------> } P + h\sqrt{} \tag{3}$$

(P = Uro-, copro- or protoporphyrine; * = electronic exited state)

RESULTS AND DISCUSSION

The peroxyoxalate-chemiluminescence is an efficient energy transfer system for indirect luminescence. This quantitative porphyrin-sensitized peroxyoxalate-chemiluminescence test porphyrrinurias may be relisable discerned by their orange or red chemiluminescence (good visible in a dark room, $\lambda_{max\ CL}$ 610-650 nm) over 10 to 15 s beginning with a total concentration of 250 µg/L urine. Other pathological or physiological components of the urine are showing- according to hitherto existing examinations- in this characteristics spectral region not any chemiluminescence. Comparing examinations with the previous porphyria searching test of cho-

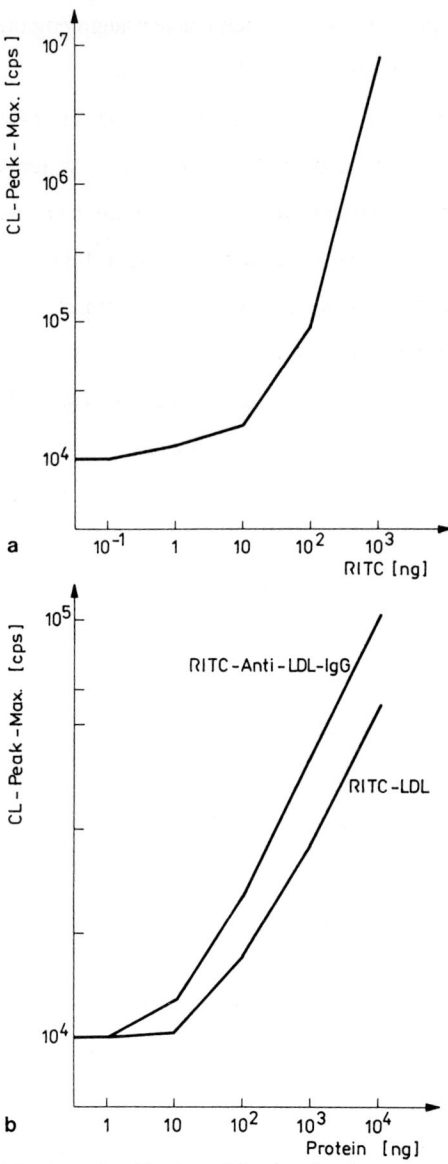

Fig. 1a, b. Sensitization of the chemiluminescent system oxalic acid/DCC/H_2O_2 by **a** free and **b** protein-bound rhodamineisothiocyanate contained in 100 μl of sample (1 mol/l oxalate solution, pH 1)

ice (the porphyrin fluorescence talc-test) resulted in an at least comparable sensitivity of the PCL-test, so that it will be possible to record beginning disorders in porphyrin metabolism resp. latent stages up to a high percentage.

Is a solution of alkalioxalate (0,1 -1 mol/L, adjusted with HCl to pH=1) used in the system carbodiimide / H_2O_2 / ethanol, a sensitization of the system by 3 orders of magnitude depending on the concentration of fluorescer can be achieved regarding the CL measuring signal. However, many of the common fluorescent dyes cannot be employed because of insufficient fluorescence quantum yield at pH=1. Apart from polycondensed carbohydrates (e.g. diphenylanthracene), brillant sulfoflavine, rhodamine and porphyrins (except complexes of metals with several stable valence states, e.g. iron and cobalt) are, among others, known as excellent sensitizers.

Covalent bounding to protein of the fluorescer will decrease CL quantum yield in comparison to an adequate quantum of free fluorescer. Yet, few nanogramms per mL can still be detected. (Fig.1)

The system oxalate /DCC / H_2O_2 now also permits a quantitative analysis of porphyrins in urine down to the concentration of about 250 µg/L, with some problems of standardization of the procedure remaining to be solved. The use of a suitable photodetector with a maximum sensitivity within a narrow range of the fluorescence (chemiluminescence) maximum of porphyrins (about 630 nm) is a basic requirement for sufficient high sensitivity or further enhancement of sensitivity.

REFERENCES

[1] H. Brandl and S. Albrecht, Nachweis pathologischer Beimengungen von Porphyrinderivaten im Harn.
PdN-Ch. 30, 17-24 (1990)

[2] S. Albrecht, H. Brandl and E. Köstler, Porphyrinsensibilisierte Peroxyoxalat-Chemilumineszenz - ein neuer sensitiver Suchtest zur Porphyriediagnostik.
Z.Klin.Med.44, 2071-2073 (1989)

PEROXYOXALATE-CHEMILUMINESCENCE OF FREE OXALIC ACID - A NEW METHOD FOR THE QUANTITATIVE DETERMINATION OF OXALATE IN URINE

S. Albrecht, W.-D. Böhm[1], R. Beckert[2], H. Kroschwitz[3], V. Neumeister and W. Jaroß

Institute of Clinical Chemistry and [1]Department of Urology of the OPC of the Medical Academy Dresden, [2]Department of Chemistry of the Friedrich-Schiller-University Jena and [3]Department of Chemistry of the Technical University Dresden, Germany

INTRODUCTION

Calciumoxalate is a most important phase in the case of stone disease in Europe. One of the risk factors is a hyperoxaluria.

We developed a new method for the determination of oxalate in urine. By using the chemiluminescence decay of monoperoxyoxalate acid very low concentrations of oxalate in protic solvents (up to 10 nmol/l) can be determined [1]. Optimized reaction conditions and preanalytical factors permit direct measurements of urinary oxalate without calcium oxalate precipitation [2,3]. The accuracy and reproducibility of this new method will be discussed in comparison to enzymatic and RP-HPLC-methods.

MATERIAL AND METHODS

Reaction principle: In 1965, Rauhut discribed bright, strongly visible CL of an oxalic acid / carbodiimide / H_2O_2 system in aprotic solvents. Our own experiments showed intensive short-time CL at pH=1 with a maximum within 0.4 s after the start of reaction also in protic enviroment (ethanol/H_2O). Monoperoxyoxalate is formed as an intermediate which, in the presence of a fluorescer, decays accompanied by photoemission via several interme-diate compounds (e.g. dioxethanedione) into CO_2 and H_2O (equation 1 -3):

$$HO-\overset{O}{\underset{}{C}}-\overset{O}{\underset{}{C}}-OH + H_2O_2 \xrightarrow[\text{diimide}]{\text{carbo-}} HOO-\overset{O}{\underset{}{C}}-\overset{O}{\underset{}{C}}-OH + H_2O \qquad (1)$$

$$HOO-\overset{O}{\underset{}{C}}-\overset{O}{\underset{}{C}}-OH + F \xrightarrow{\hspace{2cm}} H_2O + 2CO_2 + F^* \qquad (2)$$

$$F^* \xrightarrow{\hspace{2cm}} F + hv \qquad (3)$$

(F^* = fluorescer in excited electronic singlet or triplet state)

Preanalytical conditions: 9 mL fresh collected urine (24h or 8h) are added to 1 mL HCl (15 p.c.). The sample is stored at 5-7°C up to four weeks and can be directly used for the chemiluminometric and chromatographic (reference method) determination of oxalate.[3]

Chemiluminometric measurement: All chemiluminometric measurements were made on a Clinilumat LB 9502 (Fa. Berthold, Germany). Dosed by injector 300 uL H_2O_2 (0.3 p.c) and 300 uL of a solution of bis(cyclohexyl)-carbodiimide (25 g/L) in abs. ethanol containing 50 mg/L diphenylantracene were added to 100 uL of aqueous sample adjusted to pH = 1. Measurement was started immediately after injection and continued for 4 s. The integral signal was used for the oxalate determination.

Commercial urine lyophilisates with definite oxalate levels (Fa. Sigma) were used as standards. A linear calibration curve was obtained in the range of 10 to 200 mg/L oxalate concentration.

RESULTS AND DISCUSSION
The reproducibility of the presented chemiluminometric oxalate determination was in the range of 3 p.c. (intraassay precision) and 6 p.c. (interassay precision). The accurancy of the method is shown in Fig.1.
A modified RP-HPLC-determination was used as reference method, because of watery solutions of alkalioxalates are suitable as standards (Fig.2).
The correlation between enzymatical oxalate determination and the chemiluminometric method was unsatisfactory: r=0.68 using Sigma-oxalate-oxidase-method and r=0.52 by using Boehringer-Mannheim-oxalate-decarboxylase-method. Correlation coefficients in the same range were found between RP-HPLC and the enzymatical methods.
We've investigated 30 patients suffering from oxalate stone diathesis with a high recurrence rate in comparison to a control group of 30 health persons. The results are shown in Fig.3. 14 of the 30 oxalate stone formers had a hyperoxaluria (ox.-excr. > 0.70 mmol/d). The simplicity and comparable sensitivity (10 nmol/l) makes the presen-

2

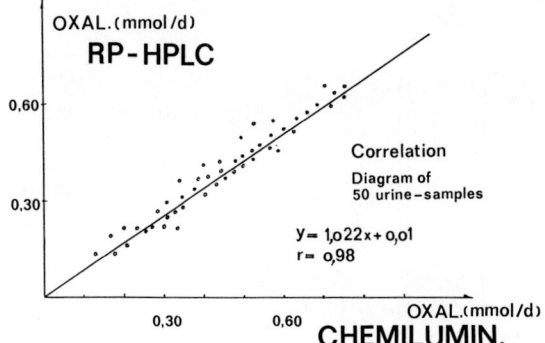

Fig.1. Correlation diagram of 50 urine samples

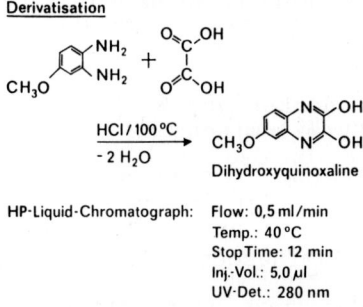

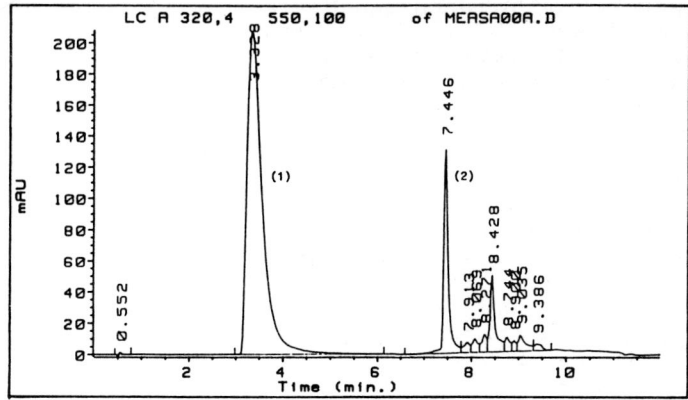

HPLC-chromatogram (Hewlett Packart, reversed phase) of a $K_2C_2O_4$-standard (78 mg/l)
(1) = methoxyphenylenediamine
(2) = methoxydihydroxyquinoxaline

Fig.2. RP-HPLC-determination of oxalate

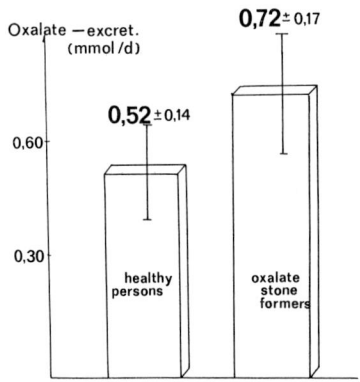

Fig.3. Oxalate excretion
of healthy persons
and oxalate stone
formers (n = 30)

ted chemiluminometric method an useful alternative to the expensive more-step enzymatical methods.

REFERENCES
[1] S. Albrecht, R. Becker and W.-D. Böhm, Zur Anwendung der Peroxyoxalate-Chemilumineszenz in der biochemischen Analytik: Bestimmung von Oxalat.
J.Clin.Chem.Clin.Biochem. 27, 451-454 (1989)
[2] S. Albrecht, H. Brandl, W.-D. Böhm, R. Beckert, E. Köstler, M. Menschikowski and W. Jaross, Application of peroxyoxalate chemiluminescence in analytical biochemistry.
Fresenius J.Anal.Chem. 337, 93-94 (1990)
[3] S. Albrecht, W.-D. Böhm and H. Kroschwitz, Peroxyoxalat-Chemilumineszenz – ein neues effizientes Verfahren zur Bestimmung von Oxalat im Harn
Z.urologie poster 2, 144-146 (1990)

CHEMILUMINESCENT ENZYME METHOD FOR DETERMINATION

OF SERUM CHOLINE-CONTAINING PHOSPHOLIPIDS

E. Bissé, J. Gissler, E. Wieland, H. Wieland

Dep. of Clinical Chemistry, Central Lab. University Hospital, Hugstetterstr. 55, D-7800-Freiburg, FRG

INTRODUCTION

Many non-enzymatic procedures for the determination of phospholipids (PL) have been reported. They require, however, solvent extraction, digestion of phospholipids and more or less specific detection of inorganic phosphorus. In this work, a sensitive simple and rapid procedure for the quantitation of serum choline-containing phospholipids is presented. The method is based on three coupled enzyme reactions[1] (1) and luminol chemiluminescence[2] :

1) PL $\xrightarrow{\text{phospholipase D (PD)}}$ phospholipase d (PD) phosphatic acid + choline

2) choline $\xrightarrow{\text{choline oxidase (COD)}}$ betaine + H_2O_2

3) H_2O_2 + Luminol (L) $\xrightarrow{\text{peroxidase (POD)}}$

   aminophtalic acid + LIGHT

MATERIALS AND METHODS

Reagents: all chemicals were obtained from Sigma (Chemical Co. St. Louis). All reagents were prepared in 50 mM Tris-HCl buffer, pH 8.5 . The enzyme stock solutions consist of:

a) PD   (from streptomyces species) 22,3 U/ml
b) COD (from alcaligenes species) 11,0 U/ml
c) POD (from horseradish)   9,5 U/ml .

These solutions were stable at 4°C for at least 10 days. A 2 mM luminol was prepared in Tris-HCl buffer and allowed to stabilize at least 10 days.
The chemiluminescent reagent (CL) was composed of:
luminol : COD : POD (1 : 1 : 1) mixed just prior to use.
Sample preparation: the sample was diluted 1:600 with Tris-HCl buffer and homogenized by 10 min. ultrasonication. The blood serum samples had been previously analyzed for phospholipids using a phospholipid kit (Boehringer Mannheim, FRG) adapted to a Wako Analyzer (Wako Chemicals GmbH, FRG).

PROCEDURE

300 μl diluted sample and 100 μl PD were incubated at 37°C. After 20 min. the tube was placed in the detector cell compartment (Biolumat LB 9500 T, Bertold, FRG), and 100 μl of CL reagent was immediately injected into the tube. The photon emission was recorded as function of time.

RESULTS AND DISCUSSION

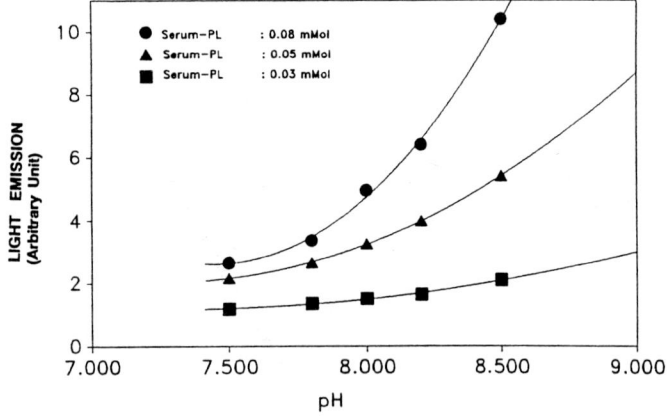

Fig. 1   Change in light emission as a function of pH at three different PL concentrations.

The pH dependence seems to be less pronounced at low PL concentrations and the reaction proceeds best at pH 8.5 .

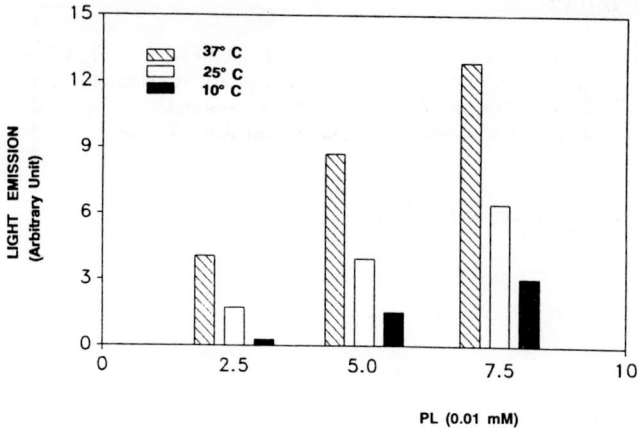

Fig. 2 Change in the light emission as a function of temperature at three different PL concentrations. As expected, the optimal temperature was 37°C.

As expected, the optimal temperature was 37°C. The light emission was linear with the increase of temperature.

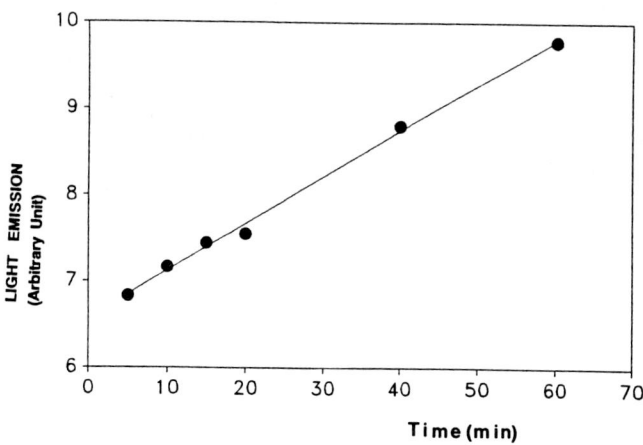

Fig. 3 Dependence of light emission upon the time of incubation of PL (serum PL : 0.15 mM) with PD (0.58 U) at 37°C.

The maximum of photons emitted was linear with time.

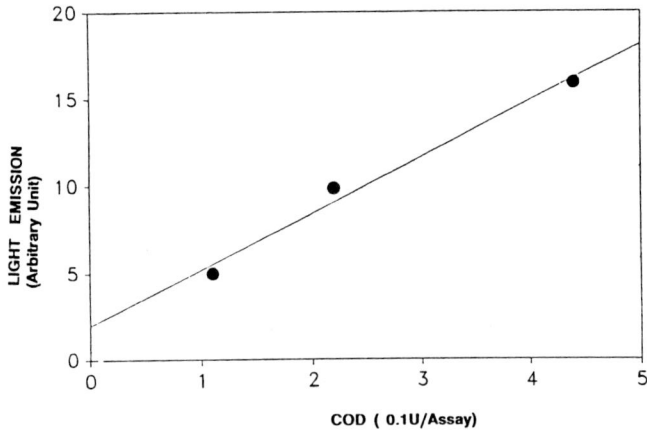

Fig. 4   Effect of COD concentration on the photon emission at 37°C, pH 8.5.

The comparative determination of PL in patients' serum by use of CL method and by the routine method (Boehringer Mannheim, FRG) in Wako-Analyzer showed reasonable correlation (r = 0.941). The regression line was: y = 0.99 X + 1.86, n = 35 . The CL values were read from a standard curve which was linear up to 5 mmol/L PL. The within-run and between-run CV varied from 2.2 % to 4.0 %.

KEY WORDS

Phospholipids, Chemiluminescence, Luminol

REFERENCES

1) M. Takayama et al. Clin. Chem. Act. 79 93-98 (1977).
2) Methods in Enzymology LVII
   M. DeLuca Ed. Academic Press N.Y. 1978

The authors wish to acknowledge the contribution of I. Mattmüller for her assistance in the preparation of this paper.

CHEMILUMINESCENCE DETECTION OF RAT LIVER OXIDASES IN PEROXISOME PROLIFERATION.

E.H.J.M. Jansen, R.H. van den Berg, W.A. van de Ham and F.X.R. van Leeuwen.

Laboratory for Toxicology, National Institute of Public Health and Environmental Protection, P.O. Box 1, 3720 BA Bilthoven, The Netherlands.

INTRODUCTION

The administation of hypolipidaemic agents and plasticizers results in a proliferation of liver peroxisomes in rodents. Besides the increase in liver weight, the induction of several enzymes involved in the β-oxidation of longchain fatty acids are good biochemical markers for peroxisome proliferation [1]. Specific for peroxisomal β-oxidation is the enzyme acyl Co-enzyme A oxidase. This enzyme can be determined by the production of hydrogen peroxide or by determination of NADPH which is produced by the enzyme enoyl CoA hydratase two steps further in the β-oxidation sequence. Other oxidizing enzymes that are known to occur in peroxisomes are glycolate oxidase, urate oxidase, amino acid oxidase, allantoinase [2].

In the present report attempts have been described to measure liver enzymes with chemiluminescence by oxidation of luminol by hydrogen peroxide using an iron-chelator as catalyst and by the determination of NADPH. A comparison will be made with spectrophotometric determinations on samples from a toxicological experiment with di(ethylhexyl)phthalate in rats.

MATERIALS AND METHODS

Chemiluminescence and bioluminescence measurements were performed with the Luminoskan microtitre plate reader manufacturered by Labsystems, Helsinki, Finland or with the Biocounter M2000 (Lumac, Landsgraaf, The Netherlands). Chemiluminescence measurements of hydrogen peroxide were performed as described for xanthin oxidase [3]. Chemiluminescence measurements of cytochrome P-450 were performed as described earlier [4]. Bioluminescence measurements of NADH were performed with the NAD(P)H bioluminescence kit from Boehringer Mannheim. Spectrophotometric measurements of NADH were performed on the Cobas Bio autoanalyzer (Hoffmann La Roche). The preperation of rat liver homogenates will be published elsewhere [5].

RESULTS AND DISCUSSION

The key-enzyme in the β-oxidation in liver peroxisomes, acyl Co-enzyme A oxidase, produces hydrogen peroxide similar to a number of other oxidases, like xanthin oxidase, glucose oxidase, etc. The production of hydrogen peroxide can be detected very sentitive with a signal reagent containing the substrate for the enzyme, an iron-chelate complex as catalyst and luminol as the luminogenic reagent [6]. The attempts to measure the enzymatic activity of palmitoyl CoA oxidase failed, however. Since the enzyme palmitoyl CoA oxidase requires FAD as cofactor, which strongly quenches the emitted blue light of luminol at 430 nm, this compound was omitted from the reaction mixture because its concentration in the

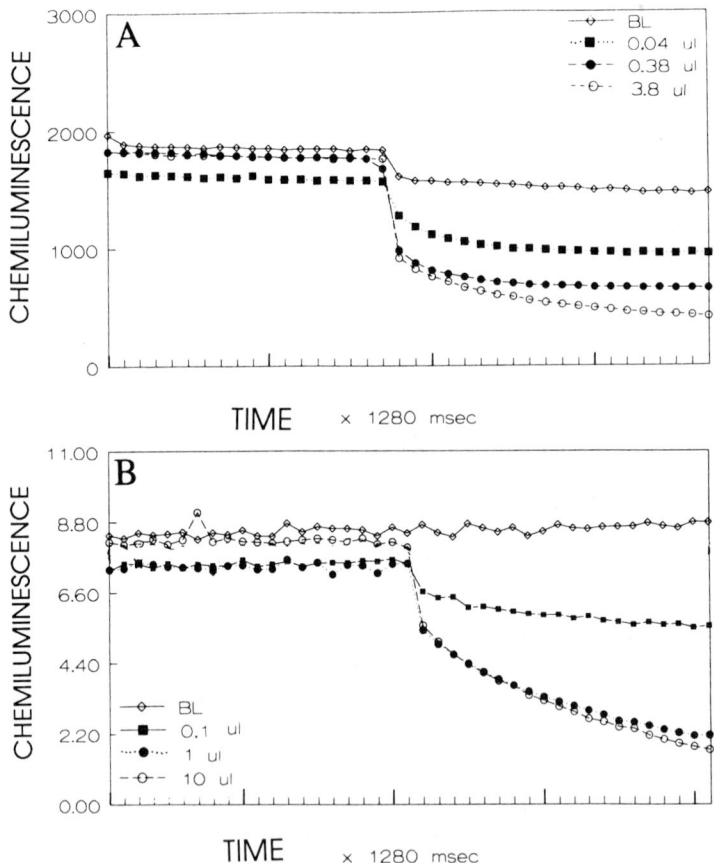

FIGURE 1: Effects of different amounts of liver homogenates on the chemilumines-cence signal of xanthin oxidase (A) and horseradish peroxidase (B).

liver was supposed to be sufficient. Xanthin oxidase also has two FAD molecules as prosthetic group, which do not need to be added in the signal reagent. Another possible explanation for the failure is the existence of catalase in the peroxisomes. Since catalase in the peroxisomes is rather unstable at room temperature, the liver homogenates were kept for several hours at room temperature, without any improvement, however. The addition of cyanide, necessary to block the mitochondrial enzyme, had also a rather disturbing effect on the chemiluminescence. It appeared, however, that also the addition of just liver homogenates quenched the chemiluminescence reaction of xanthin oxidase, which is probably a homolytic splitting of hydrogen peroxide in hydroxyl radicals by the iron-EDTA complex. The hydroxyl radicals react very efficiently with luminol to produce chemiluminescence. In Fig. 1A the effect of various amounts of liver homogenates on the long-term chemiluminescence of xanthin oxidase is shown. Also the enzyme enhanced signal of horseradish peroxidase, which is a more direct oxidation of luminol, is inhibited also by the liver homogenate (Fig. 1B).

The enzymatic activity of palmitoyl CoA oxidase was measured spectrophotometrically with an clinical chemical autoanalyser (Cobas Bio) by the detection of NADH which is produced two steps further in the enzyme sequence of the β-oxidation. With this method the enzymatic activity of palmitoyl CoA oxidase could be measured very sensitive and conveniently. The no-effect-level for di(2-ethylhexyl)phthalate in male rats was found to be even below the lowest dose level

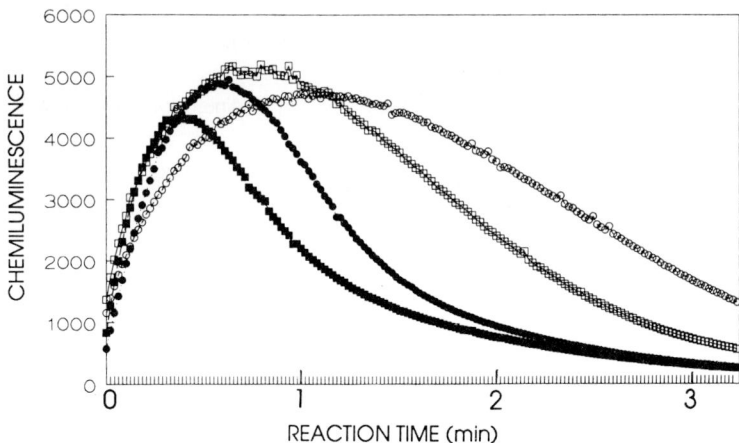

FIGURE 2: Chemiluminescence signals of NADH from liver homogenates of a controle rat (open symbols) and from a rat treated with di(2-ethylhexyl)phthalate (closed symbols). The circles represent the signal after the addition of palmitoyl Coenzyme A.

of 60 mg/kg diet. Since the measurement of NADH can be performed also with chemiluminescence by coupling to the luciferine/luciferase assay, this approach was tested using a commercially available kit (Boehringer). The results of these measurements on a sample of a control and a treated rat are shown in Fig. 2. Although at about 1 min a difference can be observed between both animals, the kinetics of the chemiluminescence signals have to be explained first.

In conclusion, it appears that, although chemiluminescence detection is inherently more sensitive than other detection methods, measurements of chemiluminescence signals in biological samples is often hindered by matrix components or other factors involved in the enzymatic reactions. Further studies on the liver oxidases are in progress.

REFERENCES

1. E.D. Barber, B.D. Astill, E.J. Moran, B.F. Schneider, T.J.B. Gray, B.G. Lake and J.G. Evans. Peroxisome induction studies on seven phthalate esters. Toxicol. Industr. Health, 3, 7-22 (1987).

2. N.E. Tolbert. Metabolic pathways in peroxisomes and glycosomes. Ann. Rev. Biochem. 50, 133-157 (1981).

3. E.H.J.M. Jansen, R.H. van den Berg and G. Zomer. Characteristics and detection principles of a new label producing a long-term chemiluminescent signal. J. Biolumin. Chemilumin. 4, 129-135 (1989).

4. E.H.J.M. Jansen, E.J.M. Reinerink and R.H. van den Berg. Chemiluminescence detection of haem-containing microsomal proteins. Anal. Chim. Acta, 227, 49-55 (1989).

5. E.H.J.M. Jansen, W.A. van de Ham, M. van Apeldoorn and F.X.R. van Leeuwen. To be published.

6. E.H.J.M. Jansen, R.H. van den Berg and J.J. Bergman. Effect of iron chelates on luminol chemiluminescence in the presence of xanthine oxidase. Anal. Chim. Acta, 227, 57-63 (1989).

CHEMILUMINESCENCE REACTION RATES OF *CYPRIDINA* LUCIFERIN ANALOGUES WITH SUPEROXIDE

S. Mashiko, N. Suzuki, *[1] K. Suetsuna,[1] B. Yoda,[2] T. Nomoto,[3] Y. Toya[4] and H. Inaba[5]

Communications Research Laboratory, Ministry of Posts and Telecommunications, Koganei, Tokyo 184, JAPAN
[1]*Shimonoseki University of Fisheries, Yoshimi, Shimonoseki 759-65, JAPAN*
[2]*Koriyama Women's University, Koriyama, Fukushima 963, JAPAN*
[3]*Department of Chemistry, Faculty of Education, Mie University, Tsu, Mie 514, JAPAN*
[4]*Aichi University of Education, Kariya, Aichi 448, JAPAN*
[5]*Research Institute of Electrical Communication, Tohoku University, Aoba, Sendai 980 and Biophoton Project, Research Development Corporation of JAPAN, Miyagino, Sendai 980, JAPAN*

INTRODUCTION

Both singlet molecular oxygen ($^1O_2$) and superoxide (O_2^-) play important roles in various biological and chemical processes. For detecting the former active oxygen species ($^1O_2$), direct spectroscopic observation of near-infrared emission at 1.27 μm is one of the best ways, which is the most reliable physical method.[1,2]

However, direct observation of $^1O_2$ in biological systems is still now extremely difficult in spite of recent advances of detection techniques for the active oxygen species by using sensitive detectors constructed using semiconductors as a result of low quantum yields of its emission ($\leq 10^{-6}$ einstein/mol).[3] On the other hand, there is no direct spectroscopic way for detecting the latter oxygen species (O_2^-).

Cypridina luciferin analogues (CLA's), 2-methyl-6-phenyl- and 2-methyl-6-(p-methoxyphenyl)-3,7-dihydroimidazo[1,2-a]pyrazin-3-ones (CLA and MCLA) were shown to be versatile tools for specific detection of $^1O_2$ and O_2^-.[4-6] In the previous reports, we have reported rate constants for [MCLA][$^1O_2$] and [CLA][$^1O_2$] by measuring the quenching constants for $^1O_2$ ($^1\Delta_g$) by MCLA and CLA together with those of [luminol][$^1O_2$], [superoxide dismutase (SOD)][$^1O_2$] and [NaN$_3$][$^1O_2$].[7,8] In this report, reaction rate constants of superoxide (O_2^-) with *Cypridina* luciferin analogues (CLA's), 2-methyl-6-phenyl-, 2-methyl-6-(p-methoxyphenyl)-3,7-dihydroimidazo[1,2-a]pyrazin-3-ones (CLA and MCLA) and 2-methyl-6-[p-[2-[sodium 3-carboxylato-4-(6-hydroxy-3-xanthenon-9-yl)phenylthioureylene]ethyleneoxy]phenyl]-3,7-dihydroimidazo[1,2-a]pyrazin-3-one (FCLA), [CLA][O_2^-], [MCLA][O_2^-], and [FCLA][O_2^-] were determined by using superoxide dismutase (SOD) as a quencher as 8.0 x 10^7, 1.82 x 10^8 and 8.5 x 10^7 mol/l^{-1}s^{-1} in pH 7.1 buffer solutions at 25℃ for the first time.

MATERIALS AND METHODS

Reagents   Superoxide dismutase (SOD II, 3,520 U/mg) was purchased from Toyobo Co. Ltd.   CLA and FCLA were from Tokyo Chem. Ind. Co.   Hypo-xanthine, xanthine oxidase (XOD) and albumin were from Sigma Chem. Co.   MCLA was prepared according to the method described in the literature.[5]   Superoxide was generated *in situ* from a reaction of hypoxanthine and XOD.[9]

CLA: R = H
MCLA: R = OMe
FCLA: R = (structure)

Quenching experiments were performed as follows;   A solution of 10 µl of CLA (final concn. 4.4×10^{-8} mol/l), MCLA (1.1×10^{-8} mol/l) or FCLA (4.8×10^{-8} mol/l); $(2.2 - y)$ ml of 25 mmol/l phosphate buffer (pH 7.1); 0.5 ml of 1 mg/ml albumin solution in the buffer solution; 50 µl of XOD (200 µl in 1.8 ml of the albumin soln); and y ml of SOD (2.54×10^{-9} mol/l for CLA's and 1.03×10^{-10} mol/l for luminol) in the buffer solution in a 18 mmφ quartz cell was placed on a photo-multiplier tube (R331, Hamamatsu Photonics) at 25°C in a dark cell chamber.   Into the solution was added a 0.2 ml of 3 mmol/l hypo-xanthine[9] in the buffer solution through an injection needle.   CL was measured through the bottom of the quartz cells in the single photon counting technique as usual.

Reaction rates for O_2^- can be given as follows,

$$d[O_2^-]/dt = E - k_1[O_2^-]^2 - k_2[Q][O_2^-] - k_3[SOD][O_2^-]$$

where E, k_1, k_2 and k_3 are production rate of O_2^-, rate constants of disappearance and reactions with the quencher and with SOD of O_2^-.

[Q], [O_2^-] and [SOD] represent concentration of quenchers used (CLA, MCLA and FCLA), of superoxide and of SOD.   At steady state, $d[O_2^-]/dt$ should be zero,

$$0 = E - k_1[O_2^-]^2 - k_2[Q][O_2^-] - k_3[SOD][O_2^-]$$
$$E = k_1[O_2^-]^2 + k_2[Q][O_2^-] + k_3[SOD][O_2^-]$$

Quantum efficiency (Φ) of CL of Q (CLA, MCLA, and FCLA) is represented as

$$\Phi = k_2[Q][O_2^-]/E = k_2[Q]/(k_1[O_2^-] + k_2[Q] + k_3[SOD])$$

When [SOD] = 0, $\Phi = \Phi_0$, $\Phi_0 = k_2[Q]/(k_1[O_2^-] + k_2[Q])$

Therefore, to determine the rate constants (k_2) for quenching O_2^- with the quencher, Q, the Stern-Volmer description of dynamic quenching is given in the following equation[10a]

$$I_0/I = \Phi_0/\Phi = (k_1[O_2^-] + k_2[Q] + k_3[SOD])/(k_1[O_2^-] + k_2[Q])$$
$$= 1 + \{k_3/(k_1[O_2^-] + k_2[Q])\}[SOD] \qquad \text{(eq. 1)}$$

where I_0/I is ratio of the emission intensity of O_2^- in the absence and presence of the quencher.   The equation 1 indicates that a plot of I_0/I vs [SOD] gives a straight line with a slope equal to $k_3/(k_1[O_2^-] + k_2[Q])$.   From these values the rate constant (k_2) for quenching O_2^- with Q can be determined if the values of k_1,[10b] k_3[10c] and [Q] are known.   The term of $k_1[O_2^-]$ is negligible, because its value is much less than those of $k_2[Q]$ in these cases.

RESULTS AND DISCUSSION

From the chemiluminescence – time curves of CLA's initiated by O_2^- in the absence and presence of SOD in pH 7.1 buffer solutions, values of I_0/I at various SOD concentration were read and plotted in Fig. 1(a). The straight lines were drawn by a least-squares fit of the data points.   For reference, a similar graph for luminol at pH 7.1 and 10.1 was shown in Fig. 1(b).   The correlation coefficients for the data points and the straight lines are 0.997 (CLA), 0.997 (MCLA), 0.997 (FCLA), 0.977 (luminol at pH 7.1) and 0.985 (luminol at pH 10.1), respectively.

Reaction rates (k_2) for $[CLA's][O_2^-]$ obtained from the equation 1 and Fig. 1 were summarized in Table 1.   The results suggest that the CLA's react and give CL with O_2^- as fast as SOD does[10c].

According to our previous papers,[7, 8] the CLA's react with $^1O_2$ at the rates 6.30×10^8 (CLA), 2.94×10^9 (MCLA) and 8.00×10^8 (FCLA) $(mol/l)^{-1}s^{-1}$ (k_q values), respectively.   The reaction rates obtained for O_2^- seem to parallel those for $^1O_2$.   As similarly as in the previous paper, the electron donating substituent MeO- on the phenyl group of CLA seems to accelerate the reaction of MCLA to electrophilic superoxide anion radical.   In the case of FCLA, the electron withdrawing thioketone, carboxylate and quinoid groups might cancel the electrophilicity.

We are also conscious that antioxidative activity of anti-oxidants such as oligopeptides and tea leaf catechins can be measured quantitatively when SOD is substituted by them in the present quenching experiment.

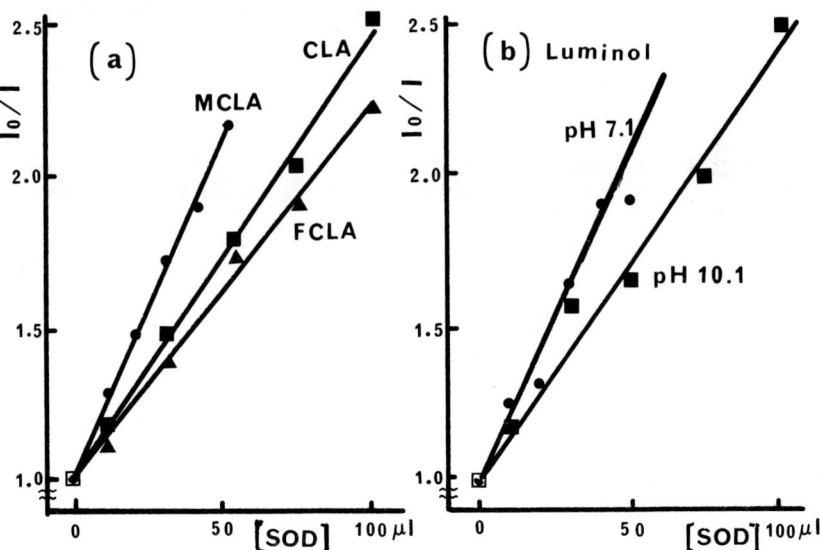

Fig.1. I_0/I – [SOD] Plots for (a) CLA, MCLA, and FCLA and (b) Luminol at 25 ℃.   [SOD] = 100 μl means 2.54×10^{-9} and 1.03×10^{-10} mol/1 for CLA's and luminol, resp.   [CLA], [MCLA], [FCLA], and [Luminol] are 4.4, 1.1, 4.8, and 300×10^{-8} mol/1, resp. $[O_2^-] \geq 1 \times 10^{-6}$ mol/1 s^{-1}.

Table 1.   Rate constants (k_2) and quantum yields (Φ) of CL of CLA's with O_2^- in 25 mmol/l buffer soln. (pH 7.1) at 25℃.

Quencher	$k_2 \times 10^{-7}$ ((mol/l)$^{-1}$s^{-1})	$\Phi \times 10^4$ einstein/mol[a]
CLA	8.0	3.8
MCLA	18.2	8.7
FCLA	8.5	22.1
Luminol	0.0037	--
	0.0092 (pH 10.1)	--

a) Relative to the Hastings standard and are corrected both to the sensitivity of the photomultiplier tube (Hamamatsu Photonics, R331: max at 420 nm) and the CL max of the standard (λ_{max} 420 nm) [J. W. Hastings and G. Weber, *Photochem. Photobiol.*, 4, 1049-1050 (1965); *J. Opt. Soc. Am.*, 53, 1410-1415 (1963)].

ACKNOWLEDGEMENT
We thank Professor Toshio Goto of Nagoya University for his helpful suggestions.

REFERENCES

1) A. U. Khan and M. Kasha, *Proc. Natl. Acad. Sci.*, *U. S. A.*, 76, 6047-6049 (1979).
2) A. A. Krasnovsky, Jr., *Photochem. Photobiol.*, 29, 29-36 (1979).
3) A. A. Krasnovsky, Jr., *Chem. Phys. Lett.*, 81, 443-445 (1981).
4) M. Nakano, K. Sugioka, Y. Ushijima and T. Goto, *Anal. Biochem.*,157, 363-369 (1986).
5) A. Nishida, H. Kimura, M. Nakano and T. Goto, *Clin. Chim. Acta*,179, 177-182 (1989).
6) K. Sugioka, H. Sawada and M. Nakano, in "Medical Biochemical and Chemical Aspects of Free Radicals," (ed. by O. Hayaishi, E. Niki, M. Kondo and Y. Yoshikawa), Elsevier Scientific Publishers, Amsterdam, 1988, pp. 899-903.
7) S. Mashiko, T. Ashino, I. Mizumoto, N. Suzuki, M. Nakano, T. Goto and H. Inaba, *Photomed. Photobiol.*, 11, 191-194 (1989).
8) N. Suzuki, I. Mizumoto, Y. Toya, T. Nomoto, S. Mashiko and H. Inaba, *Agric. Biol. Chem.*, 54 (11), 0000 (1990), in press.
9) Generation of O_2^- from the hypoxanthine/XOD system:
   G. G. Roussos, *Methods Enzymol.*, 12, 5-16 (1967); I. Fridovich, *J. Biol. Chem.*, 245, 4053-4057 (1970).
10)(a) D. M. Jameson, in "Fluorescein Hapten, An Immunological Probe," (ed. by E. Voss, Jr.), CRC Press, Boca Raton, FL., 1984, pp. 23-48.
   (b) The reaction rate constant, k_1 for $O_2^- + O_2^- \rightarrow O_2 + O_2^{2-}$ is reported as $k_1 = 10^2$ (mol/l)$^{-1}$s^{-1}.   A. A. Frimer, in "Chemistry of Peroxides" (ed by S. Patai),(1983), pp. 429-461, Wiley, New York, N. Y.
   (c) The rate constant k_3 for the reaction of SOD and O_2^-; $k_3 = 2 \times 10^9$ (mol/l)$^{-1}$s^{-1}. G. Rotilio, R. C. Bray and E. M. Fielden, *Biochim. Biophys. Acta*, 268, 605-609 (1972); E. Michael, R. A. Fox, F. Lavelle and E. M. Fielden, *Biochem. J.* 165, 71-79 (1977).

ENUMERATION OF BACTERIA IN MILK

M.W. Griffiths, L. McIntyre, M. Sully[1] and I. Johnson[1]

Hannah Research Institute, Ayr KA6 5HL, Scotland
[1] *BioTrace Ltd, The Innovation Centre, Mid Glamorgan Science Park, Bridgend CF31 3NA, Wales*

INTRODUCTION

The dairy industry has sought a rapid, reliable test for enumerating bacteria in raw milk. Ideally, the performance time for the test should be less than 10 minutes to enable creameries to assess the hygienic quality of tanker milk samples within the turn around time of the vehicle. Although such tests based on bioluminescence have been described in the past (1), their lack of sensitivity made them unsuitable for routine use in the UK. The lack of sensitivity was mainly due to high levels of non–bacterial ATP, present in milk as free ATP from the mammary gland and ATP from somatic cells (2). Improvement in the sensitivity were obtained by removal of the bacterial cells from the milk prior to extraction and assay of the microbial ATP (3,4). Incorporation of a filtration step to concentrate the bacterial cells from milk resulted in a sensitivity of 2×10^4 cfu/ml (4). However, the test described by Waes et al. (4) was cumbersome and labour–intensive. A new test is described that allows assay of about 1×10^4 cfu/ml. The test uses a novel somatic cell lysing agent and a novel filtration system to remove non–microbial ATP. A resuscitation step is incorporated to increase intracellular ATP pools in the organisms before extraction and assay. The test has been evaluated in a number of dairy microbiology laboratories.

MATERIALS AND METHODS

Milk samples Samples of tanker milks were obtained from a variety of sources. Milks with raised bacterial counts were produced by storage at 2 to 6°C for up to 72 hours.

Reagents All reagents and materials are available from BioTrace Ltd as part of the Milk Microbial ATP Kit (MMAK).

Assay protocol Milk sample (1 ml) was mixed with Somex A (1 ml) in a cuvette and incubated at 37°C for 5 minutes to remove somatic cell ATP. The mixture was then filtered through a bacteria–retaining filter pre– warmed to 37°C. The filter was washed 4 times with rinse solution (1 ml) and the filter removed to a cuvette. To the filter was then added Bactisol (200 μl) to aid recovery of stressed organisms and the cuvette allowed to stand for 2 minutes. Bactex (100 μl) was added, followed, after gentle mixing, by Enzyme–HM (100 μl). The resulting light emission was read after 1 minute in a BioTrace M3 Luminometer, or other suitable luminometer.

RESULTS AND DISCUSSION

There was a highly significant relation between log ATP count and log plate count for the milks analysed (r = 0.64; n = 240; P<0.001). There was little difference between the repeatability of the MMAK assay and plate count, the standard deviations about mean being 0.17 and 0.11 log units respectively. For fresh tanker milk samples with plate counts above 2×10^4 cfu/ml the standard deviation of plate count about the regression line ($S_{y.x}$) was 0.34. This was similar to values quoted previously (4). Geometric means of plate counts within a range of counts were calculated and the means of ATP counts from the same milk samles determined. A plate count interval of 0.25 log units was chosen. A plot of mean log plate count against mean log ATP count showed a linear relation (r = 0.97; n = 15; P<0.001) across the entire count range from 5.2×10^3 to 3.7×10^7 cfu/ml (Fig. 1). The regression equation was log plate count = 3.58 x log ATP count – 1.03.

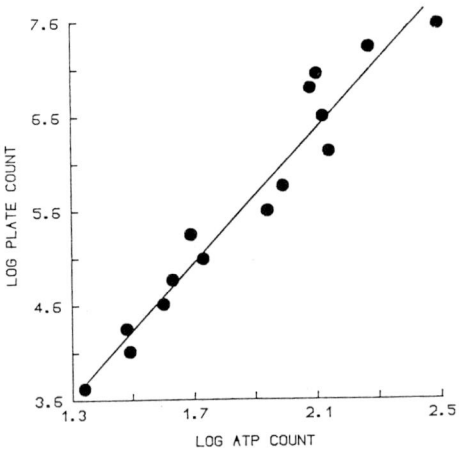

Fig. 1. Relation between mean log plate count values and mean log ATP count for 240 milk samples divided according to plate count range

The test was not primarily designed to enable precise bacterial counts to be extrapolated from ATP values but rather as a means to predict whether a milk fell within a particular plate count range. Using selective cut-offs determined using the regression data, milks could be assigned to count ranges with about 85% reliability (Table 1). However, when adjustments were incorporated to allow for experimental error in the plate count and ATP count procedures, the reliability increased to about 90% of samples correctly assigned to plate count range.

Table 1. Assignment of milk to plate count ranges using the MMAK

Selective cut-offs	% sample assigned to groups		
	Correctly identified	ATP count/an overestimate	ATP count/an underestimate
Log ATP count 1.40 (≡ 25 RLU) Log plate count 4.00	85.8 (91.2)*	2.9 (1.7)	11.3 (7.1)
Log ATP count 1.70 (≡ 50 RLU) Log plate count 5.00	80.0 (88.8)	8.3 (5.0)	11.7 (6.2)
Log ATP count 2.00 (≡ 100 RLU) Log plate count 6.00	84.2 (87.1)	5.0 (3.3)	10.8 (9.6)
Log ATP count 2 20 (≡ 160 RLU) Log plate count 7.00	84.2 (88.8)	5.8 (3.7)	10.0 (7.5)

*Figures in parenthesis allow for 2% variation in ATP count and/or 0.1 log unit variation in plate count.

The method can be used, therefore, to determine the count range of incoming tanker milks at creameries down to at least 1×10^4 cfu/ml and can also be used to monitor growth in silo milk during storage. This would allow rapid processing of milk before counts are attained that compromise product quality.

ACKNOWLEDGEMENTS
The authors would like to thank UKDIRPC and the DTI for funding. The help of the SMMB in providing samples is also appreciated.

REFERENCES
1. R. Bossuyt, A 5-minute ATP platform test for judging the bacteriological quality of raw milk. Neth. Milk Dairy J. 36, 355-364 (1982).
2. M.W. Griffiths, Microbial estimation in dairy products. In ATP Luminescence Rapid Methods in Microbiology, P.E. Stanley, B.J. McCarthy and R. Smither (Eds), Blackwell, Oxford, p. 167-173 (1989).
3. M.W. Griffiths and J.D. Phillips, Rapid assessment of the bacterial content of milk by bioluminescent techniques. In Rapid Microbiological Methods for Foods, Beverages and Pharmaceuticals, C.J. Stannard, S.B. Pettit and F.A. Skinner (Eds), Blackwell, Oxford, p. 13-32 (1989).
4. G. Waes, J. Van Crombrugge and W. Reybroeck, The ATP-F test for estimation of the bacteriological quality of raw milk. In Modern Microbiological Methods for Dairy Products, IDF Special Issue 8901, IDF, Brussels, p. 279-286 (1989).

CHEMILUMINESCENCE MONITORING OF ATP AND HEMOGLOBIN RELEASE FROM PERMEABILIZED ERYTHROCYTES

T. Köszegi, M. Kellermayer

Department of Clinical Chemistry, University Medical School of Pecs, H-7624 Pecs, Hungary

INTRODUCTION

Cellular ATP content can vary to a significant extent without appreciable diminution of certain cell functions (1). Therefore determination of total intracellular ATP content alone gives only limited information on the functional capacity and integrity of the cells. In order to get additional data on the active processes occurring in living cells, a promising approach would be to analyse the molecular interactions of cellular ATP. As early as 1962, it was hypothesized by Ling (2) that ATP might be adsorbed to intracellular proteins. In previous work we presented evidence supporting the concept of the adsorbed ATP by monitoring non-ionic detergent-induced ATP release kinetics in different cell types (3,4). The non-ionic detergent cell permeabilization model had already been successfully used to demonstrate the co-compartmentation of K^+ with intracellular proteins (5). In our studies, similarity in the observed delay of K^+, ATP and protein mobilization after permeabilization of the cells with Brij-58 indicated that these three components are not freely dissolved in the cellular water but they are co-compartmentalized inside the living cell. We also showed that partial ATP depletion results in an alteration of intracellular adsorption of ATP (3-5). However, protein release in our experiments was not directly compared with that of ATP by a continuous monitoring technique.

In the present work, the non-ionic detergent-induced protein (hemoglobin) mobilization was followed continuously by the luminol-hydrogen peroxide chemiluminescent system. Triton-X 100 or Brij-58 induced ATP release kinetics were monitored by the firefly luciferin-luciferase method, and ATP and hemoglobin solubilization were directly compared. Here, we will show that at above a critical micellar concentration (CMC) of detergent, ATP is mobilized more quickly than hemoglobin. It will also be shown that below this CMC concentration but above a critical level Triton-X 100 induces almost identical sigmoid-type ATP and hemoglobin release kinetics. Our findings give further support to the view that intracellular ATP is associated with proteins (mainly hemoglobin in the erythrocytes).

MATERIALS AND METHODS

Reagent and sample preparation. All measurements were made in an isotonic buffer solution containing 0.15mol/l Tris/acetate, 10mmol/l MgSO$_4$, 2mmol/l EDTA, pH7.5. ATP reagent was freshly prepared when required by adding 5.4mg/l crystalline luciferase and 12μmol/l

luciferin (Boehringer, Mannheim) to the above buffer. Quantitative
measurements were done by standard addition technique. For hemoglobin
detection our buffer contained 400μmol/l luminol (Boehringer, Mannheim) and 2mmol/l hydrogen peroxide. For characterization of the
light signal we used reduced and SDS-denatured hemoglobin (240ng/sam-
ple, derived from human erythrocytes). In some cases we applied the
luminol-H_2O_2 and the luciferin-luciferase systems simultaneously in
the same reagent with 500μmol/l H_2O_2 present. In permeabilization
experiments, the solutions contained Triton-X 100 (Serva, Heidelberg)
or Brij-58 (Sigma, St.Louis). For detergent treatment of the erythrocytes at above the CMC, we used lg/l detergent concentration, whereas
0.08g/l Triton-X 100 was applied in the study below the CMC values.

Human venous whole blood anticoagulated with EDTA was obtained from
healthy donors. The blood samples were prediluted with 0.15mol/l NaCl
1000fold before ATP, and 10 000fold before hemoglobin measurements.
When ATP and hemoglobin release were followed simultaneously, the
predilution was 4000fold.

Analysis of ATP and hemoglobin release. Before detergent permeabilization, total red cell ATP was determined by a Triton-X 100 method
and ATP values were referred to the packed cell volume (6). Non-
ionic detergent-induced ATP and hemoglobin mobilization was monitored
directly in the presence of the detergents. For ATP detection at a-
bove the CMC level, a final dilution of 25 000fold and for that of
hemoglobin a dilution of 250 000fold was used. Continuous monitoring
of the luminescence signal was performed at $25\pm2°C$ with an LB 9505
six channel luminometer (Lab.Prof.Dr.Berthold, Wildbad). Below the
CMC value of Triton-X 100, ATP and hemoglobin mobilization were followed in a system containing both the two types of chemiluminescent
reagents. The light signal originating from the firefly system (560
nm) was selected by an OG4 glass filter while that of the luminol
(425nm) by a BG12 filter in a photon counting instrument constructed
in our laboratory.

RESULTS AND DISCUSSION

As observed in our previous studies the presence of non-ionic detergents in the ATP reagent caused a minor quenching (Triton-X 100) or
enhancing (Brij-58) effect without affecting the constant light signal characteristics of the reaction (4). The reduced and SDS-denatured hemoglobin showed an exponential decay of the emitted light,
irrespective of the absence or presence of the non-ionic detergents.
However, lg/l Triton-X 100 or Brij-58 significantly decreased the luminescent intensity.

Above the CMC value (∼0.15g/l for Triton and∼0.09g/l for Brij,ref.
7), Triton-X 100 induced a very rapid ATP mobilization and caused an
unexpected biphasic hemoglobin release curve. After an immediate
light signal initiated by the release of hemoglobin we could observe
an additional slower forming luminescent peak. In Fig.1, the results
of a representative experiment are shown. Brij-58 at lg/l was also
effective in the solubilization of both ATP and hemoglobin but with
much slower mobilization kinetics. There was little, if any difference in the time required for maximal release of ATP and hemoglobin.
However, the release of ATP preceded that of hemoglobin (Fig.2).
Our results are summarized in Table 1.

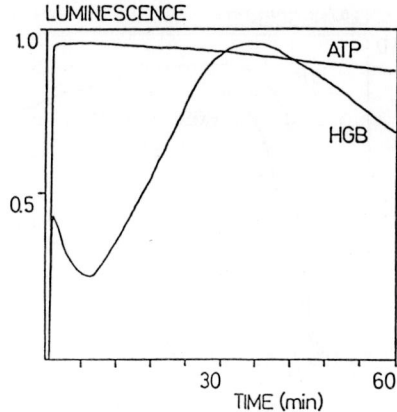

Figure 1. Triton-X 100 induced ATP and hemoglobin
(HGB) release in erythro-
cytes at above the CMC.
The curves are normalized.

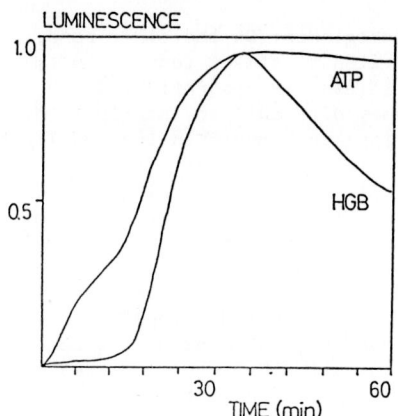

Figure 2. Brij-58 induced
ATP and hemoglobin (HGB)
release in erythrocytes at
above the CMC. The curves
are normalized.

	T_{max} (min)		
	Triton-X 100	Brij-58	Total ATP (mmol/l RBC)
ATP	<0.1	43.9 ± 10.6	1.50 ± 0.27
	(n=14)	(n=16)	(n=14)
HGB*	<0.1	38.7 ± 8.2	
	35.2 ± 12.5	(n=16)	
	(n=7)	p>0.1	

Table 1. Time required for maximal ATP and hemoglobin (HGB)
release (T_{max}) in erythrocytes at above the CMC. (RBC) red
blood cells, (HGB*) time values for the two maxima are given.

In the next series of experiments Triton-X 100 was used below the
CMC value in a reagent capable of detecting both ATP and hemoglobin.
In order to avoid severe distortion of the low intensity ATP release
curves by the high luminescent intensity of the luminol, we reduced
the H_2O_2 content of the reagent. Triton-X 100 (0.08g/l) induced delayed, sigmoid-type release kinetics of ATP and hemoglobin which were
almost identical (Fig. 3).

The main goal in our work is the direct continuous monitoring of
the non-ionic detergent-induced intracellular ATP and protein release.
Direct comparison of the release curves at high detergent concentration indicates that even the mild detergent Brij-58 desorbs intracellular ATP which diffuses out of the cells more rapidly than

hemoglobin. On the other hand, Triton treatment induces an immediate ATP desorption. It is very difficult to interpret the biphasic hemoglobin induced light signal seen at lg/l Triton content. A possible explanation is the limited access of the chemiluminescent compounds to hemoglobin during cell lysis caused by hemoglobin aggregation. The very similar ATP and hemoglobin release curves found at low concentration Triton-X 100 treatment strongly support the view that ATP and hemoglobin are co-compartmentalized in the erythrocytes.

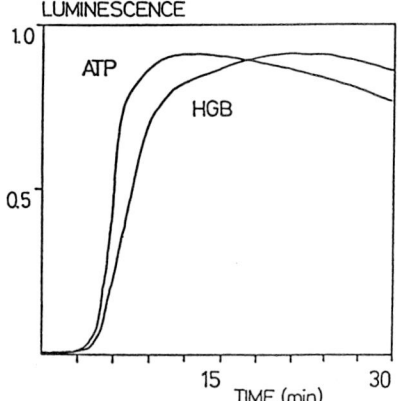

Figure 3. Triton-X 100 induced ATP and hemoglobin (HGB) release in erythrocytes at below the CMC. Simultaneous ATP and hemoglobin detection. The curves are normalized.

ACKNOWLEDGEMENTS
This work was supported by the Hungarian Academy of Sciences Grant No: OTKA 83, and the Ministry of Health Grant ETT 1/267.

REFERENCES
1. J.Palek, A.Liu, D.Liu, L.M.Snyder, N.L.Fortier, G.Njoku, F.Kiernan, D.Funk and T.Crusberg, Effect of Procaine HCl on ATP: Calcium-dependent Alterations in Red Cell Shape and Deformability. Blood 50, 155-164 (1977).
2. G.N.Ling, The Role of Metabolism in Biological Functions. A Physical Theory of the Living State, New York: Blaisdell, p. 189-212 (1962).
3. T.Köszegi, M.Kellermayer, E.Berenyi, K.Jobst and C.F.Hazlewood The Bulk of ATP is Associated to Proteins in the Living Cell: a Release Kinetics Study. Physiol.Chem.Phys.Med.NMR 19, 143-146 (1987).
4. T.Köszegi, M.Kellermayer, F.Kövecs and K.Jobst, Bioluminescent Monitoring of ATP Release from Human Red Blood Cells Treated with Nonionic Detergent. J.Clin.Chem.Clin.Biochem. 26, 599-604 (1988).
5. M.Kellermayer, A.Ludany, K.Jobst, Gy.Szücs, K.Trombitas and C.F.Hazlewood, Cocompartmentation of Proteins and K^+ within the Living Cell. Proc.Natl.Acad.Sci. USA 83, 1011-1015 (1986).
6. T.Köszegi, Rapid Bioluminescent Measurement of Human Erythrocyte ATP Content. Clin.Chem. 34, 2578 (1988).
7. A.Helenius and K.Simons, Solubilization of Membranes by Detergents. BBA 415, 29-79 (1975).

PROPERTIES AND ANALYTICAL PERFORMANCE OF IMMOBILIZED RECOMBINANT FIREFLY LUCIFERASE

A. Roda, B. Grigolo [1], S. Girotti[1], S .Ghini[1], G. Carrea[2] and R .Bovara[2]

Institute of Analytical Chemistry, University of Messina, Salita Sperone 31, 98010 S. Agata, Messina ,Italy, [1] *Institute of Chemical Sciences, University of Bologna, Via S. Donato 15, 40127 Bologna, Italy, and* [2] *Institute of Hormone Chemistry, CNR, Milano, Italy*

INTRODUCTION

Since the first report by De Luca group (1) who successfully cloned c-DNA encoding firefly luciferase in Escherichia coli and showed that it could be used as a reporter gene in studies with animal and plant cells, interest has grown in molecular biology and genetic engineering of luminescent enzymes from firefly, marine bacteria and photoproteins.

The possibility to produce recombinant luciferases or manipulate or introduce lux genes into proteins or other molecules has opened new frontiers in the analytical use of such systems and r-luciferase has recently become commercially available.

In previous papers (2,3) we reported the analytical use of immobilized firefly luciferase on various supports including nylon tubes and epoxymetacrylate beads and the development of continuous flow assay of ATP or related enzymes or analytes. One of the major drawbacks of this system and others (4,7) is the relatively poor stability of the luminescent enzymes even though this was improved by immobilization.

Specificity problems due to the presence of other enzymes in the extracted luciferase (adenylate kinase) or the reaction of the enzyme with other nucleotides (GTP) have been observed when this method is applied on a real matrix such as sea water and sediments.

In order to evaluate the analytical performance of a commercial preparation of r-luciferase we report a comparison between natural and recombinant luciferases when immobilized on nylon and epoxymethacrylate. The enzymes were assessed in terms of stability, enzymatic activity, sensitivity, precision and specificity for the assay of ATP using a continuous flow system in sea water samples, platelets and erythrocytes (8,9).

MATERIALS AND METHODS

Reagents Luciferase from Photinus pyralis (EC 1.13.12.7) specific activity 8 mU/mg and D-luciferin were purchased from Boehringer, Mannheim, GFR. r-Luciferase was kindly donated by Dr.Thomas Hardy, Amgen, Ca, USA, specific activity 30 U/mg, Nylon 6-tubes with 1 mm i.d. were obtained by Portex LTD, Kent, England, epoxymethacrylate (Eupergit C) was supplied by Sigma St. Louis, MO-USA. All other chemicals were of analytical grade and the solutions were made with apyrogenic reagent water prepared with a Milli-Q System (Millipore Corp. Bedford, MA, USA).

Luciferases immobilization Nylon coils were formed by heating tubes at 100° C for 5 min and O-alkylated with triethyloxonium tetrafluoroborate as previously reported (3). Usually 1 m of the activated nylon washed once with 0.1 mol/l K-phosphate pH = 8 was filled with solution of either luciferase or r-luciferase (0.1-1 mg/ml) in 0.1 mol/l K-phosphate pH = 8, 0.2 mmol/l DTT and 2 mmol/l EDTA and left overnight at 4° C. The tubes were washed and stored in 0.1 mol/l K-phosphate buffer pH = 7, 1 % bovine serum albumin, 1 mmol/l DTT and 0.02 % sodium azide at 4° C.

For Eupergit C immobilization 1 ml of 0.5 mol/l K-phosphate buffer pH = 8, containing 1 mg of luciferase, 1 mmol/l ATP, 1 mmol/l luciferin and 0.5 mmol/l DTT was added to 250 mg of dry Eupergit C. After equilibration overnight at room temperature the swelled gel (about 1.3 ml) was washed with 0.5 mol/l K-phosphate buffer, pH = 8, containing 0.1 mol/l ethanolamine and incubated in the same buffer for two hours. Finally, the gel was washed extensively with 0.1 mol/l K-phosphate buffer, pH = 8, and stored in 0.1 mol/l K-phosphate buffer, pH = 7, 1 % bovine serum albumin, 1 mmol/l DTT, and 0.02 % sodium azide at 4° C.

Samples. A platelet count was determined for plasma samples. After centrifugation and suitable dilution to give 100000 platelets per ml, plasma was diluted 1:200 with 0.1 mmol/l Tris buffer, pH = 8.5, containing 4 mmol/l EDTA and 0.2 % Triton X100 (TET solution) and stored at -18° C. This solution lyses the PLT membrane causing ATP release which is then determined with the immobilized luciferase/luciferin system. This procedure is similar to a previously reported method (8,9).

Erythrocytes were similarly treated except that accurate counting, they were diluted 1:5000 or 1:3000 with TET and analyzed. Waste and sea water samples were filtered through a 200 um mesh net to separate largest particles (detritus and zooplankton). Microorganisms were filtered through 0.6 um and 0.2 um filters. ATP was extracted from the microorganisms putting the filters in boiling 0.1 mol/l pH = 7.75 Tris EDTA buffer solution and then analyzed in the flow system. ATP assays were carried out the same day. Otherwise the samples were stored frozen (-20° C).

Analytical method. The manifold has been described elsewhere (2) and involved two streams: the first was the working bioluminescent solution (0.02 mol/l Tris-acetate buffer pH = 7.75, containing 0.12 mmol/l luciferin, 5 mmol/l Mg^{2+}, 0.1 mmol/l EDTA, and 1 mmol/l DTT) and the second was a continuous flow of air into which a known volume (5-50 ul) of sample was intermittently added. A multichannel peristaltic pump (Minipuls HP4, Gilson) and calibrated tubes of different diameters were used to produce different flow rates.

RESULTS AND DISCUSSION

r-Luciferase can be immobilized on Nylon-6 and Eupergit-C with recoveries giving slightly higher enzymatic activity than that obtained using natural luciferase as reported in Table 1. r- Luciferase shows a very low recovery on nylon-6 like natural luciferase while the enzymatic activity recovered on Eupergit C is much higher for both luciferases. The Km value for luciferin for the free enzyme in solution is 2.4×10^{-6} mol/l and generally immobilization slightly increased this value without significant differences between the two luciferases studied. The Km value for ATP was lower in the case of the immobilized enzymes method without any difference between the two enzymes.The stability of the nylon and Eupergit-C immobilized r-Luciferase is higher than that of natural luciferase both at 4° C and 25° C and the reactors can be used for almost two months performing more than 100 assays daily (Table 1). r-Luciferase shows a good specificity for ATP and other related nucleotides are not measured. This is extremely important when this method is applied to detect ATP in sea water which contains appreciable amounts of nucleotides such as GTP which has been documented to interfere when natural luciferase is used. The r-Luciferase preparation does not contain other enzymes which could in some way increase the background signal as previously observed for the presence of adenylate kinase and other related enzymes. When the immobilized r- Luciferase is used in the continuous flow system assay for ATP the analytical performance of the assay is slightly improved in terms of better precision and reproducibility since the coefficients of variation in the intra and inter assay studies are lower, from 6-9% to 3-5%. The assay using Nylon-6 shows the lower detection limit (0.3 pmoles ATP) when compared with Eupergit-C (3 pmoles ATP) without any differences between the two luciferases. The standard curve presents a linearity of four decades when r-luciferase is used in both systems, while the range of linearity was narrow for natural luciferase.

For normal platelets and erythrocytes samples, the reference range, using nylon or Eupergit-C, show that data for r-Luciferase immobilized were in agreement to previously data obtained i.e. 3.5-7.5 umol/10^{11} platelets and 1.1-1.6 umol/ml erythrocytes. For waste and sea water the ATP content was respectively 10^{-9} 10^{-8} and 10^{-10} 10^{-9} mol/l.

In conclusion the reported data suggest that r-Luciferase can be used for ATP analysis and the analytical performance of the assay is slightly improved in respect to natural luciferase.

TABLE 1. Recovery, stability and enzymatic properties of   immobilized luciferase from Photinus pyralis and recombinant luciferase.

	Luciferase from Photinus pyralis		r-Luciferase	
	Nylon	Eupergit	Nylon	Eupergit
% Recovery	0.6-0.8	25	1.2-1.4	40
Stability (halflife,days)				
4°C	60	20	130	35
25° C	20	3	70	10
$K_{m \, app}$ mol/lx10^{-5}				
D-Luciferin	2.8	8.9	1.4	2.2
ATP	2.2	0.66	0.8	0.73

REFERENCES

1.   D.W. Ow, K.V. Wood, M. De Luca, J.R. De Wet, D.R. Helinski and S.H. Howell.
.   Transient and stable expression of the firefly luciferase gene in plant cells and transgenic
.   plants. Science, 234, 856-859 (1986).

2.   G. Carrea, R. Bovara, G. Mazzola, S. Girotti, A. Roda and S.   Ghini. Bioluminescent
.   continuous-flow assay of adenosine 5'-triphosphate using firefly luciferase immobilized
.   on nylon tubes. Anal.Chem. 58, 331-333 (1986).

3.   G. Carrea, R. Bovara, S. Girotti, E. Ferri, S. Ghini and A. Roda. Continuous-flow .
        bioluminescent   determination   of   ATP in platelets using firefly luciferase.
        immobilized on epoxy methacrylate. J. Biolumin. Chemilumin. 3,7-11 (1989).

4.   L.J. Kricka, G.K. Wienhausen, J.E. Hinkley and M. De Luca. Automated bioluminescent.
        assays for NADH, glucose-6- phosphate, primary bile acids, and ATP.Anal. Biochem.
        129, 392-397 (1983).

5.   A. Lundin, M. Hasenson, J. Persson and A. Pousette. Estimation of biomass in growing.
        cell lines by adenosine triphosphate assay. Methods Enzymol. 133, 27-42, (1987).

6.   D.C. Vellom and L.J. Kricka. Continuous-flow bioluminescent assays   employng.
        Sepharose-immobilized enzymes. Methods Enzymol. 133, 229-237 (1987).

7.   P.J. Worsfold   and A. Nabi. Bioluminescent   assay with immobilized firefly luciferase.
        based on flow injection   analysis. Anal. Chim. Acta. 179, 307-313 (1986).

8.   S. Girotti, E. Ferri, M.L. Cascione, S. Comuzio, A. Mazzuca, A. Orlandini and A.Breccia.
        Methodological problems of direct bioluminescent ATP assay in platelets and.
        erythrocytes. J. Biolumin. Chemilumin. 4, 594-601   (1989)

9.   E. W. Chappelle, G. L. Picciolo and J. W. Deming. Determination of bacterial content.
        in fluids. In Methods in Enzymology. Academic Press. LVII 65-72 (1987).

DIRECT BIOLUMINESCENT ADP AND ATP ASSAY IN PLATELETS AND ERYTHROCYTES: APPLICATION TO QUALITY CONTROL IN TRANSFUSIONAL CENTERS

S. Girotti, E. Ferri, S. Nucci[1], F. DI Graci[1] M. L. Cascione, G. Sermasi[1], A. Orlandini[2] and L. Farina[2]

Institute of Chemical Sciences, University of Bologna, 40127 Bologna, Italy
and
[1]St. Orsola Hospital, 40127 Bologna, Italy and.
[2]Bouty SpA, 20099 Sesto S.Giovanni, Italy

INTRODUCTION

Studies during the past decade have shown that cells concentration, agitation, temperature, anticoagulants, nutrients, stabilizing media and type of container are important factors for the viability and function of stored platelets (PLT) and red cells (RBC) concentrates. Since adenine nucleotides play an important role in the fuction of these cells the relationship between ATP and ADP changes and cells viability have been estimated and a high correlation of PLT ATP with the various in vitro parameters has been found [1-2]. One of the methods most widely used to determine nucleotides concentrations is the bioluminescent assay because of its high sensitivity and specificity [3-4]. We have previously improved a group of reagents in order to obtain a direct bioluminescent method for ATP and more recently also for ADP determination. We have employed these reagents, assembled as a kit formulation [5], in the quality control of stored RBC and PLT concentrates.

MATERIALS and METHODS

Reagents All solutions were made with apyrogenic reagent-grade water. Phoshoenolpyruvate, pyruvate kinase, ATP and ADP were supplied by Boehringer, Mannheim, FRG. For luminescent analysis, reagents and kits supplied by several companies (Boehringer; Bouty, Milan, Italy, BioOrbit, Turku, Finland) were used according to the manufacturers' instructions.

Extraction solution TET extracting solution contains 0.2 % (w/w) Triton X 100, 4 mmol/l EDTA in 0.1 mol/l Tris-acetate buffer, pH 7.75.

PEP-PK solution PEP (2 mmol/l) was dissolved in the 5 mmol/l HEPES buffer, pH 7.75 containing 1 % (w/w) Ficoll-400, 300 mmol/l $MgCl_2$ and 20 mmol/l KCl and 100 μl of PK were added at each 2.5 ml of PEP solution.

Samples For stored PLT whole blood was collected into CPDA-1 anticoagulant and PLT concentrates was preparated according to Adams G.E. et al.[6] For RBC and Platelet rich plasma (PRP) whole blood was collected into citrate-EDTA.

Assay protocol ADP and ATP were extracted from cells, after accurate counting, by dilution with TET solution at a ratio 1:200 for PLT and 1:3000 for RBC. Samples could be immediately analyzed or frozen.

ATP concentration, expressed as ng/ml, was obtained interpolating the sample light emission, recorded by the automatic Berthold Auto-Clinilumat LB 952, with that of the calibration curves. ADP concentration was obtained comparing the difference between the light emission of the ADP and ATP sample with the calibration curve.

RESULTS AND DISCUSSION

We have used calibration curves instead of standard addition to determine ADP and ATP concentrations. The values obtained using the TET solution agree well with those achieved by Ethanol-EDTA extraction (y = 1.028 x + 0.067; r = 0.917, n = 37, P < 0.001) in the analysis of both normal or pathological samples. For PLT the ADP normal range was 1.9-3.7 μmol/10^{11} PLT and ATP normal range was 3.5-7.5 μmol/10^{11} PLT. RBC normal range for ATP was 3.2-3.8 μmol/g Hgb and for ADP was 0.56-0.73 μmol/g Hgb. These values agree to literature data [5].

Recently the assay was applied to direct ADP and ATP bioluminescent analysis of PLT concentrates stored in transfusional centers. The rapid evaluation of ATP/ADP ratio on a large number of samples allows to find the better conditions for preparation and storage, which have a large influence on nucleotide content.

Firstly we measured the nucleotides levels in PLT concentrates preparated by PLT apheresis and in PLT obtained by the same donors prior and after the treatment. Our results indicate that there were no alteration in nucleotides content of treated PLT and this finding was confirmed with each type of filter unit used in performing apheresis.

In a second step we focused our interest on the ATP and ADP levels in PLT concentrates stored for 7 days in plasma or in artificial media with (PCD) or without (PC) dextrose. Some examples of the time course of nucleotides concentrations are showed in Fig. 1 and Tab. 1. The results obtained till now indicate that the PLT stored in the medium with dextrose show a reduced percentage of decrease in ATP and ADP contents. Other parameters like pH, platelet count, lactate and ammonium concentrations and pCO_2 seem to support the hypothesis of an higher rate of ATP synthetizing metabolism in PCD bags, mainly as anaerobic glycolysis.

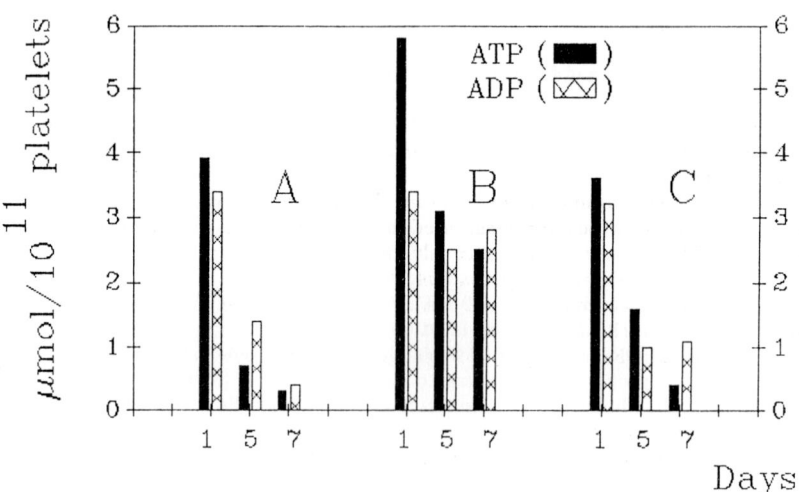

FIGURE 1:Time course of ATP AND ADP concentrations in PLT concentrates stored in plasma.

TABLE 1: Time course of nucleotides contents in PLT concentrates stored in different artificial media (concentrations expressed as $\mu mol/10^{11}$ PLT).

SAMPLE	MEDiUM	NUCLEO-TIDES	DAYS		
			1	5	7
A	PC*	ATP	4.7	1.6	0.8
		ADP	2.0	1.4	0.5
	PCD	ATP	3.6	2.6	2.0
		ADP	2.2	2.0	1.0
B	PC	ATP	7.8	3.8	3.2
		ADP	3.5	3.0	2.9
	PCD	ATP	6.2	4.2	2.6
		ADP	3.7	3.0	2.4
C	PC	ATP	3.3	1.7	0.6
		ADP	3.1	0.9	0.3
	PCD	ATP	3.6	0.8	0.7
		ADP	3.1	1.3	1.4

* PC:Sodium citrate plus Plasmalyte-A
PCD:Sodium citrate, Plasmalyte-A and 50% Dextrose

These findings suggest that the dextrose addition can slightly reduce the extent of the platelet storage lesion, at least at the nucleotides metabolism level. A previous work on the same problem led to a different conclusion : no effect was obtained when dextrose was added [6] but it must be borne in mind that it was not produced any evaluation of nucleotides content.

Compared to other techniques the direct ADP and ATP bioluminescent method offers several advantages particularly when used as an assay to measure relative variations. Ethanol-EDTA extraction [5], commonly used, is a long process and requires an ultracentrifuge which is not always available.

Also in pathological conditions in which the metabolism is altered, the direct determination of nucleotides can be applied, because it is of diagnostic value and it is increasingly used in the investigation of heart, liver and kidney diseases and transplants.

ACKNOWLEDGEMENTS
This work was supported by grants from MURST and CNR (n. 89.03414.14).

REFERENCES
1.   Holme, S., Heaton, W.A., and Courtright, M., Platelet storage lesion in second-generation containers: correlation with platelet ATP levels. *Vox Sang. 53*, 214-220, (1987).

2. Wallvik J., and Åkerblam O., Platelet concentrated stored at 22 °C need oxygen. The significance of plastics in platelet preservation. *Vox Sang.* *45*, 303-311, (1983).
3. Holmsen H., Nucleotide metabolism of platelets. *Ann. Rev. Physiol.*, *47*, 677-690, (1985).
4. Hampp R., *Luminometric method*, Methods of Enzymatic Analysis, VCH, Weinheim, Vol. VII, p.370-379, (1984).
5. Girotti S., Ferri E., Cascione M. L., Comuzio S., Mazzuca A., Orlandini A., and Breccia A., Methodological problems of direct bioluminescent ATP assay in platelet and erythrocytes. *J. Biolumin. Chemilumin. 4*, 594-601, (1989).
6. Adams G. A. and Rock G., Storage of human platelet concentrates in an artificial medium without dextrose. Transfusion 28, 217-220 (1988).

THE ATP BIOLUMINESENCE TECHNIQUE IN EVALUATING THE BACTERIOLOGICAL QUALITY OF DRIED MILK

A.Kahru and R.Vilu

Institute of Chemical Physics and Biophysics of the Estonian Academy of Sciences, Akadeemia tee 23, Tallinn, 200026, Estonia/USSR

INTRODUCTION

The luciferin-luciferase method of adenosine triphosphate (ATP) determination provides a mean for quick enumeration of microorganisms within minutes to hours depending on sample type and its microbial count. As traditional plate counts take days and weeks before the results could be interpreted the application of ATP-technique as a rapid test for counting bacteria is coming more and more popular in laboratories of food industry, environmental control, clinics etc (1). In this paper the use of ATP-technique in evaluating the bacteriological quality of dried milk is discussed. Up to our knowledge this technique has not been applied to dried milk before despite of the fact that this could save a lot of time: according to our experience the viable plate counts of dried milk take 5...6 days.

MATERIALS AND METHODS

Dry milk samples . Dried cream, whole milk, skim milk and whey samples from Paide Dairy Plant (Estonia), APV Anhydro A/S and Niro Atomizer (both Denmark) and Valio (Finland) were analyzed.

Reconstitution of dry milk. 1 g of dried milk was aseptically added to 9 ml of 0.9% NaCl solution and shaked for 30 min at room temperature.

Enrichment of dry milk. 1 ml of reconstituted dried milk was aseptically added to 2 ml of medium consisting (g/1): Tryptone (Difco) - 5; yeast exract (Difco) - 2.5; glucose - 1. Samples were incubated at 30°C for 24 h.

Total bacterial counts. Total viable counts were made in duplicate by preparation of pour plates from Plate Count Agar (Fluka). Plates were incubated aerobically 6 days at 30°C before counting.

Preparation of samples for ATP assay. For the analysis of **free ATP** content 100 ul of reconstituted dry milk or enriched dried milk was added to 900 ul of sterile Tris-acetate/EDTA (0.1 mol/1; 2mmol/1, pH 7.75). **Total ATP** was extracted by adding 150 μl of reconstituted dry milk (or enriched dry milk) sample to 150 μl of ice-cold TCA/EDTA solution (10%; 4 mmol/1). For the analysis of **bacterial ATP** 50 μl of reconstituted dry milk (or enriched dry milk) was added to 50 μl of NRS (Lumac) and 50 μl of apyrase solution (Sigma, 5.8 U/ml in the reaction mixture) and shaked at

room temperature for 30 min. Then 150 µl of ice-cold TCA/EDTA
solution (10%; 4mmol/l) was added, samples vortexed and
centrifuged.
Determination of the ATP content. Before the measurement of ATP
the samples extracted with TCA (total ATP and bacterial ATP
samples) were 50-fold diluted with Tris-acetate/EDTA buffer. Free
ATP samples were analyzed without further dilution. The ATP
content was measured at least in duplicate with
luciferin-luciferase and a Luminometer 1251 (Bio-Orbit, Finland).

RESULTS AND DISCUSSION
Lots of papers have been published on ATP analysis in milk (2-5, to
cite only few of them). The applicability of ATP-technique in
enumeration of bacteria in milk is complicated by the fact that
apart of bacterial cells ATP is also present in somatic cells and
remarkable amount of ATP in milk is so-called free ATP possibly
associated with casein micelles (2). According to (6) the ratio of
intrinsic ATP to bacterial ATP in milk is from 6 to 1500. The
insufficient removal of non-bacterial ATP before the extraction
of bacterial ATP is assumed to be the main reason for lack of the
sensitivity of the ATP-test in milk (4). We supposed that the
amount of intrinsic ATP in dry milk should be remarkably lower as
ATP is not a stable molecule and could be hydrolyzed during the
procedure of making milk powder.
Enumeration of bacteria from the reconstituted dried milk. In
this study 26 dried milk samples having plate counts from $3*10^3$ to
$3*10^5$/g were analyzed on ATP content. Our data showed that dried
milk contained from 0.1 to 2 nmol ATP/g, 70% of it being free ATP.
It means that the ATP content is practically not changed during
the milk drying procedure as the bovine milk contains from 0.13 to
0.31 nmoles ATP/ml (2) and the dried milk is around 10-times more
concentrated. No correlation between bacterial ATP and viable
plate counts was found as the bacterial ATP signal was comparable
to the backround's signal. For **milk** samples it has been shown that
the ATP-technique is applicable if the respective plate count
numbers exceed $5*10^5$/ml (3). If a concentration step by membrane
filtration is included into the assay the linear relationship
could be established between the plate count and ATP-test above a
plate count of $2*10^4$/ml (5). The standard plate count in certified
dry milk samples should not exceed $5*10^4$/g, i.e. $5*10^3$/ml in
reconstituted (diluted dry milk) samples. Therefore, the membrane
filtration procedure developed for the milk is not applicable.
**Conclusion 1: the number of bacteria cannot be quantified directly
from the reconstituted dried milk** and therefore the enrichment step
was added to the assay.
Enrichment of dry milk samples. During the enrichment (growth)
the number of bacteria increases exponentially. The ATP-test
involving the enrichment step is applicable if the growth rate of
bacteria from different dry milk samples is similar. If so, the
pre-enrichment bacterial counts could be calculated from
post-enrichment ATP data knowing the growth rate and the amount of
ATP per bacterium.

The enrichment of samples for 6 and 24 h was studied because in former case the samples could be analyzed within the same day and on the latter case – the next morning. Two parallel experiments with enrichment of dried milk samples were performed, first with 9 samples (Table 1) and second with 14 samples (not shown).

Table 1. Enumeration of bacteria in dried milk by viable plate counts and ATP-technique after enrichment of samples in nutrient broth.

Dry milk	μ,[a] h^{-1}		fg ATP per cell[b]	Bacteria*10^4 per g	
	Time of enrichment at 30°C			Viable plate count data[c]	ATP-test data after 24-h enrichment[d]
	6 h	24 h			
1	n.t.	0.69	0.55	0.3	1.5
2	0.42	0.59	2.50	2.9	5.7
3	0.45	0.68	0.49	1.4	4.6
4	0.53	0.35	28.10	0.25	0.002
5	0.31	0.59	0.51	5.0	1.9
6	0.78	0.60	0.47	0.7	0.4
7	0.68	0.66	0.79	0.5	1.8
8	0.74	0.59	2.74	1.1	2.2
9	0.80	0.50	1.67	16.0	2.3

a – calculated from plate count data at 0, 6 and 24 h of enrichment.
b – measured by dividing the amount of bacterial ATP per ml to viable plate count per ml determined after 24-h enrichment
c – measured before enrichment
d – calculated from post-enrichment ATP data for $u=0.6$ h^{-1} and ATP per cell value of 1 fg.
n.t. – not tested

The growth rate of bacteria of different dry milk samples determined from 24-h enrichment experiments varied less than 6-h enrichment data (Table 1). From 0 to 6 h the number of bacteria in enrichment broth increased by two orders of magnitude and the ATP content per cell after 6 h enrichment was around 100 fg (not shown) and not 1 fg as described in the literature (1) which means that the number of bacteria determined from the ATP data is overestimated in comparison with the plate counts. As the errors of estimation of both (growth rate value and ATP per cell value) contribute to final error of estimation of initial bacterial count by ATP (see above) we reached the **conclusion 2: 6-h enrichment is not appropriate in determination of bacterial counts by ATP in dry milk.**

During 24 h of incubation the bacterial count in the enrichment broth increased by 6 orders of magnitude, from 10^3/ml to 10^9/ml. The amount of free and somatic ATP after 24-h enrichment was negligible (not shown). The average growth rate of bacteria in the first experiment was 0.6 h^{-1} and the average ATP content per cell was 1 fg (Table 1). For the second experiment u was 0.46 h^{-1} and the ATP content was 6.3 fg per cell. Using the data on bacterial ATP concentration in 23 reconstituted dry milk samples after 24 h of enrichment and taking into account the growth rate and ATP per cell values the initial pre-enrichment bacterial counts were calculated as shown in Table 1. Statistically significant correlation (r=0.64, n=23) between predicted and actual plate count was found. In case 19 samples out of 23 the ATP count did not differ from plate count more than 10 times. The number of false positives was equal to false negatives. As already mentioned, in case of dry milk samples the agar plates ought to be incubated at 30°C for 5...6 days: the number of colonies counted after 3 days of incubation was only 57% from that counted after 6 days of incubation. Thus, the ATP test which involves the 24-h preincubation step still saves time. As the precision of viable plate counts is not much bigger than order of magnitude the accuracy of these two methods is comparable. Whether this analysis time and precision is acceptable in the industrial practice is to be decided by the further experiments and economic calculations. However, the following **conclusion 3** could be drawn: **the data obtained by us showed that the ATP-test involving the 24-h enrichment step is applicable at least for the qualitative analysis of the total bacterial count in dried milk.**

ACKNOWLEDGEMENTS
We wish to thank Anneli Room for skillful technical assistance.

REFERENCES
1. P.E. Stanley, A review of bioluminescent ATP techniques in rapid microbiology. J. Biolumin. Chemilumin.4, 375-380 (1989).
2. T. Richardson, T.C.A. McGann, R.D. Kearney, Levels and location of adenosine 5´-triphosphate in bovine milk. J. Dairy Res. 47, 91-96 (1980).
3. T. J. Britz, J. J. Bezuidenhout, J.M. Dreyer and P.L. Steyn, Use of adenosine triphosphate as an indicator of the microbial counts in milk. S. Afr. J. Dairy Technol. 12, 89-91 (1980).
4. D.P. Theron, B.A. Prior and P.M. Lategan, Determination of bacterial levels in raw milk. Selectivity of non-bacterial ATP hydrolysis. J. Food Protection 49, 4-7 (1986).
5. J. Van Crombrugge, J., G. Waes and W. Reybroeck, The ATP-F test for estimation of bacteriological quality of raw milk. Neth. Milk Dairy J. 43, 347-354 (1989).
6. A. N. Sharpe, M.N. Woodrow and A.K. Jackson, Adenosinetriphosphate (ATP) levels in foods contaminated by bacteria. J. Appl. Bact., 33, 758-767.

MEASUREMENT OF ADENYLATE KINASE (MYOKINASE) IN HUMAN NEUTROPHILS AND ITS RELEASE BY BACTERIA. EFFECT OF LIPID A ON THE ACTIVITY OF THE ENZYME.

I. Tarnok and Zs. Tarnok

*Bacteriology Division, Research Institute Borstel,
2061 Borstel, Federal Republic of Germany.*

INTRODUCTION.

Adenylate kinase (ADKi; E.C. 2.7.4.3.) has been detected in 1942 by Kalckar (1) in muscle and it catalyses the formation of ATP from ADP according to the equation:

$$ADP \rightleftharpoons ATP + AMP.$$

ATP is measurable using the luciferase-luciferin reagent (LULU). Recently we were able to measure it in supernatants of PMN lysed by ultrasonication.

The question arose whether ADKi could only be released by ultrasonication or perhaps by mechanical agitation or even by contact with bacteria. An increased enzyme activity in supernatants of PMN plus bacteria may represent a measure for such an interaction. On the other hand, it was of interest whether lipid A, the toxic principle of endotoxin of Gram-negative bacteria, or the chemotactic peptide N-formylmethionyl-leucyl-phenylalanine (fMLP) could be able to induce the release of ADKi, for both compounds influence PMN in many respects e.g. they stimulate superoxide output.

Furthermore, characteristical data of ADKi such as Km, pH- and temperature optimum, denaturation temperature, stability in the frozen state, and units per PMN (u/PMN) have been collected.

MATERIAL AND METHODS

Reagents: PBS: phosphate balanced salt solution according to Dulbecco; PBSAG: PBS supplemented with 0.1 % human albumin and 1 % glucose; fMLP: a 10 μmol/l solution of the chemotactic peptide N-formylmethionyl-leucyl-phenylalanine, in PBSAG;
phosphate buffer 1/15 m/L of different pH-values according to Sörensen (2);
Salmonella abortus equi lipid A Sae TEN H 1178: 1 mg/ml in water. Salmonella minnesota R7 lipid A: 750 μg/ml in water.
ATP standard: concentrations between 10 and 100 μmol/l in PBS.
ADP solution: concentrations between 0.2 and 1 mmol/l in PBS.
LULU (luciferin-luciferase; Sigma Chemie GmbH, Germany) 10 mg of the freeze-dried material/ml.
M. bovis BCG SN 2008: suspension of 20 mg bacteria/ml in water.
Neutrophils: isolate PMN according to (7). Determine the number of living cells using the Trypan blue exclusion test (7).
Definition of ADKi units: 1 unit will form 1 μmol ATP from 2 umoles of ADP per min at pH 7.2 and 20° C.

Assay protocol: ADKi in cell-free homogenate of ultrasonicated PMN: chill a PMN suspension (2 x 10^6 cells/ml) in ice. Treat it 3 times in a Branson ultrasonicator for 10 seconds following by a break for 20 seconds each. Remove cell debris by centrifugation. Use supernatant (referred to as "cell-free homogenate") immediately or keep it at -20°C until use. Calibration: use 5 µl of the LULU reagent and 20 µl of the respective ATP standard.

Michaelis constant (Km): prepare ADP dilutions (from 1 µmol/l to 2 mmol/l). Mix ADP dilutions and freshly prepared cell-free homogenates. Take aliquots after intervals and add LULU to each. Measure counts resp. ATP and calculate Km.

pH-Optimum: mix aliquots of phosphate buffer of different pH-values, that of ADP solution and cell-free homogenate. Keep the solution at 20°C. Take aliquots at intervals, add LULU to each one and calculate the amount of ATP formed. Measure the final pH of the mixture.

Temperature stability of ADKi: Keep aliquots of a freshly prepared cell-free homogenate for 10 min in a water bath of the desired temperature. Chill it in ice. Add ADP solution. Take aliquots at intervals, add LULU to each and determine counts resp. amounts of ATP formed and calculate enzyme units.

Stability at -20°C: dispense aliquots of a freshly prepared cell-free homogenate in Eppendorf tubes and refrigerate them (-20°C). Measure ADKi units in thawed aliquots immediately and after some days.

Release of ADKi by stirring of a PMN suspension: take aliquots at intervals from a stirred PMN suspension and add ADP solution to each. Take aliquots from this mixture at intervals, add LULU to each and calculate enzyme units. Continue both stirring and activity determinations for about 3 hours. Count number of living PMN in each sample.

Release of ADKi by BCG: mix BCG (= 20 mg bacteria/ml) and PMN suspensions. Incubate the mixture at 20° C. Remove bacteria and PMN by centrifugation. Add ADP solution to aliquots of the resulting supernatant; take samples from the mixture at intervals, add LULU to each, and calculate the ATP concentration and enzyme units. Perform controls using neutrophils without BCG and vice versa.

Release of ADKi by different amounts of BCG: incubate a mixture of aliquots of an ADP solution, and of BCG and PMN suspensions at room temperature. Take samples at intervals, add LULU to each and determine the amount of ATP formed. Repeat the experiment using decreasing amounts of BCG. Perform controls without BCG.

Viability of PMN after treatment with BCG: mix PMN and BCG suspensions. Take samples at intervals and determine the number of viable PMN using the Trypan blue exclusion test.

Phagocytosis of BCG by PMN: use the Ziehl-Neelsen staining in order to make intracellularly located bacteria visible.

Influence of lipid A on ADKi release: mix aliquots of a lipid A solution (1 mg/ml in water) and a PMN suspension; incubate the mixture at room temperature. Take samples after 0, 10, 20 and 30 min. Add ADP solution to each. Take samples from this second solution after 0, 2, 4 and 6 min, add LULU to each and calculate the ATP amount. Perform controls using PMN and water instead of lipid A.

Influence of fMLP on ADKi release: mix a 10 µmol/l fMLP solution and a PMN suspension; incubate the mixture at room temperature. Take samples at intervals (up to 2 or 3 hours). Add ADP solution to each. Take samples from this second solution after 0, 2, 4 and 6 min, add LULU to each and determine the amounts of ATP formed. Perform controls using PMN and PBSAG instead of fMLP.

RESULTS AND DISCUSSION

ADKi can be released from PMN by ultrasonication (Fig.1), stirring or contact with bacteria. ADKi from   ultrasonicated PMN of different donors showed the following characteristical data:

$$K_m: 1.05 \times 10^{-6} - 8.9 \times 10^{-4} \text{ mol};$$
$$\text{enzyme units/PMN}: 1.0 - 2.4 \times 10^{-12};$$

temperature stability: no loss of activity at 50°C for 10 min; about 20 and 80 percent loss of activity after 1 month at -20°C; no defined pH-optimum.

The increase of ADKi activity in stirred PMN suspensions was accompanied by a decrease of PMN number (Fig.2).

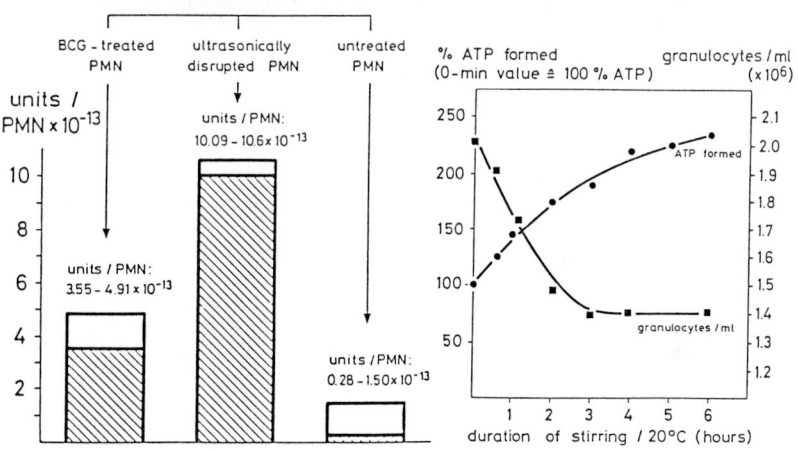

Fig.1. Units of adenylate kinase (u/PMN) released a) by contact with M.bovis BCG; b) after ultrasonication and c) in the untreated control.

Fig.2. Increase of ADKi activity and decrease of the number of granulocytes in stirred PMN suspension. Values at 0-minute: 100 %.

The increase of ADKi activity, caused by contact with Mycobacterium bovis BCG, is a function on the number of bacteria applied (Fig.1, 3). PMN, incubated without BCG, showed a significantly lower ADKi release. As a rule, the contact with BCG did not influence the number of living PMN within the short contact time applied (0-15 min). Ziehl-Neelsen staining of the PMN for engulfed bacteria yielded negative results during the incubation period indicating that ADKi release occurred earlier than microscopically visible phagocytosis.

If PMN had been incubated with lipid A, the toxic principle of lipopolysaccharide of Gram-negative bacteria, a significant repression of ADKi release (or its activity) could be observed (Fig.4). However, we

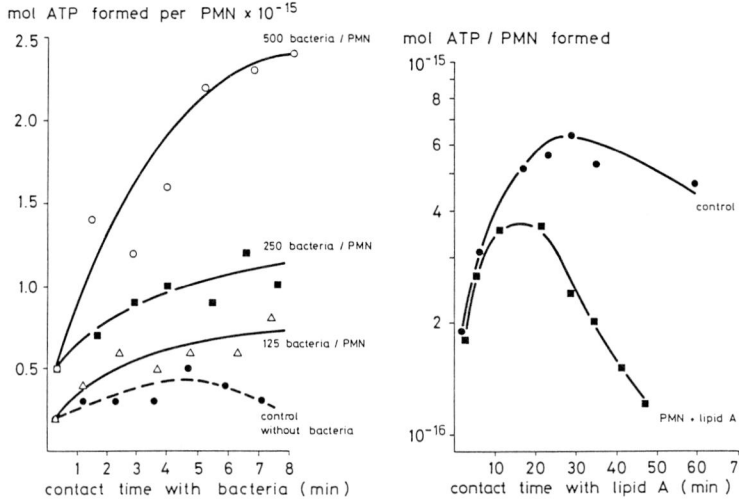

Fig.3. ADKi activity in PMN released by different amounts of M.bovis BCG. Contact time versus mol ATP formed after addition of ADP.

Fig.4. ADKi activity in PMN with and without S.minnesota R7 lipid A. Contact time versus mol ATP formed after addition of ADP.

need more data on this topic, for PMN of different donors reacted irregularly with lipid A. Similarly, further experiments are necessary to clear the question whether lipid A might influence the release procedure or rather the free enzyme itself.

In regard to the chemotactic peptide fMLP, it was not able to induce an enzyme release or inactivate ADKi in cell-free solutions.

The physiological role of ADKi is still controversial and thus, we can just speculate about its significance in vivo. The enzyme might be released into the extracellular fluid during an inflammation caused by microorganisms. Its measurable and most striking effect could be a dramatic increase of the ATP level in the inflamed tissue, provided that ADP is available. Whether an increased ATP concentration could influence the inflammation is not subject of this paper.

Details will be published elsewhere.

ACKNOWLEDGEMENT
We are very much indebted to Dr. H. Brade for making the lipid A preparations available.

REFERENCES
1. J.Kalckar, J. Biol. Chem. 143, 299 (1942). Cited in: R. Ammon and W. Dirscherl, Transferasen. Fermente, Hormone, Vitamine. Georg Thieme Verlag, Stuttgart, p. 343-344 (1959).
2. J. Stauff and R. Jaenicke, Physikalische Chemie der Lösungen. In Biochemisches Taschenbuch, H.M. Rauen and R. Kuhn (Eds), Springer-Verlag, Berlin, 1964, p. 90 ff.

BACTERIOLOGICAL TESTING OF RAW MILK WITH FIREFLY LUCIFERASE

E. Schram and A. Weyens-van Witzenburg

Vrije Universiteit Brussel, 1640-Sint-Genesius-Rode, Belgium

INTRODUCTION

Since the introduction of commercial reagents for the assay of biomass with firefly luciferase, the bacteriological control of milk has been one of the important goals pursued with this method. The speed and simplicity of the technique make it indeed particularly attractive for the rapid screening of bulk milk reaching dairy plants. For many years the assay protocols did not involve fractionation of the milk samples, chiefly to shorten the duration of the tests. However, practical experience has shown that the non-bacterial ATP contained in milk is so much higher than bacterial ATP that the detection limit for bacteria can hardly be lowered below 500,000/ml (1, 2), which is more than an order of magnitude higher than the desired limit of 10-20,000. Concentration of the bacteria by centrifugation has proved inapplicable to milk and filtration seems therefore the only way out. Several difficulties are however associated with this technique, the first one being the poor filtrability of milk. The second one has to do with the fact that casein micelles, which bind part of the non-bacterial ATP, should be eliminated while retaining bacterial cells on the filter. According to literature the maximum diameter for casein particles is around 300 nm, which is but little less than the diameter of some bacteria.

MATERIALS AND METHODS.

Milk samples were kindly supplied to us by the Comité Provincial de la Qualité du Lait in Soignies (Belgium), together with data on CFU's, fat and protein content, and number of somatic cells. The reagents used have been described and commented in two of our previous publications (3, 4). As far as filtration is concerned conventional bacteriological filters were all found inadequate in our working conditions. The filters used in the present work were supplied to us by Hamilton (Switzerland) as 8 mm diameters discs, suitable for introduction into the standard cuvettes used for the luminescence measurements.

Standard procedure (see Figure 1) : 0.5 ml of milk is mixed with 0.5 ml of buffer containing Triton and apyrase to remove somatic ATP, and incubated at 40°C for 5 minutes. The mixture is then filtered in a prewarmed manifold conceived to accomodate six samples in parallel, and washed three times with 1 ml of prewarmed buffer. The filters are subsequently transferred to luminescence vials, suspended in 200 ul luciferin/luciferase reagent, and the vials introduced in the six-sample holder of a LUMICON luminometer (Hamilton, Switzerland). After counting of residual ATP, 100 ul of bacterial extractant is automaticallly injected and countings are resumed. The assay is finalized by the automatic injection of 100 ul of standard ATP.

Specific software was developped that takes care of the automatic injection and of the calculation of the results, inclusively the dilution effect produced by the addition of bacterial extractant and ATP standard (see Table 1).

RESULTS AND DISCUSSION

Milk contains currently some 400,000 somatic cells/ml whose ATP content is three orders of magnitude higher than that of bacteria. In order to be able to detect 10,000 bacteria/ml, non-bacterial ATP, which also includes ATP bound to casein micelles, must therefore be eliminated as quantitatively as possible. Lysis of somatic cells with a non-ionic detergent and hydrolysis of the freed ATP with apyrase proved to be insufficient and a filtration step has therefore been proposed as an improvement to the standard procedure (2, 5, 6). Filtration shows an additional advantage in that apyrase is eliminated from the medium prior to the extraction of bacterial ATP. However, filtration efficiency is quite critical and depends, beside the choice of the filter itself, on parameters as temperature, dilution of the milk, addition of detergent, etc. The use of positively charged filters with a diameter of 47 mm has been described by Griffiths (5) and by Waes et al. (6). These filters allow the filtration of 5 ml of milk. However, in order to miniaturize the technique we had to select filters with better hydrodynamic properties. These filters were specially adapted for us by Hamilton (Switzerland) and allow the filtration of 0.5 ml of milk through a 8 mm diameter disk. The use of such minifilters, which fit inside regular luminescence vials, avoids the need for an extraction step in a separate vial and allows the control of residual ATP before the extraction of bacterial ATP (see standard procedure). Furthermore, working with small size filters significantly reduces the volumes of reagents needed for extraction and filtration. Incubation of the milk samples with trypsine or lipase did not improve filtrability to a significant extent. Repeated washing of the filters with buffer, on the contrary, is most important for the elimination of residual ATP. The efficiency of our filters was checked for both the retention of bacteria and for the elimination of proteins, namely casein. Protein analysis showed that less than 10% of the proteins remained on the filters while over 95% of the bacteria were retained as indicated by plate count before and after filtration. Addition of luciferin/luciferase reagent prior to the extraction of bacterial ATP offers the advantage that residual ATP can be controlled. This implies, however, that the extraction occurs in the presence of lecithin added for the protection of luciferase; extraction is therefore somewhat slower and extractant concentration has to be increased.

Using the procedure described above, over 150 milk samples were analyzed whose bacterial contents ranged from 15,000 to 1.5 million bacteria/ml. The correlation coefficient between CFU's (colony forming units) and luminescence was 0.80. This correlation coefficient was found to be lower when the figures for residual ATP were not taken into account. No significant effect was observed on the filtration properties of milk due to the number of somatic cells (0.7×10^5 - 1.5×10^6 cells/ml), fat content (3.3 - 6.7%), protein content (2.8 - 3.5%) or aging for up to three days. The standard deviation of repeated measurements was found to be 10% at the level of 10^5 bacteria/ml (see Table 1). This was further checked by showing that the addition of 20,000 bacteria to such samples could easily be detected.

Further standardization of our procedure and a more stringent control of temperature during filtration would probably still improve results, but the possibility now exists to detect 10-20,000 bact./ml in raw milk with the procedure described in this paper.

REFERENCES

1.   D.P. Theron, B.A. Prior and P.M. Lategan, Determination of bacterial ATP levels in raw milk: selectivity of non-bacterial ATP hydrolysis. J. Food Prot. 49, 4-7 (1986).

2.   J.A.J. Webster, M.S. Hall, C.N. Rich, S.E. Gilliland, S.R. Ford and F.R. Leach, Improved sensitivity of the bioluminescent determination of numbers of bacteria in milk samples. J. Food Prot. 51, 949-954 (1988).

3.		E. Schram and A. Weyens-van Witzenburg, Control of experimental factors involved in luminescent ATP assays. In <u>ATP Luminescence, Rapid methods in Microbiology</u>, P.E. Stanley,		B.J. McCarthy and R. Smither (Eds), Blackwell, Oxford, p. 37-43 (1989).

4.		E. Schram and A. Weyens-van Witzenburg, Improved ATP methodology for biomass assays. <u>J. Biolumin. Chemilumin.</u> 4, 390-398 (1989).

5.		M.W. Griffiths, Microbial estimation in dairy products. In <u>ATP Luminescence, Rapid Methods in Microbiolgy</u>, P.E. Stanley, B.J. McCarthy and R. Smither (Eds), Blackwell, Oxford, p. 167-173 (1989).

6.		G. Waes, J. Van Crombrugge and W. Reybroeck, The ATP-F test for estimation of the bacteriological quality of raw milk. In <u>Modern Microbiological Methods for Dairy Products</u>, IDF Special Issue 8901, IDF, Brussels, p. 279-286 (1989).

ACKNOWLEDGEMENTS

The authors wish to thank Mr. G. Mary and Mr. Y. Wouters of the "Comité Provincial de la Qualité du Lait et de la Crème de la Province de Hainaut" in Soignies (Belgium) for their kind help in supplying the milk samples and data concerning CFU's, protein and fat content, and number of somatic cells.

	Pos. 1 (cpm)	Pos. 2 (cpm)	Pos. 3 (cpm)	Pos. 4 (cpm)	Pos. 5 (cpm)	Pos. 6 (cpm)
Res. ATP (a)	28901	27214	34732	26620	31759	28153
Bact. ATP (b)	45683 47638 48154	46391 46747 47249	58888 59056 61260	45190 46967 48965	53531 55702 55045	41310 42650 41993
Int. stand.(c)	842514	844899	895045	882452	800390	727889
ATP conc. $(10^{-10}$ M) (d)	1.44	1.39	1.77	1.52	1.79	1.66
Bact./ml (10^3) (e)	90	87 mean = 100	111	95 s.d. = 10.1 %	112	104

Table 1 - Reproducibility test on milk containing 10^5 bacteria/ml

d = (b-a)/(d-c) x concentration factor

e = d x 10^2 / 1.6

Preincubation (40°C):
    0.5 ml milk + 0.5 ml somatic
    extractant and apyrase

Filtration (40°)

Transfer of filters and addition of
    luciferin/luciferase reagent

Counting in LUMICON luminometer: (20°C)
    a. residual ATP
    b. + bacterial extractant
    c. + standard ATP

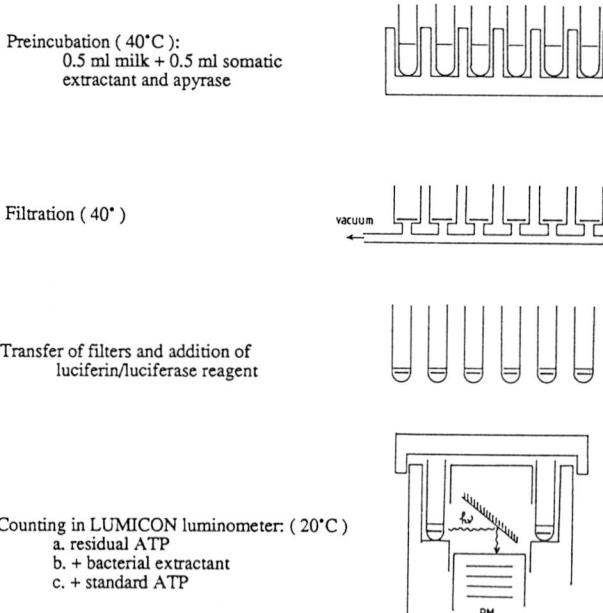

Fig. 1 - Standard procedure for the bacteriological assay of milk

BIOLUMINESCENT ASSAY
FOR CHOLINE-CONTAINING PHOSPHOLIPIDS AND LUCIFERIN

A.F.Dukhovich, N.Yu.Philippova and N.N.Ugarova

Chemistry Department, Lomonosov Moscow State University, 119899, Moscow, USSR

INTRODUCTION

Recent years, lipids have been shown to play an important role in the regulation of bioluminescent systems [1, 2]. Fundamental studies of the lipid dependence of luciferin--luciferase firefly system afforded the application of bioluminescent assay in some novel domains. This work is devoted to new bioluminescent methods and preparations for assaying phospholipids and luciferin and to their possible applications in various fields of microassay. Highly sensitive and rapid procedures for assaying phospholipids in biological samples and foods are extremely necessary. Most of the known physico-chemical methods for assaying phospholipids are laborious, rather time-consuming and costly. The known enzymatic methods for assaying phospholipids [3] take 1.5-2 h and are based on a few enzymatic reactions, which lowers the reproducibility of the method and rises its costs. So, the application, in phospholipid assays, of bioluminescent systems featuring the high sensitivity, simplicity and cheapness of luminescence equipment is of great interest. The importance of production of new bioluminescent, highly sensitive and inexpensive preparations for luciferin assays depends on possible application of various luciferin derivatives in enzyme immunoassay and in genetic engineering studies [4].

MATERIALS AND METHODS

Reagents: firefly *Luciola mingrelica* luciferase (Immunobiotechnology Institute, U.S.S.R); firefly *Photinus pyralis* luciferase (Sigma, USA); luciferin synthesized by Dr. A.A.Shiogolev (Moscow State University, U.S.S.R.); phosphatidyl choline (Kharkov plant for bacterial preparations, U.S.S.R.); sphingomyelin and lysophosphatidyl choline (Sigma, USA). Bioluminescence intensity was measured on a LKB-1250 and 1251 luminometers (Vallac, Finland). Assay of choline-containing phospholipids. Place 1 ml of detergent mixture + 0.1 ml 0.1 mol/l magnesium sulphate + 0.15 ml 0.01 mol/l ATP into a luminometer cuvette. Add 10 μl luciferase solution and incubate the mixture for definite period of time. Add 0.15 ml 1 mmol/l luciferin and measure bioluminescence intensity (I_0). Repeat the experiment but, before adding luciferase into

the reaction mixture, add 10 μl choline-containing phospholipids or their mixture and measure bioluminescence
intensity (I) to obtain the dependence of the sample bioluminescence intensity (I_s = I - I_o) at various concentrations of phospholipids. Define the unknown concentration of choline-containing phospholipids in the sample
either from the calibration plot or by the Internal Standard method. Repeat all measurements a few times and
define the concentration of choline-containing phospholipid from the mean values with an error no greater than
5%. Luciferase reagent and luciferin assay. Extract
luciferase from firefly lanterns [5] and purify the
former from luciferin admixture by gel filtration. After
the addition of stabilizing agents luciferase is liophi-
lized. To assay luciferin, place 0.78 ml 0.05 mol/l Tris-
-acetate buffer (pH 7.8) + 2 mmol/l EDTA + 0.1 ml 0.1
mol/l magnesium sulphate + 0.1 ml 0.01 mol/l ATP + 20 μl
reagent into a cuvette to define the background intensity
(I_o). Add 10 μl luciferin and measure the light intensity
of the sample (I). Obtain the bioluminescence intensity
values (I_s = I - I_o) at various luciferin concentrations.
The luciferin in the unknown sample was quantified either
from the calibration plot, or by the Internal Standard
method.

RESULTS AND DISCUSSION

We showed that under definite conditions in luciferase-
-detergent-phospholipid system, enzyme is reactivated and
stabilized only in the presence of choline-containing
phospholipids: lysophosphatidyl choline, phosphatidyl
choline and sphingomyelin. Phosphatidyl inosite, phosphatidyl glycerol, phosphatidyl ethanolamine, cardioli-
pine and lysophosphatidyl ethanolamine almost do not
reactivate luciferase. We revealed a quantitative dependence of the degree of firefly luciferase reactivation on
concentration of the added choline-containing phospholipid. When phospholipid was added into the reaction mixture
before introduction of luciferase, the observed light
intensity was likewise proportional to phospholipid
concentration. We assumed namely these data as a basis of
the above bioluminescent assay of choline-containing
phospholipids. The calibration plot for quantifying phosphatidyl choline (Fig.1) shows the linear dependence
between the light intensity and phospholipid concentration in a wide concentration range (0.1-70 μg/ml phospholipid). The detection limit for phosphatidyl choline is
0.1 μg/ml. Compared to the conventional enzymatic methods
[3], the bioluminescent assay of phospholipids is 50
times as sensitive, which permits one to work with diluted
samples and thus almost obviates the effect of admixtures
which can be present in the test samples. In addition,
the assay procedure is much simplified due to a reduced
number of steps. This rises its significance and reduces
the assay time a few times.

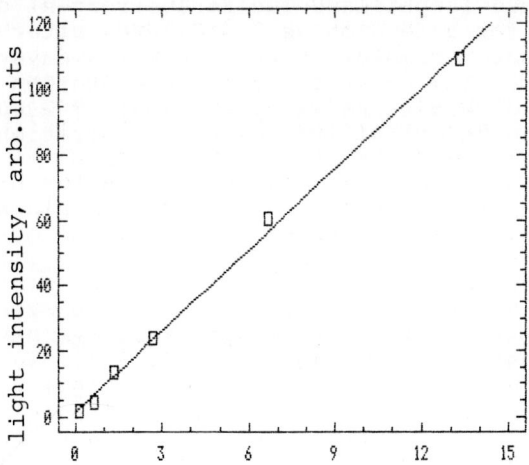

[phosphatidyl choline], µg/ml

Fig.1. Calibration plot for phasphatidyl choline assay

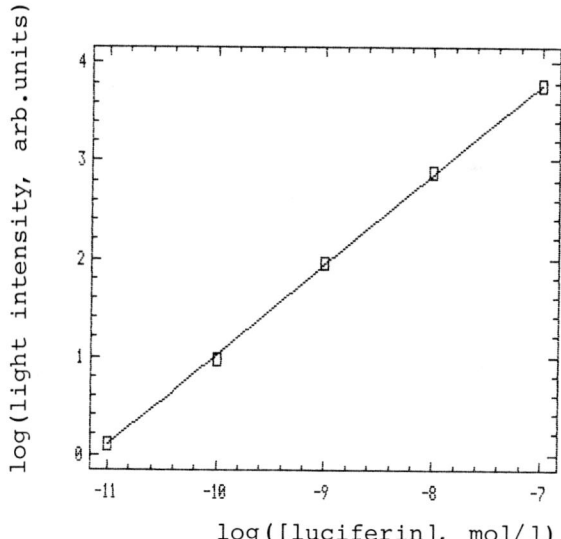

log([luciferin], mol/l)

Fig.2. Calibration plot for luciferin assay

To obtain luciferase preparation for luciferin assay, we
also took into account the lipid-dependence of bioluminescence firefly luciferin-luciferase system. To extract
the enzyme, we used the method of differential centrifugation with stabilizing agents [5]. Optimization of gel
filtration conditions helped to lower luciferin admixture
content more than 10.000 times by one purification step.

The obtained lyophilized luciferase reagent does not lose
activity for a few months (+4°C) and, after reconstitution with water remains active for a few days even at room
temperature. The activity of the preparation upon its use
in standard conditions was approximately 2,000 mV (LKB-
-1251 luminometer) at luciferin concentration 0.01 mmol/l
with linearity in the range 0.01 nmol/l-1 µmol/l (Fig.2).
Further studies of lipid-bioluminescence dependence of
firefly luciferase helps to expand possible applications
of the bioluminescent microassay.

REFERENCES
1. A.F.Dukhovich, N.N.Ugarova and I.V.Berezin, Firefly
   luciferase as a protein-lipid complex. Dokl.Acad.Nauk
   SSSR, 298, 231-233 (1986).
2. A.F.Dukhovich, N.Yu.Philippova, A.I.Yefimov, N.N.Uga-
   rova and I.V.Berezin, Choline-containing phospholipids
   as specific activators and stabilizers of firefly luciferase. Dokl.Acad.Nauk SSSR, 298, 1257-1260 (1988).
3. H.U.Bergmeyer, Methods in Enzymic Analysis, Verlag
   Chemie Weinheim; Academic pree, N.Y., (1974).
4. W.Miska and R.Geiger, Synthesis and characterization
   of luciferin derivatives for use in bioluminescence
   enhanced enzyme immunoassays. J.Clin.Chem.Clin.Bio-
   chem. 25, 23-30 (1987).
5. N.Yu.Filippova, A.F.Dukhovich and N.N.Ugarova, New
   approaches to the preparation and application of
   firefly luciferase. J.Biolum.Chemilum., 4, 419-422
   (1989).

BIOLUMINESCENT ASSAY OF β-GALACTOSIDASE USING D-LUCIFERIN-o-β-GALACTOSIDE

N.N.Ugarova, Ya.V.Vozny, G.D.Kutuzova and E.I.Dementieva

Chemistry Department, Lomonosov Moscow University Moscow 119899 USSR

INTRODUCTION

DNA probes are widely applied in molecular biology and genetic engineering. Among non-radioactive methods of detection, the most promising are the luminescent methods for detection of DNA probes with the use of enzymes.These methods are almost as sensitive as the radioactive methods. Recent years, the methods for detection of DNA probes were offered using the coupled reactions involving firefly luciferase (1, 2). D-luciferin derivatives were described both in -COOH and -OH groups, which, under the action of specific enzymes (alkaline phosphatase, carboxypeptidase, arylsulfatase, etc.), form free luciferin detectable by firefly luciferase. We obtained a new bioluminogenic substrate for β-galactosidase, β-D-galactopyranoside D-luciferin and developed the method for detection of the enzyme, using firefly luciferase. The bioluminescence detection limit for β-galactosidase was 0.25 fmol/l (0.02 amol).

MATERIALS AND METHODS

Reagents. Alkaline phosphatase from marine bacteria and β-galactosidase from *E.coli* were used. Firefly *Luciola mingrelica* luciferase purified by gel filtration and ion exchange chromatography had a specific activity $1\cdot10^8$ - $-1\cdot10^9$ arb.u/mg (1 arb.u corresponds to $1\cdot10^9$ quanta/s). D-luciferin phosphate was courteously presented by Prof. Geiger R.; D-luciferin β-galactoside and luciferin were both synthesized in our laboratory and quantified spectrophotometrically ($\mathcal{E}_{327}$ = 18 000 $mol^{-1}.l.cm^{-1}$). Bioluminescence was measured on a LKB-1250 luminometer.
Assay protocol. The activity of alkaline phosphatase was measured by spectrophotometry, fluorimetry and bioluminometry usiny 0.05 mol/l diethanol amine buffer, pH 9.8, containing 1 mmol/l $MgSO_4$. Spectrophotometry. To 2 ml 1.0 mmol/l p-nitrophenyl phosphate in the buffer, there was added 2-20 µl alkaline phosphatase. The mixture was incubated at 37°C for 100 min and A_{405} was measured.
Fluorimetry. To 2 ml o.1 mmol/l 4-methylumbelliferyl phosphate in the buffer, there was added 2-20 µl alkaline phosphatase. The mixture was incubated at 37°C for 100min

and the fluorescence was measured at 450 nm (excitation at 360 nm). <u>Bioluminometry</u>. To 1 ml 10 µmol/l luciferyl phosphate in the buffer, there was added 2-20 µl alkaline phosphatase. The mixture was incubated at 37°C for 100min. Luciferin was quantified before and after incubation in the presence and absence of phosphatase by bioluminescent method using firefly luciferase. β-Galactosidase activity was likewise measured by the three methods using 0.1mol/l phosphate buffer, pH 7.3, containing 0.1 mol/l NaCl. <u>Spectrophotometry</u>. To 2 ml 1 mmol/l o-nitrophenyl galactoside in the buffer, there was added 2-20 µl β-galactosidase. The mixture was incubated at 37°C for 100 min and A_{420} was measured. <u>Fluorimetry</u>. To 2 ml 0.4 mmol/l 4-umbelliferyl galactoside in the buffer, there was added 2-20 µl β-galactosidase. The mixture was incubated at 37°C for 100 min and fluorescence intensity was measured at 450 nm (excitation at 360 nm). <u>Bioluminometry</u>. To 1 ml 0.8 mmol/l D-luciferin β-galactoside, there was added 2-20 µl β-galactosidase. The mixture was incubated at 37°C for 100 min and luciferin in solution was quantified before and after incubation (as well as before and after incubation in the enzyme absence) by bioluminescent method using firefly luciferase.

<u>Bioluminometry of luciferin</u>. Tris-acetate buffer (0.7 ml, 0.1 mol/l), pH 7.8, containing 2 mmol/l EDTA + 10 mmol/l $MgSO_4$ + 0.3 ml 3.3 mmol/l ATP in the same buffer + 10 µl luciferase was placed into a cuvette of luminometer. The background light (I_o) was recorded. Luciferin (20-50 µl) of unknown concentration was added and luminescence intensity (I_1) measured. Then, 10 µl luciferin solution of known concentration ("internal standard") was added and luminescence intensity (I_2) measured. At luciferin concentration in the test solution less than 0.1 µmol/l, a steady light was observed for 2-3 min. The unknown luciferin concentration was calculated by the formula:

$$[LH_2] = \frac{[LH_2]_{st} \ (I_1 - I_o)}{(I_2 - I_1)} \times \frac{N_1}{N_2}$$

where $[LH_2]_{st}$ is the concentration of the standard luciferin solution; N_1 is the dilution of luciferin solution of unknown concentration in the cuvette; and N_2 is the dilution of the standard solution.

RESULTS AND DISCUSSION

Bioluminescent assay of luciferin is simple and highly sensitive. Using *Luciola mingrelica* luciferase not containing luciferin admixture (3), we assayed luciferin with the detection limit 0.8 fmole. This is why the bioluminescent substrates, firefly luciferin derivatives, become increasingly applied for detection of marker enzymes.

Application of luciferyl phosphate for detection of alkaline phosphatase has been described (2). We synthesi-

zed a new luciferin derivative, luciferin galactoside, a
specific substrate for β-galactosidase, which has proved
itself to be a good marker enzyme, and compared the
sensitivity of bioluminescent assay of two marker enzymes,
alkaline phosphatase and β-galactosidase, using firefly
luciferase. For comparison, the calibration curves were
constructed to assay the same enzyme preparations spectrophotometrically and fluorimetrically. Table shows that
the bioluminescent assay of alkaline phosphatase is
little more sensitive than spectrophotometry. This is due

Table. Comparison of various methods for detection of
alkaline phosphatase and β-galactosidase

Enzyme	Assay*	Substrate	Detection limit, fmol/l
Alkaline phosphatase	Sp	p-Nitrophenyl phosphate	75
	Fl	4-Methyl-umbelli-feryl phosphate	75
	Bl	D-Luciferin--o-phosphate	30
β-Galactosidase	Sp	o-Nitrophenyl galactoside	65
	Fl	4-Methyl-umbelli-feryl galactoside	0.65
	Bl	D-Luciferin β-D-galactoside	0.25

*
Sp is for spectrophotometry; Fl is for fluorimetry;
Bl is for bioluminometry

to the fact that the initial luciferyl phosphate
preparation contained an appreciable amount of free
luciferin (about 4%). During incubation at 37°C, the
substrate was further spontaneously hydrolyzed. The
substrate instability created a notable background which
was a hindrance for achieving a high sensitivity in
detection of alkaline phosphatase. Luciferyl phosphate
decomposed in solution during storage at +4°C. In 10 days,
in 10 mmol/l luciferyl phosphate, the amount of free
luciferin increased from 4 to 13%. Thus, we see that
luciferyl phosphate has notable drawbacks confining its
application as substrate for detecting microquantities of
alkaline phosphatase.
Unlike luciferyl phosphate, luciferin β-galactoside is
very stable, does not decompose in solution during

storage and incubation at 37°C, contains minute admixture of free luciferin (0.02-0.1%) and is easy to purify by HPLC chromatography. Luciferin galactoside at a concentration higher than 1 mmol/l samewhat inhibited β-galactosidase. The optimal concentration for assay was 0.8 mmol/l. On the other hand, upon luciferase-assisted detection of luciferin, the luciferin galactoside was present at a concentration of 10-20 µmol/l and did not inhibit firefly luciferase.

Luciferin galactoside is a good substrate for β-galactosidase (k_{cat} = 2.6·10^3 s^{-1} and K_m = 0.43 mmol/l; (for comparison, K_m = 0.11 mmol/l for the routine susbtsrate o-nitrophenyl galactoside). The detection limit of β-galactosidase was 0.25 fmol/l. Hence, the bioluminescent method of detection is a few times as sensitive as fluorimetry (Table).

Thus, using a new bioluminogenic substrate, luciferin galactoside, we succeeded in sensitizing the marker enzyme detection more than a 100 times (30 fmol/l for alkaline phosphatase compared to 0.25 fmol/l for β-galactosidase).

REFERENCES

1. W.Miska and R.Geiger, Synthesis and characterization of luciferin derivatives for use in bioluminescence enhanced enzyme immunoassays, new ultrasensitive detection systems for enzyme immunoassays. I. J.Clin. Chem.Clin.Biochem. 25, 23-30 (1987).

2. R.Geiger and W.Miska, Bioluminescence detection systems for enzyme immunoassay, II. J.Clin. Chem.Clin. Biochem. 25, 31-39 (1987).

3. E.I.Dementyeva, G.D.Kutuzova and N.N.Ugarova, Biochemical properties and stability of homogeneous firefly luciferase. Vestnik Moscow Univer. 30, Ser. 2, 601-606 (1989).

BIOLUMINESCENCE ASSAY FOR PSYCHROTROPH PROTEASES

M.T. Rowe,[1,2] J. Pearce,[1,2] L. Crone,[1] M. Sully[3] and I. Johnson[3]

[1]The Queen's University of Belfast, [2]Department of Agriculture
for Northern Ireland, Newforge Lane, Belfast BT9 5PX and
[3]Biotrace Ltd., The Innovation Centre, Mid Glamorgan Science
Park, Bridgend CF31 3NA, Wales

INTRODUCTION

Refrigeration of raw milk supplies create conditions which are
selective for the growth of psychrotrophic bacteria.
Psychrotrophs, particularly strains of *Pseudomonas* spp., elaborate
heat-stable extracellular proteases which can survive
pasteurization and UHT heat treatments (1) even though the
producing organisms are heat-sensitive. These enzymes are of
importance to the dairy industry because they have been shown to
cause a reduction in cheese yield (2) and gelation in UHT milk (3).
Rapid methods of bacterial enumeration are of limited value because
of the lack of correlation between bacterial numbers and level of
enzyme (4). The dairy industry has need therefore for a rapid,
sensitive and simple protease assay which could enable laboratory
staff at creameries to detect incipient spoilage by psychrotroph
proteases. The work reported here describes the development of such
an assay based on the use of luciferase as a protease substrate.
The resulting reduction in light output, after the remaining
components of the bioluminescence reaction are added in excess, is
used as an index of protease activity.

MATERIALS AND METHODS

Reagents: All reagents and materials are available from Biotrace
Ltd. as part of the Bioluminescence Protease Assay Kit.

Assay Protocol: Commercial protease originating from *Streptomyces
griseus* (Sigma P5147) was dissolved in UHT milk previously diluted
1:10 with Dilution Buffer. A range of protease concentrations were
incubated separately at 25° C for 5 minutes in a reaction mixture
(1 ml total volume) comprising; 100 l protease solution; 200 l
Enzyme-L and co-factors and 400 l of Dilution Buffer. After this
incubation 300 l of Substrate-L was added and the resulting light
output (integrated over 10 seconds) was measured in an LKB Wallac
luminometer (model 1250).

Milk Trial: Fifteen individual raw cow's milk samples were
obtained from milk recorder jars. Samples were taken on arrival at
the laboratory and after 2 and 4 days incubation at 7° C and 60 rpm
in an orbital incubator for the performance of a psychrotroph count

(Plate count agar; 6.5° C/10 days) and a protease assay. Prior to assay only the milk samples were subjected to laboratory pasteurization (63.5° C/35 min) and diluted 1:10 with Dilution Buffer. This was used to replace the commercial protease solutions in the assay.

Inhibitor: A number of protease inhibitors including soyabean trypsin inhibitor and phenylmethyl sulphonyl fluoride were screened to determine if they inhibited the luciferase/psychrotroph protease reaction. Each inhibitor was added at various concentrations to assay reaction mixtures containing as a psychrotroph protease source a stationary phase culture in UHT milk of a *Pseudomonas fluorescens* strain with the genetic competence to produce protease. One inhibitor, Milk Protease Deactivator (MPD) was further tested for its ability to inhibit the luciferase/intrinsic protease reaction by using as a source of intrinsic protease 4 samples of late lactation milk.

Luciferase Susceptibility to Different Pseudomonad Proteases: Stationary phase cultures in UHT milk of 4 strains of *Ps. fluorescens*, known to produce proteases, were diluted with uninoculated UHT milk previously diluted 1:10 with Dilution Buffer to give a range of protease concentrations. Each protease concentration was subjected to a protease assay and the slopes of the resulting regression lines compared statistically.

RESULTS AND DISCUSSION
A highly significant linear relationship ($r = 0.91$; $P<0.001$) was found between commercial enzyme concentration and light output (Fig. 1) over the protease concentration range 0 to 0.05 units protease/ml.

In order to evaluate the assay further, raw milk samples were subjected to a psychrotroph count and, after laboratory pasteurization, a protease assay at intervals during refrigerated storage (7° C and 60 rpm) in an orbital incubator. As shown in Fig. 2 the following trends were seen. Initially at the beginning of the storage period i.e. when the psychrotroph count was approximately 10^4 cfu/ml a number of samples demonstrated comparatively high protease levels. This was followed by a phase of declining protease activity reaching a minimum at approximately 10^5 cfu/ml and thereafter the activity increased with increasing count. It was considered that the high initial protease activity was due to the inability of the laboratory pasteurization treatment to inactivate completely the intrinsic protease which is present in raw bovine milk. The subsequent decline in protease activity was assumed to be due to gradual inactivation of this intrinsic protease, perhaps by autolysis. The final rise in protease activity is consistent with synthesis and secretion of the enzyme by the psychrotrophic microflora present.

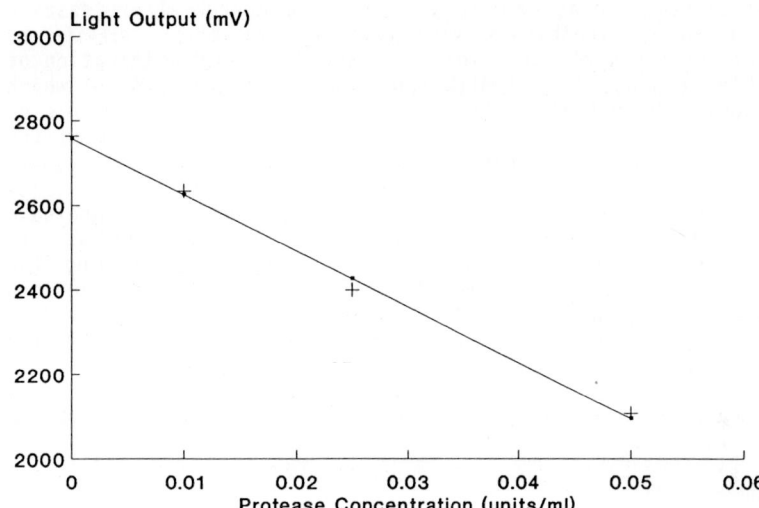

Fig. 1 Relationship between commercial protease concentration and light output. Each point represents the mean of 10 replicate values.

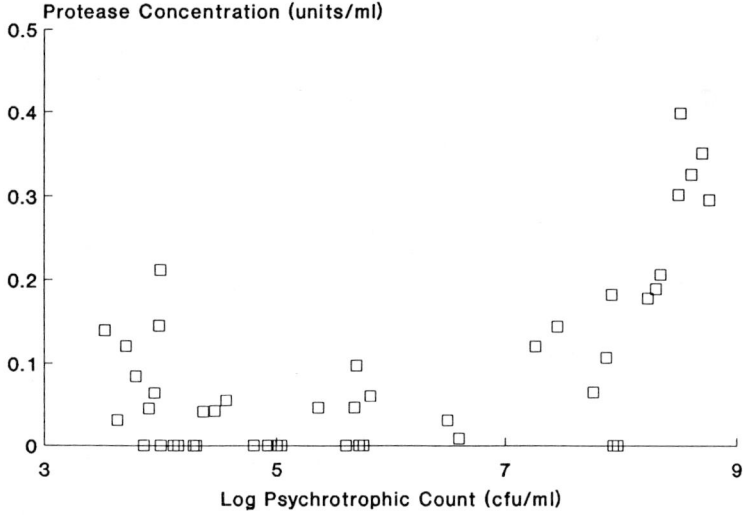

Fig. 2 Relationship between $\log_{10}$ psychrotrophic count (on raw milk) and protease concentration (after laboratory pasteurization) for 15 milk samples during storage at 7° C.

In order to make the assay more selective for microbial proteases a number of enzyme inhibitors were screened for their ability to inhibit intrinsic protease. This resulted in the identification of a suitable reagent, termed Milk Protease Deactivator (MPD), which was found to inhibit the intrinsic protease present in samples of late lactation milk by between 100 and 63% without affecting the psychrotroph protease/luciferase reaction.

Finally, the susceptibility of luciferase, used as a protease substrate, to proteases from 4 strains of *Ps. fluorescens* was tested. The slopes of the 4 regression lines representing the relationship between protease activity and light output for the individual strains were found to be similar (P>0.05) demonstrating that the susceptibility of the luciferase as a substrate for psychrotroph protease was constant.

Commercially produced assay kits containing the reagents described are now available.

ACKNOWLEDGEMENTS
The authors would like to thank UKDIRPC and the DTI for funding and Dr D. Kilpatrick, Biometrics Division, The Queen's University of Belfast for the statistical analyses.

REFERENCES
1.   B. Law, Review of the progress of Dairy Science: Enzymes of psychrotrophic bacteria and their effects on milk and milk products. J. Dairy Sci. 46, 573-588 (1979).
2.   C.L. Hicks, C. Onuorah, J. O'Leary and B.E. Langois, Effect of milk quality and low temperature storage on cheese yield - A summation. J. Dairy Sci. 69, 649-657 (1986).
3.   A. Gilmour and M.T. Rowe, Microorganisms associated with milk. In Dairy Microbiology, Vol. 1, R.K. Robinson (Ed), Elsevier Applied Science, London, p. 37-75 (1990).
4.   M.T. Rowe and A. Gilmour, The present and future importance of psychrotrophic bacteria. Dairy Ind. Internat. 50, 14, 15, 17, 19 (1985).

RAPID HYGIENE MONITORING USING ATP BIOLUMINESCENCE

A.L. Kyriakides, S.M. Costello, G. Doyle[1],
M.C. Easter and I. Johnson[2]

Grand Metropolitan Foods Europe, South Ruislip, HA4 OHF, UK
[1]Express Foods Group, South Ruislip, HA4 OHF, UK
[2]Biotrace Limited, Bridgend, CF31 3NA, UK

INTRODUCTION

Hygiene monitoring of production areas in factories has traditionally relied on the enumeration of micro-organisms present on a surface using plate count techniques. The dependance of such techniques on microbial growth necessitates a 2-7 day wait for results making the exercise retrospective, a fact consistently criticised.

Hygiene monitoring, normally performed after cleaning, should ideally indicate the presence of micro-organisms and food debris both/either of which being present on an improperly cleaned surface.

Traditional plate count techniques are limited to the detection of microbial contaminants only, but the presence of ATP in both food debris and viable micro-organisms allow the dual detection of these sources of contamination using ATP bioluminescence. The greatest advantage conferred by the application of an ATP bioluminescence assay is in providing instant assessment of the hygiene status of production equipment. ATP bioluminescence has previously been demonstrated to be applicable for hygiene assessment of production areas (1,2,3), but the uptake of the technique has been very slow, primarily due to the high capital cost of laboratory based instrumentation (>£10k). The development of cheaper (£3-5k) instruments suitable for portable, on-site analysis and more cost effective, stable reagents has stimulated recent interest in such applications (4,5).

The aim of this evaluation was to assess the application of a commercially available hygiene monitoring kit and portable luminometer for rapid determination of surface hygiene at a cheese manufacturing site.

MATERIALS AND METHODS

Reagents and Instrumentation

ATP bioluminescence hygiene monitoring kits (Biotrace HM Kit) consisting of Enzyme HM (luciferin/luciferase), Swab Diluent XT (microbial extractant) and sterile swabs together with ATP free cuvettes and pipette tips were obtained from Biotrace Ltd., Bridgend, U.K. All reagents were reconstituted and stored in accordance with the manufacturers instructions. Adenosine triphosphate (disodium salt) was obtained from Sigma, UK and used to prepare a 200nmole/l standard in sterile water. An M3 Hygiene Monitor (Biotrace Ltd.) was used to measure the light output for the ATP assays.

Assay Protocol

ATP Bioluminescence Hygiene Monitoring
     One millilitre of Swab Diluent XT was dispensed into an ATP free cuvette and used to moisten a sterile swab prior to swabbing a 10 cm^2 area of the test surface. Any ATP (microbial or non-microbial) captured by the swab was released by placing the swab into the original Swab Diluent XT, mixing and leaving for one minute. 300 μl of this suspension was then pipetted into a clean cuvette followed by 100 μl Enzyme HM. The cuvette was inserted into the luminometer and light output (relative light units, RLU) was recorded over a ten second period.
     The cuvette was then removed, 10 μl (2pmole) of the ATP standard added, mixed and re-inserted into the luminometer and a second determination of light output recorded. This internal standardisation was performed to determine if the light signal was being quenched.
     Control assays were performed by following the procedure described above using fresh unused swabs. Blank readings of instrument light output were recorded by taking light readings in the absence of any reagents.

Traditional Method for Hygiene Assessment
     A sterile cotton swab (Sterilin, UK) was moistened with sterile thiosulphate Ringer solution (Unipath, UK) and used to swab a 10 cm^2 area of the test surface adjacent to the area sampled for the ATP bioluminescence assay. The swab was placed into a test tube containing 10 ml sterile thiosulphate Ringer solution and vortexed to remove adhering micro-organisms. 1 ml of this suspension was then used to prepare a pour plate using milk agar (Unipath, UK). Plates were incubated at 30°C for 3 days prior to enumeration of colonies.
     Production surfaces tested included milk silos, cheese vats, cheese moulds, conveyors and other process equipment. Areas were swabbed both before and after cleaning to provide a wide distribution of results and the visual appearance of the test surface was also noted (i.e. dirty or clean).

RESULTS AND DISCUSSION

A total of 179 production surfaces were swabbed by the two techniques during the evaluation (raw data available on application). Criteria for grading the results are shown in Table 1.

TABLE 1

GRADES	PASS	<-------- FAIL -------->		
	(-)	(+)	(+2)	(+3)
PLATE COUNT (cfu/swab)	< 100	100-999	1000-9999	> 10000
ATP ASSAY (RLU/swab)	< 27	28-49	50-149	> 150

The pass level for the ATP assay was calculated by the sum of the mean instrument light output and four times the mean control swab light output based on 20 control assays.
     The number of production areas failed by the plate count technique and ATP assay were 110 and 122 respectively.

Table 2 below shows the agreement found between the results of the two methods. Figures represent number of results in each category. Total agreement = 77%.

TABLE 2		ATP ASSAY	
		Pass	Fail
Plate Count	Pass	42	27
	Fail	15	95

The greater number of areas deemed unacceptable by the ATP assay probably reflects the ability to detect product debris present on a surface where microbial numbers were either very low or absent. This has been observed in other studies (3). As previously mentioned this represents a significant advantage over traditional plate count assessments since product debris, as well as indicating poor cleaning, provides a source of nutrients to micro-organisms enabling growth of low numbers to unacceptable levels in a short period of time. Detection of this debris can therefore allow remedial action to be taken directly, thereby preventing such scenarios.

The 15 areas failed by the plate count assay and passed by the ATP technique probably resulted from the presence of low numbers of stressed micro-organisms, sufficient to fail the plate count criteria (>100/swab) but containing insufficient ATP to be detected by the bioluminescence assay.

The overall percentage agreement between the two techniques of 77% (Table 2) represents an excellent relationship when consideration is given to the experimental variables in existence such as sub sample variation in the swab area and differences in experimental procedures and parameters assessed. A similar relationship is observed in Fig. 1, results being compared using grades described in Table 1.

Fig. 1 : Relationship between ATP Assay and Plate Counts for Hygiene Assessment

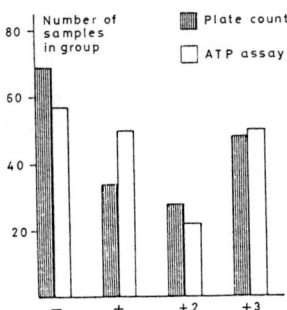

Acceptable relationships between ATP bioluminescence and plate counts for assessment of surface hygiene have been observed in other studies (2,3,5) and have also been reproduced in a similar evaluation by authors ALK, SMC and GD of the Lumac ATP Bioluminescence Hygiene Kit (Sonco Limited, Batley, UK). Results from this evaluation are available on request to ALK.

Comparison of results obtained by the two methods to visual assessment of surface cleanliness (Table 3) indicates a poor agreement between surfaces visibly clean and microbiologically clean, as determined by ATP and Plate Count assays, illustrating the necessity for a rapid indicator of hygiene.

TABLE 3: Visual Assessment of Hygiene vs ATP/Plate Count Assessment

Visibly clean surfaces		119
Clean by:	ATP Assay	42
	Plate Count	47

Visibly dirty surfaces		45
Dirty by:	ATP Assay	36
	Plate Count	33

An internal standard was included in our experiment because it is often necessary, when performing ATP assays, to compensate for any quenching (physical or chemical) of the light signal by components of the sample material. Our studies indicate that this was not necessary, since few of the samples were subject to these problems.

The running costs associated with ATP bioluminescence is often an important obstacle limiting its use and therefore a cost comparison was made between it and the plate count technique. The total labour and material cost of the ATP bioluminescence and plate count swabbing techniques have been calculated as £1.33 and £0.87 respectively. It is interesting to note that labour accounts for 70% of the total plate count cost whilst material represents 90% of the ATP bioluminescence costs. The obvious cost disadvantage of the ATP technique should be readily balanced by labour saving advantages and particularly by the generation of instant results.

In conclusion, the results from this evaluation have shown that ATP bioluminescence is clearly applicable for hygiene monitoring of production equipment where the generation of instant, meaningful results can at last provide factory personnel with information enabling direct action. The provision of cheap instrumentation (Biotrace M3 Hygiene Monitor), stable reagents (Biotrace HM Kit) and simple protocols are additional reasons why such systems should become commonplace in factory environments.

REFERENCES

1.   S. Shaw, Rapid Methods of Estimating Numbers of Organisms on Surface Swabs. In Rapid Microbiological Methods in the Food Industry. J.M. Woods and P.A. Gibbs (Eds), Leatherhead Food R.A. Tech. Note No. 4, p29-30 (1983).

2.   S. Barnett, Rapid Hygiene Screening. In Proceedings of the ATP Bioluminescence Brewing Workshop May 1988, Lumac bv, PO Box 31101, 6370 AC Landgraaf, Netherlands.

3.   W.J. Simpson, J.R.M. Hammond, P.A. Thurston and A.L. Kyriakides, Brewery Process Control and the Role of 'Instant' Microbiological Techniques. In Proceedings of the European Brewery Convention, Zurich 1989, IRL Press, Oxford, 663-674.

4.   P. Thompson, Rapid Hygiene Analysis using ATP Bioluminescence. European Food and Drink Review, Spring Issue 1989, 42-48.

5.   W.J. Simpson, Instant Assessment of Brewery Hygiene using ATP Bioluminescence. Brewers' Guardian 118, 20-22 (1989).

CHEMICAL AMPLIFICATION BY COUPLING TO ENERGY-PROVIDING REACTIONS

S.E. Brolin, P. Naeser[1] and G. Wettermark
Department of Medical Cell Biology and [1]Department of Ophthalmology,
Uppsala University, Uppsala, Sweden

Keywords: Bioluminescence analysis; coupled reactions;
enzymatic cycling

INTRODUCTION

In contrast to non-destructive photometric assays where the outgoing signal is a measure of concentration the luminescence intensity usually represents the rate of degradation of the substance being measured(1). Consequently this rate determines the fading of the emitted light and, neclecting other factors, high intensity implies a short duration of the emission. The total amount of light emitted i.e. the integral of intensity vs. time can be increased by arranging the analytical system where the concentration of the substance to be measured is reconstituted by a chemical cycle. The cycle may be used to give a chemical amplification, e.g. by storage and counting of the number of photons emitted. The intensity stays in principle constant and the number of cycles and thus the degree of amplification becomes determined by the collection of time.

LUMINESCENCE INTENSITY REPRESENTS THE DECAY OF A LIGHT INDUCER

The situation in bioluminescence analysis is entirely different from that experienced in other kinds of photometry, where the intensity is a direct measure of the concentration of the substance to be measured. Repeated read outs are here required for determining a possible degradation. In bioluminescence the intensity represents a rate and reflects directly the decomposition of a light inducting compound. The amount of a light inducer at the start of the luminescence is to be determined by the analysis.

The rate of decomposition of the light inducer and thus the intensity of the light may be varied by changing the conditions of the assay but the total amount of available light i.e. the total integral of intensity vs. time is in principle the same for a given amount of light inducing substance. By arranging the analytical conditions so that the luminescence reaction occurs with a high rate it is feasible to record the entire emission (=measure the total integral) as the light now is given off as an isolated flash of light.

RECONSTITUTION OF LIGHT INDUCER THROUGH ENZYMATIC CYCLING YIELDS A CONSTANT HIGH INTENSITY LIGHT AND CHEMICAL AMPLIFICATION

Chemical amplification is obtained when the number of emitted photons exceeds the efflux obtained during a straight complete decomposition of the light inducer. Such a situation is possible to create by ensuing that the light inducer is reconstituted in a cyclic process. Storage of photon pulses in an electronic memory may cover several cycles in order to enhance sensitivity. Moreover the cycling makes it possible to obtain a "true" steady light emission close to the intensity level reached in a flash, i.e. keeping the conversion rates of the same order as during "flash analysis". Reconstitution of the light inducer proceeds simultanously with the luminescence reaction in the same solution and supplies energy for continuous emission. The reconstitution implies that a product of lumine-scence, i.e. the inactivated state of the light inducer is recharged via another route and is brought back to provide new light inducer molecules. The light inducer as well as the product and the enzyme catalyzing the reconstitution become determinable. The constant light level obtained in cycling is entirely different from the durable light intensity obtained by slowing down the reaction rate. In this case due to the possibility of

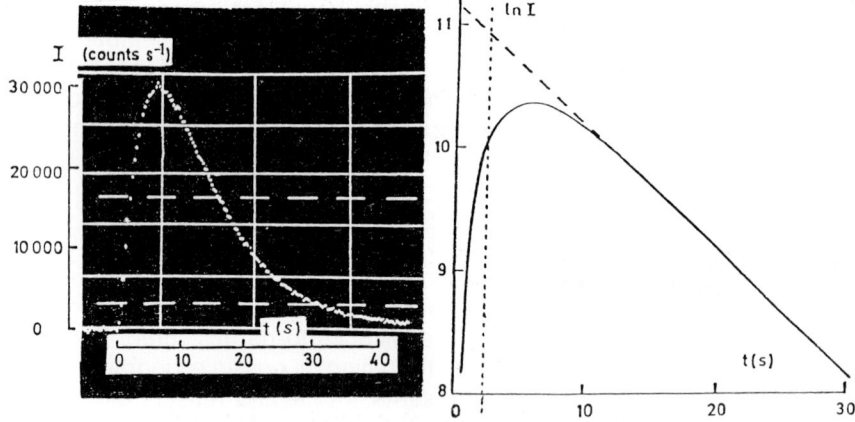

Fig.1 Light emission obtained by injecting 5 pmole NADH in a reaction solution in the abscence of sorbitol dehydrogenase. Recording of the emission takes place by photon counting. Each dot represents the number of photons collected during 0.3 s. The intensity recording is seen to the left and a logarithmic representation of the data to the right.

measuring even very weak emissions it may at instances be permissable to select such a low decomposition rate that only a small percentage of light inducing substance is decomposed during the time of analysis and measure a light intensity which stays essentially constant during the time of measurements. Using this technique a very small portion of the totally available light is recorded.

APPLICATION TO ANALYSIS OF NADH

The recycling principle discussed above is here applied to analysis of NADH using the bacterial luciferase system for the production of light and the sorbitol dehydrogenase reaction for the reconstitution. The reaction is thus

$$NAD^+ + sorbitol \xrightarrow{\text{sorbitol dehydrogenase}} NADH \qquad (1)$$

$$NADH + reactants \xrightarrow{\text{oxidoreductase - luciferase}} light + NAD^+ \qquad (2)$$

The rates of the enzym reactions employed in bioluminescence usually follow Michaelis-Menten kinetics and may be written as

$$-d[S]/dt = k_{ES}[E]_0 \, [S]/(K_m+[S]) \qquad (3)$$

Analytical conditions are usually arranged so that all other reactants except the cycled species are present in excess. NADH and NAD^+ are thus held at concentrations well below the respective K_m values of the two reactions. This means that equation (3) may be reduced to that of a first order reaction with rate constants k_1 and k_2 for reaction (1) and (2) respectively. With b as a constant the light intensity I becomes:

$$I = b \, k_2 \, [NADH] \qquad (4)$$

Without cycling (no sorbitol dehydrogenase present) the emission should essentially be an exponential decay

$$I = b \, k_2 \, [NADH]_0 \, e^{-k_2 t} \qquad (5)$$

which, except for the initial build-up of intensity, describes fairly well the experimental results, cf. Fig. 1. $[NADH]_0$ is the concentration to be determined through the analysis. The slope of the logarithmic plot yields $k_2 = 0.10 \text{ s}^{-1}$. Hence by integration of the emission one obtains b $[NADH]_0$ = 510 000 counts, which is the possible emission for a complete decomposition of the light inducer:

$$A_\infty = \int_0^\infty I\,dt = b\,[NADH]_0 = 510\ 000 \text{ counts.} \qquad (6)$$

Multiplying by the value of k_2 yields an intercept value of 10.85 in the logarithmic plot,cf. eq(5). Extrapolating to this value (dashed portion of the straight line) yields the start-up time for an instantaneous build-up of emission (dotted vertical line).

With cycling (sorbitol dehydrogenase present) a constant light intensity level, I_∞ , is obtained when steady state is reached. In the experiments shown in Fig. 2 I_∞ = 8 200 counts s^{-1}. The value 510 000 counts given by eq. (6) represents the emission during one cycle and may be used to estimate the number of cycles the system goes through, which is indicated on a separate scale below in Fig.2. It is seen that the recorded emission covers about 22 cycles.

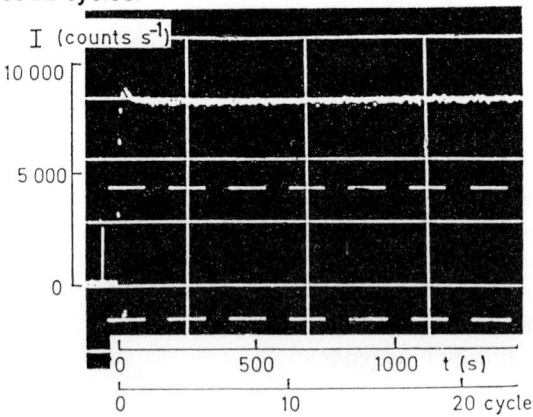

Fig.2 Light emission obtained by injecting 5 pmole NADH in a reaction solution identical with that of Fig.1, but where sorbitol dehydrogenase has been added. Same recording principle was used as in Fig.1, but collection was instead made for 3 s.

ACKNOWLEDGEMENTS
Supported by the Swedish Medical Research Council (12x-07476), the Swedish Diabetes Association, the Swedish Hoechst Diabetes Fund, the Nordic Insulin Foundation, The Family Ernfors Fund, and Clas Groschinsky's Memorial Fund.

REFERENCE
1.   S.E.Brolin, P. Naeser and G.Wettermark, Design of coupled reactions for simplification of bioluminescence analysis. J.Biolumin. Chemilumin. 4, 446-453 (1989)

INDUSTRIAL SEMINAR:
ELA TECHNOLOGIES, INC.

APPLICATIONS OF RECOMBINANT BIOLUMINESCENT PROTEINS IN DIAGNOSTIC ASSAYS

D.F. Smith[1,3], N.L. Stults[1], H. Rivera[1], W.D. Gehle[2], R.D. Cummings[3], and M.J. Cormier[3]

ELA Technologies, Inc.[1], Athens, GA 30604, USA, Dynatech Laboratories, Inc.[2], Chantilly, VA 22021, USA, and University of Georgia[3], Athens, GA 30602, USA

INTRODUCTION

Enzyme-linked immunosorbent assays (ELISAs) have been used extensively for the detection of antigens and antibodies in basic research and in assays that are considered diagnostic for both human and animal diseases. ELISAs have replaced many immunoradiometric assays and afford greater sensitivity because many molecules of a colored product can be generated from a single molecule of enzyme-labeled antibody. However, very low concentrations of chromophore result in only small changes in incidence light, and the sensitivity of ELISAs is therefore limited as the plate reader is required to measure small changes in relatively large numbers. On the other hand, samples generating large absorbance values must be appropriately diluted since absorbance measurements are linear over a limited range (2 to 3 decades) of chromophore concentration. The introduction of bioluminescent proteins as antibody labels in bioluminescent immunoassays or BLIAs will overcome these disadvantages of the ELISA by providing light energy as a signal which can be quantified by more sensitive instrumentation over 6 to 8 decades of bioluminescent protein concentration.

There are a variety of applications where greater sensitivity than is obtained by ELISA would be desirable. For example, the food producer, using even rapid and sensitive ELISA tests, must hold inventory for a 48 hour culturing period required to enrich and select for pathogens prior to testing. A sensitive assay that would reduce the time required for pathogen detection to 24 hours is considered an industry goal. The development of a bioluminescent immunoassay (BLIA) for salmonellae antigen and its comparison to ELISAs using colorimetric as well as chemiluminescent detection systems are described in this report.

MATERIAL AND METHODS

Bioluminescent proteins ELA Technologies, Inc. produces a recombinant form of the photoprotein Aequorin by overexpression in E. coli of the cDNA which was cloned from the jellyfish *Aequoria victoria* (1). Coelenterate luciferin was synthesized as previously described (2) or was obtained from Dr. Frank McCapra, The University of Sussex, Brighton, U.K. The recombinant apoAequorin was converted to **RecAcquorin** by addition of coelenterate luciferin as described previously (1). The biotinylated derivative of **RecAequorin** is a product of ELA Technologies, Inc., and this derivative produces light with essentially the same photon yield as the underivitized **RecAcquorin**. Buffered solutions of the biotinylated **RecAcquorin** lose very little activity when stored at 4° C for up to 30 days. These solutions are stable to freezing and thawing and to lyophilization.

Luminescence measurement Luminescence was measured in a Microtiter™ plate luminometer from Dynatech Laboratories, Inc., Chantilly, VA (DLI). This instrument has two operating modes, the Cycle Mode for measuring light from reactions that exhibit glow kinetics, and the Integrate Mode for reactions with flash kinetics. In the Cycle Mode, wells may be read up to 99 times with a variable pause between readings. Twenty 10 msec readings of each well are averaged, internally corrected for noise, drift and aging of the photomultiplier tube (PMT), and the corrected value which is proportional to the photons striking the PMT is reported as relative light units (RLU). In the Integrate Mode, a single well is moved under the PMT and reagent(s) are injected to initiate the luminescent reaction. The plate carrier shakes to ensure proper mixing, and as the reaction occurs, the flash of light is quantified. The peak of light, the time at which the peak occurred, a before-peak and after-peak integral, and a total integral are recorded as RLU after appropriate correction.

Assays An ELISA for detecting *Salmonella* antigen developed by Dynatech Laboratories Inc., the Q-TROL^{TM} Salmonella assay (3), was used to compare both colorimetric and chemiluminescent substrates for the enzymes horseradish peroxidase (HRP) and alkaline phosphatase (AP) with a BLIA using the bioluminescent label, biotinylated-**RecAequorin**. Monoclonal antibodies to salmonellae were absorbed to Microlite$^{®}$ 2 Removawell$^{®}$ (DLI) strips which were specially designed to be used as solid-phase supports for the absorption of antigens and antibodies in luminescent ELISAs (4). Samples (100 ul of boiled suspensions of *Salmonella typhimurium* for the positive control and *Proteus vulgaris* for the negative control) were prewarmed to 35° C, added to the wells and incubated for 25 min with gentle agitation. The wells were then washed 5 times with PBS (pH 7.4) containing 0.0025% to 0.05% Tween 20 to remove unreacted antigen. Prewarmed biotin-labeled antibody (100 ul) to salmonellae antigens (Kirkegaard and Perry Laboratories, Inc. (KPL), Gaithersburg, MD) were added to the wells and incubated at 35° C for 20 min, then washed five times to remove unreacted derivatized antibody. Unconjugated streptavidin (KPL), HRP-streptavidin conjugate (KPL), or AP-streptavidin conjugate (KPL) were added to the wells and incubated at 35° C for 10 min, then washed five times to remove unreacted conjugate before addition of substrate. For the BLIA the final wash contained 5mM EDTA.

For bioluminescent detection, biotinylated-**RecAequorin** in 0.01 mol/l Tris, 0.15 mol/l NaCl, 2 mmol/l EDTA, pH 8.0 containing 0.1 mg/ml bovine serum albumin was added to wells and incubated for 30 min at room temperature, then washed 5 times with the same buffer without BSA. The wells were placed in the luminometer and 100 ul of 0.1 mol/l CaCl$_2$ were injected to initiate the flash reaction. The emitted light was quantified as its full integral during 1 s in the Integrate Mode. The colorimetric substrate for the HRP was 3,3',5,5', tetramethylbenzidine (TMB) and for the AP was phenolphthalein monophosphate (PMP). Substrates were added and incubated at room temperature until color appeared in the negative control wells (10 min and 15 min, respectively). The reactions were stopped and read at 450 nm and 550 nm, respectively. The chemiluminescent substrate for HRP was enhanced luminol: 2.3 mM 3-aminophthalhydrazide, 0.9 mM 4-iodophenol (Aldrich Chem. Co.) and 0.0006% v/v H$_2$O$_2$ in 0.1 M Tris, pH 8.5. The chemiluminescent substrate for AP was the phosphorylated 1,2-dioxetane (AMPPD^{TM}, Tropix, Inc. Bedford, MA or Lumi-Phos^{TM}, Lumigen, Inc., Detroit, MI). After addition of substrate, the wells were placed in the luminometer and the emitted light was quantified in the Cycle Mode with measurements taken every 60 s during 20 min.

RESULTS AND DISCUSSION

With the Microtiter plate luminometer in the Integrate Mode, the **RecAequorin** flash reaction can be initiated by injecting CaCl$_2$. This protein can be detected in the attomol range, and the instrument response is linear between 10^{-18} and 10^{-11} mol of protein. The BLIA for salmonellae antigen was linear over several log units of antigen dilution. The detection limit for the BLIA was defined as the dilution of positive control antigen that produced a RLU value twice that of the negative control antigen. This definition was also used for comparing the detection limits of colorimetric assays (absorbance) and the chemiluminescent assays (RLU). As shown in Table 1 the most sensitive colorimetric assay was obtained using HRP with TMB as substrate. The sensitivity of the assay with AP and HRP was increased approximately 10-fold by using the chemiluminescent substrates AMPPD and enhanced luminol, respectively. The BLIA in the configuration used for these studies was 5-6 fold more sensitive than the chemiluminescent based detection systems and at least 50-fold more sensitive than the most sensitive colorimetric detection system (Table 1).

TABLE 1

RELATIVE SENSITIVITIES FOR SALMONELLAE DETECTION

ASSAY	*DILUTION AT DETECTION LIMIT*
Alkaline Phosphatase with PMP	16
Alkaline Phosphatase with AMPPD	250
Horseradish Peroxidase with TMB	25
Horseradish Peroxidase with Enhanced Luminol	250
BLIA with RecAequorin	1450

The chemiluminescent based assay systems required monitoring the individual reactions in the Cycle Mode of the instrument for a sufficiently long period of time to determine the maximum ratio between positive and negative control samples. For example, serial 10-fold dilutions of *Salmonella milwaukee* containing 2×10^7 colony forming units (CFU)/ml were assayed using HRP and enhanced luminol. As shown in Table 2, maximum ratios between the positive samples and negative control varied with the time at which light measurements were taken.

TABLE 2

RATIOS OF SAMPLE RLU TO NEGATIVE CONTROL RLU AT VARIOUS TIMES AFTER INITIATING ENHANCED LUMINOL REACTION

TIME (MINUTES)	*Salmonella milwaukee* CFU/ml		
	200	2000	20,000
1	2.0	5.8	2090
2	2.1	9.5	1160
4	2.6	25	
5	3.0	35	
7	3.7	36	690
9	3.9	29	
10	4.0	29	
11	4.0	25	
12	3.8	21	
13	3.5	15	125

The higher the concentration of antigen in the sample, the sooner the maximum was obtained. At low antigen levels, higher test sensitivity could be obtained upon longer incubation in the presence of substrate. The data reported for this assay were collected under maximal signal to background ratios. In contrast, the AP with phosphorylated 1,2-dioxetane, generated maximum ratios in less than 1 min after addition of substrate and these were maintained for 20 min. Thus, the incubation period for the AP substrate was not critical in determining test sensitivity when the measurement was based on ratios between test and negative control samples.

The BLIA for salmonellae antigen is sufficiently sensitive to be potentially useful in developing an assay that could detect salmonellae contamination in food within 24 hours of sampling. As shown in Fig. 1, the assay is linear over several log units of *Salmonella typhimurium* concentrations.

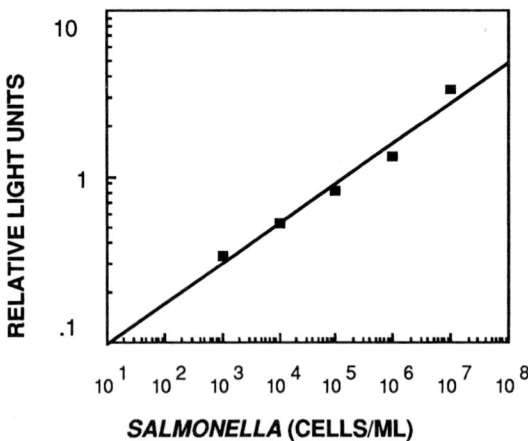

SALMONELLA (CELLS/ML)

Fig. 1   *Detection of Salmonella typhimurium using a bioluminescent immunoassay (BLIA).*
S. typhimurium was grown in M-Broth to a concentration of 10^8 cells/ml as determined by optical
density.   The culture was boiled, and 10-fold serial dilutions were assayed as described in
"Materials and Methods".

The limit of detection in this assay configuration for *Salmonella typhimurium* was between 10^3
and 10^4 cells/ml.   The BLIA, based on **RecAcquorin** detection, is simple to use as it only
requires the injection of $CaCl_2$ and is extremely rapid since the reactions in a 96 well microtiter
plate are complete and data can be recorded within 2 min.   The chemiluminescent and
colorimetric detection systems in the ELISAs require significantly longer incubation times after
addition of substrates to reach end points.   The total assay time from sample addition into wells
until data collection can be decreased to less than 90 min using a conjugate of **RecAcquorin** and
the polyclonal anti-salmonellae antibody.   Preliminary data indicate that this conjugate-based
assay is as sensitive as the streptavidin-biotin "sandwich" assay reported here.

ACKNOWLEDGEMENTS

This work was supported by Small Business Innovative Research Grants AI28628 to D.F.S. and
GM45093 to N.L.S. from the National Institutes of Health.

REFERENCES

1. Prasher, D., McCann, R.O. and Cormier, M.J. Cloning and expression of the cDNA for
Aequorin, a bioluminescent calcium-binding protein. *Biochem. Biophys. Res. Commun.*, 126,
1259-1268 (1985).

2. Inoue, S., Sugiura, S., Kakoi, H. and Hasizume, K., Goto, T. and Iio, H. Squid
Bioluminescence II. Isolation from *Watasenia scintillans* and synthesis of 2-(p-hydroxybenzyl)-6-
(p-hydroxyphenyl)-3,7-dihydroimidazo[1,2-a]pyrazin-3-one. *Chem. Lett.* 141-144 (1975) .

3. Flowers, R.S., Klatt, M.J., Keelan, S.L., Swaminathan, B., Gehle, W.D. and Chandonnet, H.E.
Fluorescent enzyme immunoassay for rapid screening of *Salmonella* in foods: Collaborative study.
J. Assoc. Off. Anal. Chem., 72, 318-325 (1989).

4. Gehle, W.D. and Lazar, B.S. An enhanced-response system for performing chemiluminescent
and bioluminescent tests. *Am. Biotech. Lab.*, 8, 42-45 (1990).

APPLICATIONS OF RECOMBINANT BIOLUMINESCENT PROTEINS AS PROBES FOR PROTEINS AND NUCLEIC ACIDS

N.L. Stults, N A. Stocks, [1]R.D. Cummings, [1]M.J. Cormier, and D.F. Smith

ELA Technologies, Inc., Athens, GA 30604, USA and [1]Dept. of Biochemistry, University of Georgia, Athens, GA 30602 USA

INTRODUCTION

We have investigated the application of the recombinant bioluminescent proteins, RecAequorin and RecLuciferase for the detection of protein and nucleic acid targets immobilized on membrane supports. RecApoAequorin is the recombinant product of the cloned gene for the protein component of aequorin obtained from the luminescent jellyfish *Aequoria victoria* (1). Native aequorin is a 20,000 dalton protein which emits blue light (λ_{max}=469 nm) in the presence of calcium ion. Each mol of aequorin contains one mol of coelenterate luciferin which is required for light production (2). Aequorin produces all of its light in a single step within 10 s and thus can be considered to catalyze a 'flash reaction.' RecApoAequorin is readily converted to RecAequorin in high yield upon incubation with synthetic coelenterate luciferin, dissolved oxygen, and a thiol agent. In addition, RecAequorin exhibits light producing properties which are essentially identical to native aequorin. Luciferase from the sea pansy, *Renilla reniformis*, is a 36,000 dalton protein which also catalyzes the oxidation of luciferin with concomitant sustained light production (λ_{max}=480 nm) (3). Luciferase has recently been cloned, and the recombinant protein, RecLuciferase exhibits properties similar to the native protein (4).

Since RecAequorin and RecLuciferase can be detected in the attomol range using photon counting luminometers, their use as biochemical tags should facilitate very sensitive detection of target molecules. Both recombinant bioluminescent proteins have been successfully derivatized and found to be stable when stored at 4°C or when frozen or lyophilized. We have examined the feasibility of using different derivatives in solid-phase assays including those conducted in microtiter wells and on membrane supports. Use of the biotinylated derivatives, in combination with streptavidin, have resulted in the detection of subnanogram quantities of biotinylated proteins and nucleic acid targets immobilized on Western and Southern blots using both polaroid instant and x-ray films. In addition, a goat anti-rabbit antibody-RecAequorin conjugate has proven useful for immunodetection of specific protein antigens. Advantages of bioluminescent detection are that it is sensitive, rapid, nonradioactive and it does not involve the use of toxic substrates. Furthermore, a permanent, non-fading record of the blots is obtained.

MATERIALS AND METHODS

Reagents: Human transferrin, Polaroid 612 ASA 20,000 instant film, and Kodak X-OMAT AR x-ray film were obtained from Sigma Chemical Co (St. Louis, MO). Rabbit anti-human transferrin was from Dakopatts A/S (Denmark). Nitrocellulose, gelatin, Tween-20, biotinylated-protein standards, goat anti-rabbit-alkaline phosphatase, biotinylated-goat anti-rabbit antibody, bromochloroindoyl phosphate (BCIP) and nitroblue tetrazolium (NBT) were obtained from Bio-Rad (Richmond, CA). Streptavidin was from Infergene (Benicia, CA). Nytran nylon membrane was from Schleicher and Schuell (Keene, NH). Biotinylated RecAequorin, biotinylated-RecLuciferase, synthetic coelenterate luciferin, and goat anti-rabbit IgG-RecAequorin conjugate were provided by ELA Technologies. The pTZR-Luc1 plasmid containing the 1.5 kB *Renilla* luciferase cDNA insert was provided by Dr. Walter Lorenz (Univ. of GA). Photoprobe Biotin was obtained from Vector Laboratories (Burlingame, CA).

Protein Detection: The sample of interest was applied to the membrane in a dot blot or by electrophoretic transfer following separation in SDS-PAGE. Nonspecific binding sites were blocked by incubating the membrane in 3% gelatin or 0.5% casein for 30 min. After exposing the blots to the appropriate primary and biotinylated-secondary antibody for 30 min they were sequentially incubated for 30 min with streptavidin (10 µg/ml) and biotinylated-RecAequorin (1 µg/ml). Alternatively, a goat anti-rabbit antibody-RecAequorin conjugate was substituted for the biotinylated-secondary antibody/streptavidin/biotinylated-RecAequorin detection system. All reagents were prepared in TBS/E/BSA (Tris-buffered saline, pH 8.0 containing 2 mmol/l EDTA and 0.1 mg/ml bovine serum albumin) or 3% gelatin in TBS/E. Washes between incubation steps were carried out with TBS/E. For the experiments conducted with biotinylated-alkaline phosphatase or goat anti-rabbit-alkaline phosphatase, the reagents were prepared in TBS containing 0.05% Tween-20. For development, the blots were rinsed with TBS and BCIP/NBT were added according to the vendor's instructions.

DNA detection: Linearized pTZR-Luc1 plasmid was immobilized onto nitrocellulose or nylon membranes using standard protocols for dot blotting or after Southern transfer from an agarose gel. The target DNA was probed with the *Renilla* luciferase cDNA fragment which was generated by restriction enzyme digestion with Sma I and EcoR I and biotinylated with Photoprobe Biotin according to the vendor's instructions. After incubation in prehybridization buffer (6X SSC, 0.5% SDS, 5X Denhardt's, 0.1 mg/ml salmon sperm) for 4-6 h at 52°C, the biotinylated-*Renilla* cDNA fragment was added in hybridization buffer (6X SSC, 0.5% SDS, 5X Denhardt's, 0.1 mg/ml salmon sperm, 10 mmol/l EDTA) at 100 ng/ml and incubated overnight at 52°C. Following hybridization, the membranes were washed two times with 2X SSC/0.1% SDS for 15 min at room temperature and then one time with 0.1X SSC/0.1% SDS for 1 h at 52°C. The membrane was then blocked as described above and incubated sequentially with streptavidin and biotinylated-RecAequorin or biotinylated-RecLuciferase.

Development of protein and DNA blots: The streptavidin or antibody detecting complex of RecAequorin or RecLuciferase was visualized by developing the blots against polaroid instant or x-ray film by addition of calcium or luciferin, respectively. This was accomplished by placing the membrane in sandwich assembly against the shutter of the Dynatech camera luminometer (Chantilly, VA) containing the instant film or directly against x-ray film. The sandwich consisted of a piece of plastic wrap to prevent wetting of the film, the moist membrane with the complex facing the film, and a moist piece of filter paper to prevent the blots from drying out. Light production was initiated by saturating a second piece of thicker, dry filter paper which lies on top of the sandwich with either 0.1 mol/l $CaCl_2$ or 1 µmol/l luciferin. In the case of RecAequorin, calcium was added with a syringe in the darkroom to allow immediate exposure of the film. With RecLuciferase, the membrane was dipped into a solution of luciferin on the bench, the sandwich assembled, and then placed against the film in the darkroom. The films were exposed for 10-15 s and 10 min using RecAequorin or RecLuciferase, respectively and then developed according to the manufacturer's instructions.

RESULTS AND DISCUSSION

In order to pursue the development of protein and nucleic acid detection on membranes using bioluminescent probes, it was first necessary to define conditions where they could be detected at low levels on different types of films. Serial dilutions of biotinylated-RecAequorin and biotinylated-RecLuciferase were prepared and placed in the wells of a clear plastic microtiter plate. The plate, situated in a metal holder which prevents crosstalk between the wells, was placed in the darkroom against ASA 20,000 film in the Dynatech camera luminometer or Kodak X-OMAT AR x-ray film and then either calcium or luciferin was added to initiate light production. Six ng (272 fmol) of biotinylated-RecAequorin was detected on the instant film after a 10 s exposure. Significantly less biotinylated-RecLuciferase, about 20 pg (<1 fmol) was detected after a 10 min exposure. In terms of molar concentration, this represents about a 300 fold difference in sensitivity. These results were anticipated since RecLuciferase catalyzes the oxidation of numerous luciferin molecules, while RecAequorin catalyzes a single turnover event. Relative to the instant film, about 2-3 fold less of each protein was detectable on x-ray film. Preflashing the x-ray film was found to lower the detection limit of the bioluminescent proteins by about 2 fold, thereby making it's sensitivity comparable to the instant film.

Immunodetection of human transferrin in dot blots with biotinylated-RecAequorin and biotinylated-RecLuciferase is shown in Fig. 1. Different amounts of transferrin (1-60, 2-6; 3-.6, 4-.06, 5-.006 ng) were spotted on nitrocellulose and nylon membranes and then incubated sequentially with rabbit anti-human transferrin, biotinylated goat anti-rabbit antibody, streptavidin, and biotinylated-bioluminescent protein. Using biotinylated RecAequorin, 60 pg (fourth spot) of transferrin was visualized on both membranes. Using biotinylated-RecLuciferase, 6 pg (fifth spot) was seen on the nitrocellulose. In this particular experiment, only the 3rd spot (600 pg) was visible on the nylon membrane. In general, with RecLuciferase detection, the different membranes give more similar results. The increased sensitivity obtained on the blots relative to that seen with the dilutions of the bioluminescent proteins in the microtiter plate described above is due to antibody amplification and because the blots are immediately juxtaposed to the film.

Immunodetection of human transferrin in dot blots using a goat anti-rabbit-RecAequorin conjugate, which has the advantage of simplifying and speeding up the detection procedure, also proved successful. The conjugate detected 6 pg of transferrin on both nitrocellulose and nylon membranes. The sensitivity compared favorably with that obtained with a goat anti-rabbit antibody conjugate of alkaline phosphatase using the substrate pair, BCIP/NBT.

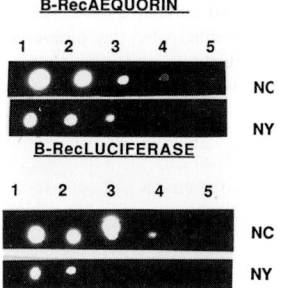

Bioluminescent detection of proteins in the Western blot format was also investigated. In a preliminary experiment, different amounts of commercially available biotinylated protein standards were separated by SDS-PAGE and then transferred to nitrocellulose. The blots were probed by sequential incubation with streptavidin and biotinylated-RecAequorin and developed against instant and x-ray film. As expected, the instant film was more sensitive where the best results were obtained when 10-30 ng of total protein were loaded onto the gel. The best pattern on the x-ray film corresponded to 80 ng total protein.

Fig. 1. Bioluminescent immunodetection of human transferrin

The question of whether the bioluminescent proteins could be used to detect different antigens in a complex mixture after Western blotting was addressed. Rabbit serum and human serum were separated by SDS-PAGE, transferred to nitrocellulose and probed using the appropriate antibody reagents in combination with streptavidin and biotinylated-bioluminescent proteins for the presence of immunoglobulin G and transferrin, respectively. The heavy and light chains of the immunoglobulin were visualized in lanes containing both purified rabbit IgG and rabbit serum using both biotinylated-RecAequorin and biotinylated-RecLuciferase. A band corresponding to transferrin was observed in lanes loaded with purified transferrin and human serum using biotinylated-RecAequorin. A second transferrin blot was probed in parallel with biotinylated-alkaline phosphatase and BCIP/NBT. Biotinylated-RecAequorin detection showed comparable, if not greater, sensitivity to the results obtained with the enzyme detection system.

Detection of DNA using the biotinylated bioluminescent protein probes was accomplished in dot blot and Southern blot formats. Fig. 2 shows an example of a dot blot hybridization experiment on nitrocellulose where different amounts (1-500, 2-250, 3-125, 4-63, 5-31, 6-15 ng) of the linearized pTZR-Luc1 plasmid containing the *Renilla* luciferase cDNA insert were probed with the biotinylated- *Renilla* cDNA fragment generated by restriction enzyme digestion of the plasmid. After hybridization, the blot was incubated sequentially in blocking protein, streptavidin and biotinylated-RecAequorin or

biotinylated-RecLuciferase. With biotinylated RecAequorin, although it is quite faint, the fifth spot at 31 ng of target DNA is detectable. The sixth spot corresponding to 15 ng of DNA is visible at about the same intensity using biotinylated-RecLuciferase following a 10 min exposure to the film. Biotinylated-RecAequorin was also used in Southern blot analysis of linearized pTZR-Luc1 plasmid DNA following fractionation by agarose gel electrophoresis and transfer to nitrocellulose and nylon membranes. A single band of target DNA corresponding to about 5 kB was seen on both membranes. The limit of detection was approximately 8 ng of target DNA and was comparable to that obtained with biotinylated-alkaline phosphatase and BCIP/NBT in a parallel Southern blot.

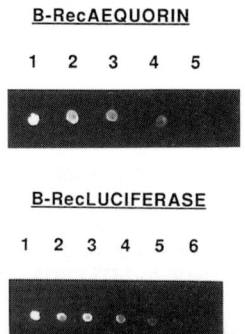

In summary, derivatives of the bioluminescent proteins, RecAequorin and RecLuciferase, are stable and can be used for the detection of protein and nucleic acid targets immobilized on membrane supports using either instant or x-ray film. These reagents offer a rapid and sensitive alternative to radioactive and enzyme detectionsystems. It is anticipated that bioluminescent detection systems will become even more simple to use and more sensitive with further development and optimization.

Fig. 2. Bioluminescent detection of
DNA dot blot hybridization

ACKNOWLEDGEMENTS
Supported by Small Business Innovative Research Grants AI28628 (DFS) and GM45093 (NLS) from the National Institutes of Health.

REFERENCES
1. Prasher, D., McCann, R.O., and Cormier, M.J. (1985) Cloning and expression of the cDNA for aequorin, a bioluminescent calcium-binding protein. Biochem. Biophys. Res. Commun. 126, 1259-1268.

2. Cormier, M.J. (1978) Comparative Biochemistry of Animal Systems. In Bioluminescence in Action. Herring, P. (ed.) , Academic Press, London, p. 75-108.

3. Matthews, J.C., Hori, K., and Cormier, M.J. (1977) Purification and properties of *Renilla reniformis* luciferase. Biochemistry, 16, 85-91.

4. Lorenz, W.W., McCann, R.O., Longiaru, M., and Cormier, M.J. (1990) Isolation and expression of the cDNA encoding for *Renilla reniformis* luciferase. Submitted for publication.

A SOLID-PHASE ASSAY FOR THE EVALUATION OF THE GLYCOSYLATION STATUS OF SERUM GLYCOPROTEINS USING RECOMBINANT AEQUORIN

P.F. Zatta[1,2], D.F. Smith[1,3], M.J. Cormier[1], and R.D. Cummings[1]

University of Georgia[1], Athens, GA 30602, USA, University of Padua[2], Padova, Italy, ELA Technologies, Inc.[3], Athens, GA 30604

INTRODUCTION

Serum contains numerous glycoproteins, such as immunoglobulins, fibrinogen, transferrin, and orosomucoid, many of which are synthesized by the liver and plasma cells. Most of these glycoproteins contain numerous complex-type Asn-linked oligosaccharides consisting of outer branches with the non-reducing sequence N-acetylneuraminic acidα2,6galactoseβ1,4N-acetylglucosamineβ1,-mannose---R (1). Ashwell and Morrel (2) first demonstrated that the liver is also the major site for the clearance and catabolism of serum glycoproteins through the action of the hepatic asialoglycoprotein receptor, which recognizes the penultimate galactosyl residues of these glycoproteins after the systemic removal of terminal N-acetylneuraminic acid. However, the biological functions of these carbohydrate chains on serum glycoproteins, other than to facilitate clearance, is not well understood. There is increasing evidence, nevertheless, that disease conditions can result in alterations in the glycosylation status of the serum glycoproteins. Fibrinogen in patients affected by dysfibrinogenemia, has 1.3 to 3.5 more N-acetylneuraminic acid per mole with respect to fibrinogen from healthy controls (3). Decreased amounts of both N-acetylneuraminic acid and galactose have been found on serum IgG from patients affected with rheumatoid arthritis (4-6). The glycosylation of serum transferrin is significantly altered in patients with hepatocellular carcinoma (7). Other factors can also indirectly influence the glycosylation status of serum glycoproteins; for instance, heavy alcohol consumption is correlated with increased levels of desialylated forms of serum transferrin (8).

Many different approaches have been utilized to examine the glycosylation status of serum glycoproteins, with the idea that a quantative evaluation of the status might be reflective of specific diseases and might serve as useful diagnostic and/or prognostic indicators. Recent approaches, involving serum IgG in patients with rheumatoid arthritis, have been based on difficult, semi-quantitative techniques. For example, the Asn-linked oligosaccharides of IgG were liberated chemically by hydrazinolysis, radiolabeled, and examined by charge and size-exclusion chromatographic techniques (5). Clearly, more rapid and sensitive methods are required before the clinical utility of serum glycosylation status can be thoroughly evaluated. Because of the high degree of specificity some plant lectins exhibit in their interactions with glycoconjugates (9), they have recently been employed in solid phase techniques to assess the glycosylation status of IgG, with notable success (6). Along similar lines, we explored the development of solid-phase assays for assessing glycosylation status of serum glycoproteins using different plant lectins as probes for the glycoprotein oligosaccharides. To enhance the sensitivity and speed of these assays we have used the derivatized recombinant bioluminescent protein, RecAequorin (10), as the label. RecAequorin produces light in a flash upon addition of calcium ions and is easily detectable in the attomol range. Our results indicate that serum glycoproteins can be adsorbed by a variety of methods to the surfaces of microtiter plates allowing the evaluation of the glycosylation status of the immobilized glycoprotein. These new assays employing RecAequorin are sensitive and reliable and should significantly enhance the possibility of using bioluminescent labels in future diagnostic and prognostic assays based on serum glycoprotein glycosylation.

MATERIAL AND METHODS

Reagents - Streptavidin, biotinylated *Ricinus communis* agglutinin (RCA-I), fibrinogen, bovine thyroglobulin, and anti-human transferrin from goat were purchased from Sigma Chemical Company. Immunlon-2 microtiter plates were obtained from Dynatech Laboratories. Other reagents for ELISA procedures were purchased from Boehringer-Mannheim. The biotinylated derivative of RecAequorin was supplied by ELA Technologies, Inc., Athens, Georgia.

Immobilization of Serum Glycoproteins and Antibodies to Microtiter Wells and Detection with Plant Lectins Coupled to RecAequorin - Sera collected from healthy donors was serially diluted in Tris-buffered saline (TBS) (0.1 mol/l Tris-HCl, .15 mol/l NaCl, pH 7.5). One-hundred microliter portions of each diluted serum samples were placed in wells of a removalwell microtiter plate from Dynatech. The coating was carried out at 4°C overnight or at room temperature for 1 h. In other experiments transferrin, thyroglobulin, albumin or goat antibody to the human transferrin diluted in TBS were immobilized in plates at concentrations ranging from 0 to 10 μg/ml using similar conditions. In all cases after immobilization of glycoproteins, the microtiter wells were washed with TBS containing 0.05% Tween 20. The wells were blocked by incubating for 30 minutes at room temperature with the ELISA-blocking reagent from Boehringer-Mannheim, according to the suggestions of the manufacturer. After three washings a biotin-lectin solution (20 μg/mL) was added into each well and incubated at room temperature for 2 h followed by washing three times. Streptavidin (15 μg/mL) was added to each well, followed by two washings. The biotinylated derivative of RecAequorin (0.18 μg/mL) was then added to each well and allowed to incubate for 30 min followed by 5 washings.

Luminescence Measurement - After washing, wells were removed from the microtiter plate, placed in a larger test tube, and placed within the chamber of a Berthold CliniLumat luminometer. Calcium was injected automatically into the enclosed microtiter well within each tube and the amount of light produced was recorded as counts.

RESULTS AND DISCUSSION

In our experiments we have examined glycoproteins for the presence and level of terminal β-galactosyl residues using the plant lectin *Ricinus communis* agglutinin-I (RCA-I), which binds with high affinity to such residues (11). Using an biotin-streptavidin system it is possible to couple the biotinylated RCA-I to the biotinylated RecAequorin through a streptavidin bridge. As indicated in the section on "Materials and Methods", components are added sequentially to microtiter wells with washing steps in between each addition. Thus, it is possible to include Ca^{++} at all steps in the assay up to the final step in which biotinylated RecAequorin is introduced; at that step EDTA is added to chelate free Ca^{++}. Within the luminometer Ca^{++} can then be introduced at concentrations sufficient to overcome the EDTA.

As a first step in evaualting the sensitivity and reliability of this assay we examined the binding of biotinylated RCA-I to two different glycoproteins, thyroglobulin and fibrinogen, and to a non-glycosylated protein, bovine serum albumin (Fig. 1). There was no significant binding of RCA-I to albumin, whereas the binding to thyroglobulin and fibrinogen was significant and detectable in the range of 10-100 ng of glycoprotein per well, with maximal binding observed for both glycoproteins at approximately 1 μg/ml. In all cases binding was obliterated by the inclusion of 0.5 M lactose in the assay, which is a competitive inhibitor of RCA-I binding to glycoconjugates. These results demonstrate that RCA-I binding to glycoproteins is specific and allows the sensitive detection of terminal β-galactosyl residues on adsorbed glycoproteins.

We then investigated the binding of RCA-I to glycoproteins contained within total serum samples. Five samples of diluted serum, obtained from pin pricks of five fingers of a volunteer, were incubated in microtiter wells at increasing dilutions of serum and the adsorbed glycoproteins were detected by biotinylated RCA-I (Fig. 2). The results indicate that up to 10,000-fold dilutions of serum contain enough glycoprotein to be detected in this assay. (It should be noted that at such high dilutions it is not feasible to determine the actual amount of adsorbed glycoprotein within the wells.)

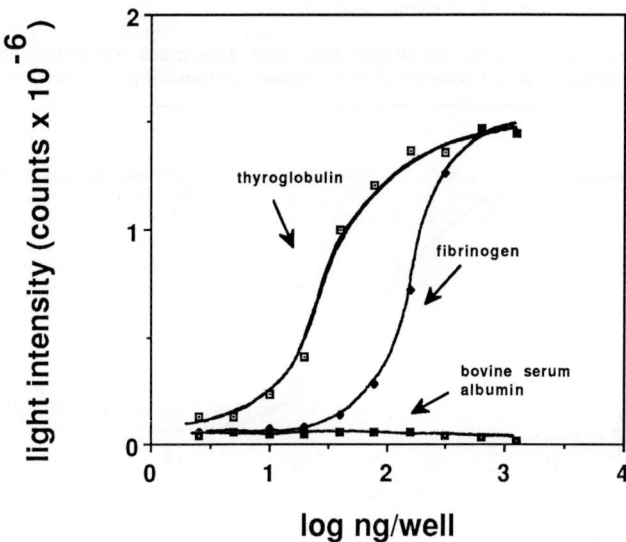

Fig. 1 RCA-I binding to Standard Glycoproteins

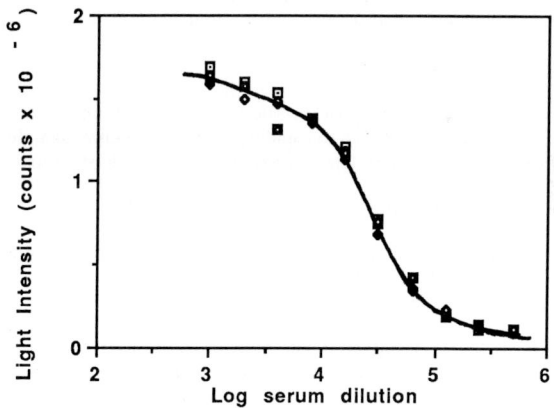

Fig. 2 RCA-I binding to Serum Glycoproteins

Finally, we assessed another method of immobilization to allow evaluation of the glycosylation status of a specific serum glycoprotein, transferrin. Microtiter wells were pretreated with antibody to human transferrin. Diluted serum samples were applied and the glycosylation status of the bound transferrin was assessed by biotinylated RCA-I (Fig. 3.). The RCA-I detection assay allowed assessment of β-galactosyl residues on serum transferrin in ranges of 100 to 10,000-fold serum dilutions.

Taken together, these results demonstrate the feasibility of using solid-phase assays to study the glycosylation of soluble glycoproteins. The assay has several key features which make it attractive. The assay is rapid and can be performed within 90 minutes using readily available plant lectins. Most importantly, the use of RecAequorin allows the rapid detection of bound materials over several log units of sample dilution.

In recent studies Sumar, et al., (6) utilized the lectins *Bandeiraea simplicifolia* and *Ricinus communis* agglutinin to detect terminal β-N-acetylglucosaminyl and β-galactosyl residues,

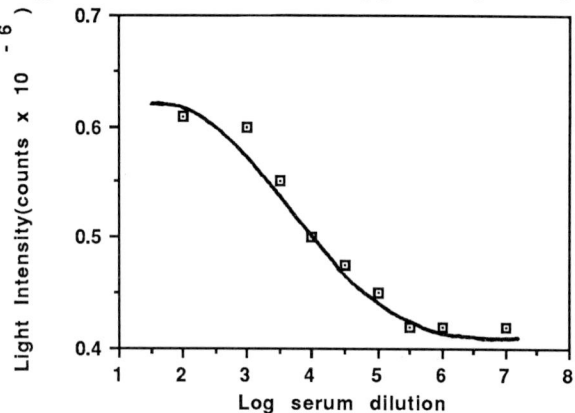

Fig. 3 RCA-I Binding to Serum Transferrin

respectively, in serum IgG from patients with rheumatoid arthritis. Their assay employed a similar biotin-streptavidin system, except that horseradish peroxidase was used as the label. However, in terms of sensitivity their assay required μg quantities of IgG. Our results indicate that the use of RecAequorin can significantly enhance the sensitivity and speed of these types of assays and may permit a more quantitative assessment of the glycosylation status of IgG and other serum glycoproteins. Although we have used the Berthold Clinilumat, we have found that the assay has similar sensitivity in a Dynatech Microtiter® plate luminometer (data not shown). Using the latter instrument it is possible to rapidly process many more samples, thus increasing the potential feasibility of using these types of assays in the clinical situation.

ACKNOWLEDGEMENTS

Supported by NIH grants IT4 RR05351 (RDC) and SBIR Grants AI28628 and GM43025 (DFS).

REFERENCES

1. Kornfeld, R. and Kornfeld, S. (1985) Annu. Rev. Biochem. 54, 631.
2. Ashwell, G. and Morrell, A.G. (1974) Adv. Enzymol. Relat. Areas Mol. Biol. 41, 99.
3. Martinez, J., MacDonald, K.A. and Palascak, J. (1983) Blood 61, 1196.
4. Mullinax, F. (1975) Arthritis Rheum. 18, 417.
5. Parek, R.B., Dwek, R.A., Sutton, B.J., Fernandes, D.L., Leung. A., Stanworth, D., Rademacher, T.W., Muzuochi, T., Tamiguchi, T., Matsuta, K., Takeuchi, F., Nagano, Y., Miyamato, T. and Kobata, A. (1985) Nature 316, 452.
6. Sumar, N., Bodman, K.B., Rademacher, T.W., Dwek, R.A., Williams, P., Parekh, R.B., Edge, J., Rook, G.A.W., Isenberg, D.A., Hay, F.C. and Roitt, I.M. (1990) J. Immunol. Methods 131, 127.
7. Yamashita, K., Koide, N., Endo, T., Iwaki, Y. and Kobata, A. (1989) J. Biol. Chem. 264, 2415.
8. Storey, E.L., Anderson, G.J., Mack, U., Powell, L.W., and Halliday, J.W. (1987) Lancet 1, 1292.
9. Cummings, R.D., Merkle, R.K. and Stults, N.L. (1989) Meth. Cell Biol. 32, 141.
10. Prasher, D., McCann, R.O. and Cormier, M.J. (1985) Biochem. Biophys. Res. Commun. 126, 1259.
11. Baenziger, J.V. and Fiete, D. (1979) J. Biol. Chem. 254, 2600.

INDUSTRIAL SEMINAR: BERTHOLD INSTRUMENTS

RECENT ADVANCES AND PROSPECTS FOR USE OF BEETLE LUCIFERASES AS GENETIC REPORTERS

Keith V. Wood

Promega Corporation, 2800 Woods Hollow Rd., Madison WI 53711

INTRODUCTION

Where it is applicable, luminescence can be a highly sensitive reporter of biological events. Beetle luciferases have long realized this potential in the assay of ATP. Current assays, which can detect 10^{-16} moles of ATP, provide the most sensitive means for such measurements. More recently, cloning of their cDNAs has made beetle luciferases useful for measuring genetic events. As with measurements of ATP, applications as a genetic reporter have been dominated by luciferase of the North American firefly, *Photinus pyralis*.

First reports of firefly luciferase as a genetic reporter were published in 1986 for plant cells (1) and 1987 for mammalian cells (2). For these uses luciferase was shown to offer extreme sensitivity with little required effort. In a recent review, the cDNA coding luciferase was cited as probably being "the most versatile reporter gene yet (3)." Other publications have further displayed a diversity of studies benefitting from use of a luminescent reporter. A brief sampling of this diversity is presented in this "mini-overview."

Also presented are developments in the methodology of these applications. Recent insights into the chemistry and structure of luciferase are advancing our ability to use this enzyme. Some advances have both greatly increased the yield of light and simplified the assay. Other advances also may offer improvements in the physical and enzymological properties of the enzyme. Adoption of these improvements into common use will make firefly luciferase more favorable as a genetic reporter.

APPLICATIONS OF GENETIC REPORTERS

The most common use of genetic reporters is in the analysis of promoter and enhancer activity. An example is described by Economou, *et al.*, for analyzing regulation of the gene coding tumor necrosis factor (TNF) α/cachectin (4). Previous results had shown this gene to be inducible by the phorbol ester, PMA. By coupling a region of DNA upstream of the TNF-α/cachectin gene to the luciferase gene, PMA could similarly cause a 12-fold induction of luminescence. Analysis of deletions allowed identification of the smallest DNA region still capable of inducing luminescence. Analysis of RNA structure and processing is also a common use of reporter genes. A study by Callis, *et al.* showed that intron processing could increase expression of the luciferase gene in maize (5). Luminescence was increased 6- to 16-fold depending upon the source of promoter and intron used.

Reporter genes can often be used to study the structure and function relationships of proteins that act upon genetic material. In a study reported by Waterman, *et al.*, the luciferase gene was used to examine the genetic activity of estrogen receptor (6). A system was established where estrogen receptor expressed from one plasmid could regulate expression from the luciferase gene, which was coupled to an estrogen responsive element (ERE) on another plasmid. With this arrangement, a 3- to 7-fold increase in luminescence was caused by the presence of estrogen. Point mutations to the DNA binding region of the receptor were shown to abolish the induction of luminescence. Another experiment showed that truncation of the steroid binding region caused constitutively high level expression of luminescence. These results help to map functional domains within the structure of the receptor.

Some studies have used reporter genes to reveal the presence of other genes or organisms. Rodriguez, *et al.* used the luciferase gene to monitor growth of vaccinia virus in mice (7). By inserting the gene into the virus, its relative propagation into various tissues could be easily monitored. Using this method they showed that an attenuated strain of virus would propagate more slowly than wild type, but with the same tissue distribution. Schwartz, *et al.*, described an assay for growth of HIV virus in cell culture by using trans activation of the luciferase gene (8). In this example, luciferase was coupled to the HIV LTR and transformed into the cells prior to viral infection. Upon infection, the *tat* gene product of HIV caused up to 50-fold induction of luciferase expression from the plasmid. Protective action of anti-viral drugs could be rapidly tested with this system.

Development of techniques that involve genetic material can often be benefitted by a rapid and sensitive reporter gene. Malone, *et al.*, demonstrated this benefit when they used the luciferase gene to develop a method of cationic liposome-mediated RNA transfection (9). RNA coding luciferase was synthesized *in vitro* and used to optimize conditions for its transfer into cells. Expression of luciferase activity could be detected within 30 minutes of its addition to cells. The authors also showed effects of RNA structure on protein expression. For example, addition of a 5' cap structure onto the RNA increased luminescence 40-fold.

ADVANCES IN THE LUMINESCENCE CHEMISTRY

Upon mixing of enzyme with substrates, the reaction of firefly luciferase has been characterized by a brief flash followed by rapid decay of the light intensity. In a 1 minute measurement, 50% of the light is emitted during the first 15 seconds (Fig. 1). After 1 minute only about 10% of the peak intensity remains. For this reason, assays of luciferase were best done using a luminometer equipped with an injection device to allow the light of the flash to be collected for measurement. Since the magnitude of the flash is dependent upon the rate of mixing, reproducibility of the injection device has been important for good results.

This description of the luciferase assay may soon be obsolete due to discoveries in the underlying chemistry. I recently have been developing a new assay for luciferase based on the action of coenzyme A (see "The Origin of Beetle Luciferases" this volume). This new assay largely overcomes the self-inhibition that has been characteristic of luciferase activity (Fig. 1). Mixing of enzyme and substrates under the new conditions yields a luminescence intensity that remains at a high and steady level. With optimization, the steady level is maintained for about 30 seconds, followed by a slow decay with a half-life of roughly 5 minutes.

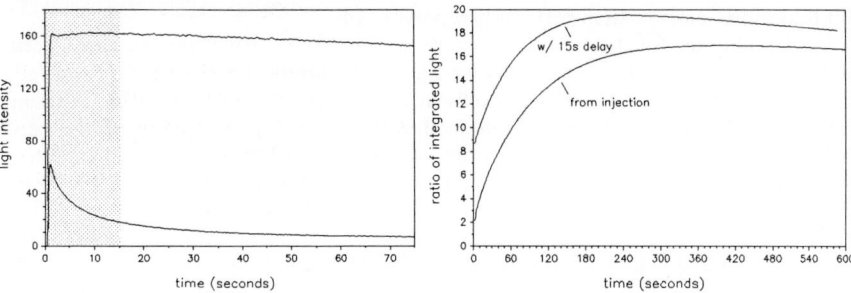

Fig. 1   On the left is shown luciferase activity measured from NIH 3T3 cells transformed with pRSVL (2) using the new assay (upper curve) or the conventional assay (lower curve) (1, 2). Grey area shows luminescence typically lost when using a scintillation counter for assay. On the right is shown the relative efficiencies of the two assays for luminescence integrated over the time indicated on the abscissa. The upper curve shows the relative efficiencies expected for measurements using a scintillation counter (i.e., without luminescence of the initial 15 seconds).

The new assay, therefore, produces substantially more light than the conventional assay. In a 1 minute measurement, the new assay yields 10-fold more luminescence. Maximal improvement over the conventional method is approximately 17-fold, achieved after about 5 minutes of measurement (Fig. 1). The benefits of improved light output are especially evident if measurements are made using a scintillation counter. Since many molecular biology laboratories do not have a luminometer, scintillation counters are a common method of quantifying luminescence. By this method, approximately the first 15 seconds of luminescence are typically lost due to the time required to place the sample into the instrument after mixing of substrates. Because the new assay produces a nearly steady luminescence intensity, it suffers little from loss of the initial reaction. Thus, in a 1 minute measurement the new assay yields 15-fold more luminescence, rising to nearly a 20-fold increase after 4 minutes (Fig. 1).

Probably more important than the increase in total luminescence is the constancy of the intensity, which eliminates the advantage provided by an injection device. This simplifies the assay since enzyme and substrates may be mixed by usual means without significant loss of sensitivity. Precision between replicate measurements using normal mixing have a coefficient of variation typically less than 1.5%. Since most injection devices require a minimum liquid volume, the new assay also can allow measurements to be made using substantially less reagent.

The sensitivity and linearity of the new assay was shown in a titration of luciferase activity. For dilutions of enzyme covering a 10^8-fold range, linearity proved to be excellent (Fig. 2). The sample of lowest enzyme concentration contained 3×10^{-21} moles (2000 molecules). This is

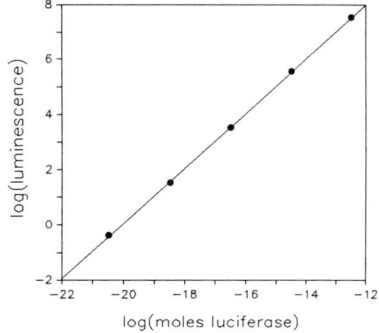

Fig. 2   Titration of luciferase activity using the new assay. Purified luciferase was diluted using the buffer for cell lysis.

100-fold more sensitive than previously reported (10).

Conditions for lysis of cells expressing luciferase have also recently been improved (11). In a modification of those conditions developed at Promega, enzymatic activity was released from cells in 2 to 5 minutes, and was stable at room temperature for several hours. Combined with the ease and rapidity of the new luciferase assay, results from the reporter can be obtained in less than 10 minutes. The optimized reagents for cell lysis and the luciferase assay will be available from Promega.

ADVANCES IN LUCIFERASE STRUCTURE

Changes to the structure of the gene coding luciferase offer potential for its improvement as a genetic reporter. There has been particular interest concerning the peroxisomal targeting sequence at the carboxy terminal of luciferase. Though relationship of this sequence to peroxisome localization has been well studied, the effect of localization on the expression of luciferase has not been thoroughly investigated. Experiments are underway to decide whether luciferase without a targeting sequence may be preferable as a reporter. Addition of other targeting sequences also may prove to be of value.

Luciferases that emit different colors of light present especially intriguing possibilities. It may be possible, for example, to assay two genetic events simultaneously by associating with each a different color of luminescence. Results from a study of click beetle luciferases show that the color of luminescence can be changed with substitutions of specific amino acids. With further research, enzymes may be developed with similar physical and enzymological properties, but widely differing wavelengths of emission. Mutagenesis of the luciferases also may yield other properties beneficial to applications in genetic reporting.

REFERENCES

1. D.W. Ow, K.V. Wood, M. DeLuca, J.R. de Wet, D.R. Helinski, and S.H. Howell, Transient and Stable Expression of the Firefly Luciferase Gene in Plant Cells and Transgenic Plants. *Science*, **234**, 856-859 (1986)
2. J.R. de Wet, K.V. Wood, M. DeLuca, D.R. Helinski, and S. Subramani, Firefly Luciferase Gene: Structure and Expression in Mammalian Cells. *Molec. Cell. Biol.*, 7, 725-737 (1987).
3. J. Alam and J.L. Cook, Reporter Genes: Application to the Study of Mammalian Gene Transcription. *Anal. Bioch.*, **188**, 245-254 (1990).
4. J.S. Economou, K. Rhoades, R. Essner, W.H. McBride, J.C. Gasson, and D.L. Morton, Genetic Analysis of the Human Tumor Necrosis Factor α/Cachectin Promoter Region in a Macrophage Cell Line. *J. Exp. Med.*, *170*, 321-326 (1989).
5. J. Callis, M. Fromm, and V. Walbot, Introns Increase Gene Expression in Cultured Maize Cells. *Genes and Development*, 1, 1183-1200 (1987).
6. M.L. Waterman, S. Adler, C. Nelson, G.L. Green, R.M. Evans, and M.G. Rosenfeld, A Single Domain on the Estrogen Receptor Confers Deoxyribonucleic Acid Binding and Transcriptional Activation of the Rat Prolactin Gene. *Mol. Endocr.*, 2, 14-21 (1988).
7. D. Rodriguez, J.R. Rodriguez, J.F. Rodriguez, D. Trauber, and M. Esteban, Highly Attenuated Vaccinia Virus Mutants for the Generation of Safe Recombinant Viruses. *Proc. Natl. Acad. Sci. USA*, **86**, 1287-1291 (1989).
8. O. Schwartz, J.L. Virelizier, L. Montagnier, and U. Hazan, A Microtransfection Method Using the Luciferase-encoded Reporter Gene for the Assay of Human Immunodeficiency Virus LTR Promoter Activity. *Gene*, **88**. 197-205 (1990).
9. R.W. Malone, P.L. Felgner, and I.M. Verma, Cationic Liposome-mediated RNA Transfection. *Proc. Natl. Acad. Sci. USA*, 86, 6077-6081 (1990).
10. S.J. Gould and S. Subramani, Firefly Luciferase as a Tool in Molecular and Cell Biology. *Anal. Bioch.*, **175**, 5-13 (1988).
11. A.R. Brasier, J.E. Tate, and J.F. Habener, Optimized Use of the Firefly Luciferase Assay as a Reporter Gene in Mammalian Cell Lines. *Biotech.*, 7, 1116-1122 (1989).

LABELS TO MONITOR LIGAND BINDING ASSAYS VIA LUMINESCENCE

Hartmut R. Schroeder

ExOxEmis, Inc., 18585 Sigma Rd., San Antonio, Texas 78258 USA

INTRODUCTION

Luminescence technologies have been developed using various labels to monitor immuno- and nucleic acid hybridization assays. Some assets of these label systems are stability, ease of conjugation to ligands, sensitivity, rapid readout, versatility and flexibility. Operating at a pH of 5-13, these systems may be coupled to the generation of NAD(P)H, ATP or peroxide and used for homogeneous or heterogeneous procedures. The complexity of detection equipment ranges from a simple photographic apparatus to an automated photon counter. The selection of chemistry and format depend on analyte concentration, interference from the sample matrix, and ultimately, the nonspecific binding of the labeled conjugate to the solid phase employed. This review will give perspective to the choice of luminescent labels, highlight new developments and evaluate the status of commercial applications and instrumentation.

LUMINESCENCE MEASUREMENTS

The kinetics of various luminescence systems differ dramatically and can affect both precision and sensitivity. Enzyme labels such as peroxidase have historically been monitored by accumulating a colored dye product over 20-60 minute readout time. The greater sensitivity of chemiluminescence (CL) with luminol as substrate permits much more rapid readout, usually less than 0.5 minutes. However, the benefit of lengthy integration is lost because photons can only be accumulated in front of the photodetector. By contrast, the cyclic hydrazides and acridinium ester labels are measured at extremely low concentration via peak light intensity in less than a second. But this requires precise control of injection and highly sensitive photon counting, which has been achieved in the Berthold instruments.

Ultimately, the detection limit is based on the statistical confidence limit of the precision. However, for comparative purposes, choosing an arbitrary limit of twice the total background mitigates excessive claims achieved only under special circumstances. By such criterion and using a standard chemistry, such as the luminol-microperoxidase (1) or ATP-luciferase (2) system as reference, various chemistries or instruments may be compared. Although the detection of NAD(P)H at 1 nM(3) and ATP at 0.5 pM(4) is excellent for chemistries, their endogenous presence in the sample limits their usefulness as luminescence labels to monitor homogeneous immunoassays. By contrast, cyclic hydrazides and acridinium derivatives are absent in biological preparations. The cyclic hydrazides are very stable and detectable at less than 1 pM at both pH 8.6 and 13.0 (Table 1). They are also useful in homogeneous immunoassays based on enhancement of CL of the antibody bound fractions (5) and an energy transfer system (6). The greater CL yield of the naphthalhydrazide ABENH (>0.04) allows a lower detection limit near 10^{-14}M. Its longer wavelength emission (515 nm) eliminates the quenching sometimes observed with isoluminol derivatives covalently coupled to proteins. Present acridinium derivatives are stable and display similar sensitivity as ABENH, having about half the quantum yield and about half the chemical background in the detection system. An acridinium ester label's unique ability to intercalate into double strand nucleic acid hybrids protects it from selective base hydrolysis which is part of the light yielding step and, therefore, makes a homogeneous readout possible in this special case (7).

TABLE 1

PROMINENT LABELS AND THEIR DETECTION
LIMITS VIA LUMINESCENCE[a]

LABEL	pH	Time[b] (sec)	Vol (mL)	mol	Limit M	Ref
Cyclic Hydrazides						
AHEI	$\leq$ 8.6	2	0.15	1.5×10^{-16}	1×10^{-12}	1
ABENH	13.0	10	0.6	1.8×10^{-17}	3×10^{-14}	1,6
Acridinium						
Ester	13.0	5	0.3	$\sim 10^{-17c}$	$\sim 10^{-14c}$	8
Firefly						
Luciferase	7.8	10	1.5	2×10^{-17}	1.3×10^{-14}	2
Aequorin	8.0	10	0.35	5×10^{-19}	1.4×10^{-15}	9
Horseradish Peroxidase						
Unenhanced	9.5	420	0.5	7.5×10^{-15}	1.5×10^{-11}	10
Enhanced	8.5	420	0.5	5×10^{-16}	1×10^{-12}	10
Alkaline Phosphatase						
dioxetane-P	9.5	1200	0.1	1×10^{-21}	1×10^{-17}	11
dioxetane-P	9.5	1200	0.5	1.5×10^{-18d}	3×10^{-15d}	12
luciferyl-P	9.8	1800	0.1	$\sim 10^{-19e}$	$\sim 10^{-15e}$	13

a Arbitrarily defined as twice the mean of all background.
b Actual luminescence measurements were all 30 seconds or less, but for comparative purposes total readout time is shown.
c These values are extrapolated. 2.2×10^{-18} mol or 7.3×10^{-15}M is claimed via statistical cutoff.
d 10^{-20} mol detection is claimed via statistical cutoff.
e These values are extrapolated. 2×10^{-20} mol or 2×10^{-16}M is claimed, presumably via statistical cutoff.

The enzyme labels (Table 1) also display excellent sensitivity, but generally require a longer readout time. In an elegant approach alkaline phosphatase can be detected at ~10^{-15}M by accumulating the intermediary luciferin product over time, followed by rapid measurement with the firefly system. Finally, the ultimate claim is 10^{-17}M alkaline phosphatase using a dioxetane-phosphate substrate with enhancers that provide a hydrophobic environment for the emitter.

PERFORMANCE OF LUMINESCENCE IMMUNOASSAYS

Translating the lower detection limits of the above labels into an immunoassay is a difficult task. In comparing the alkaline phosphatase systems with the enhanced peroxidase, an improvement of nearly 3 orders of magnitude might be expected. However, when a dioxetane-phosphate is substituted for a colorimetric substrate to monitor the alkaline phosphatase label in the Hybritech immunometric assay for thyrotropin (TSH) for example, the detection limit improves only 4-fold (14). Clearly, other factors such as the non-specific adsorption of the label conjugate predominate when high assay sensitivity is required. The lower detection limit (zero standard + 2 SD) of ~0.2 milliinternational unit/L is about the same for the Ciba-Corning and Amersham assays (14,15). The luminescence immunoassays correlate well with RIA, have similar precision and accuracy and are generally more sensitive.

Commercial luminescence binding assays are basically focused on the thyroid, pregnancy, tumor markers, infectious disease and allergy (Table 2). Expanded menus and additional suppliers will consolidate these areas shortly. Hence, a greater emphasis on special market niches which can take advantage of unique features of a particular luminescence system is advisable for the future. Application of luminescence to nucleic acid detection are elaborated by Dr. Weeks in this issue.

APPROACHES AND CHOICES

The cornerstone objectives of a clinical assay are low cost, convenience, speed and sensitivity. Yet the heart is the specificity, at times taken for granted. The assay is usually a compromise of these factors, but the specificity must be maintained. Successful development demands an integrated system approach in which the label, binder, format and readout are all considered together. Desirable characteristics in the label are high sensitivity, stability, low non-specific binding and lack of interfering substances in the sample. Specificity of the assay resides in the antibody specificity and selectivity of the format with its solid phase and wash steps. The format and instrumental readout provide the convenient packaging the clinician expects.

Several solid phases have been adopted for heterogeneous assays. The microtiter plate offers unsurpassed efficiency in wash, high convenience and throughput, but is ultimately limited by low surface area. Amersham's "Amerlite" provides an excellent example of an integrated system based on this approach and the enhanced peroxidase luminescence chemistry. The plateau light production permits initiation outside the instrument and

TABLE 2

COMMERCIAL LUMINESCENCE BINDING ASSAYS

1	Amersham International	Amersham, UK
2	Mast Immunosystems	Mountainview, CA, USA
3	Ciba-Corning	Medfield, MA, USA
4	Gen-Probe	San Diego, CA, USA
5	London Diagnostics	Eden Prairie, MN, USA
6	Byk-Sangtec Diagnostica	Dietzenbach, Germany
7	Hoechst AG	Frankfurt, Germany
8	Henning-Berlin	Berlin, Germany
9	Bouty Diagnostici	Milan, Italy

	1	2	3	4	5	6	7	8	9
THYROID									
Thyrotropin	+		+		+	+	+	+	
Thyroxine Binding Globulin	+								
Thyroglobulin						+			
Thyroxine	+		+			+			
Free Thyroxine	+		+				+	+	
Triiodothyronine	+		+			+	+	+	
Free Triiodothyronine	+							+	
T_3 Uptake	+		+						
Parathyroid Hormone			+						
PITUITARY									
Lutropin	+		+		+				
Follitropin	+		+						
Choriogonadotropin	+		+		+	+	+		
Prolactin	+					+	+		
STEROIDS									
Estradiol	+								
Progesterone	+								
Estrone-3-glucuronide									+
Progesterone-3-glucuronide									+
Cortisol	+								
TUMOR MARKERS									
Carcinoembryonic antigen	+					+	+		
∝-Fetoprotein	+					+	+		
CA 19-9						+			
CA 125						+			
CA 15-3						+			
INFECTIOUS DISEASE									
Chlamydia Trachomatis			+	+					
Neisseria Gonorrhoeae				+					
Rubella	+								
8 Mycobacteria and Fungal Confirmation				+					
ALLERGY									
Total Immunoglobin E		+	+		+				
35 allergen-specific IgE's		+	+						
OTHERS									
Digoxin	+								
Ferritin	+				+				
Creatine kinase MB			+						

simplifies luminescence measurement. Although the sensitivity
of the chemistry is only moderately high, the system covers the
need for a broad range of analytes. Furthermore, a constant
light signal is easily monitored by photographic film
instruments where only semiquantative data is needed.   As
demonstrated during this meeting, photographic detection is also
very useful to monitor luminescent reactions linked to nucleic
acid hybrids on membranes or in various blotting techniques.

Other formats based on coated tubes have more consistent protein
binding, but are a bit less convenient than microtiter plates.
The large surface area of magnetizable beads, typically 50-fold
that of wells, greatly speed up binding reactions and their
handling permits random access automation currently explored at
Ciba-Corning. The versatile Berthold luminometers are state-of-
the-art and excellent for automatic readout with fast
luminescence reactions in tubes ($\pm$ beads). Currently more than
25 companies provide luminometers with increasing options and
automation.

An area of great potential for luminescence is the detection of
whole cells. We have designed an integrated system to capture
bacterial cells rapidly on antibody coated magnetizable beads.
The sandwich is completed by multiple binding of a second
antibody labeled with glucose oxidase (GO), followed by wash
steps. On addition of glucose, the enzyme produces peroxide as
an intermediate which accumulates.   This amplification is
performed at pH5 because it is the GO pH optimum, most stable
for peroxide and inhibitory for cellular enzymes that could
degrade peroxide. Finally, the luminescence readout is carried
out with luminol, myeloperoxidase and bromide at pH5 where there
is no metal ion catalyzed background luminescence and
interfering cellular enzymes are inhibited.   Thus, maximum
sensitivity and specificity are achieved. Preliminary results
indicated detection of 10 bacteria per mL at a signal/background
ratio of 10.

In my view, the versatility and flexibility of luminescence
measurement is perhaps its most attractive feature. Numerous
chemistries, many labels, a variety of solid phases and formats
and several instrumental approaches are available. By taking
advantage of market niches and integrated system design,
luminescence will indeed become the methodology of the 90's in
the clinical laboratory.

REFERENCES
1.   H.R.Schroeder and F.M.Yeager. Chemiluminescence yields and
      detection limits of some isoluminol derivatives in various
      oxidation systems. Anal.Chem. 50, 1114-1120 (1978).
2.   K.Wulff, H.P.Haar and G.Michal. Constant light signals in
      ATP assays with firefly luciferase: a kinetic explanation.
      In Luminescence Assays, M. Serio and M.Pazzagli (Eds),
      Raven Press, New York, 47-52 (1982).
3.   H.R.Schroeder, R.J.Carrico, R.C.Boguslaski, and
      J.E.Christner, Specific binding reactions monitored with
      ligand-cofactor conjugates and bacterial luciferase.
      Anal.Biochem. 72, 283-292 (1976).

4.   R.J.Carrico, K-K.Yeung, H.R.Schroeder, R.C.Boguslaski,
      R.T.Buckler, and J.E.Christner. Specific protein-binding
      reactions monitored with ligand-ATP conjugates and firefly
      luciferase. Anal.Biochem. 76, 95-100 (1976).
5.   G.Messeri, A.L.Caldini, G.F.Bolelli, M.Pazzagli, A.Tommasi,
      P.L.Vannuchi and M.Sergio. Homogeneous luminescence
      immunoassay for total estrogen in urine. Clin.Chem. 30,
      653-657 (1984).
6.   A.Patel, C.J.Davies, F.McCapra, and A.K.Campbell.
      Chemiluminescence energy transfer: a new technique
      applicable to the study of ligand-ligand interactions in
      living systems. Anal.Biochem. 129, 162-169 (1983).
7.   L.J.Arnold, P.W.Hammond, W.A.Wiese and N.C. Nelson. Assay
      formats involving acridinium-ester-labeled DNA probes.
      Clin.Chem. 35, 1588-1594 (1989).
8.   I.Weeks, I.Beheshti, F.McCapra, A.K.Campbell and
      J.S.Woodhead. Acridinium esters as high-specific-activity
      labels in immunoassay. Clin.Chem. 29, 1474-1479 (1983).
9.   A.Patel and M.J.Cormier. Calcium-sensitive photoproteins
      as bioluminescent labels in immunoassay, In Analytical
      Applications of Bioluminescence and Chemiluminescence,
      L.J.Kricka, P.E.Stanley, G.H.G.Thorpe and T.P.Whitehead
      (Eds), Academic Press, Inc., Orlando, 273-276 (1984).
10.   M.A.Motzenbocker. Sensitivity limitations encountered in
      enhanced horseradish peroxidized catalyzed
      chemiluminescence. J.Biolum. and Chemilum. 2, 9-16 (1988).
11.   A.P.Schaap, H.Akhaven and L.J.Romano. Chemiluminescent
      substrates for alkaline phosphatase: application to ultra
      sensitive enzyme-linked immunoassays and DNA probes.
      Clin.Chem. 9, 1863-4 (1989).
12.   I.Bronstein, B.Edwards and J.C.Voyta. 1,2-Dioxetanes:
      Novel chemiluminescent enzyme substrates. Application to
      immunoassays. J.Biolum. and Chemilum. 4, 99-111 (1989).
13.   W.Miska and R.Geiger. Luciferin derivatives in
      bioluminescence-enhanced enzyme immunoassays. J.Biolum. and
      Chemilum. 4, 119-128 (1989).
14.   I.Bronstein, J.C.Voyta, G.H.G.Thorpe, L.J.Kricka and
      G.Armstrong. Chemiluminescent assay of alkaline
      phosphatase applied in an ultrasensitive enzyme immunoassay
      of thyrotropin. Clin.Chem. 35, 1441-1446 (1989).
15.   M.G.McConway, R.S.Chapman, G.H.Beastal, et al. How
      sensitive are immunoassays for thyrotropin. Clin.Chem. 35,
      289-291 (1989).

CHEMILUMINESCENT ASSAYS BASED ON ACRIDINIUM LABELS

I. Weeks and J.S. Woodhead

Department of Medical Biochemistry
University of Wales College of Medicine
Heath Park, Cardiff CF4 4XN, Wales, UK

The synthesis and study of the chemistry of acridinium compounds was reported during the last century. The synthesis of acridine-9-carboxylic acid, which is a useful precursor for the synthesis of chemiluminescent acridinium salts, was described by Lehmstedt and Wirth as early as 1928 (1). The chemiluminescence of bis(methylacridinium nitrate) was first described by Gleu and Petsch in 1935 (2) and has subsequently been better known as lucigenin. More detailed studies of the chemiluminescence of acridine compounds were carried out by Rauhut and co-workers (3) and by McCapra et al (4).

It is established that a wide range of organic molecules are potentially capable of being used as end-points for binding assays such as immunoassays and oligonucleotide probe hybridisation assays. It is possible to make use of these by coupling the chemiluminescent molecule itself into the binding reaction by attachment to one of the components thereof. Alternatively a label can be chosen which is itself not chemiluminescent but rather is one which can initiate the reaction of a chemiluminescent molecule. Examples of the latter include the use of horseradish peroxidase or xanthine oxidase labels to initiate luminol chemiluminescence and the use of alkaline phosphatase to cleave phosphorylated stable dioxetanes.

There are far fewer examples of different chemiluminescent molecules being used in the "direct" labelling mode. In the past, phthalhydrazide molecules have been most widely used for this purpose, particularly isoluminol derivatives such as aminobutylethylisoluminol (ABEI).

Such phthalhydrazides can be stimulated to emit chemiluminescence under a variety of conditions. Most commonly in the present application initiating reagents consist of alkali, hydrogen peroxide and a "catalyst". A wide variety of oxidising species can yield phthalhydrazide chemiluminescence and many chemical species are capable of "catalysing" or otherwise

modulating the reaction. Transition metal cations whether
free, chelated or part of metalloproteins display a range
of "catalytic" activities in this respect.

By contrast, certain acridinium salts exhibit
chemiluminescence in the presence of dilute alkaline
hydrogen peroxide without the need for a catalyst. Such
an oxidation system is very simple and robust and also
yields low chemiluminescence backgrounds. Most commonly,
chemiluminescent acridinium salts are aryl esters,
thioesters, active aliphatic esters and active amides.
Derivatives of these can be synthesised to enable them to
be coupled to a component of the required binding
reaction to form the labelled reagent.

Phthalhydrazide molecules frequently exhibit loss of
chemiluminescence activity when coupled to the relevant
species, particularly proteins. Acridinium salts possess
inherently high quantum yields which are maintained when
the molecules are coupled in the appropriate manner. In
part this is due to the fact that under such
circumstances the emitting species, N-methylacridone, is
dissociated from the rest of the labelled molecule and is
thus less subject to microenvironmental quenching effects
(Fig. 1).

The chemiluminescent salt most widely demonstrated for
immunoassay purposes possesses an N-succinimidyl group.
Here the N-succinimidyl ester moiety enables the molecule
to be smoothly incorporated into species containing
primary and secondary aliphatic amines, though other
active groups can be incorporated into the synthesis of
the acridinium salt if required.Many immunoassays have
been described which are based on this label (Table 1).
Of historical interest is an immunochemiluminometric
assay for thyrotrophin (TSH) based on acridinium labelled
antibodies (5). As early as 1984 this system
demonstrated the high sensitivity achievable relative to
conventional techniques. The use of high affinity
monoclonal antibodies labelled to high specific activity
with the chemiluminescent acridinium ester enabled a ten-
fold improvement in sensitivity to be achieved over
state-of-the-art methods.

Table 1.   Some immunoassays based on acridinium salts

Labelled Antibody	Labelled Antigen
Alphafetoprotein	Progesterone
Ferritin	Calcitonin
Complement C9	Growth hormone
Tamm-Horsfall glycoprotein	releasing factor
Thyrotrophin	Albumin
Growth hormone	
Parathyroid hormone	
Thyroxine	
Fluphenazine	

Recently, acridinium labelled reagents have been adapted
to full automation. Ciba Corning Diagnostics have
developed an instrument (the ACS:180) capable of running
their range of "Magic Lite" immunoassays in an automated,
random access mode. Sample throughput is 180 samples per
hour and time to first result is 15 minutes. The system
has on-board reagent capacity to run 13 different
analytes. A further development has been to make use of
new instrumentation to configure simple, robust screening
assays using acridinium labelling technology. Microtitre
plates are particularly well-suited for this purpose but
the constraints of existing instrumentation have not made
it possible to exploit the advantages of of acridinium
labels in this format. The introduction of microtitre
plate luminometers with "on-board" reagent injection
facilities now permits the high performance of acridinium
based assays to be combined with an easy-to-use
microtitre plate format. This system has been used in our
laboratory to develop screening assays for thyrotrophin
in 3mm neonatal blood spot discs and also for
alphafetoprotein in a 50ul maternal serum sample. These

assays have a one hour incubation, following which the plate is washed and inserted into the luminometer. This end-point detection is more rapid and robust than enzyme-driven systems which involve addition of substrate followed by a delay period prior to quantitation. These assays are described in greater detail elsewhere in this volume.

The acridinium salt described earlier has also been successfully used as a label for oligonucleotide probe hybridisation assays marketed by Gen Probe Inc as the PACE System. In such applications, labelled probe (typically 20-30 bases) forms duplexes with target DNA or RNA if present. The duplexes are then selectively bound to magnetisable particles which permits separation and quantitation of chemiluminescence, the intensity of which is proportional to the amount of target nucleic acid present.

A novel approach to hybridisation assays also developed by Gen Probe involves the development of a homogeneous assay which does not require the use of a separation step prior to measurement (6). This is based on the chemical properties of the acridinium label which permits its hydrolysis under alkaline conditions, the hydrolysis products being non-chemiluminescent. However, when the labelled probe is involved in duplex formation with target nucleic acid, the acridinium ester is resistant to hydrolysis and thus retains its chemiluminescent activity. In this way, labelled probe can be added to the sample and, if target is present, duplexes will form. The mixture can then be exposed to alkaline conditions at elevated temperature and then chemiluminescence emission quantified in a luminometer. The intensity of chemiluminescence is thus proportional to the amount of target nucleic acid present.

This overview thus demonstrates the ability of acridinium salts to yield sensitive yet simple end-points for monitoring binding reactions.

REFERENCES
1.   K. Lehmstedt and E. Wirth. Ber. Dtsch. Chem.
        Ges. 61, 2044 (1928).
2.   K. Gleu and W Petsch. Angew. Chem. 58, 57 (1935)
3.   M.M. Rauhut, D. Sheehan, R.A. Clarke and
        A.M. Semsel. J. Org. Chem. 30, 3587 (1965).
4.   F. McCapra and D.G. Richardson. Tetrahedron Lett
        43, 2167 (1964).
5.   I. Weeks, M. Sturgess, K. Siddle, M.K. Jones and
        J.S. Woodhead. J.Endocrinol. 20, 489 (1984).
6.   L.J. Arnold, P.W. Hammond, W.A. Wiese and
        N.C. Nelson. Clin. Chem. 35, 1588 (1989).

Index